Green
Business

Green Business

An A-to-Z Guide

The SAGE Reference Series on
Green Society
Toward a Sustainable Future

NEVIN COHEN, GENERAL EDITOR
The New School

PAUL ROBBINS, SERIES EDITOR
University of Arizona

$SAGE | reference

Los Angeles | London | New Delhi
Singapore | Washington DC

Los Angeles | London | New Delhi
Singapore | Washington DC

FOR INFORMATION:

SAGE Publications, Inc.
2455 Teller Road
Thousand Oaks, California 91320
E-mail: order@sagepub.com

SAGE Publications Ltd.
1 Oliver's Yard
55 City Road
London EC1Y 1SP
United Kingdom

SAGE Publications India Pvt. Ltd.
B 1/I 1 Mohan Cooperative Industrial Area
Mathura Road, New Delhi 110 044
India

SAGE Publications Asia-Pacific Pte. Ltd.
33 Pekin Street #02-01
Far East Square
Singapore 048763

SAGE Publications
Publisher: Rolf A. Janke
Assistant to the Publisher: Michele Thompson
Senior Editor: Jim Brace-Thompson
Production Editors: Kate Schroeder, Tracy Buyan
Reference Systems Manager: Leticia Gutierrez
Reference Systems Coordinator: Laura Notton
Typesetter: C&M Digitals (P) Ltd.
Proofreader: Rae-Ann Goodwin
Indexer: Joan Shapiro
Cover Designer: Gail Buschman
Marketing Manager: Kristi Ward

Golson Media
President and Editor: J. Geoffrey Golson
Author Manager: Ellen Ingber
Editors: Jill Coleman, Mary Jo Scibetta
Copy Editors: Anne Hicks, Barbara Paris

Copyright © 2011 by SAGE Publications, Inc.

Printed in the United States of America.

Library of Congress Cataloging-in-Publication Data

Green business : an A-to-Z guide / general editor Nevin Cohen.

p. cm.—(The Sage references series on green society: toward a sustainable future)
Includes bibliographical references and index.

ISBN 978-1-4129-9684-6 (cloth)—ISBN 978-1-4129-7379-3 (ebk)

1. Business enterprises—Environmental aspects.
2. Management—Environmental aspects. 3. Green products. 4. Green marketing. I. Cohen, Nevin.

HD30.255.G734 2010 658.4′083—dc22 2011002870

11 12 13 14 15 10 9 8 7 6 5 4 3 2 1

Contents

About the Editors

Green Series Editor: Paul Robbins

Paul Robbins is a professor and the director of the University of Arizona School of Geography and Development. He earned his Ph.D. in Geography in 1996 from Clark University. He is General Editor of the *Encyclopedia of Environment and Society* (2007) and author of several books, including *Environment and Society: A Critical Introduction* (2010), *Lawn People: How Grasses, Weeds, and Chemicals Make Us Who We Are* (2007), and *Political Ecology: A Critical Introduction* (2004).

Robbins's research centers on the relationships between individuals (homeowners, hunters, professional foresters), environmental actors (lawns, elk, mesquite trees), and the institutions that connect them. He and his students seek to explain human environmental practices and knowledge, the influence nonhumans have on human behavior and organization, and the implications these interactions hold for ecosystem health, local community, and social justice. Past projects have examined chemical use in the suburban United States, elk management in Montana, forest product collection in New England, and wolf conservation in India.

Green Business General Editor: Nevin Cohen

Nevin Cohen is an assistant professor of Environmental Studies at The New School, in New York City, where he teaches courses in urban planning and sustainable food systems. He serves as co-chair of the Tishman Environment and Design Center, The New School's interdisciplinary environmental research and education center, and home to the university's innovative bachelor program in Environmental Studies, which emphasizes urban ecosystems, sustainable design, and public policy. He has a Ph.D. in Urban Planning from Rutgers University, a Masters in City and Regional Planning from the University of California, Berkeley, and a B.A. from Cornell University.

For the past two decades, Dr. Cohen has worked with Fortune 500 companies on corporate sustainability initiatives. Prior to joining the faculty of The New School, he served as managing principal for GreenOrder, Inc., a consulting firm specializing in sustainable business practices, and held senior research positions at Rutgers University's Center for Environmental Communication, Environmental Defense, the World Resources Institute, Tellus Institute, and INFORM, Inc. As a policy analyst and planner in New York City, Dr. Cohen advised local planning boards and real estate developers on green development strategies. He was also responsible for developing landmark municipal recycling, water conservation, and clean fuel laws in New York City as a policy analyst for the City Council and Manhattan borough president.

Introduction

Over the past 40 years, there has been a dramatic transformation in the way businesses have addressed environmental issues. In response to the enactment of national legislation to control the pollution of the air, water, and land, corporations responded by cutting the most egregious end-of-the-pipe emissions and increasing attention to environmental compliance by implementing environmental management systems and best management practices. By the mid-1980s, companies began to shift from a focus on pollution control to one on pollution prevention, attempting to eliminate waste before it is created. Greener businesses began to embrace the idea that designing manufacturing systems that prevented waste offered a competitive edge by not only improving efficiency but also reducing the financial risks associated with the use and disposal of toxic chemicals. Leading businesses, nongovernmental organizations, trade groups, and socially responsible investment firms began to demonstrate the possibility of achieving a "triple bottom line," improving environmental and social performance simultaneously with profit by adopting proactive and preventive strategies and a commitment to corporate social responsibility.

Throughout the 1980s and 1990s, as regulatory agencies in the United States and Europe began to embrace market-based policies, including requiring information disclosure to drive green business practices, the "greening" of business took off as companies began to identify ways in which their business goals could be aligned with the social objectives of cleaner production and civic responsibility. These efforts were guided by global management standards developed by the International Organization for Standardization, various environmental management principles developed by nongovernmental organizations (such as the Ceres Principles), sustainability reporting guidelines, and sustainability standards for investors, developed by companies such as Dow Jones.

In the 1990s and early 2000s, firms moved from pollution prevention to a focus on product stewardship. As firms ratcheted down their end-of-the-pipe pollution through increased efficiency and the use of best available control technology, regulators and advocacy groups, together with leading green businesses, began focusing on the entire product life cycle, from the practices of suppliers to the environmental impacts associated with the use of products and with their disposition at the end of their lives. Producer responsibility regulations in Europe included product take-back requirements, particularly for waste electronics. Green businesses began to explore various forms of techniques to minimize the life cycle effects of products, including design for the environment—a process by which firms proactively design products that are durable, easily repairable, upgradable, and safely and economically recyclable; cradle-to-cradle design, which attempts to eliminate waste by designing goods so that they can be used as raw materials for other products; and the precautionary principle, which aims to avoid products and practices for which there are insufficient data on the health and environmental risks. Some firms have embraced service design, finding ways to provide the functions that consumers want without selling physical goods.

The evolution of green business has required increasingly sophisticated methodologies for measuring the consequences of green business practices, leading to the maturation of fields such as life cycle analysis, quantitative risk assessment, and environmental accounting and auditing. To capitalize on the salience of environmental issues and consumer interest in green products, particularly among the demographic groups most attuned to environmental issues (e.g., Lifestyles of Health and Sustainability), businesses have attempted to use green marketing techniques, conveying the benefits of greener products through ecolabels and environmental indicators, including easy-to-communicate concepts such as the ecological footprint. Although this has left some firms open to the charge of exaggerating green claims—or greenwashing—others, such as GE, have developed green marketing campaigns with efficiency metrics aimed at both increasing sales and identifying clean technologies to support through research and development and other investments.

The clean technology field has grown in large measure because businesses anticipate that the future will be carbon-constrained as nations adopt policies to curb greenhouse gases, and water constrained as climate change alters weather patterns and the availability of freshwater. Firms that have incorporated sustainability into their business models have been leaders in developing innovative green products, such as advanced solar panels, electric vehicles, and carbon-neutral green buildings that provide the functionality of conventional products (e.g., mobility, thermal comfort) while dramatically reducing greenhouse gas emissions. These new technologies offer new business opportunities and potentially large competitive advantages as environmental challenges grow.

The recent global recession has dampened consumption and new business investments, green or otherwise. Yet in the United States and abroad, government economic stimulus programs have focused on green economic development based on alternative energy technologies and the infrastructure (e.g., a "smart energy grid") to support the transmission of distributed sources of power, as well as low-technology innovations such as building weatherization. Proponents of green technology believe that public stimulus dollars and private investments have the potential to create large numbers of green collar jobs for working-class people, as well as to increase the overall efficiency of the economy. Action at the national level is mirrored by policies at the state and local level, where governments are revising their building codes to require green building, leading the development and construction industry to source greener building materials and products.

Notwithstanding the recession, green business innovations continue. Large retailers, from IKEA to the supermarket chain Tesco, are eliminating unnecessary packaging and providing product information (in Tesco's case, carbon footprint labels) designed to encourage sustainable purchasing. Manufacturers are investing in green chemistry techniques to reduce the hazards of their products and are using biomimicry, employing design principles found in the natural environment as models of industrial processes and manufactured products, to create environmentally preferable products and structures. In addition, the concepts of green business are being applied in sectors beyond manufacturing and retail, including agriculture, where the growth in demand for organic products and food sourced from growers who are paid a fair wage and use sustainable farming techniques continues apace.

Businesses increasingly recognize their capacity to help solve global environmental and social challenges, and the most innovative understand the business case for addressing issues such as climate change, water scarcity, pollution, poverty, hunger, and inequality. This volume, *Green Business*, in *The SAGE Reference Green Series: Toward a Sustainable Future*, provides an overview of the key principles, approaches, strategies, and tools that businesses have used to reduce their environmental impacts and contribute to sustainability, as well as describing some of the firms that have taken steps to green their operations and products. Each entry reflects the expertise of scholars and practitioners from a variety of fields of study and provides references to other entries in the volume in addition to citations for further reading. Together, the entries provide an understanding of green business practices that will be valuable for managers, policy makers, scholars, and citizens interested in the complex relationship between businesses and the environment.

Nevin Cohen
General Editor

Reader's Guide

Business Organizations, Movements, and Planning

Balanced Scorecard
Best Available Control Technology
Best Management Practices
Ceres Principles
Certification
Closed-Loop Supply Chain
Compliance
Core Competencies
Corporate Social Responsibility
Cost-Benefit Analysis
Demand-Side Management
Discounting
Dow Jones Sustainability Index
Ecoeffectiveness
Ecoefficiency
Ecoindustrial Park
Ecological Economics
Economic Value Added
Emissions Trading
Energy Performance Contracting
Energy Service Company
Environmental Accounting
Environmental Assessment
Environmental Audit
Environmental Economics
Environmental Impact Statement
Environmental Indicators
Environmentally Preferable Purchasing
Environmental Management System
Environmental Marketing
Environmental Risk Assessment
Environmental Services

Equator Principles
Extended Producer Responsibility
Extended Product Responsibility
Externalities
Factor Four and Factor Ten
Fair Trade
Genuine Progress Indicator
Global Reporting Initiative
Global Sullivan Principles
Industrial Ecology
Industrial Metabolism
Industrial Nutrients
Informational Regulation
Integrated Bottom Line
International Organization for
    Standardization
ISO 14000
ISO 19011
Leadership in Green Business
Life Cycle Analysis
Material Input per Service Unit (MIPS)
Maximum Achievable Control
    Technology
National Priorities List
Natural Capital
New Source Review
Quantitative Risk Assessment
Recycling, Business of
Reverse Logistics
Service Design
Social Return on Investment
Steady State Economy
Stewardship
Supply Chain Management
Value Chain

Business Profiles

Body Shop, The
British Telecom
General Electric (GE)
Herman Miller
IKEA
Lifestyles of Health and Sustainability
  (LOHAS)
Natural Step
Newman's Own Organics
Patagonia
Sainsbury's
Stonyfield Farm
SunEdison
Tesco
Toyota
Waitrose

Green Business Challenges

Bottom of the Pyramid
Carbon Footprint
Ecological Footprint
E-Waste Management
Exposure Assessment
Greenwashing
Persistent Pollutants
Toxics Release Inventory
Transparency
Triple Bottom Line

Green Business Solutions

Abatement
Appropriate Technology
Bio-Based Material
Biofuels
Biological Resource Management
Biomimicry
Bioremediation
Biotechnology
Blended Value
Brownfield Redevelopment
Carbon Neutral
Carbon Sequestration
Carbon Trading

Cause-Related Marketing
Clean Fuels
Clean Production
Clean Technology
Cogeneration
Conservation
Coopetition
Cradle-to-Cradle
Deposit Systems
Distributed Energy
Ecolabels
Ecosystem Services
Ecotourism
Environmental Justice
Green Building
Green Chemistry
Green-Collar Jobs
Green Design
Green Retailing
Green Technology
Gross National Happiness
Integrated Pest Management
Organic
Pollution Offsets
Pollution Prevention
Precautionary Principle
Remanufacturing
Resource Management
Responsible Sourcing
Restoration
Right to Know
Seventh Generation
Six Sigma
Smart Energy
Social Entrepreneurship
Socially Responsible Investing
Social Marketing
Superfund
Sustainability
Sustainable Design
Sustainable Development
Systems Thinking
Take Back
Upcycle
Voluntary Standards
Waste Reduction

List of Articles

List of Contributors

Ackom, Emmanuel Kofi
University of British Columbia

Alexis, Gwendolyn Yvonne
Monmouth University

Amer, Wafaa Mahrous
Cairo University

Andrews, Mitchell
University of Sunderland

Arney, Jo A.
University of Wisconsin–La Crosse

Ballamingie, Patricia
Carleton University

Beder, Sharon
University of Wollongong

Bled, Amandine J.
Université Libre de Bruxelles

Boslaugh, Sarah
Washington University in St. Louis

Brohé, Arnaud
Université Libre de Bruxelles

Büscher, Bram
Institute of Social Studies, the Netherlands, University of Johannesburg

Büscher, Stacey
HIVOS, the Netherlands

Carveth, Rodney Andrew
Fitchburg State College

Cidell, Julie
University of Illinois

Cooley, Amanda Harmon
North Carolina A&T State University

Cullari, Francine
University of Michigan–Flint

Daniels, Peter
Griffith University, Australia

de Bakker, Frank G. A.
Vrije Universiteit Amsterdam

de Souza, Lester
Independent Scholar

Denault, Jean-François
Université de Montréal

Duffy, Lawrence K.
University of Alaska Fairbanks

Emery, Barry
University of Northampton

Ertel, Jürgen
Brandenburg University of Technlogy

Farley, Heather M.
Northern Arizona University

Finley-Brook, Mary
University of Richmond

Flanagan, Patrick
St. John's University

Fritz, Anju
University of Maryland

Gachechiladze, Maia
Central European University

García-Olmedo, Belén
University of Granada

Grover, Velma I.
*United Nations University–Institute for
    Water, Environment and Health
    (UNU–IWEH)*

Harper, Gavin D. J.
Cardiff University

Hasselmann, Franziska
*Swiss Federal Institute of Forest, Snow,
    and Landscape Research*

Helfer, Jason A.
Knox College

Howes, Michael
Griffith University, Australia

Jarvie, Michelle E.
Independent Scholar

Johnson, Sherrill
Independent Scholar

Kirkham, Elaine D.
University of Wolverhampton

Kte'pi, Bill
Independent Scholar

Lanfair, Jordan K.
Knox College

Leinaweaver, Jeff
Fielding Graduate University

Lippert, Ingmar
University of Augsburg

Lord, Richard Alastair
University of Teesside

Martin, Diane M.
University of Portland

Merskin, Debra
University of Oregon

Mullaney, Emma Gaalaas
Pennsylvania State University

Mulvaney, Dustin
University of California, Santa Cruz

Nash, Hazel
Cardiff University

Newell, Josh
University of Southern California

Nieuwenhuis, Paul
Cardiff University

Nuñez, Maura Troester
University of Colorado, Boulder

Orsini, Marco
*Institut de Conseil et d'Études en
    Développement Durable (ICEDD)*

O'Sullivan, Robin
University of Texas at Austin

Palmer, Daniel E.
Kent State University, Trumbull

Paolini, Federico
University of Siena

Parker, Jonathan
University of North Texas

Patnaik, Rasmi
Pondicherry University

Persaud, Nadini
University of the West Indies,
   Cave Hill Campus

Philipsen, Dirk Peter
Virginia State University

Pojasek, Robert B.
Harvard University

Ponting, Cerys Anne
Cardiff University

Poyyamoli, Gopalsamy
Pondicherry University

Putnam, Heather R.
University of Kansas

Rands, Gordon P.
Western Illinois University

Rands, Pamela J.
Western Illinois University

Reed, Matt
Countryside and Community Research
   Institute

Roby, Claire
Independent Scholar

Roy, Abhijit
University of Scranton

Roy, Mousumi
Penn State University, Worthington
   Scranton

Salimath, Manjula S.
University of North Texas

Schouten, John W.
University of Portland

Schroth, Stephen T.
Knox College

Sekerka, Leslie E.
Menlo College

Smith, Alastair M.
ESRC Centre for Business Relationships,
   Accountability, Sustainability & Society

Stancil, John L.
Florida Southern College

Stoll, Mary Lyn
University of Southern Indiana

Tsoi, Joyce
Brunel University

Van Leuven, Nancy
Bridgewater State College

Vos, Robert O.
University of Southern California

Watson, Derek
University of Sunderland

Weaver, Susan H.
Independent Scholar

Woodward, David G.
University of Southampton School of
   Management

Zaccai, Edwin
Université Libre de Bruxelles

Green Business Chronology

c. 1530: Commercial whaling begins as the Basques start the pursuit of right whales in the North Atlantic, taking an estimated 25,000 to 40,000 whales over the next 80 years.

1892: British reformer Henry S. Salt, a socialist, pacifist, and vegetarian, publishes a landmark work on animal welfare, *Animal Rights Considered in Relation to Social Progress*.

1905: Upton Sinclair publishes his novel *The Jungle* in serial format in the socialist magazine *Appeal to Reason*. Public outcry over the filthy conditions of the meatpacking industry portrayed in this novel lead to passage of the Pure Food and Drug Act in 1906.

1918: The Save the Redwoods League forms in the United States to purchase the remaining redwood forests, which have been extensively harvested for lumber.

1920: In response to the perceived failures of existing federal laws to deal with the mining of coal and oil resources, the United States passes the Mineral Leasing Act to regulate mining on public lands. The law governs deposits of coal, oil, gas, oil shale, phosphate, potash, sodium, and sulfur.

1932: American president Franklin D. Roosevelt uses the phrase "bottom of the pyramid" in a radio address, which refers to "the indispensable units of economic power . . . the forgotten man at the bottom of the economic pyramid." The term is used today to refer to the 4 billion or so people who live on less than US$2 per day.

1934: The Taylor Grazing Act establishes grazing districts on public lands (which have formerly been unregulated) and grants leases to individuals to use them. The fees are deliberately set low and in general remain lower today than the fees charged for comparable private grazing land.

1936: In an early example of cost-benefit analysis, the U.S. Flood Control Act requires that flood control projects must demonstrate that their benefits outweigh their costs.

1946: The International Convention for the Regulation of Whaling holds its first meeting in Washington, D.C., and sets quotas for whaling that are intended to allow the whaling industry to continue at reduced levels so that whales are not hunted to extinction. The convention establishes the International Whaling Commission in 1949, which is intended to regulate whaling but has been beset by conflicts between nations with traditional whaling industries (e.g., Japan, Iceland, and Norway) and those that wish to impose a moratorium on all whaling.

1948: The Federal Water Pollution Control Act creates comprehensive programs to regulate interstate waters in the United States and provides financing for state and local governments.

1954: Introduction of highly industrialized fishing vessels into the codfish industry in the United States and Canada greatly increases the annual catch, causing a steady decline in the cod population, so that in 1977 it was determined that the number of spawning codfish off the coast of Newfoundland had decreased by 92% since 1962.

1956: The first victims of Minamata disease, caused by alkylmercury poisoning from eating contaminated fish, are identified in Japan. Victims suffer from neurological impairments including retardation, contorted limbs, and sensory disturbances. The source of the contamination is traced to the Chisso Company, which had been dumping industrial pollutants into Minamata Bay on the west coast of Kyushu island.

1965: The Water Quality Act requires U.S. states to establish water quality standards or accept those set by the Federal Water Pollution Control Administration.

1970: Dr. Norman Borlaug, father of the "green revolution," which is credited with substantially increasing crop yield in the Third World, wins the Nobel Peace Prize. Although few question that the green revolution saved millions of people from starvation, many—particularly in more recent years—will criticize Borlaug's reforms because they rely heavily on chemical fertilizers and irrigation and on seeds that must be purchased annually from multinational corporations, thus increasing corporate control of Third World agriculture.

1971: The U.S. Public Health Service sets the level of "undue lead absorption," that is, the level above which medical intervention is recommended, at 40 μg/dL. The level is lowered to 30 μg/dL in 1975, 25 μg/dL in 1985, and 10–15 μg/dL in 1991. Some cities implement campaigns to remove lead paint from housing because this is a primary source of lead exposure for children.

1972: The London Dumping Convention, the formal name of which is the International Convention on the Prevention of Marine Pollution by Dumping of Waste and Other Matter, prohibits dumping wastes in the ocean from ships, aircraft, platforms, or any other man-made structure in the sea, as well as from land-based sources.

1973: Ernst Friedrich Schumacher's *Small Is Beautiful: Economics as if People Mattered* criticizes the assumption that economic development requires adoption of large-scale Western technologies and a lifestyle based on acquisition of consumer goods.

1975: The Convention on International Trade in Endangered Species of Wild Fauna and Flora (CITES), signed in 1973, comes into force. The convention's purpose is to end international trade in endangered or threatened animal and plant species or products made from them, and it requires signatories to implement domestic laws (such as the Endangered Species Act in the United States) to carry out conventional principles.

1976: Anita Roddick founds the Body Shop, which sells beauty products made from natural ingredients and manufactured in ways that do not harm the environment. Body Shop products are not tested on animals, and many are obtained through a Fair Trade program that provides suppliers with sufficient income to maintain themselves.

1976: Wes and Dana Jackson found the Land Institute in Salina, Kansas, to develop sustainable agriculture practices. The institute includes both a school and an agricultural research station that investigate alternatives to industrial agricultural practices such as monoculture and extensive use of pesticides and herbicides.

1976: In the Godkin Lectures at Harvard University in Cambridge, Massachusetts, Charles Schultze argues that taxes and incentives are a more efficient method for controlling pollution than are laws and regulations. The lectures are later published as *The Public Use of the Private Interest* and argue that taxing pollution would encourage those industries that can easily reduce it to do so, and new businesses to develop methods to pollute less while allowing other industries to remain in business by paying a tax or fine equivalent to the damage caused by their pollution.

1977: The U.S. Consumer Product Safety Commission bans the use of lead in house paint and restricts lead in paints used on toys and furniture to 0.06%.

1979: A series of equipment failures and human error cause a partial core meltdown at a nuclear plant at Three Mile Island, Pennsylvania. There is no major steam or hydrogen gas explosion, although it is uncertain how much radiation escapes. This near-catastrophe draws attention to the dangers involved in nuclear power generation and halts construction of new nuclear power facilities in the United States.

1980: Alex Pacheco and Ingrid Newkirk found People for the Ethical Treatment of Animals (PETA) to protect animal rights. Philosophers such as Tom Regan and Peter Singer heavily influence the PETA founders, and in particular their belief that "speciesism," or the belief that man is superior to all other species, is incorrect. PETA goes on to become perhaps the best-known animal rights organization in the world and protests many common commercial practices involving animals, from using laboratory animals for product research and testing to raising meat animals in factory-like conditions and using animal fur for human garments.

1980: Paul Ehrlich and Julian Simon make a famous bet about the future supply of natural resources as human population increases. Ehrlich, an ecologist at Stanford University, predicts that because there is a finite supply of natural resources, the price of basic commodities will increase over time, whereas Simon, an economist from the University of Maryland, predicts that commodity prices will fall as a result of increasing supply and substitution. Ehrlich chooses five metals to track—copper, chrome, nickel, tin, and tungsten—and they agree to evaluate the change in prices 10 years later, calculated in constant dollars. Ehrlich loses the bet despite increased world population because he failed to consider that consumption patterns often change in response to prices and that new sources of commodities may be discovered that could lower the price.

1983: American Express coins the phrase "cause-related marketing" to describe their campaign to raise money to restore the Statue of Liberty. The company raises over $2 million in four months by donating a penny every time someone uses their card and a dollar for each new card issued. Card usage and the number of new cards also increase, establishing the principle that this type of marketing can be beneficial to both the company and the cause.

1984: An accident at a Union Carbide pesticide-producing plan in Bhopal, India, releases a toxic cloud of methyl isocynate, causing over 2000 deaths immediately and at least 2000 more in ensuing years. Some estimates of deaths run as high as 10,000, with estimates of the number injured running from 200,000 to 500,000. In 1989, Union Carbide pays $470 million in settlements to the Indian government, with the money to go to survivors of the leak. This disaster brings closer scrutiny to other Union Carbide plants, including a plant in Institute, West Virginia, which also handles methyl isocynate.

1985: A group of activists found the Rainforest Action Network in San Francisco, California, with the purpose of protecting the world's rainforests and the people who live in them from environmental destruction. Their first major action is a boycott of the American fast-food chain Burger King, which at that time imported much of its beef from Central and South America, where rainforest destruction was hastened by the economic incentive of clearing the forest and turning it into grazing land for cattle. In 1987, Burger King announces that it is no longer importing beef from rainforest areas.

1987: American activist Dave Foreman publishes *Ecodefense: A Field Guide to Monkey-wrenching*, which advocates sabotage to prevent environmentally destructive development and other commercial activities. Many of Foreman's suggested tactics are illegal, including driving metal spikes into trees to prevent their being logged, sabotaging earth-moving equipment such as bulldozers, removing surveyor's stakes, and pulling down power lines. The book's title refers to *The Monkey Wrench Gang*, a 1975 novel by Edward Abbey that calls for individuals to take direct action to halt the destruction of wilderness.

1988: Negative publicity about polluted beaches in New York and New Jersey leads to passage of the Ocean Dumping Ban Act, which takes force in 1991. The act prohibits dumping industrial waste and sewage sludge in the ocean and is more stringent than previous laws, which allowed some dumping as long as it did not seriously degrade the marine environment.

1988: Ron Arnold and Alan Gottlieb publish *The Wise Use Agenda*, which states the opposition of the Wise Use Movement in the Western United States to increased government regulation of commercial uses of public lands. In opposition to many spokespeople for the environmental movement, the Wise Use Movement advocates clear-cutting in national forests, petroleum drilling in the Arctic National Wildlife Refuge in Alaska, and placing individual states rather than the federal government in charge of regulating water resources on public lands.

1989: The Coalition for Environmentally Responsible Economics (CERES), which includes environmental organizations, socially responsible investors, labor unions, and other groups, forms to encourage environmentally responsible and sustainable business practices. The Ceres principles are also introduced in this year: they ask corporations to commit to a number of principles (initially called "the Valdez Principles"), which include protection of the biosphere, sustainable use of natural resources, waste reduction, conservation of energy, and environmental restoration.

1989: The worst oil spill in American history occurs when the supertanker ship *Exxon Valdez* grounds on a reef and spills over 11 million gallons of oil into Prince William Sound near Valdez, Alaska. The resulting oil slick extends 50 miles and is estimated to kill 10% of the region's bird population. In 1990, Joseph Hazelwood, captain of the *Exxon Valdez*, is convicted on one misdemeanor count related to the spill, and in 1991 Exxon pled guilty to four misdemeanor counts and paid about $1 billion in fines and environmental damage payments.

1990: The U.S. Food and Drug Administration approves the Pfizer product Chy-Max chymosin, a genetically modified version of rennet used in cheese. At this time, 60% of U.S. cheeses are produced using genetically modified chymosins.

1992: The Flavr Savr tomato becomes the first genetically engineered vegetable approved by the Food and Drug Administration for sale in the United States. The Flavr Savr will be introduced commercially in 1994 and withdrawn from the market in 1996 because of lack of consumer interest.

1992: British Telecom begins publishing their Society and Environment Report.

1996: Monsanto plants the first commercial fields with the Roundup Ready soybean, the first commercially genetically modified crop in the United States. The beans are engineered to resist the common herbicide glyphosate, which can therefore be sprayed on the fields without damaging the soybean crop. By 2002 about 70% of soybeans grown in the United States are engineered to be herbicide resistant.

1997: The Kyoto Protocol, an international agreement linked to the United Nations Framework Convention on Climate Change, which aims to reduce or prevent global warming, is adopted. Under the protocol, which goes in effect in 2005, most industrialized countries agree to reduce their emissions of greenhouse gases: some were set specific targets, some were given the goal of reducing their emissions to 1990 levels, and others were allowed to reduce their levels. The protocol also allows countries to trade carbon emissions to meet their goals.

1998: Great Britain announces that by 2008, over 60% of new housing starts will be on brownfield grounds; that is, they will reuse previously developed land, such as abandoned commercial developments, that is underused or neglected.

1999: A study by Cone Millenial Cause finds that most (89%) of Americans age 13–25 years would be willing to switch to a brand associated with a "good cause" and that many would prefer to work for a socially responsible company.

2002: The Center for Food Safety publishes *Fatal Harvest: The Tragedy of Industrial Agriculture*, a collection of essays criticizing the effects of industrial agriculture, which is the norm in the United States, on the environment and on human health. Specific criticisms include overuse of pesticides, loss of topsoil, use of genetically engineered species, loss of agricultural diversity, and loss of family farms.

2002: William McDonough and Michael Braungart popularize the term "cradle to cradle," which was introduced by Walter Stahel in the 1970s. "Cradle to cradle" refers to the principle that companies should be responsible for recycling the materials from their products after they are discarded.

2006: The New Oxford American dictionary selects "carbon neutral" as its word of the year. Ironically, there is no single accepted definition of the term, which refers in general to the achievement of net zero greenhouse gas emissions through reducing emissions and purchasing carbon offsets: the question is the scope of the emissions included in the calculations. For instance, should a company include the emissions related to raw materials that they purchase, or is that part of someone else's business?

2008: Honda offers the first fuel cell car, the FCX Clarity, for lease.

Sarah Boslaugh
Washington University in St. Louis

ABATEMENT

Abatement involves the reduction or elimination of pollutants in the environment. Abatement is used to reduce a variety of pollutants, such as noise or dust. Abatement also involves adjusting tax rates and rebates that businesses and other organizations pay or receive for their contribution to the environmental protection or degradation of certain areas. Abatement can be tied to specific types of pollution, such as air, water, or noise. Abatement is multifaceted, as it involves governmental regulation that requires or encourages certain actions, technology that can reduce the occurrence of certain pollutants, and incentives and penalties that support or punish those who produce or eliminate pollution. Pollution abatement typically refers to the reduction in pollution after it is generated but before it is released into the environment. It is often contrasted with pollution prevention, which seeks to eliminate the generation of pollution in the first place.

Abatement is often used in response to air pollution. Air pollution is highly problematic, particularly in cities where emissions released from motor vehicles, factories, and other sources create environmental conditions such as smog. Air pollution may exacerbate preexisting health conditions of individuals and can cause higher rates of respiratory problems such as asthma. Air pollution can create a toxic environment that contributes to such problems as acid rain and the destruction of organisms reliant on a set pH balance in water. Since population growth continues unabated, some maintain that pollution and waste disposal issues will remain a necessary by-product of our current lifestyle. Abatement advocates maintain that societal behaviors that cause environmental toxins can be reduced so as to decrease pollution while maintaining current standards of living.

The need for abatement comes from a history of tragic pollution-related events, such as 60 deaths in 1930 in Belgium that were related to air pollution, 20 respiratory-related deaths in Pennsylvania in 1948, and approximately 4,000 lives in London in 1952 due to "black fog." The cause for these events has been identified as an inability for atmospheric dispersion due to high levels of pollutants in the environment. Much abatement technology and regulation in the United States stems from the Clean Air Act of 1963 (subsequently amended in 1970, 1977, and 1990) and the Air Quality Act of 1967. This legislation led to the ultimate creation of the U.S. Environmental Protection Agency (EPA) in 1970, which was charged with establishing and regulating national standards to reduce pollution through a variety of environmental laws, in consultation with state, tribal, and local governments. Early research found that a majority of Americans lived in metropolitan areas where air pollutants and other toxins were most prevalent. The effects of this continue to be felt and were a primary factor in the passage of the Air Quality Act, which established programs that would reduce air pollutants and thus increase the safety of citizens in local areas. The EPA has pursued four different tactics to establish cleaner air: direct regulation, taxation, facilitation for private litigation, and tradable pollution permits. These options arise in a time when air pollution is becoming linked to drastic environmental problems and the release of toxins into the atmosphere is contributing to an increased climate change.

Regulation

Direct regulation has become the staple of policy initiatives as it provides a level of transparency and a coherent guideline for appropriate and inappropriate practices. Direct regulation is dependent on policy formation and consequent enforcement and monitoring. Regulation has led to the implementation of technology that reduces toxic and harmful emissions. The most common form of air pollution abatement technology is the use of electrostatic precipitators, baghouses, and baghouse filters. Baghouses and baghouse filters use fabric filters and air flow to capture potential pollutants before they are released into the atmosphere. In addition to the fabric filters, other porous materials such as ceramics, glass, and fibers are used to reduce pollutants. Baghouses and baghouse filters normally capture more than 99 percent of fine particles. Electrostatic precipitators are also used to remove particles from flowing gases by using an electrostatic charge that removes fine particulate matter such as dust and smoke from the air stream. Baghouses, baghouse filters, and electrostatic precipitators have all been deemed highly effective at removing fine particles from an air stream. Unfortunately, all three devices lack the ability to prevent gas emissions from escaping, and also fail to prevent these emissions from forming acidic particles once in contact with the environment.

Taxation is another option available for air pollution abatement. Through a series of taxes for emissions and rebates for emission reduction, potential polluters are provided with financial incentives to reduce emissions and to invest in technology and other pollution abatement strategies. Taxation also is attractive insofar that it can provide the EPA and other government agencies with revenue needed for enforcement, research, and innovation. Private litigation is another strategy that allows for air pollutant abatement. Private litigation creates a cause of action that allows third parties to take legal action against polluters who violate certain standards regarding the amount of pollutants to be generated—if this amount is exceeded, a third party is allowed to directly sue the polluter to recover monetary damages. Private litigation has been touted as particularly effective when used in conjunction with other strategies, since polluters in otherwise highly profitable endeavors are left with no incentives to stall cleanups.

Tradable permits are another tactic used to attain air pollution abatement. Tradable permits combine some of the financial incentives provided manufacturers that are similar to those provided by tax policies while also maintaining control over the total amount of emissions allowed. When tradable permits are used, governmental agencies such as the EPA set a limit, or cap, on the total amount of a pollutant that can be emitted. Companies or other groups are then issued emission permits that allow for a certain level of release of that pollutant. Companies that anticipate producing emissions are required to hold an equivalent number of allowances or credits that represent the right to emit a specific quantity of a given pollutant. All manufacturers' total amounts of allowances and credits cannot exceed the overall limit, thereby restraining total emissions to that cap. Tradable permits allow users to buy or sell permits, thereby creating an incentive for those who pollute less. Over time, the limit or cap of a given pollutant would be lowered to decrease the level of emissions released.

Water pollution abatement authority stems from the Federal Water Pollution Control Act of 1948, which created a comprehensive set of water quality programs for interstate waters that also provided some financing for state and local governments, and the Water Quality Act of 1965, which mandated that states issue water quality standards and authorized the Federal Water Pollution Control Administration to set those standards if states did not take action. The overarching goal of both pieces of legislation was reclamation and restoration of national water supplies. Federal funds have been provided to deter pollutants from public water sources as well as to help restore areas that support aquatic wildlife. Individual states are permitted to work to protect their water sources and to reduce water pollution. Water pollution abatement is a key part of federal and local response to reducing water contamination.

Point and Nonpoint Sources

Water pollution abatement addresses point and nonpoint sources of water pollution. A point source of water pollution is a single, localized, and identifiable source of contamination, such as wastewater from an oil refiner. Point sources are so identified because in mathematical modeling, they can be approximated as a mathematical "point" to simplify analysis. Agencies such as the EPA have dealt quickly and severely with point sources of water pollution as these sources are easily identifiable, often traceable to a single entity, and relatively easy to eliminate, at least in terms of identifying the responsible party. Nonpoint sources, on the other hand, result in the creation of water pollution through indirect environmental changes and result in contamination of a body of water from diffuse sources. Nonpoint sources can stem from a variety of causes, including polluted agricultural or sewer runoff into a river or debris that is blown out to sea. Although the majority of the contamination to bodies of water comes from nonpoint sources, it is more difficult to prevent because its many different sources make it difficult or impossible to regulate. Nonpoint source water pollution is the leading cause of water contamination in the United States. Nonpoint source water pollution is also tied to other forms of contamination, such as air pollution, where acidic rain can change the pH and safety of bodies of water.

Water pollution's primary sources can be categorized as municipal, agricultural, and industrial. Municipal pollutants are those from homes and commercial areas, such as cleaning supplies or sewage. Agricultural pollutants include livestock waste, commercial fertilizers, and other contaminants related to the raising of livestock or the production of crops. Industrial pollution varies depending on the industry involved, but generally relates to contaminants that result from industrial discharge, such as secretions from a chemical plant. Municipal pollutants are abated primarily through treatment facilities, with the removal of suspended solids, oxygen-demanding materials, and harmful bacteria. In recent years, however, there has been more emphasis placed on the removal of solid residue, which results in a reduction of 25 percent to 50 percent of the operational costs of a treatment plant.

Agricultural pollution tends to contain dirt and other crop surplus in addition to animal remains and waste, which creates the need to treat agricultural waste on land and through the adoption of environmentally friendly farming practices. Industrial pollutants, which are distinctive based on their dissimilar characteristics, are made more difficult to deal with due to organic and inorganic substances present in their composition. Industrial wastewater is treated in differing ways, either by individual manufacturers or collectively, which makes enforcing standards a difficult task for local governments attempting to adhere to federal standards. Pollutants can be treated at individual facilities, before they are discharged to municipal sources, or before they are discharged directly into receiving bodies of water. The Clean Water Act has provided over $50 billion to cities and states building wastewater treatment plants, facilities that focus chiefly on the purification of surface water. States' ability to choose those measures that they would take to comply with federal mandates have caused certain problems, as states have focused primarily on providing fishable and swimmable waters. As a result, bacteria levels have been used as the chief indicator of water quality.

See Also: Appropriate Technology; Compliance; Green Technology; Industrial Metabolism; Pollution Offsets.

Further Readings

Couach, O., et al. "A Development of Ozone Abatement Strategies for the Grenoble Area Using Modeling and Indicators." *Atmospheric Environment* (March 1, 2004).
Moslener, U. and T. Requate. "The Dynamics of Abatement Strategies for Multiple Pollutants—An Illustration in the Greenhouse." *Ecological Economics*, 68/5:1521–34 (2009).
Oxley, T., M. Valientis, A. Elshkaki and H. M. Apsimon. "Background, Road and Urban Transport Modeling of Air Quality Limit Values (The BRUTAL Model)." *Environmental Modeling & Software*, 24/9:1036–50 (2009).
Viessman, Warren, Jr., et al. *Water Supply and Pollution Control*, 8th Ed. Upper Saddle River, NJ: Prentice Hall, 2008.

Stephen T. Schroth
Jason A. Helfer
Jordan K. Lanfair
Knox College

APPROPRIATE TECHNOLOGY

At the most basic level, proponents of appropriate technology attempt to clarify which technologies are appropriate to a given context. What is appropriate is, of course, a highly problematic issue about which it is difficult to form consensus. What is generally characteristic about appropriate technology is an effort to mold technologies to be context sensitive; that is, rather than developing a technology without regard to the cultural and environmental situations in which the technology will be used, an advocate of appropriate technology strives to develop technologies that can be enmeshed into a specific place with due consideration given to the impact on the cultural traditions and ecology of those people and ecosystems where the technology will be implemented.

When the term *appropriate technology* is currently invoked it is often used to refer specifically to technologies in developing countries. Appropriate technology in Western contexts generally refers to alternatives to dominant existing technologies that are seen as inappropriate to cultural ideals and values and/or unsustainable. In this context such appropriate technology is sometimes referred to as alternative or intermediate technology.

Appropriate technology advocates seek to develop technologies that are low energy/energy efficient, small scale, minimal to no pollution output, sensitive to the local ecological,

Appropriate technology advocates, such as the NGO Light Up the World, seek small-scale, energy-efficient technologies. Here, a Light Up the World beneficiary in Costa Rica, Daniel Zuniga, holds up a light-emitting diode lamp.

Source: Christoph Schultz/LUTW.org

cultural, and economic context, and sustainable. In some instances there is an effort to build appropriate technologies through local and available resources, in some cases using recycled materials. Caution must be taken, however, not to confuse appropriate technology with "low" technology. An often-cited example of this is the acclaimed nongovernmental organization (NGO) Light Up the World Foundation's efforts to introduce white light-emitting diodes (WLED) lighting to poor and remote rural areas. WLEDs offer alternative safe, healthy, and affordable lighting that is based on renewable energy sources. This is a high-tech solution to kerosene or wood-burning lighting which can pose health and safety risks. There is often implicit if not explicit care taken for the impact the technology will have on future generations and their ability to lead a good life.

Political Dimensions

There is often implicit, if not explicit, care taken for the impact the technology will have on future generations and their ability to lead a good life. There is a political dimension to appropriate technology that recognizes that technologies can have social and political dimensions built into them. Adoption of one or another technology may require the adoption of structural changes to social, economic, and political conditions. Adoption of a given technology may also increase a society's dependence on that technology and perhaps thereby on a power beyond that society. Appropriate technology, therefore, attempts to meet the basic needs of a society—for example, "cooking, sanitation, heating, cooling, and shelter"—while allowing the greatest degree of self-determination. Examples of appropriate technologies include solar cookers, solar water heaters, fuel-efficient wood stoves and ovens, hayboxes, alternative building techniques, dehydrators, and biodiesel. Not only innovative new technologies are sought after, but also improvements on frequently used existing technologies that often are ignored by technologists but are central technologies used on a daily basis for many people around the world, such as wheelbarrows and carts.

Philosophical Background

The second half of the 20th century witnessed the rise of the environmental movement in Western countries, with the 1960s and 1970s giving birth to the field of environmental ethics in North America. Some of the first alarms raised in the field called attention to the increasing technological power humankind had achieved and the need to limit our actions due to this new ability achieved through modern technology to radically alter ecosystems on a local and global level. Also in the mid-20th century, the philosopher Martin Heidegger, in *The Question Concerning Technology*, critiqued modern technology and its tendency to frame the natural world in such a way that it becomes perceived as natural resources there for human exploitation and use; in his terms, modern technology frames nature as "standing reserve."

The environmental movement and the critique of modern technology can be seen as converging in the concept of appropriate technology. The text most often cited as a founding text for the idea is Ernst Friedrich Schumacher's (1911–77) *Small Is Beautiful: Economics as If People Mattered* (1973). Schumacher, who was a student of economics, visited Burma as an economic adviser and was struck by the different way of life he encountered there. This experience forced him to revise his understanding of what a good life is and the role economics and technology play therein. He advocated a more spiritual understanding of humanity, the reduction of material desires, and highlighted the importance of good work for a good life.

Drawing upon Schumacher's insights, appropriate technology advocates challenge the development agendas of developing nations that can be interpreted as striving to enter the global economy with its attendant consumptive lifestyle. They recommend the adoption of culturally and ecologically appropriate technologies instead of advanced Western technologies. The underlying goal of such efforts is the introduction of technologies that increase the ability of people to pursue a good life while not imposing cultural changes upon communities, but rather allow for community and individual self-determination. Such efforts, however, can be critiqued as paternalistic, raising questions as to the appropriateness of Westerners' determining what technologies are "appropriate" to a given place, and thus what forms of life people in developing countries are justified in striving for or not.

What these discussions reveal is the impact technologies have on ways of life. Langdon Winner raised the question of whether artifacts have politics built into them. For example, if a society adapts nuclear power, such a technology will impose an authoritative political structure upon that society due to the hierarchical structures needed to maintain and safeguard nuclear technology, whereas solar and wind technologies require much less, if any, centralized authority, and thus are more conducive to alternative forms of political structures. Appropriate technology operates on this kind of understanding of technology and the politics inherent to them. One question that attends such a conception is whether it fails to do justice to the dynamic nature of technology, and the fact that different users will make different uses of a given technology and, hence, that any technology cannot be understood as having a fixed meaning. Andrew Feenberg points this out in his treatment of France's Minitel phones, which were designed with one function in mind but redefined and socially reconstructed in their use by actual users, thus transforming their meaning and function.

Current Manifestations

Contemporary philosophy of technology exhibits a shift from speaking about "technology" in an essentializing manner, to speaking about "technologies." That is a turn away from talking about technology in universal and deterministic fashion to engaging particular technologies and attempting to come to terms on how specific technologies will engender certain ways of life and prohibit others. There are roughly three ways to do this; these can be called, for convenience, downstream, midstream, and upstream models.

The downstream model has enjoyed a longer tenure; in this version, technologies are developed and, only after the fact, philosophers of technology and ethicists are handed a technology ready-made to make of it what they will. They can argue against it or in favor of it, but little can be done at this point to change it. A midstream model suggests that the role of philosophers of technology and ethicists is to engage scientists and engineers in the laboratories through the development process of a given technology. In this stage, the technology is still relatively flexible, and the input of an ethicist can help to shape a technology to be more sensitive to its impact on society. An upstream model argues for deliberation on a proposed technology before the development process gets under way.

Experiments in Scandinavian countries in particular have attempted to bring in diverse members of society to participate in the discussion of a proposed technological path at the early stages. This process could bring various values to the surface and thus discourage or encourage certain technologies, resulting in better, in the sense of more democratic technologies, or otherwise technologies more in line with the values of a society. While appealing, the problem with this model is that it is difficult to set a fixed meaning to a technology before the fact. Its meaning will likely shift through the development process, and it is more or less open how society will take up, adapt, and use that technology. For certain technologies, then, the meaning ascribed by society in its use is more or less flexible and difficult to determine from far upstream. However, there remains value in facilitating discourse on a given technology from the beginning to attempt, as best possible, to shape a technology so that it will be in accordance with the goals and values of society.

This is very much the spirit of appropriate technology—the realization that humans have an active role in the development technology will take, and we need not be bound by a technological determinism that holds that technology follows its own internal logic and we are powerless over it. Appropriate technology holds out for the possibility that communities can determine their own fate and meet their own needs in a sustainable way by actively playing a role in the developmental process of new technologies, or revisions of already-existing technologies. There is a strong emphasis on the role of local knowledge and participation of local community members in the development process. For there to be genuine participation, the input of community members must be solicited from the early stages of discussion regarding a new technology. Thus, appropriate technology tends to operate on the upstream model outlined above. Diverse voices from within the community are necessary at the outset to help model the technology to bolster and not to undercut community values and practices.

There is a recent trend in countries like the United States and the Netherlands, where science-funding agencies are requiring the incorporation of "users" in the formulation and early stages of innovative research projects. The role of these users is to provide feedback to increase the relevance and appropriateness of a given technology. The *Appropriate Technology Sourcebook* emphasizes that a way to make technologies more relevant to everyday problems of people is "to involve people who will use the fruits of research in the research process itself, and in decisions about research content."

The *Appropriate Technology Sourcebook* further provides four guidelines to improve appropriate technology development efforts. They are taken from the sourcebook as follows:

1. Change the criteria for "good" research. Good research should be that which is likely to reduce poverty.

2. Seek to understand the viewpoint of the poor—their perceptions of problems and opportunities.

3. Actively include the poor, especially small farmers and crafts people, in both decisions about research content, and in the research itself.

4. Offer basic relevant science education geared to the challenges of local problems, with curricula adapted to employ available materials and common devices to illustrate principles, and to provide young people and farmer-inventors with a more scientifically sound basis for their innovation efforts. Offer related courses on simple machines, how they work and how to fix them.

At work in much of appropriate technology efforts is a redefinition of the good. We see this in principle number 1 above, redefining good research. There is also an effort to redefine the good life. Appropriate technology often attempts to avoid technologies that fit with the adoption of a consumptive lifestyle that actively pursues more technological devices, which leads to an ultimately unsustainable lifestyle. Efforts must be made to avoid the paternalism of outsiders' defining what a good life is for another, however, and thereby limiting which goods and technologies they may or may not pursue. This is why involvement of the local peoples themselves and considerations for their self-determination are essential.

Occasionally appropriate technology discourse can be seen to lean toward two apparently distinct poles—either more toward the needs of poor people or toward the needs of the environment. The above principles focus on poor people and the goal of appropriate technology is understood as an effort to alleviate their poverty and facilitate their ability to meet their basic needs through appropriate technology. In other instances, appropriate technology is construed as being not so much about poor people as about the impact of technology on the environment and the need to protect the health and stability of ecosystems. However, in general, appropriate technology discourses are able to avoid strict nature/culture binary opposition. While some proponents may lean more toward emphasizing the environment and others toward vocalizing on behalf of poor people, ultimately proponents of appropriate technology recognize the interrelation between people and the environments they inhabit, such that the health and sustainability of a community is deeply related to the health and stability of their surrounding environment. Consider the following from the mission statement of the Development Center for Appropriate Technology: "Appropriate technology is that which strives to minimize negative consequences to all life, and connects people with each other and the Earth."

See Also: Conservation; Cradle-to-Cradle; Green Design; Green Technology; Precautionary Principle; Sustainability.

Further Readings

Aprovecho. "Appropriate Technology (AT)." http://www.aprovecho.net/pg/at.htm (Accessed April 2009).
Development Center for Appropriate Technology. "Mission, Vision, Values." http://www.dcat.net/about_dcat/mission.php (Accessed April 2009).
Feenberg, Andrew. *Questioning Technology*. London: Routledge, 1999.
Heidegger, Martin. *The Question Concerning Technology and Other Essays*. New York: Harper & Row, 1977.
Light Up the World Foundation. http://lutw.org/home.htm (Accessed April 2009).
Schumacher, E. F. *Small Is Beautiful: Economics as If People Mattered*. New York: Harper Perennial, 1973.
Village Earth. "Introduction to the Appropriate Technology Sourcebook." http://www.villageearth.org/pages/Appropriate_Technology/ATSourcebook/Introduction.php (Accessed April 2009).
Winner, Langdon. "Do Artifacts Have Politics?" *Daedalus*, 109/1 (1980).

Jonathan Parker
University of North Texas

B

BALANCED SCORECARD

The balanced scorecard (BSC) was popularized by Robert Kaplan and Greg Norton in the 1990s. The BSC is a strategic tool that seeks to align objectives of the enterprise with its vision and strategy. As such, it focuses on four perspectives: financial, customer, internal business processes, and learning and growth. Measurable goals within each objective are set, monitored, and evaluated. It is balanced because it looks at financial and nonfinancial measures in reaching the company's objectives. It is integrated, as success in meeting goals of one perspective will result in improved performance in other perspectives.

With an enhanced interest in environmentalism and sustainability, many have suggested incorporating sustainability into the BSC. Any attempt to measure success in sustainability must meet four criteria: (1) it must address the triple bottom line of sustainability— economic, social, and environmental performance; (2) it must have performance measures that can be clearly understood and communicated; (3) it must add value and be integrated into the value-adding systems of the organization; and (4) it must be supported by the existing management tools and resources. The BSC meets these criteria.

There is no agreement on how to incorporate sustainability into the BSC. Three approaches have been utilized. The first approach integrates environmental and social aspects into the existing perspectives. This approach considers sustainability as another strategic aspect of that perspective, and is seen as particularly appropriate in cases where sustainability is already integrated into the market system. This approach does not appear to have gained much use in practice.

The second approach adds an additional perspective to the existing four in order to incorporate environmental and social aspects. Referred to as the "nonmarket perspective," this approach is seen as necessary because environmental and social issues are not fully integrated into the market exchange process. Adding a nonmarket perspective allows inclusion of sustainability in the organization's strategy.

Those opposed to this approach argue that this leads to a wrong conclusion, missing a fundamental understanding of the BSC. Sustainability or environmental strategy is a theme of the strategy that spans the existing perspectives and is not separate from the other perspectives. Opponents observe that this approach lumps anything associated with sustainability into a single perspective whether it involves learning and growth, internal business, customer, or financial issues.

The third approach develops a separate scorecard, which is referred to as the sustainability balanced scorecard (SBSC). This approach evaluates each of the four perspectives along the triple bottom line of economic, social, and environmental performance.

The financial perspective would incorporate at least three factors. The first of these factors is the cost of compliance, innovation, and the cost and risk of noncompliance. The cost of compliance should be easy to determine as the organization would be dealing in an area where costs are known or knowable. The cost of innovation parallels research and development costs as applied to sustainability. Thus, a methodology for estimating these costs is readily available. The cost/risk of noncompliance is not difficult to determine. It involves fines for lack of compliance or the extreme measure of closing the business if the cost of compliance becomes too excessive.

The second factor includes the benefits of environmental processes, products, and services that create revenue. Sustainability carries a number of intangible benefits such as a conservation of resources and a healthier climate. Placing a price tag on these benefits is difficult, if not impossible. However, the focus is on actions the organization takes that create revenues. Sales of hybrid vehicles, compact fluorescent lightbulbs, and energy-efficient appliances are sustainable actions that create revenue streams.

The third factor in the financial perspective is the increase in the market value of the company due to its sustainable activity. This amount is not known, but can be projected using cost of capital measures and added revenue streams.

The customer perspective identifies four types of customers in regard to sustainability. First are actual customers that buy the organization's products or services. This customer is normally the object of consideration in a traditional balanced scorecard.

The community that would be affected by the actions taken by the organization is the second group of "customers." This group bears social costs when an organization avoids its social responsibility. The third group is the government that levies fines for noncompliance with environmental regulations. Finally, investors are "customers" who may be influenced to lower the cost of capital when a company actively promotes sustainability.

Internal business processes include processes for monitoring potential legislation affecting the organization as well as compliance with the regulations. Additionally, the organization may implement an effective sustainability compliance reporting process. This includes environmental auditing skills and tracking costs of reporting compliance. An additional process is the enhancement of the organization's research and development capability. This involves incorporating sustainability issues in the organization's existing research and development process.

Learning, the final perspective, involves training of personnel at all levels of the company in the triple bottom line of sustainability—economic, social, and environmental.

This third approach, a separate SBSC, seems to have gained a larger following than the other two perspectives, primarily due to the realization that sustainability is not something to be isolated as a part of the organization's strategy, but as an overarching issue that spans the existing balanced scorecard perspectives. Additionally, it focuses attention on sustainability rather than burying it in other strategies.

See Also: Cost-Benefit Analysis; Economic Value Added; Environmental Audit; Environmental Management System; Triple Bottom Line.

Further Readings

AP Institute. "Balanced Scorecards and Sustainability." http://www.ap-institute.com/Balanced%20scorecards%20and%20sustainability.html (Accessed September 2009).

Gminder, Carl and Thomas Bieker. "Managing Corporate Social Responsibility by Using the 'Sustainability-Balanced Scorecard.'" Göteborg, Sweden: Paper presented at the 10th International Conference of the Greening of Industry Network (June 23–26, 2002).

Kaplan, Robert and Greg Norma. *The Balanced Scorecard: Translating Strategy Into Action.* Cambridge, MA: Harvard Business School Press, 1996.

Moller, Andreas and Stefan Schaltegger. "The Sustainability Balanced Scorecard as a Framework for Eco-Efficiency Analysis." *Journal of Industrial Ecology* (August 2004).

Sidiropoulos, Michalis, et al. "Applying Sustainable Indicators to Corporate Strategy—The Eco-Balanced Scorecard." *Environmental Research, Engineering and Management*, 1 (2004).

John L. Stancil
Florida Southern College

BEST AVAILABLE CONTROL TECHNOLOGY

Best available control technology (BACT) is a standard established under the U.S. Clean Air Act (CAA) that requires regulated activities to incorporate the use of currently available technology to reduce their emissions of criteria pollutants (i.e., particulate matter, carbon monoxide, nitrogen oxides, sulfur oxides, and volatile organic compounds) to the maximum degree possible while taking into consideration economic, energy, environmental and other related costs. Under the U.S. federal New Source Review (NSR) program, BACT applies to construction of major new sources and major modifications (including relocation) of existing facilities within areas that meet U.S. Environmental Protection Agency (EPA) ambient air quality standards.

For the purposes of the NSR program, a major project is defined as one that has the potential to significantly increase emissions of pollutants, ammonia, or any factor that contributes to depletion of the ozone layer. This potential is considered relative to the prevention of significant deterioration (PSD) increment in any given area. The PSD increment is the difference between the baseline level of ambient air pollutants (carbon monoxide, lead, nitrogen dioxide, PM_{10} and $PM_{2.5}$ [particulate matter smaller than 10 micrometers or 2.5 micrometers], ozone, and sulfur dioxide) and the maximum concentration for those pollutants allowed under the U.S. National Ambient Air Quality Standards (NAAQS) established by the CAA. BACT only applies to projects in areas that comply with the NAAQS (attainment areas) and criteria subject to PSD. Major new or modified facilities in areas that do not meet NAAQS (nonattainment areas) must meet the stricter standard of lowest achievable emissions rate (LAER), and control for criteria for which the region is out of attainment. Existing facilities located in nonattainment areas are required to use reasonably available control technology (RACT).

Permits to construct (PTCs) and permits to operate (PTOs) are issued either by state, regional, or local air quality regulatory agencies. Due to regional differences in air quality, the thresholds for project review under NSR vary. Furthermore, because the extenuating circumstances of environment, energy, and economics vary from place to place, the technology that constitutes BACT for any given application may vary on a case-by-case basis. BACT is a time dependent determination; for example, given two similar facilities, the BACT required when the second one seeks a permit may be substantially different from the technology approved for the first if a more effective or efficient technology becomes available after the first facility was permitted.

A project proponent initiates the process of identifying BACT for a particular project with an application for a PTC or a PTO. Any activity that will result in increased emissions of pollutants is considered a project. The first step is to identify all available control options, including those designated LAER. Applicants are expected to consider technologies in use both in the United States and abroad. They are directed to investigate the possibility of transferring technology from other similar facilities, and encouraged to consider adopting innovative control technology. BACT determinations in recently awarded permits for similar facilities are often helpful in expediting the search process. The EPA maintains the RACT/BACT/LAER Clearinghouse, an online, searchable database that proposers can use to identify control technologies for potential use in their own facilities. Many state, regional, and local agencies maintain similar databases.

The second step in the process is to analyze each available control option to determine whether its application is feasible. There are two aspects to feasibility: the first is availability. Any licensed and commercially available control technology is considered available. The second aspect is applicability, that is, certainty that a given control can be installed and will operate effectively on the proposed facility. If a technology is installed and operating on a similar source-type facility, it is demonstrably feasible, and it will be presumed that the proposed facility will be able to achieve comparable pollution reduction by employing it unless solid evidence to the contrary is presented. In order to eliminate a control option, the applicant must substantively document that it is not available, could not be installed, or would not satisfactorily provide the desired level of control. Technically infeasible options require no further review. The decision on whether a control option is technically feasible rests with the review authority.

At this juncture, if the control option being proposed is deemed feasible and satisfies LAER, no further analysis is required. The applicant need only document that the option is indeed the LAER technology and review the environmental impacts associated with this option. If the preferred option is not LAER, then all remaining control options must be scrutinized to determine how efficiently they will remove pollutants, what the anticipated emissions rate will be for the facility where the control mechanism installed, and what the estimated emissions reduction would be, following installation. Other environmental impacts, energy profiles, and economic implications are also taken into account. Environmental impacts that must be considered include influence on air quality (reduction in visibility, odor, release of toxins or other pollutants, or visible emissions), water quality, solid waste impacts, and noise. Given two technologies equally effective in removing a pollutant of key concern, the one that is more effective in removing additional pollutants will be preferred. Reliability of the control technology must also be considered by examining the potential for accidental releases and breakdowns. At a minimum, any BACT must meet the New Source Performance Standards (NSPS), although, given technological improvements, BACT is not limited by NSPS. Any BACT must also meet the National Emission Standards for Hazardous Air Pollutants (NESHAP).

Preliminary analysis of candidate control options includes consideration of energy efficiency (energy use per unit of pollution reduction), as well as type of fuel used, fuel characteristics, availability issues, and source reliability.

Finally, the economic factors are identified. Cost effectiveness is an important issue in this regard, with those technologies that minimize the cost of reducing emissions being preferred, all else being equal. The ability of the applicant to afford a particular technology is not a consideration in determining BACT. The BACT requirement is mandatory, and is therefore as much a component of construction and operating costs as any other crucial equipment.

Following the feasibility analysis, the available control options are rank ordered by their effectiveness in pollution reduction, that is, the extent to which they remove pollutants compared to the required control level. The second order ranking for controls that are equally effective at reducing pollutants is based on their economic costs, in other words, total annualized costs, average cost effectiveness and incremental cost effectiveness, ancillary environmental impacts (toxics impacts, other adverse impacts), and the incremental increase of use over the baseline year. Normally, only direct energy impacts are considered, but if unusual or significant indirect impacts are identified and can be quantified, they can be included in the analysis.

The fourth step in the process of determining BACT is an in-depth analysis of the top-ranked option. Each impact must be examined and supporting information must be presented to document both beneficial and adverse impacts. In the absence of exceptional circumstances, only direct impacts need be analyzed.

At this stage, the applicant is essentially substantiating the rationale for designating the top choice as BACT. The in-depth analysis is intended to determine whether there are extenuating circumstances, such as accompanying, unavoidable environmental impacts, that dictate against the top choice. If no such circumstances emerge, the analysis is concluded, and the top-ranked control option is determined as BACT.

If some issue does dissuade the use of the top choice, the applicant must make a solidly documented case for elimination of that option, and then the next most effective option must be fully analyzed. This process continues iteratively until a control measure is finally chosen by the applicant and deemed BACT by the regulating authority.

While BACT determinations have generally not been subject to litigation, there is a large body of heavily nuanced, interpretive memos. Recent EPA comments on BACT analyses for proposed power generating facilities have been seen by some as a change in direction away from taking the nature of the project as a given toward an approach that modifies the project. The concern is really one of scale, as basic equipment substitutions for cleaner technologies have often been required, particularly when increased capital costs are offset by reduced operating costs. In point of fact, project proponents are increasingly being forced to consider alternate basic technology, that is, best available technology (BAT) that addresses pollution reduction on the front end rather than through backdoor controls.

The designation of greenhouse gas (GHG) emissions as regulated pollutants, anticipated in the second quarter of 2010, is also causing concern. Carbon dioxide is released in much greater quantities than the 250 tons per year threshold established for trace pollutants. The EPA will purportedly adopt interim thresholds for GHG to address the discharge concerns, but as of early 2010, there was no BACT available, although at least one technology was being tested.

See Also: Appropriate Technology; Best Management Practices; Clean Technology; Corporate Social Responsibility; Cost-Benefit Analysis; Environmental Risk Assessment; Green Technology; Leadership in Green Business; Maximum Achievable Control Technology; Pollution Prevention; Voluntary Standards.

Further Readings

Schnelle, K. B and C. A. Brown. *Air Pollution Control Handbook*. Boca Raton, FL: CRC Press, 2002.

U.S. Environmental Protection Agency (EPA). "New Source Review Workshop Manual: Prevention of Significant Deterioration and Non-Attainment Area Permitting." (Draft October 1990). Washington D.C.: EPA, 1990. http://www.epa.gov/nsr/ttnnsr01/gen/wkshpman.pdf (Accessed February 2010).

U.S. Senate Committee on Environment and Public Works. U.S. Code Title 42, Chapter 85 § 7401-7671q. http://epw.senate.gov/envlaws/cleanair.pdf (Accessed February 2010).

Susan H. Weaver
Independent Scholar

BEST MANAGEMENT PRACTICES

Best Management Practices (BMPs) comprise a range of different concepts and definitions, with varying emphases depending on the primary interests or perspectives of the definers. Environmental BMPs, however, share in common that they are geared toward real (or apparent) reduction or elimination of detrimental impact on the natural environment (such as pollution or habitat destruction). Though references to the general idea of BMPs can be found as far back as the late 19th century, the concept did not gain serious traction until it appeared in the revised 1977 Clean Water Act (though it lacked clear definition), and was originally concerned only with water quality. Since then, BMPs have found a variety of uses in agriculture, manufacturing, land maintenance, and animal care, covering the use and protection of everything from water to air, soil, and natural resources.

Common to BMP *regulations* is that they are primarily concerned with so-called nonpoint source pollution, that is, pollution that is an indirect result of production or consumption, and that does not have a clearly defined or definable source of origin. But BMPs have by now also found a much more common usage, as in "best practices" to minimize detrimental environmental impact of one's actions, whether they are personal or communal, in production or consumption. Typical for the kinds of rationales given for the creation and implementation of BMPs, the U.S. Department of the Interior states that BMPs are significant in the pursuit of "enhancing quality of life for all citizens through balanced stewardship of America's public lands and resources." In short, the most fundamental goal of all BMPs is what can best be summarized as pollution prevention, resource conservation, and environmental stewardship.

Basic Uses for Best Management Practices

Ever since the industrial revolution, business interests and environmental concerns seem to have been on a coalition course. The reason is simple: virtually all conventional forms of extraction, manufacture, and disposal are destructive to the natural environment.

Government agencies frequently find themselves caught in the middle of the competing interests of environmentalists and industrialists, seeking to mediate, and attempting to find (or at least appear as attempting to find) some middle way agreeable to both parties.

BMPs are a classic example of a political answer to a thorny problem—in this case the problem of how to assist virtually indiscriminate growth in the economy without unduly risking to destroy the natural habitat upon which all life (and business) ultimately depends.

A wide array of increasingly pressing environmental problems—ranging from meat contamination to species extinction, toxic runoffs, smog, and global warming forced the government to intervene in the presumably free market regulating the commercial interactions of citizens in the United States.

While historians and economists can argue about when exactly the point was reached, there is little doubt that by the time of the first Clean Air (1963) and Clean Water (1972) Acts, had the government not begun to provide comprehensive guidelines, regulations, and enforceable restrictions on the use of resources and our natural environment, much of the country today would look like the unregulated slaughterhouses of Chicago vividly portrayed by Upton Sinclair in his 1906 book *The Jungle*, or the 1969 city of Cleveland where the Cuyahoga River self-ignited due to its toxic contamination with industrial waste.

As the term *best management practices* already suggests, the challenge is to "manage" the environment (such as water, land, animals, forests, coral reefs) in a way that allow for use and exploitation of habitat and resources, but at the same time minimize consequences on both nature and people. As such, the format of BMPs ranges from *general guidelines* (as in, for instance, attempts "to minimize the disturbance of wetland fauna" or "make every effort to restrict toxic runoff") to enforceable *regulations* (as in, for instance, the U.S. Environmental Protection Agency [EPA]–enforced "new technology-based effluent limitation standards for wastewater discharges associated with operation of aquatic animal production facilities"). Indeed, the Federal Water Pollution Control Act Amendments of 1972, Public Law 92-500 (and as amended by Sec. 319, 1986), for instance, require the management of nonpoint sources of water pollution from sources including forest-related activities.

It is an early example for the development of federal BMPs, in this case for forest landowners, other land managers and timber harvesters, yet compliance with the act was voluntary, and the BMPs merely provided guidance. The central goal of the law was to accomplish maintenance of water quality to provide "fishable" and "swimmable" waters. Whether the sources of pollution are clearly identifiable (point pollution) or have multiple, not clearly identifiable sources (nonpoint pollution), the EPA has over the years established an array of BMPs—both as guidelines and as enforceable regulations—as an acceptable method of reducing pollution.

Increasingly, BMPs are also used as recommendations for individuals or families in an attempt to combat things like erosion, toxic water runoff, or greenhouse gas emissions. Most states have such BMP recommendations, and they urge citizens to take measures such as the following:

- Reduce the number of necessary herbicide applications by applying herbicides when they will be most effective against target weeds.
- Bring oil, antifreeze, and other contaminants to the Household Hazardous Waste disposal collection sites instead of pouring them onto driveways, streets, or into storm drains.
- Conserve electricity and set air conditioners to 78 degrees.
- Share a ride to work, use public transportation, bicycle, or walk whenever possible.
- If you must drive, use your newest car or truck since new vehicles pollute less.

In the United States today, all commercial activities that have a potential effect on the environment are captured by one or more sets of BMPs, ordinarily issued by the federal, state, or local governments. Whether in the fields of construction, manufacture or agriculture, pharmaceutical or materials' research, waste or water management, energy generation or transportation, commercial activities today are all subject to a continuously increasing range and number of BMPs.

The effectiveness of these BMPs in reaching their own stated goals, however, is a much-researched and much-debated question. Studies by several public and private groups, often in association with the government, routinely find that evaluating effectiveness of BMPs continues to be difficult due to a variety of problems, ranging from a lack of clear evaluation standards or centralized, easy-to-use, scientifically sound tools for assessment, as much as a lack of reliable and standardized data, and clear reporting and performance evaluation protocols. Nevertheless, there is strong evidence in a wide range of applications that environmental protection outcomes have improved since the introduction of BMP guidelines and regulations (though it is often impossible to decipher conclusively as to what specific BMPs yielded what kind of results). Results, however, differ a great deal depending on application and time frame.

A recent report found, for instance, that certain Great Lakes communities, after realizing how economically significant the cleanliness of the lake water was to their tourism industry, implemented stringent BMPs and succeeded in reducing bathing water quality advisories from 66 percent of days during the swimming season in 2000 to 5 percent or less in four consecutive years (2005–08). Even a cursory search, however, also yields a great number of less than satisfactory outcomes. Failures in implementation as well as successful accomplishment of stated goals of BMPs abound, for instance, in the field of industrial and roadway wastewater runoff, aggravated by increasingly overwhelmed communal wastewater treatment facilities and an aging sewage system.

A comprehensive evaluation of BMP outcomes, in other words, is far from possible at this point. Evaluation attempts in almost all applications are ongoing, however, and will continue to generate additional insights. What is possible to state, unequivocally, is that BMPs have become more sophisticated with new incoming research findings on how best to prevent harm to the natural environment, and that BMPs have provided essential tools, without which environmental harm across a wide range of applications undoubtedly would have been far greater.

See Also: Best Available Control Technology; Corporate Social Responsibility; Resource Management; Stewardship; Sustainability.

Further Readings

Epstein, Marc J. and John Elkington. *Making Sustainability Work: Best Practices in Managing and Measuring Corporate Social, Environmental, and Economic Impacts.* San Francisco: Berrett Koehler, 2008.

Hall, Ridgway M., Jr. "The Clean Water Act of 1977." *National Resources Law,* 343 (1978).

Ice, George. "History of Innovative Best Management Practice Development and Its Role in Addressing Water Quality Limited Waterbodies." *Journal of Environmental Engineering,* 130/6:684–89 (June 2004).

Sharpley, A. N., et al. *Best Management Practices to Minimize Agricultural Phosphorus Impacts on Water Quality.* Washington, D.C.: USDA-ARS 163 (2006).

U.S. Environmental Protection Agency. "The Clean Water Act and Best Management Practices." http://www.epa.gov/watertrain/forestry/forestry3.htm (Accessed January 2010).

Wilson, W. Gary and Dennis R. Sasseville. *Sustaining Environmental Management Success: Best Business Practices From Industry Leaders*. Hoboken, NJ: John Wiley & Sons, 1998.

Dirk Peter Philipsen
Virginia State University

BIO-BASED MATERIAL

A bio-based material is an organic material that is derived from a live, renewable resource, usually plant, and occasionally animal. Common examples of bio-based materials are corn, soy, wood, linoleum, straw, humus, manure, bark, cotton, spider silk, chitin, fibrin, and bone. Bio-based materials are primarily used as biofuels, but they are also used for bio-plastics and consumer goods. Other common uses include packaging, coatings, foams, cleaners, and adhesives. Common examples of bio-based products include detergents, plywood/particleboard, soy-based plastics, or a variety of bio-based packaging products, the most common of which is the paper bag. Bio-based materials are becoming ever more significant as an alternative to fossil-based products because they come from renewable resources, have the potential to reduce pollution, and potentially decrease pressure on the food chain. Although an increasing number of consumers are interested in bio-based products (due to growing green consumerism), their commercial viability continues to be limited since they often cost more to produce than comparable standard goods. As the price of standard goods never reflects the *real* cost of the good (much of the cost of extraction, production, pollution is carried by society, not the consumer), many bioproducts are dependent on direct government subsidies to be competitive in a market favoring fossil fuel–based industrial products.

Raw Materials

The main raw materials used to create bio-based materials are either conventional crops, such as corn or soybeans, or cellulosic-based sources such as wood. The foremost advantage of conventional feedstock is that growth technologies are highly advanced and efficient. The major drawback is that crops such as corn or soy are also used for feed and food production: the more they are used as a basis for bio-based materials, the more prices will increase in their use as food. Furthermore, and perhaps even more importantly, turning plant matter into fuel has also been found to be highly energy inefficient, as it routinely uses more energy to produce than it yields. The use of cellulosic sources, on the other hand, does not create any direct pressure on the food chain, but conversion methods and technologies are not yet available at a commercially viable level. In addition, the use of wood can lead to further deforestation with all its related negative side effects. Other raw materials used include recycled paper, animal by-products, and raw rubber.

Uses for Bio-Based Materials

As rising levels of pollution and resource depletion generate rising economic, political, and social demands for sustainable alternatives to fossil fuel–based energy and goods, bio-based materials, chemicals, and energy sources are likely to transform the materials industries in fundamental ways. As demands for sustainability and environmental regulations will grow, the materials industry will increasingly have to heed the principles of ecoefficiency, industrial ecology, green chemistry, and engineering.

While one of the main uses of bio-based material is still biofuels, other uses are rapidly increasing in both variety and number, ranging from its most common usage, bioplastics (for tractor parts, water bottles, and so on), to clothing and building materials. Other common usages of biomaterials include pharmaceuticals, coatings, packaging materials, foams, adhesives, cleaning products, and insulation. As environmental consciousness is advancing, consumers increasingly demand bio-based products, particularly for short-term, single-use purposes such as food or drink containers, or a range of packaging uses.

Advantages of Bio-Based Products

Compared to a fossil fuel-based economy that inevitably depletes nonrenewable resources, pollutes the environment, creates dependency on oil-rich regions, and threatens our health, a bio-based economy yields a multitude of significant advantages.

Renewable Resource

Bio-based materials are, by definition, renewable, though the resources required to grow some raw materials—water, fuel, fertilizers—may be finite. As demand for fossil fuels continues to skyrocket, the world is rapidly approaching—or, as some argue, has already reached—what is commonly known as "peak oil," that is, the point of largest possible extraction. With rapidly growing economies such as China and India generating a voracious new appetite for fossil fuels, bio-based fuels and energy sources are not only becoming economically viable but a necessity. As fossil fuels will inevitably become scarcer and more expensive, bio-based materials and energy sources will become essential to the continued existence of developed consumer societies.

Pollution

While the growth, harvest, storage, transportation, and use (both industrial and private) of bio-based materials is not without harmful effects for the environment, researchers widely agree that the impacts may be far less significant than a fossil-fuel based economy, depending on how the bio-based feed stocks are produced. Considering that some 85 percent of major atmospheric pollutants today—O_3, CO_2, SO_2, NO_x—directly result from burning fossil fuels, or that some of the most devastating forms of environmental pollution of soil, air, and water are generated by fossil-based chemicals, such as oil spills or pesticide runoffs, bio-based materials generally represent a far less toxic alternative.

Resource Dependency

Policies to reduce our dependence on fossil fuels have not quelled increasing demand, with all of the negative consequences of such dependence. The increased substitution of sustainably raised bio-based materials for fossil fuels provides a possible alternative.

Health

Potential health benefits of an increasingly bio-based economy are too numerous to mention. Scientists are merely beginning to have a glimpse of the aggregate health effects of decades of emitting fossil-based toxic chemicals into the soil, air, and water, and adding them to the vast majority of goods produced and sold. For one, our essential food supplies are exposed to toxic chemicals from the point of origin to the point of consumption. The breast milk of the average American mother, the most essential diet for newborn babies, for instance, contains a staggering number of toxic chemicals. The list includes chemicals used in pesticides, fire retardants, by-products of manufacturing and incineration, solvents, lubricants, wood preservatives, cosmetics, and cable insulation, as well as metals such as lead, mercury, and cadmium—by far the most of which are fossil based, and many of which are easily replaceable with far less toxic bio-made materials.

It should also be noted that the emergence of cellulosic bio-based materials could lead to reduced demand in crops traditionally used for food supply, reducing the "food or fuel" dilemma which forces farmers to choose between growing a crop for food or for industrial production. Finally, one of the main attractions of bio-based materials is their renewability. Since many bio-based sources are agricultural crops, they can be harvested annually, making them an ideal renewable source. Furthermore, these crops renew themselves on a predictable time frame, from annual renewal (grains and grasses) to years (like managed forests). The fact that these materials can be farmed in lands sometimes unsuitable for traditional agriculture also reduces the "fuel or food" dilemma.

Commercial Viability

To encourage research and commercial viability of bio-based products, many government agencies have implemented purchasing programs that encourage the purchase of bio-based products. This provides a much-needed boost for bio-based products, as they are in many instances, following current pricing structures, more expensive than standard goods. In the United States, the Department of Agriculture (USDA) has compiled a list of products and the recommended composition of biological components that has to be attained for the product to be certified as "bio-based." For example, a cleaning product should have at least 74 percent biological components to be recognized as bio-based, while a laundry product might only need 34 percent. This is done to support existing programs (such as subsidies and tax credits) to encourage local industry to develop and market these products, and to ensure that a credible end buyer exists once the product reaches the market.

Another key concern for bio-based materials, and especially for biofuels, is their total environmental impact over their entire life cycle (for instance, from farm to finished fuel to use in the vehicle). To fully assess the environmental impact of biofuel, scientists are increasingly calling for an honest account of the "carbon intensity" of a particular bioproduct, that is, a measure that accounts for the full life cycle emissions per unit of energy delivered. While such accounts pose a variety of analytical challenges (kind of feedstock, production process, model inputs and assumptions), they are essential for any kind of realistic evaluation of the environmental benefits and drawbacks of each bio-based material, and especially biofuel.

As with all green products, the bio-based material economy finds itself in a conflict between environmental sustainability and economic viability. Not all bio-based materials or fuels are environmentally sound or sustainable, and many cannot compete economically as long as the pricing structure of standard goods effectively "externalizes" many of the costs of extraction, production, and consumption (such as depletion, pollution, sickness). In addition, those who promote bio-based materials for the purpose of environmental protection routinely apply different criteria than those who see business opportunities in bio-based materials.

A great deal of research is going into the creation and possible applications of bio-based materials. New uses are developed on an almost daily basis—from building materials to packaging and fuels—and there is no doubt that bio-based materials will make up a growing percentage of goods consumed by Americans in the future.

See Also: Biofuels; Externalities; Green Technology; Life Cycle Analysis.

Further Readings

Deffeyes, Kenneth S. *Hubbert's Peak: The Impending World Oil Shortage.* Princeton, NJ: Princeton University Press, 2001.

Dorsch, R. "Sustainable Materials and Chemicals for the Next Generation." In A. Eaglesham, et al., eds., *The Bio-Based Economy of the Twenty-First Century: Agriculture Expanding Into Health, Energy, Chemicals, and Materials.* Ithaca, NY: National Agricultural Biotechnology Council, 2000.

Hatti-Kaul, R., et al. "Industrial Biotechnology for the Production of Bio-Based Chemicals—A Cradle-to-Grave Perspective." *Trends in Biotechnology* (2007).

Miller, S. A., A. E. Landis and T. L. Theis. "Environmental Trade-Offs of Bio-Based Production." *Environmental Science & Technology,* 41:5176–82 (2007).

Weber, Claus. "Biobased Packaging Materials for the Food Industry." Food Biopack Project, November 2000. http://jobfunctions.bnet.com/abstract.aspx?docid=119683 (Accessed January 2010).

Jean-François Denault
Université de Montréal

BIOFUELS

Research into biofuels began in earnest soon after the oil crisis of 1973. By the early to mid-1990s, biofuels emerged as a serious alternative to fossil fuels, and were touted as an Earth-friendly answer to the need for lower fossil fuel carbon emissions and a reduction of foreign dependency on overseas oil. In essence, biofuel is an energy resource that is derived either from living organisms (called biomass) or from waste biomass (also called biowaste). It can be produced in solid, liquid, and gaseous form, but since it is used mostly for transportation, it is primarily found in liquid form.

In the public debate, biofuels generally refer to ethanol or biodiesel, the former favored by the Americas (Brazil and the United States produce 90 percent of the world's ethanol between them); the latter preferred by Europe (accounting for 89 percent of global biodiesel production in 2005). While biofuels can derive from a large number of crops, most ethanol in the world today comes from corn and sugar cane, whereas most biodiesel emanates from vegetable oils or animal fats, including waste cooking oil.

Biofuels, such as sugar cane at this distillation plant near São Paulo, Brazil, are believed to be an important pollution-reduction alternative and a renewable source of energy.

Source: Shell

While biofuels undoubtedly have several benefits over fossil fuels, it has become increasingly apparent in recent years that they do not represent the environmental panacea many hoped they would. The United Nations Commission on Sustainable Development in a recent report acknowledged, for instance, that biofuels are more carbon dioxide (CO_2) neutral than fossil fuels, but may have harmful effects if growing the feedstock is done in a way that causes deforestation, competes with the food supply and drives up the cost of food, or consumes water in water-scarce regions.

First-Generation Biofuels: Agricultural Assets (Agrofuels)

First-generation biofuels use biomass generated by agricultural assets; hence, they are derived from sources such as starch, sugar, animal fats, and vegetable oil. The main crops employed to produce biofuels are corn, sugarcane, soybeans, and canola.

There are many different types of first-generation biofuels that are produced. One of the most common is biodiesel. It is produced by mixing transformed biomass with methanol. This fuel is commonly found as it can be used in diesel engines after being mixed with a mineral diesel. There are also vehicles that are designed to use regular diesel or biodiesel without distinction. Some countries promote biodiesel in an attempt to reach economic independence, while others promote it as a way to reduce vehicle pollution emissions. Biodiesel mix can be anywhere from 5 percent to 20 percent (which means 5 percent to 20 percent biodiesel is added to regular diesel). Other biofuels include biogas (which is produced by anaerobic digestion) and bioalcohol (produced through fermentation of starches and sugar).

The main advantage of first-generation biofuels is that the technology has been extensively researched and developed. As such, many transformation facilities are almost commercially viable (excluding subsidies). The main disadvantage is that converting crops into biofuels creates pressures on the food chain, as there is a competitive relationship between traditional food industries and energy industries to purchase agricultural assets.

Second-Generation Biofuels: Waste Fuels

Second-generation biofuels are cellulosic ethanol created by using nonagricultural crops or waste. This waste is a by-product of regular agricultural activities. Some of the waste used includes agricultural scraps, crop residue, woody waste, and organic waste.

Second-generation biofuel emphasizes optimizing the transformation process of a non-useful waste into a useful resource. For example, some second-generation biofuels are made out of wood waste, low-yield food crops, or recycled newspapers. Nonetheless, processes and technologies available at this time are not commercially viable and inefficient. Even with the inherent advantages that they may have (for example, little to no cost for the waste itself or for government subsidies), the transformation process is often uncompetitive with first-generation biofuels or traditional energy sources.

Advantages Linked to Biofuels

There are a number of reasons why biofuels have been heralded as holding great promise for both the environment and economic progress. On a fundamental level, all developed economies depend on a vast and steadily increasing amount of fossil fuels, yet the planet is either at or close to peak oil. Biofuels provide relatively simple and readily available alternatives that, in many cases, do not even require large-scale structural changes in the economy. Some diesel-powered cars, for instance, can just as easily run on biodiesel; ethanol is already mixed into regular fossil-based fuels at gas stations across the country.

From a more purely environmental standpoint, biofuels have been celebrated as being carbon neutral, meaning that they sequester as much CO2 in plant form as they later release when burned as biofuel. More accurately, however, the potential greenhouse gas emission benefits of biofuel plants vary significantly, from an estimated high of 40 percent reduction of emissions compared to fossil fuels (in the case of corn), to an estimated 110 percent reduction (in the case of switchgrass). But such figures do not fully consider various other factors that potentially reduce the benefits significantly (see below).

Proponents of biofuels point to yet another important benefit: developing an alternative to fossil fuels can reduce dependence on international energy providers, and particularly oil-producing nations in the Middle East. As over 60 percent of crude oil and petroleum products used in the United States today come from other countries, finding domestic alternatives to fossil fuels is becoming ever more important. Others have argued that dependence on foreign oil also entangles the United States in foreign conflicts and wars, which cost the American taxpayers trillions of dollars—money that might be saved if the American economy was energy independent.

Savings are significant: Brazil is said to have substituted over $60 billion of oil imports since its large-scale implementation of biofuel production (mostly ethanol). Most European nations, as well as Japan, have negligible amounts of domestic oil reserves, and thus depend on either foreign oil or domestic alternatives even more than the United States. Given the drastically growing demand for oil in large expanding economies such as China and India, and the resulting inevitably fierce global competition for the finite supply of oil, in short, finding domestic fuel alternatives is no longer just a benefit—it has become a necessity.

Issues and Challenges

Few products reveal as clearly as biofuels do that there is never an easy answer to the resource and environmental crisis. The vast majority of biofuels are based on plants, and thus require land, air, water, and nutrients to grow. Even though biofuels today make up only about 1 percent of fuel consumption for transportation, an estimated 12 million hectares have already been turned over for their production—land that is no longer used for the growth of food, thus driving up food prices. Land has also been clear-cut to allow for the growth of biofuel designated plants, leading to further deforestation and a net increase in CO_2 emissions. Furthermore, according to Sweden's Stockholm Environment Institute, even if we were to replace merely 50 percent of the fossil fuels with biofuels for the estimated fuel consumption in transportation and energy in 2050, we would need an extra amount of water that is almost equal to the total annual flow of all of the world's rivers. This would be a serious problem in the United States, but would be an insurmountable problem in rapidly growing countries where water supplies are already scarce, such as China or India. There is also the problem of the amount of fertilizers used on biofuel crops, which, according to Nobel-winning scientists, could wipe out all of the biofuel carbon savings, as fertilizers emit large amounts of nitrous oxide—a heat trapping gas almost 300 times more harmful than CO_2.

A recent study found that whether biofuels offer carbon savings or not depends on how they are produced. Converting rainforests, peat lands, savannas, or grasslands to produce food crop–based biofuels in Brazil, southeast Asia, and the United States creates a "biofuel carbon debt" by releasing 17 to 420 times more CO_2 than the annual greenhouse gas (GHG) reductions that these biofuels would provide by displacing fossil fuels.

Some of these issues may be partly remedied by using less fertilizers, pesticides, and energy, and above all, by utilizing lands with low agricultural value. According to a report by the U.S. Agricultural Research Service, for instance, planting native switchgrass on the prairies of Nebraska and North and South Dakota could yield cellulosic ethanol that contains 5.4 times more energy than what went into its production.

Another challenge linked to biofuel is its economical viability. When oil prices are high, investors are interested in projects related to biofuel, but when world oil prices plunge, it becomes uneconomical to produce biofuel. Analysts estimate that a price point of roughly $70 per barrel of crude oil is the limit below which biofuels become uneconomical. This assumes, of course, that the regulatory and pricing structures will not change. If the U.S. government began to tax carbon emissions, ended subsidies for fossil fuels, or internalized pollution expenses into the price of oil and gasoline, biofuels would become far more economically viable.

Another concern is that growing biofuel crops, particularly in poor countries, leads to both monoculture and agribusiness of scale, undermining biodiversity, and the livelihoods and independence of small farmers, and leading to depletion of soil and deforestation. Additionally, to the extent biofuels become economically attractive, biofuel crops will begin to extract higher prices than comparable food crops, steering farmers away from growing food, thus further aggravating the food crises in many countries.

In the end it remains unclear whether biofuels are a direct benefit to the environment. When one accounts for the costs and energy of growing the crops, making (and using) fertilizers, as well as processing the biofuel, many biofuels arguably yield less energy than is needed to produce them. And since many industrial complexes manufacturing biofuel burn coal or natural gas in the process, significant amounts of pollution are created during the manufacturing process.

Finally, critics have pointed out that even if all of the crops currently producing food were shifted to biofuel production, there would still not be enough nonfossil fuels to substitute for the current oil consumption. At the current rate of consumption (which is much less than the predicted rates of consumption in the future), biofuels can never fully replace fossil fuels. As an intermediate source of fuel, or even as a long-term solution in a much less energy intensive economy, however, biofuels most likely can and will play an increasingly significant role.

See Also: Bio-Based Material; Carbon Neutral; Clean Fuels; Externalities.

Further Readings

Doornbusch, R. and R. Steenblik. "Biofuels: Is the Cure Worse Than the Disease?" Organisation for Economic Co-operation and Development's Round Table on Sustainable Development, 2007. http://www.cfr.org/publication/14293/oecd.html (Accessed January 2010).

Hill, Jason, et al. "Environmental, Economic, and Energetic Costs and Benefits of Biodiesel and Ethanol Biofuels." *Proceedings of the National Academy of Sciences,* 103/30:11206–10 (2006).

Mousdale, David M. *Biofuels: Biotechnology, Chemistry, and Sustainable Development.* Boca Raton, FL: CRC Press, 2008.

Pahl, Greg. *Biodiesel: Growing a New Energy Economy*, 2nd Ed. New York: Chelsea Green, 2008.

Ruth, Laura. "Bio or Bust? The Economic and Ecological Cost of Biofuels." *European Molecular Biology Organization Reports*, 9/2 (2008).

Jean-François Denault
Université de Montréal

BIOLOGICAL RESOURCE MANAGEMENT

Biological resource management (BRM) refers to any programmatic science-based intervention designed to preserve, restore, or enhance the value of a biological resource. A BRM plan (BRMP) can focus narrowly on a single species, or, as is now more commonly the case, on the wider issue of biodiversity. BRM and habitat conservation and restoration are closely related, but BRM frequently involves economic considerations that habitat conservation efforts generally do not. Furthermore, given climate change, world population growth and its concomitant impacts, the need for BRM is becoming increasingly more apparent.

The fisheries industry was among the first to recognize the need for BRM in the mid-20th century when the problem of coastal eutrophication lead to dramatic declines in food fish production. The presence of elevated concentrations of organic matter in the water that reduced water clarity, lowered oxygen levels in deep waters, and killed sea grasses, was traced mainly to nitrogen and phosphorus from agricultural water runoff. Nitrogen introduced to the atmosphere by fossil fuel combustion was also implicated. Faced with the loss of their livelihood, fisheries around the world began to address the issue, engaging with their governments to devise strategies to address and reverse the problem, an effort that required the involvement of the agricultural industries as well. Though eutrophication was not the first sign of an ecological system failure, it was the first to be recognized as putting a whole industry at risk. This was due in large part to the fact that, though ecological systems are integral to human welfare, their contribution to economic production has been taken for granted. They are poorly handled by traditional economic models, which address scarcity through price mechanisms but fail to account for resource degradation and depletion. Consequently, natural and environmental resources are undervalued, misallocated and misused. The issue of eutrophication helped raise awareness of the environment as an asset that requires management, but it also highlighted the complexity of management issues and the need to think of the environment as a system rather than focusing on isolated elements.

Despite significant advances in our understanding of ecological systems, their complexity can be overwhelming. Incidents like the mysterious decline of the honeybee populations in Europe and North America that began in 2004 brought the issues of economics and the ecosystem together on the nightly news, as commentators discussed how food prices would be affected and whether certain fruits might disappear completely without bees to pollinate them. Scientists, unable to pinpoint a definite cause, were unable to prescribe a definitive course of corrective action.

An understanding of what is happening ecologically is critical to establishing a BRMP, determining how to proceed, and at what scale. As the examples of coastal eutrophication and declining honeybee populations illustrate, some problems must be dealt with on an international level. Local BRMPs are also needed, but even for these, the monitoring efforts must transcend several political boundaries, for example, municipalities, watersheds, and airsheds. Like the habitats they seek to fortify, BRMPs must be hierarchically nested, which raises a host of social, political, and economic issues that require balancing the short-term needs of multiple stakeholders against the long-term interest of community sustainability and species survival. Individuals tend to worry about livelihood, commodity affordability, and private property rights. Industry groups and individual businesses tend to be defensive and inflexible, frequently focusing on short-term financial considerations rather than on the long-term sustainability of the resources they consume. Governmental policies tend to be inconsistent and difficult to amend. Delegating responsibility for formulating, financing, implementing, and monitoring a BRMP can be a lengthy, fractious, and often litigious process.

How aggressively to intervene in an ecosystem is also a point of contention. For some, BRM should simply attempt to restore a system to a preexisting condition. Others see this as a futile effort, given climate change, and argue for building resilient ecosystems. Advocates of the latter call for adaptive management, seeing change as inevitable and proposing a proactive scientific approach involving strategic interventions and sophisticated statistical analysis to find out exactly how systems operate in order to better promote and maintain resilience. There are also those who place an extreme degree of confidence in science, anticipating that technology will advance quickly enough to produce human-made substitutes for the services now provided by the environment.

In practice, BRMPs have to date taken a relatively conservative approach of restoration and mitigation more often than active management. But as the effects of global climate change intensify, there is a growing realization the BRM must be undertaken more widely—managing reserved or otherwise protected areas is not enough. As a result, there is greater support for BRMPs that engineer resilience so as to inure ecosystems to disturbances (e.g., droughts or mismanagement) that might otherwise cause them to collapse. Biodiversity is critical to resilience because it serves to provide functional redundancy; resiliency is the key to sustainability. The more conservative efforts at BRM have often focused solely on species that are able to thrive in many habitats, but now more attention is being paid to the importance of those that are peculiarly adapted to specific habitats. And since habitats are hierarchically nested, new generation BRMPs also take species with large ranges into consideration. Even when they are transients, they still play crucial roles in the health of the ecosystem.

Though many industries currently engage in BRM, their level of engagement is likely to increase over the next decade. Architects and land developers are increasingly required to design their projects not only to avoid environmental degradation but to complement natural systems. The forestry, fisheries, and agricultural industries have all incorporated BRM practices, but they, too, will have to increase their efforts, coordinate with one another, and work with the chemical and biotech industries to effectively address watershed issues, integrated pest management, and other biocontrols. Society as a whole will have to grapple with the issue of genetically modified agricultural products, one of the most aggressive forms of BRM short of geoengineering.

Overall, our track record of managing our biological resources is rather poor, despite some local successes. History offers multiple examples of societies that outstripped their natural resources, and either failed to take the collective action that might have prevented their demise or used BRM methods not up to the task. With better knowledge and tools at our disposal, both businesses and individuals will need to consider their activities as integral elements of the ecosystem and enact strategies to best manage our future.

See Also: Conservation; Ecological Economics; Ecosystem Services; Sustainability.

Further Readings

Brand, Stewart. *Whole Earth Discipline: An Ecopragmatist Manifesto*. New York: Viking, 2009.

Grimble, R. and M. Laidlaw. "Biological Resource Management: Integrating Biodiversity Concerns in Rural Development Projects and Programs." Washington, D.C.: World Bank, 2002. http://www-wds.worldbank.org/external/default/WDSContentServer/WDSP/IB/2002/09/24/000094946_02090504023464/Rendered/PDF/multi0page.pdf (Accessed February 2010).

Grumbine, R. "What Is Ecosystem Management?" *Conservation Biology*, 8/8 (March 1994). http://www.life.illinois.edu/ib/451/Grumbine%20%281994%29.pdf (Accessed February 2010).

Holling, C. S. "Understanding the Complexity of Economic, Ecological and Social Systems," *Ecosystems*, 4 (2001). http://www.tsa.gov/assets/pdf/PanarchyorComplexity.pdf (Accessed February 2010).

Rockwood, L., R. E. Stewart and T. Dietz, eds. *Foundations of Environmental Sustainability: The Coevolution of Science and Policy*. New York: Oxford University Press, 2008.

Smajgl, A. and S. Larson. *Sustainable Resource Use: Institutional Dynamics and Economics*. London: Earthscan, 2007.

Susan H. Weaver
Independent Scholar

BIOMIMICRY

The term *biomimicry* is made up of two Greek words—*bios*, meaning life, and *mimesis*, meaning imitation—and is a method of innovation that aims to provide sustainable solutions to humankind's problems in a variety of areas such as the production of new materials, manufacturing processes, construction, and new product development by copying or adapting and employing ideas from nature. The principles of biomimicry have been made popular by the influential work of Janine Benyus, an American natural scientist and author who was the first to identify this emerging area of knowledge and to propose that sustainable design and technology of the future should take its inspiration from nature. Having originally studied forestry, Benyus was fascinated to see how wildlife adapted to its habitat, managing its environment and living in its environment without causing damage or creating waste and pollution. The logic behind biomimicry is therefore quite simple. For over 3.8 billion years, the animal and plant life of the planet have adapted and survived, facing essentially the same challenges as the human race—we can surely learn from the track record of adaptation in our struggle for water, food, space, and shelter, and not only incorporate this natural know-how into scientific disciplines to improve the quality of life, but come up with ways to support human life sustainably.

Biomimicry is based on realizing that there is much more to discover from nature and, therefore, less for humankind to have to invent. Proponents of biomimicry view nature as the ultimate research and development laboratory designed to produce viable sustainable solutions to ensure the continuation of life on the planet. The term *biomimicry* can be found in scientific, engineering, business, and marketing literature to indicate the process of understanding and applying biological principles to human designs in order to avoid the ills associated with the conventional practice of industry: depletion of nonrenewable resources, waste of energy, pollution, and destruction.

Biomimicry is centered on learning from nature as model, measure, and mentor. Nature as model entails the understanding that nature is the best source of problem-solving ideas, as it contains billions of organisms that have developed successful strategies for adaptation and growth. For example, Columbia Forest Products developed a super-strong, toxin-free glue by copying the secretions that mussels use to attach themselves to surfaces under water. Nature as measure refers to the basic idea that nature can show us what is possible, what the boundaries and limitations are, and what is or is not sustainable. Following the principles of biomimicry, for instance, industry would have to avoid standard so-called heat, beat, and treat processes, taking vast amounts of raw materials, heating them, beating them, treating them with chemicals, and placing them under enormous amounts of pressure to produce relatively small amounts of finished product in comparison to the waste and toxicity generated (not to mention the energy used in the process).

For example, super-strong Kevlar® is used in bulletproof jackets and is made in a heated, pressurized vat of concentrated sulfuric acid. In comparison, spider silk is tougher than Kevlar and is made by spiders on a diet of insects without a heavy industrial process. The spiderweb is also totally biodegradable and can be eaten by its maker to build a new one. Last but not least, nature as mentor refers to the need to change our relationship with nature. Rather than viewing nature as something to be dominated and managed, something to be exploited for its wealth of resources, something to be extracted, used, abused, and then discarded and polluted, as if it were something separate from human existence, nature should rather be viewed as a guide and tutor. Mother Earth is an apt metaphor, for while we are separate and distinct from our mothers, we nevertheless originate from her, receive nurture and education from her, and share most genetic/biological features with her. Abusing or exploiting or depleting or discarding her would, even from a most selfish perspective, make little sense. A good example for the practical nature of this insight comes from scientists at the University of Southern Mississippi who, by learning how crabs auto-repair their shells, have developed a paint that when scratched, responds to ultraviolet light by forming chemical chains that bond with other materials in the paint, eventually smoothing and repairing the scratch.

One of the more well-known examples of the application of biomimicry is that of the glue-free adhesive tape inspired by the feet of the gecko. The gecko holds on to surfaces by using the millions of tiny bristles on its feet that carry positive and negative charges known as van der Waals forces. Van der Waals force is one of the weakest attractive forces known to humans; however, when used through the millions of bristles of the gecko, it creates one of the strongest adhesive powers available. The gecko-inspired adhesive tape is just as strong under water and in the vacuum of space as it is under normal conditions. While the applications of the tape are limitless, the true benefits derive from its sustainable credentials. It is not toxic, does not contaminate, and can therefore be used safely. The bond also completely separates at an angle of 30 degrees, allowing for simple disassembly and easy recycling of components.

The shark, for example, has evolved over millions of years and is built for swimming efficiency. Mimicking the tiny, ribbed scales of a shark's skin, scientists have found a number of beneficial applications. Besides providing increased speed through water by reducing drag, and therefore reducing the amount of fuel needed, the scales better resist the buildup of contaminants on the hulls of boats, reducing the need for toxic chemicals to be used to clean them. In fact, at a constant speed of 4–5 knots, the shark scale–inspired hull is self-cleaning. Synthetic shark skin made a significant impact at the 2008 Beijing Olympics, with Speedo's Fastskin® FSII swimsuits helping swimmers to break a number of world records.

By imitating the strategies of life, in short, biomimicry generates solutions designed for sustainability, with fewer impacts on nature and businesses that are cleaner, leaner, and more energy efficient. It is not a stretch to argue that a successful adoption and integration of nature's innovations may be the most important, if not the only, key to a sustainable world.

See Also: Bio-Based Material; Biotechnology; Green Design; Sustainable Design.

Further Readings

Ask Nature. http://www.asknature.org (Accessed April 2009).
Benyus, Janine. *Biomimicry*. New York: Harper Collins, 2002.
Biomimicry Guild. "The Biomimicry Guild Product and Service Reference Guide 2009." http://www.biomimicryguild.com/guild_product_service_reference_09.pdf (Accessed April 2009).
Hawken, Paul, Amory Lovins and Hunter Lovins. *Natural Capitalism: Creating the Next Industrial Revolution*. London: Earthscan, 1999.

Barry Emery
University of Northampton

BIOREMEDIATION

Synthetic chemicals and heavy metals are major soil, water, and air pollutants. Most of these pollutants are long lived and sustained in the environment for hundreds of years.

Some of these compounds, such as pesticides, herbicides, and dyes, are toxic to living organisms and cannot be decomposed by natural decaying processes. Pollution accumulation poses short-term and long-term threats on living organisms and their environment, often indicated by pollutant "action" and "trigger" levels that denote the potential danger associated with the pollution concentration.

In recent decades, biological organisms, particularly microbes, have been utilized to assist in the removal of environmental pollutants from the surface, soil, and atmosphere in a process known as *bioremediation*. The ability of biological organisms to remove a specific pollutant is controlled by many factors, such as the nature of the organism, the pollutant type, and the prevalent environmental conditions. Oil spills and tar are among the long-lived pollutants; however, both can be removed from the environment using bacterial species and higher plants (tracheophytes that include ferns and conifers). Genetically modified bacteria have been designed to remove target pollutants such as petroleum, hydrocarbons, or radioactive wastes. Heavy metal pollutants of freshwater bodies are being removed using submerged aquatic plants known as hydrophytes, plants that include sago pondweed (*Potamogeton pectinatus*), Eurasian water milfoil (*Myriophyllum spicatum*), rigid hornwort (*Ceratophyllum demersum*), curly pondweed (*Potamogeton crispus*), common water hyacinth (*Eichhornia crassipes*), and the common reed (*Phragmites australis*). Although bioremediation cannot readily remove all pollutants (e.g., heavier metals such as lead), the process generally provides a cost-effective and sustainable green business practice for addressing environmental toxins.

Nature and Sources of the Environmental Pollutants

Environmental pollution refers to any substance, whether natural or synthetic, that has an adverse effect on the well-being of living organisms and their habitat. The amount and complexity of synthetic chemicals has increased dramatically during the past century and has been a large source of environmental pollution. Huge amounts of industrial waste have been liberated into the environment, impacting water supplies, soils, and the atmosphere. Synthesized compounds, in particular, cause drastic pollution problems as nearly up to 50 percent of the raw materials used are eventually liberated as contaminants. In addition to synthetic chemicals, gasoline, petroleum products, and heavy metals are significant environmental pollution contributors worldwide.

Pollutant Concentrations and Their Effect on the Environment

An environmental pollutant may be presented in an "action level," which generally indicates a danger, or in a "trigger level," which needs further investigation to verify whether the concentration implies a danger. Trigger and action levels of a pollutant change according to the site usage and the pollutant nature. For example, the trigger level of highly dangerous chemicals such as aldrin and benzo(a)pyrene must be not exceed 2 mg/kg of dry matter in playgrounds; however, in industrial areas, trigger levels as high as 12 mg/kg of dry matter are permissible. In cases of less dangerous chemicals, such as lead, the trigger concentration in playgrounds is about 200 mg/kg of dry matter, rising up to 2,000 mg/kg of dry matter in industrial areas. Bioremediation and other similar technologies (e.g., phytoremediation) are being utilized to reduce the pollution concentrations to below site-specific trigger levels.

Bioremediation

Bioremediation uses microorganisms and their enzymes to decompose, degrade, and enhance removal of soil, water, and air pollutants in order to transform an environment to its pre-contaminated state. In contrast to other remediation methods, bioremediation is entirely based on heterotrophs, which are organisms that require organic carbon rather than light (or autotrophs) for growth. Bioremediation technologies, such composting, bioventing, land farming, and bioreactors among others, may involve treating the hazardous material on site (in situ) or removing the material to another location for processing (ex situ).

Bioremediation plays an important role in several processes such as the transformation of complex biounreactive chemicals to bioreactive ones; the removal of slow-decomposing chemicals such as plastics; the biotransformation of pollutants to environmentally safe chemicals, or at least unreactive compounds; and the chemical breakdown of some complex pollutants that cannot be otherwise chemically treated (e.g., tar). The major advantages of bioremediation are the limitation of additional chemicals; the preservation of environmental resources (fauna, flora, water, soil, and air); and the reuse of contaminated sites after complete treatment. In some cases (e.g., hydrocarbon spills), bioremediation may be significantly more effective in both cost and toxin removal than traditional cleanup methods (e.g., excavation or incineration).

Phytoremediation

While bioremediation focuses on the use of heterotrophs for toxin removal, phytoremediation utilizes higher (or vascular) plants, algae, and some fungi to remove containments or accelerate the chemical waste breakdown by microorganisms in the rhizosphere (the zone surrounding the roots of plants). Phytoremediation primarily uses photoautotrophic plants (utilizes sunlight for food production) and saprophytic fungi (uses enzymes for digesting decaying organic matter) for managing waste materials and stabilizing toxic chemicals. Phytoremediation includes several methodologies such as phytodegradation, where the plant enzymes break down the pollutants into more benign constituents; phytoextraction, where the toxins are stored within the plant material itself; or volatilization, where the chemicals are transpired from the leaves. In comparison to bioremediation, phytoremediation technologies are relatively new and can often have a longer remediation time, but offer the advantages of being solar powered and able to remove and recycle heavier metals.

Whether employing heterotrophs or autotrophs, microbial species are essential for the natural decomposition and degradation of environmental pollutants. However, some investigations must be carried out before a bioremediation (or phytoremediation) process can begin, among them are to (1) identify the chemical pollutant; (2) determine the presence of the pollutant as individual chemicals or in mixture; (3) test for pollutant toxicity; (4) examine microbial resistance to the identified pollutants; (5) assess microbial and environmental toxicity of the bioremediation products; (6) detect site pH (a measure of acidic or alkaline levels), moisture, and temperature; (7) select suitable microorganisms; and (8) study the effect of the pollutant mixture on the microorganism's growth rate. In the case of higher plant remediation, short-lived plant species (e.g., beans, cabbage, cress, or lettuce) can be used to study the pollutant toxicity. Phytotoxicity tests in higher plants are indicated by the inhibition of root growth, seed germination, and vegetative growth of test plants. The efficiency of the remediation process depends on several factors, some of them related to the organism used, such as the organism utilization rate, nature of the organism, and the presence of other organisms. Moisture content, temperature, pH levels, and toxin levels also impact remediation efficiency. In some cases, the accumulation of some petroleum biodegradation by-products, such as fatty acids, limits the degradation process. Bioremediation and phytoremediation have been successful in the removal of hydrocarbon containments from soils using bacteria and higher plants, and metals from fresh water bodies and saturated soils using various hydrophytic plant species.

Remediation of Tar-Polluted Soils Using Bacteria

Oil spills, mostly composed of hydrocarbons, are extremely harmful to the environment and threaten human and animal life. Hundred species of bacteria, yeasts, and molds are known to attack hydrocarbons in temperate regions. In Prince William Sound, Alaska, the site of the *Exxon Valdez* oil spill, natural populations of bacteria, stimulated by fertilizers, grow and metabolize the polycyclic aromatic hydrocarbons (carcinogenic chemical compounds commonly found in oil, tar, and coal by-products) of the spilled oil. After a few years of application, the level of polycyclic aromatic hydrocarbons fell five times faster in the bioremediated areas than through untreated areas. Genetically modified organisms are also being designed for remediation of certain pollutants, including hydrocarbons. In fact, the first genetically engineered bacterium patented in the United States was to degrade petroleum hydrocarbons. Other bacteria (e.g., *Deinococcus radiodurans*) have also been devised to remediate radioactive nuclear waste.

Phytomediation has also been effective in dealing with tar-polluted soils. Several xerophytic species can grow and tolerate soil with 10–20 percent tar content, such as *Aizoon canariense, Anabasis setifera, Atripex leucoclada, Fagonia indica, Salsola imbricata, Senecio glaucus, Sporobolus arabicus, Suaeda aegyptiaca,* and *Zygophullum quatarense.* There are many factors controlling the use of these plants for remediating tar piles, including the age of the tar, the pile disturbances and tar moisture content, the structure of the plant communities in the polluted site, the size of the disposal area, the method of dumping, and the prevailing environmental conditions.

Remediation of Freshwater Bodies and Hydrosoils

Pollution of freshwater bodies is an extremely important environmental topic, as waterborne pollutants can adversely impact several sectors of the environment (e.g., human, plant and animal well-being, soils, and the atmosphere). To remediate pollutants from freshwater bodies, several submerged hydrophytic species have been adopted to remove heavy metals and other toxins. Sago pondweed (*Potamogeton pectinatus*) is used for remediation of highly polluted freshwater bodies characterized by elevated levels of temperature, pH, conductivity, sulfates, and chloride concentration; this plant is used also for remediation of hydrosoils containing very high organic matter content. Eurasian watermilfoil (*Myriophyllum spicatum*) is used for remediation of nitrates, nitrites, and sulfates from freshwater bodies rich in organic matters, as this species can tolerate up to 30 percent organic matter. Known as coon's tail or rigid hornwort, *Ceratophyllum demersum* is used for remediation of freshwater with high carbohydrate and high chemical oxygen demands, as well as for hydrosoils with excessive potassium, sulfates, and alkalinity levels. Curly pondweed (*Potamogeton crispus*) is used for remediation of freshwater with very high dissolved oxygen and sulfate concentrations and can also be applied in hydrosoils that have high alkalinity, low phosphates, and organic matter. *Eichhornia crassipes,* the common water hyacinth, is a free-floating freshwater plant that can survive also in saline water. It has the ability to absorb and accumulate high levels of heavy metals, such as cadmium, copper, iron, manganese, lead, and zinc in waterways (e.g., the Nile) and hydrosoils. Water hyacinths are also effective for water bioremediation of lead contaminants.

Recently, international attention has been directed toward the capacity of constructed reed wetlands to control water pollution and to treat municipal and industrial wastewater. The common reed (*Phragmites australis*) is a perennial grass species that grows in shallow lakes and wetlands in the boreal and tropical parts of the world. The plant can survive a long-term drought, and rhizomes remain viable in dry soil for a period of eight to nine years. Common reed plants also grow in lead-polluted areas, where the lead accumulates in crystal-like form in the stem cortical cells. In lakes in northern Egypt, *Phragmites australis* is able to grow and be used for bioremediation of soil with high alkalinity levels (pH values of 6.9–9.3) and conductivity up to 12 milli-Siemens per centimeter (mS cm-1).

See Also: Biological Resource Management; Biotechnology; Environmental Management System; Persistent Pollutants.

Further Readings

Ahamd, Hegazy. "Plant Succession and Its Optimization on Tar-Polluted Coasts in the Arabian Gulf Region." *Environmental Conservation*, 24/2:149–58 (1997).

Magdi, Ali and M. E. Soltan. "Heavy Metals in Aquatic Macrophytes, Water and Hydrosoils From the River Nile, Egypt." *Journal of Union Arab Biologists*, 9(B):99–115 (1999).

Maha, Ali, et al. "Biophysical Measurements of Lead in Some Biosensors." *Egyptian Journal of Biophysics and Biomedicinal Engineering*, 4:1–13 (2003).

Magdi, Ali, et al. "Submerged Plants as Bioindicators for Aquatic Habitat Quality in the River Nile." *Journal of Union Arab Biologists*, 9(B):403–18 (1999).

Mamdouh, Serag. "Ecology and Biomass of Phragmites australis (Cav.) Trin. Ex. Steud. in the Northeastern Region of the Nile Delta, Egypt." *Ecoscience*, 3/4:473–82 (1996).

Wafaa M. Amer
Cairo University

BIOTECHNOLOGY

The Organisation for Economic Co-operation and Development (OECD) was the first international organization to have proposed a definition of biotechnology in 1982. They defined biotechnology as "the application of science and technology to living organisms as well as parts, products and models thereof, to alter living or nonliving materials for the production of knowledge, goods and services." This technical definition includes a broad range of products and reveals in practice that biotechnology applications are part of our everyday life. Indeed, one of the oldest applications of biotechnology is the use of microorganisms for the production of foods such as yogurt, cheese, beer, and pharmaceuticals, including antibiotics. Since 1980, modern biotechnology has emerged to complement these initial techniques, this time by using genetic manipulation.

The study of agro-biotechnology is meant to improve crop production. Nobel Prize–winning agro-biotechnologist Norman Borlaug (second from left) consults with local leaders near wheat plots in Kenya.

Source: Kay Simmons/U.S. Department of Agriculture

As part of our everyday life, the issue of the potential impacts of biotechnology, and in particular of modern biotechnology, on the environment is important. Like any other kind of technical innovations, biotechnology applications present some risks as well as promises in the field of sustainable development.

Modern biotechnology is more of a tool—the manipulation of genes—than a product in itself. The industrial sectors that are primarily employing biotechnology as an innovative tool are agriculture/agribusiness and pharmaceutical companies. There are also several small and medium-sized enterprises that have developed innovative products based on biotechnology for specialized markets, such as soil decontamination or biorecycling. While such specialized uses are still rare, biotechnology makes up an increasingly substantial proportion of the pharmaceutical industry, as well as of the agricultural/food industry. The popular buzz word, of course, has been "genetically engineered" food, crops, plants, and seeds. Hailed by the few remaining multinational agricultural biotechnology giants, such as Monsanto, as the great hope of the future, consumers and environmentalists around the world have raised multiple concerns, even leading to major international trade disputes between the United States and the European Union. While genetic/biological manipulation of seeds and plants is nothing new, both possibilities and extent of such manipulations have virtually exploded since the 1990s, and now include forms of direct genetic engineering such as cloning and gene splicing that directly alter the structure and characteristics of genes. Both promise and potential dangers of such advances in biotechnology are as scientifically complex as they are relatively easy to summarize: the promise is "better" (more nutritious, more resilient, faster growing, less in need of pesticides and fertilizers) and "more" (greater yield, more resistant against pathogens and parasites); the dangers (real or perceived) are "loss of control" over our collective food supply (over 90 percent of genetically engineered seed, for instance, comes from one multinational corporation), "safety" (such as unwanted genetic transfers, drug resistance, immune deficiencies), and "destruction/alteration of natural environment" (such as loss of biodiversity, killer bugs, or infertile terminator seeds). The applications derived from modern agricultural biotechnology have consequently been the subject of many intense debates.

Agricultural Biotechnology

In practice, the development of agricultural biotechnology applications has been uneven. On the one hand, the total surface of genetically modified crops has increased between 1995 and 2006 from a handful of field trials to more than 110 million hectares—with

13 percent growth between 2005 and 2006. On the other hand, most of the genetically modified plants concern only four different varieties and two processes: soy resistant to weeds (60 million hectares); cotton resistant to caterpillars (10 million hectares); corn varieties resistant to weeds and/or insects (25 million hectares); and canola seeds also potentially modified to resist weeds or pests (2 million hectares). These varieties are cultivated in a handful of countries: the United States (55 million hectares), Argentina (18 million), Brazil (11.5 million), India (3.8 million), China (3.5 million), Canada (6.1 million), Paraguay (2 million), and South Africa (1.8 million). The United States represents half of the worldwide production of transgenic plants.

As their properties show, the early development of modern agrobiotechnology products was initially meant to improve crop production, while using innovations that were more environmentally friendly. Indeed, as genetically modified varieties were resistant to pests or weeds, their production was supposed to require less pesticides or insecticides. However, the concern regarding the environmental impact of agricultural varieties grew over time. Progressively, the risks of development of "super bugs" or "super weeds" resistant to the new crop varieties emerged. The potential harmful effects of genetically modified species on the wild fauna were also mentioned, as well as the possible flows of genetic material to wild natural species. The main controversies started in 1999, when a U.S. experiment demonstrated that one wild species of butterfly, the monarch butterfly, was disappearing in direct correlation to the introduction of genetically modified corn in the environment. The modified pollen was suspected of killing the monarch caterpillars. At the end of the same year, a Hungarian researcher, Arpad Pusztai, published in *The Lancet* his results concerning the dangers of modified potatoes on mice. Perhaps the most publicly visible outcry was over the so-called terminator gene, which is a gene sequence directly inserted into a seed's DNA and then activated in order to make future generations of the seed sterile. As such biogenetic manipulation has no benefit other than to protect the intellectual property rights and profits of the manufacturer, but directly threatens centuries of agricultural traditions as well as the livelihood of small farmers, its introduction led to a wave of public protest. Initially patented by the U.S. Department of Agriculture (USDA) and Delta and Pine Land Co., the current owner, Monsanto, eventually agreed to withhold application of terminator seeds. Biological manipulations of this kind, however, will and should continue to be an essential part of the public debate.

Moreover, environmental concerns were soon coupled with fears concerning the socioeconomic impacts of biotechnology in general, and genetically modified varieties in particular. While genetic engineering was propagated as a possible solution to the world's growing demand for food, the result in poorer countries more often was destruction of small-scale local farming, increased dependence on agroconglomerates with headquarters in the United States or Europe, and an environmentally destructive push for monocultural cash-crop farming methods. In Europe, moreover, a higher level of environmental awareness coupled with a bigger political presence of environmental groups and parties translated into a widespread public resistance to genetically modified food. Significant sectors of both poor and rich countries, in short, began to resist the introduction and spread of advanced forms of biotechnology, at least when it came to fundamental questions about how food was grown and consumed, who controlled and benefited from the process, and what long-term consequences it had on the environment and people's health. Today, both uncritical advocacy of, and wholesale resistance to, genetically modified seeds/food has mostly given way to a range of more balance views, taking into account potential benefits as well as dangers.

International Moves

At the international level, the Cartagena Protocol was adopted in 2000 under the Convention on Biological Diversity—an international regime related to the United Nations Environment Programme—and currently has 103 member states. This international agreement aims at managing the environmental and socioeconomic risks related to agrobiotechnology applications. More precisely, the Cartagena Protocol deals with the transboundary movements of genetically modified organisms, and recognizes the need for a precautionary approach to handle these organisms. It requires the documentation of genetically modified crops to be kept during their transport in order for countries to make informed decisions prior to the release of transgenic varieties into the environment.

At the national and the regional levels, several countries have adopted different regulations to balance the possible benefits of biotechnology applications with their risks. For example, the Indian government adopted in 2003 a special clause that requires that every new transgenic variety should respond to the Indian population's need in order to be accepted for commercialization nationally. In Europe, the European Commission has developed a strong regulatory policy concerning the deliberate release of genetically modified organisms into the environment. This legislation requires a full risk assessment and the agreement of numerous governments and agencies before the use of any novel transgenic variety. In countries where regulations have traditionally been favorable to the spread of agrobiotechnology applications, such as the United States, several practical measures have been introduced in order to manage the possible environmental damages of modified varieties. These measures can include the establishment of buffer zones between conventional crops and modified plantations or the need to mix modified varieties with a small percentage of nonmodified varieties in order to avoid the development of resistant insects or weeds. The growing demand for organic foods and products also encourages countries to differentiate the different food supply chains for conventional and transgenic as well as organic productions.

Biotechnology firms also have made several steps toward the improvement of their products. Transnational corporations, such as Monsanto, Syngenta, and Bayer, have put several international nongovernmental organizations in place—such as CropLife International—in order to inform the public about the benefits of agricultural biotechnology applications. These companies are also working on the elaboration of new plant varieties that could better answer the needs of developing countries. For instance, in 2000, the private sector started to develop a new variety of rice, known as "golden rice," which has been enriched in vitamin A and was created to answer food deficiency issues in African countries. Research is also being conducted on modified crops that could produce drugs, on transgenic trees, and on biofuel plantations. These innovations are often considered the second generation of transgenic crop applications.

However, such developments are progressing at a slow pace, and developing countries have just started to engage in research in agricultural biotechnology. In fact, whereas biotechnology has long been perceived as a technological tool in the hands of richer nations, several developing countries today are encouraging their national firms to find a compromise between national competitiveness and sustainable development objectives. This is particularly the case for emerging countries such as India, China, and Brazil.

These countries also aim to develop their skills in biotechnology research, pharmaceutical products, and specialized industrial applications.

Even though the international debate on agricultural biotechnology has tarnished the image of biotechnology, its applications continue to spread and grow. As in other areas of the economy, when it comes to the question of which research to push forward, and which biotechnological findings to apply, the pursuit of potentially enormous profits has to be responsibly balanced with important concerns for the health and well-being of people and environment.

See Also: Corporate Social Responsibility; Green Technology; Greenwashing; Precautionary Principle.

Further Readings

Bail, Christopher, Robert Falkner and Helen Marquard, eds. *The Cartagena Protocol on Biosafety: Reconciling Trade in Biotechnology With Environment & Development?* London: Earthscan, 2002.

Falkner, Robert, ed. *The International Politics of Genetically Modified Food: Diplomacy, Trade and Law*. Basingstoke, UK: Palgrave Macmillan, 2007.

Grace, Eric S. *Biotechnology Unzipped: Promises and Realities*, 2nd Ed. Washington, D.C.: Joseph Henry Press, 2006.

Lacy, Peter. "Deploying the Full Arsenal: Fighting Hunger With Biotechnology." *SAIS Review*, 23/1 (2003).

Mitchell, C. Ben, et al. *Biotechnology and the Human Good*. Washington, D.C.: Georgetown University Press, 2007.

Amandine J. Bled
Université Libre de Bruxelles

BLENDED VALUE

Blended value most fundamentally describes two closely related notions: (1) that all organizations, whether for-profit or not, create value that consists of the three major components—economic, social, and environmental values; and (2) to the extent that organizations realize that they affect all three areas of value production—economic, social, and environment—and then try to integrate and promote all three, they will be more successful.

Attributed to a professor of business as well as founder and fellow of numerous foundations, Jed Emerson, the term *blended value* has found wide resonance in the business world, and has even made some of the top five lists of future trends in business. Integrating economic, social, and environmental values brings with it a wide range of efforts, all driven by the idea of value creation that is more holistic (i.e., not segregated into multiple strands of artificial separation). This is applicable to nonprofit and for-profit as well as governmental and nongovernmental organizations. By recognizing these three different value components, for instance, investors can be more focused on the appropriate mix of funds in companies and the amount of value that matches their own value.

Usually people distinguish between "doing well" and "doing good." For-profit firms ordinarily concentrate on maximizing economic value, for example, profit for the shareholders, while not-for-profit groups create social and environmental values for their stakeholders. However, it is important to note that not-for-profit firms create economic value, while for-profit firms obviously have significant social and environmental impact. For example, the Girls Scouts create economic value by selling several million boxes of cookies each year. On the other hand, Wal-Mart creates a social impact by providing a particular kind of employment for almost one and a half million individuals, and it creates environmental impact by building superstores, employing a fleet of thousands of trucks, or by purchasing goods only from particular farming and manufacturing businesses.

The blended value proposition mandates that all organizations, whether not-for-profit or for-profit, create the tri-component sets of values—economic, social, and environmental—and that investors (for-profits, charitable, or some combination of the two), simultaneously generate all three forms of value by providing capital to organizations. The ideal of this model would be to do well by doing good, and to do good by doing well.

Blended value strategies ideally do not allow the interests of any constituency to dominate—shareholder values are not allowed to supersede all other concerns, but neither are customer or environmental or larger social concerns given preference. In that sense, organizations that follow the blended value paradigm redefine what "success" is: One does not succeed over or in competition with other concerns; one succeeds by taking all concerns into consideration. Satisfying the basic concerns of all stakeholders, blended value strategies seek to approximate what is known as "sustainable" development or success.

All organizations, whether for profit, nonprofit, educational, or governmental, that are interested in maximizing a blended version of economic, social, and environmental values, according to Emerson, ordinarily pursue work that is largely aligned with one of the following five categories of activity. He calls each field of activity a "silo," for he argues that in practice, they usually remain quite isolated from each other:

- Corporate social responsibility
- Social enterprise
- Social investing
- Strategic/effective philanthropy
- Sustainable development

However different and separate these pursuits are, though, they have in common that they all face, again according to Emerson, the following set of challenges:

- The need for the creation of more efficient capital markets
- A lack of commonly defined performance metrics to assess both organizations and capital returns
- The evolution of new approaches to entrepreneurial leadership and organizational development
- The need to advance a more effective supporting environment through the introduction of new policy and regulations
- In order to establish an environment more conducive to the maximization of blended values, Emerson has articulated a range of recommendations that primarily fall under the rubrics of "Networking," "Collaboration," and the "Building of an International Infrastructure," all of which seem intended essentially toward bringing together people interested in blended values

Challenges to Optimizing Blended Value

Blended value maximization faces several obstacles, some systemic and some cultural, but most deeply ingrained. The arguably biggest obstacle is that little in our society—educationally, legally, politically, economically—supports holistic thinking or integrated approaches. On the contrary, most of what we learn and do is specialized, separate, disconnected. Consequently, there are virtually no performance measures—from the personal to the national—that integrate the three major components of blended values—economic, social, and environmental. The most important single indicator, in the United States and around the world, for instance, is an indiscriminate measure of totality of goods and services, the so-called gross domestic product (GDP), with absolutely no consideration for social or environmental benefits or costs. Second, but perhaps of equal impact, as long as the infamous "bottom line," defined by nothing other than profit, determines the articulation of most business plans, social and environmental concerns will never be more than an afterthought. Nonprofit organizations, on the other hand, are notorious for their lack of performance, efficiency, and accountability standards.

A further complication comes from widespread disagreement about what matters, and hence what to measure. Stakeholders would not only have to be able to agree on desirable outcomes, but also be in a position to agree on performance measures, guarantee transparency and access to the necessary information, and find mutually agreeable enforcement mechanisms in order to hold everyone accountable.

The legal and tax structure that demarcates a for-profit from a not-for-profit enterprise also presents complications for those pursuing and investing in blended value. This causes significant operational inefficiencies in its management and the problem of raising capital. Creating blended value investments and administering successful blended value ventures is much more difficult than running a purely commercial venture. It is critical to balance the decisions that best support financial objectives versus those that best support the social and environmental mission.

Investors still live in a divided world, focusing their professional attention primarily on financial investments, and attending to social or environmental issues only as an afterthought or a personal concern. Blending the value proposition between donations and maximizing profits, for instance, typically leads to confusion and anguish among investors.

An important concept that provides a number of useful tools, blended value maximization continues to face an uphill struggle despite a growing willingness in the business world to embrace the basic idea of blended values: that all three aspects of our lives—economics, society, and the environment—are equally important.

See Also: Best Management Practices; Body Shop, The; Corporate Social Responsibility; Environmental Marketing; Integrated Bottom Line; Social Entrepreneurship; Sustainability; Triple Bottom Line.

Further Readings

Baldwin, W. "Nice-Guy Economics." *Forbes*, 183/5:10 (2009).

Bradley, Bill, Paul Jansen and Les Silverman. "The Non-Profit Sector's $100 Billion Opportunity." *Harvard Business Review* (May 2003).

Emerson, Jed. "The Blending Value Proposition: Integrating Social and Financial Returns." *California Management Review*, 45/4:35–51 (2003).

Emerson, Jed and Sheila Bonini. "The Blended Value Map: Tracking the Intersects and Opportunities of Economic, Social and Environmental Value Creation." http://www .blendedvalue.org/publications (Accessed October 2009).

Epstein, Marc and M. J. Roy. "Making the Business Case for Sustainability: Linking Social and Environmental Actions to Financial Performance." *Journal of Corporate Citizenship* (2003).

Gertner, Jon. "Jed Emerson Wants to Change the World." *Money*, 31/12:118–26 (November 2002).

Goleman, Daniel. *Ecological Intelligence*. New York: Broadway Books, 2009.

Isaak, Robert. "The Making of the Ecopreneur." *Greener Management International*, 38:81–93 (Summer 2002).

Nicholls, A. "We Do Good Things, Don't We? 'Blended Value Accounting' in Social Entrepreneurship." *Accounting, Organizations and Society*, 34/6/7:755–67 (2009).

Rubinstein, Robert. "Environment, Social and Governance Supplement: What Is SRI? Really Responsible or Just Band-Wagon Jumping?" *EPN: European Pensions & Investments News*, 1 (November 2006).

Social Edge. http://www.skollfoundation.org/socialedge (Accessed October 2009).

Sustainability. http://www.sustainability.com (Accessed October 2009).

Abhijit Roy
University of Scranton

BODY SHOP, THE

Employing about 10,000 people in some 2,500 stores in over 60 countries, and selling well over 1,200 different beauty products around the globe, The Body Shop International is the second largest cosmetics franchise in the world. Founded by Anita Roddick in 1976, the company's corporate vision was greatly influenced by its founder, who wrote in her book *Body and Soul,* "There is no power on earth apart from business capable of making the changes we desperately need for the continued survival of the planet." Since its inception, the company has pushed hard to become known as a "green company," avoiding animal testing, seeking fair trade practices, cutting carbon emissions, using only recycled packaging materials, among many initiatives geared toward a sustainable business model. After spectacular growth during its first decade and a half, The Body Shop went through several ups and downs in its attempts to expand its business model to countries like the United States and Germany, and experienced growth rates that were significantly below expectations.

Serious losses forced the company, in the late 1990s, to give up on production of cosmetics, and instead concentrated merely on sales. Shortly before the death of the founder, but following a renewed period of growth, the company was sold in 2006 to the L'Oréal Group. In order to retain its value-based identity, it continues to be headquartered in the United Kingdom and operates independently within the L'Oréal Group.

The company seeks to be a leader not only in promoting greater corporate transparency, but also in representing a force for positive social and environmental change.

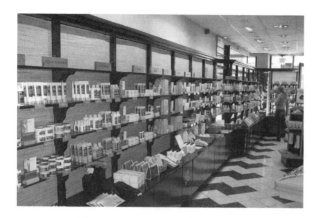

The Body Shop is a proponent of greater corporate transparency and positive social and environmental change. Here, a Body Shop store in Oslo, Norway.

Source: Nenad Bumbić/Wikipedia

As such, the company pursues campaigns around the world that are ground in five core values:

- Opposing animal testing
- Supporting community fair trade
- Defending human rights
- Activating self-esteem
- Protecting the planet

The Body Shop's key distinctive proposition is that it is against animal testing—every product is vegetarian and animal-cruelty free. It was the first cosmetics brand to be recognized under the Human Cosmetics Standard against Animal Testing Policy. The company also believes in fair trade policies—in the 1990s, it set up a fair trade program called "Community Trade" that works with 30 suppliers in more than 20 countries, providing over 25,000 people around the globe with essential incomes to sustain themselves. A charitable wing, The Body Shop Foundation, was started in 1990 to aid pioneering front line organizations that otherwise could not find funding. In 1997, the company's founder launched a global campaign to raise self-esteem in women and against the media stereotyping of women, with a particular focus on the socially pathological emphasis on skinny models, leading to massive increases in bulimia and anorexia victims. It has also focused on human and civil rights issues and environmental and animal protection.

Historical Milestones

The history of The Body Shop is the rather unlikely story of a mom-and-pop-sized store, started in Brighton on the south coast of England in 1976, with quaint ideas about socially and environmentally responsible business practices having the potential to change the world. Within less than two decades, it grew into a multinational corporation spanning the globe. And while there is ongoing debate about how true The Body Shop remained to its founding ideals, or whether the underlying motivation was not always essentially the same as in traditional businesses, namely to make money, there is no question that The Body Shop managed to create a fundamentally different business model with greater concern for the well-being of its suppliers, consumers, employees, as well as for the environment. Indeed, the corporate history of the The Body Shop is organically inseparable from its many social and environmental initiatives.

By 1982, The Body Shop was expanding at the rate of two new shops every month. In 1985, the company went public and began sponsoring several Greenpeace projects, and produced its first Community Trade product, a Footsie Roller from a supplier in southern India.

The period of 1976 to the early 1990s saw the greatest success and expansion of The Body Shop. The mid to late 1990s was a far more difficult period, as The Body Shop began to face competitors, such as Bath & Body Works, which imitated some of its strategies. This period also witnessed legislation limiting animal testing for nonpharmaceutical purposes (thus undermining its distinctive profile), and publications questioning the truthfulness of some of the company's claims. Despite reduced profit margins and slower-than-expected expansion into markets such as the United States and Germany, The Body Shop not only continued to grow, but actually expanded fair trade contracts, initiated new charities funding human rights initiatives, and launched Roddick's 1997 global campaign for women. In 1998, The Body Shop launched a joint worldwide campaign with Amnesty International to focus on the plight of human rights advocates around the globe. The related "Make Your Mark" campaign garnered over 3 million signatures. In 1999, the company decentralized its business operations to create four new business units in the United Kingdom, Europe, the Americas, and Asia.

The Body Shop continued to increase its focus on environmental practices in the 21st century. By 2001, it began using energy from renewable resources and converting it to green electricity. In 2004, it was the first global retailer to join the board of the Roundtable for Sustainable Palm Oil, working with nongovernmental organizations (NGOs) and plantations to protect tropical rainforests and to improve the human rights of workers and indigenous people. The company opened stores in Jordan and Russia in 2005, and in the same year joined the Campaign for Safe Cosmetics, and was commended by Greenpeace and the Breast Cancer Foundation for its responsible chemicals policy.

In 2006, The Body Shop was sold to the L'Oréal Group, but continued to operate with a large degree of independence. Rather than dropping its social and environmental initiatives, the company continued to embrace new social responsibility projects on issues such as violence against children, AIDS and HIV, sex trafficking, and fair trade/fair labor practices. On September 10, 2007, the founder and inspiration for the company, Anita Roddick, died at the age of 64.

The Future

In 2008, Sophie Gasperment became the chief executive, and the company unveiled a brand-new look. The company has also introduced 100 percent postconsumer recycled (PCR) bottles, with a target to convert all bottles to 100 percent recycled material by 2010. The Community Trade Program continues to create sustainable trading relationships with disadvantaged communities around the globe and provides essential income to more than 25,000 people worldwide. The Body Shop continues to be committed to trading ethically, and only trades with suppliers who are committed to its Code of Conduct for Suppliers. Finally, it has committed to becoming carbon neutral by 2010, ensuring that carbon dioxide from its core retail business worldwide is reduced.

Overall, The Body Shop has succeeded in drawing a particular, and loyal, customer base of socially and environmentally conscious people, most of whom are not only familiar with founder Anita Roddick, but appear to share her ideas about sustainability and social responsibility. Despite questions about the degree to which The Body Shop's business model is sustainable, it has undoubtedly been a trailblazer among corporations seeking to generate a profitable business model without compromising social and environmental responsibility. In so doing, The Body Shop has effectively destroyed several business truisms, chief among them the idea that prices are always determined by supply and demand. The Body Shop has been paying fair trade prices that are significantly higher than market prices, and still manages to make a profit. Also, the notion that any large cosmetics company would have to engage in animal testing was effectively dismissed by The Body Shop's pledge to forgo selling any such products.

See Also: Best Management Practices; Carbon Neutral; Cause-Related Marketing; Corporate Social Responsibility; Environmental Indicators; Environmental Marketing; Fair Trade; Integrated Bottom Line; Leadership in Green Business; Responsible Sourcing; Voluntary Standards.

Further Readings

The Body Shop. http://www.thebodyshop-usa.com/bodyshop/ (Accessed May 2009).
Coomber, Steve. "Radical Roddick." *Business Strategy Review,* 16/2:80–83 (2005).
Dennis, Bryan, Christopher P. Neck and Michael Goldsby. "Body Shop International: An Exploration of Corporate Social Responsibility." *Management Decision,* 36/10:649–53 (1998).
Hartman, Cathy L. and Caryn L. Beck-Dudley. "Marketing Strategies and the Search for Virtue: A Case Analysis of The Body Shop, International." *Journal of Business Ethics,* 120/3:249–63 (1999).
Kent, Tony and Dominic Stone. "The Body Shop and the Role of Design in Retail Branding." *International Journal of Retail & Distribution Management,* 35/7:531–40 (2007).
Livesey, Sharon M. and Kate Kearins. "Transparent and Caring Corporations? A Study of Sustainability Reports by The Body Shop and Royal Dutch/Shell." *Organization & Environment,* 15/3:233–58 (2002).
Mirvis, Philip. "Can You Buy CSR?" *California Management Review,* 51/1:109–16 (2008).
Nijssen, Edwin J. and Susan P. Douglas. "Consumer World-Mindedness, Social-Mindedness, and Store Image." *Journal of International Marketing,* 163:84–98 (2008).
Pless, Nicola M. "Understanding Responsible Leadership: Role Identity and Motivational Drivers: The Case of Dame Anita Roddick, Founder of The Body Shop." *Journal of Business Ethics,* 74/4:437–56 (2007).
Sillanpaa, Maria. "The Body Shop Values Report—Towards Integrated Stakeholder Auditing." *Journal of Business Ethics,* 17/13:1443–56 (1998).
Sinclair, Annette and Barbara Agyeman. "Building Global Leadership at The Body Shop." *Human Resource Management International Digest,* 13/4:5–8 (2005).
Wycherley, Ian. "Greening Supply Chains: The Case of The Body Shop International." *Business Strategy and the Environment,* 8/2:120–27 (1999).

Abhijit Roy
University of Scranton

Bottom of the Pyramid

The *bottom of the pyramid* (BoP) as a term and concept has seen a recent renaissance through the work of two business professors, C. K. Prahalad of the University of Michigan, and Stuart L. Hall of Cornell University. Of the roughly 6.5 billion people in the world today, according to Prahalad and Hall, more than 4 billion live at the bottom, or, by some of their varying definitions, at below $2 a day or below a purchasing power parity of $1,500 a year. Due to their inherent lack of money, businesses around the world have largely ignored this vast collection of poor people. Such neglect, according to Prahalad and Hall, constitutes a missed opportunity of vast proportions, for the multitudes at the bottom represent an almost-limitless and promising pool of future consumers. In their words, "as multinational firms search for avenues for profitable growth and radical innovation in the new millennium, they may find a unique, counter intuitive opportunity—the 4 billion poor that are at the bottom of the economic pyramid. Converting the very poor into active consumers will foster innovations in technologies and business models." In recent statements, Prahalad further elaborated that since two or three or four billion additional consumers and producers will inevitably put a tremendous strain on the environment, and particularly water and energy resources, the issue of what he calls "inclusive growth of poor people" and the issue of corporate sustainability are inextricably linked.

Sociologically, the pyramid has been used as a metaphor to describe the hierarchical nature and inequality of modern societies since the 18th century. The Industrial Workers of the World (IWW) disseminated a poster of a social pyramid in 1911 that later became famous, depicting the masses of people at the bottom, working for and feeding all, and at the top rulers and capitalists and clergy, ruling and fooling all. Then, in the midst of the Great Depression, shortly after he announced his candidacy for president, Franklin D. Roosevelt popularized the term *bottom of the pyramid* in his famous April 1932 speech "The Forgotten Man." For far too long, Roosevelt argued, those at the top had enjoyed all the benefits of economic progress. Now it was time for the government to support society's disadvantaged—"the bottom of the pyramid"—those who form "the infantry of our economic army." Roosevelt of course later became president, and his New Deal programs not only helped the poor and unemployed, but it also realized what was, at the time, a new economic theory: the basic Keynesian insight that, particularly during severe economic downturns, the federal government had a large role to play, namely to boost demand by providing people with money to spend.

During the post–World War II era, the idea that only economic growth as defined by the totality of goods and services could lift people out of poverty became an article of faith—among capitalists and communists alike. The first cracks in this axiom only appeared when development theorists and environmentalists began to raise serious questions about the consequences of unfettered and indiscriminate growth. Resource depletion, environmental destruction, and growing levels of pollution were only the tip of the iceberg. Industrial growth also destroyed centuries of independent farming traditions, turned hundreds of millions of people into wage laborers or no-wage unemployed, and created widespread dependence. Perhaps most ironically, historically unprecedented economic growth did not seem to make its primary beneficiaries any happier, and it barely (if at all) reduced the magnitude of extreme poverty.

How to define, how to think about, and then how to address people at the so-called bottom of the pyramid, in short, has become a question of survival not just for those who inhabit the bottom, but also for environmental sustainability, political stability, and with it for growth-based business and development models. Does it makes sense, for instance, to define "the bottom" by dollars earned, even when many people still live in cultures in which much economic activity is not priced, and when a dollar has such vastly different purchasing power in different locations? Does it make sense to promote indiscriminate growth, even when it leads to destruction and depletion of resources, r/>

undermining livelihoods of future generations? Or, finally, does it make sense to reduce people to "consumers" rather than promoting a process by which especially poor people can become producers and guardians of their own, largely independent communities?

The BoP concepts of Prahalad and Hall can be commended for shifting focus on those who ordinarily live behind a veil of poverty and powerlessness, and for raising awareness about the multiple challenges that are part of any kind of development that serves all *and* is sustainable. Critics have pointed to many remaining problems, chief among them that without a fundamental shift in goals and performance indicators, there is little incentive for businesses to pay attention to the economically poor, and there is no reason to think, based on traditional business models, that bringing hundreds of millions into the existing consumer society will in any way advance social or environmental sustainability. A well-known competing model that squarely focuses on the world's poor is microlending, or microcredit. A current version was popularized by Nobel Peace Prize winner Muhammad Yunus and his Bangladesh-based Grameen Bank, which has a successful record extending very small loans to those in poverty. Unlike BoP models, however, microcredit emphasizes entrepreneurship and economic independence, not enrichment opportunities for existing businesses by turning poor people into consumers.

See Also: Corporate Social Responsibility; Ecological Economics; Environmental Economics; Socially Responsible Investing.

Further Readings

Chen, Shaohua and Martin Ravallion. "How Have the World's Poor Fared Since the Early 1980s?" *World Bank Research Observer,* 19/2:141–69 (2004).
Hammond, Al and C. K. Prahalad. "Selling to the Poor." *Foreign Policy* (2004).
Hart, Stuart L. *Capitalism at the Crossroads.* Philadelphia, PA: Wharton School Publishing, 2005.
Ignatius, David. "Profits, a Penny at a Time." *Washington Post* (July 6, 2005).
Prahalad, C. K. *Fortune at the Bottom of the Pyramid: Eradicating Poverty Through Profits.* Philadelphia, PA: Wharton School Publishing, 2004.
Ravallion, Martin. "The Debate on Globalization, Poverty and Inequality: Why Measurement Matters." *International Affairs* (2003).
Sen, Amartya. *Development as Freedom.* New York: Anchor Books, 2000.

Dirk Peter Philipsen
Virginia State University

BRITISH TELECOM

British Telecom (BT Group) is one of the most renowned companies in the United Kingdom, as well as a leading global telecommunications company. Headquartered in London, BT is considered to be one of the world's oldest telecommunications companies and, for the last two decades, has repeatedly been recognized for its leadership role among transnational corporations in its sustainability or sustainable development efforts. Indeed, until dislodged from its position by Telefónica Spain in 2009, BT had been the leading telecommunications company on the Dow Jones Sustainability Index (DJSI) for eight consecutive years. As a leading industry index on sustainability, the DJSI assesses more than 2,500 companies worldwide, looking at corporate governance and ethical practices, investor relations, environmental management and climate change, digital inclusion, community investment, human rights, health and safety, diversity, and supply chain and risk management.

The history of BT goes back to the mid-19th century, when the Electric Telegraph Company was introduced in 1846 to provide telegraph service in the United Kingdom. Through this century and the first decades of the 20th century, as a telecommunications industry was developing in the richest and most industrialized country in the world at that time, different commercial companies were unified and organized under the General Post Office. In the 1960s and 1970s, the Post Office went through several reforms, changing its condition from a department of the central government to a nationalized company and public corporation, and splitting its services into Post and Telecommunications. This process culminated in the creation of BT in 1981. As with other large national industries, BT was privatized in 1984 during the Thatcher era in British politics. This also allowed BT to expand its operation globally.

According to recent data, BT operates in more than 170 countries, with a presence in Europe, North and Latin America, the Middle East and Africa, and Asia-Pacific. It employs more than 100,000 people in the United Kingdom alone. BT continues to be the main telecommunications supplier in Britain, and its financial performance indicators have remained strong: in 2007, BT profit reached £2,495 million, with a total turnover of £20,223 million.

One of the earliest to provide a publicly accessible sustainability report, BT's strategic sustainability goals range from the broadly vague of "seeking out new long-term commercial opportunities to create a more sustainable world" to the more concrete of "reducing carbon footprint," "influence suppliers to produce products that emit fewer emissions," and "influence customers by providing lower carbon solutions." A closer look at BT's Sustainability Performance Indicators in turn suggests that the company has made progress on a wide front of issues: cutting carbon emissions, funding research and school projects, reducing waste, and many more. How serious and far ranging such efforts are, however, is not always clear. In 2002, for instance, BT, as the Corporate Social Responsibility (CSR) leader in its industry, made a seven-figure donation to sponsor a science fair for Irish school children. That very same year, it laid off several hundred employees at a company just acquired by BT in Germany in order to meet publicly stated profit goals, providing fodder for skeptics who argue that sustainability programs by transnational corporations are little more than "greenwash," intended to cover up business practices that routinely despoil the environment, aggravate social inequities, and endanger human communities.

If available indexes are a reliable measure, however, BT certainly seems to be a leader among its peers. In addition to its past ratings in the DJSI, the company was recently recognized as the only global telecommunications carrier to have made the top ten in a new Corporate Sustainability Index Benchmark Report for 2009. The index, the first by Technology Business Research (TBR), is informed by a survey of 40 companies within the computer hardware, professional services, and network and telecommunications sectors, and apparently based on an array of sustainability metrics such as emissions, energy, renewable energy, annualized emissions reductions, water utilization, recycling rates, and public environment commitments.

BT is organized into four main lines of business: BT retail (focused on retail costumers), BT global services (which serves multisite organizations worldwide), BT wholesale (focused on other communications companies), and Openreach (in charge of networks that link homes and business to the networks of communications providers). BT Group offers local and long-distance phone services and it also has a wide range of products, such as broadband, television, cell and home phones, for retail consumers; IT services, web hosting, e-mail, mobile, and phone lines for small and medium businesses; and knowledge centers and conferencing facilities for large businesses.

In its 2008 Sustainability Report, BT defines telecommunications as an environmentally sound technology that is cleaner than many traditional industries. The goal, as defined by BT's strategic mission, is "to manage our business to minimize negative impacts to the environment and to maximize the benefits we bring." In order to fulfill this objective, BT relies on an environmental management system that complies with the international standard ISO 14001 of 2004 as a mean to reduce environmental impact. UK operations are certified according to this standard since 1999. With this general purpose in mind, the company has identified the following risk areas: fuel, energy and water, waste, transport, emissions to air, procurement and the environment, product stewardship, and local environmental impacts.

According to BT's official statements, the company is particularly concerned with the production of carbon dioxide due to the significant amounts of energy used to power the telecommunications network and energy needed to run and cool the buildings. In order to reduce this production of greenhouse gases, the BT Group has established a target of reducing its carbon emission intensity by 80 percent against 1996–1997 levels by December 2020. The group continuously improves its energy efficiency, buys green electricity whenever possible, and installs on-site renewable technologies. As an example of this corporate policy, BT signed in 2004 what was the largest renewable energy purchase deal at the time. The company has also set up a computer-based energy consumption monitoring system in order to control consumption and possible misuse of energy.

As a member of the Dow Jones Sustainability Index World and Dow Jones Sustainability Index, BT is continuously compared to all other transnational corporations seeking the badge of sustainability. In all major categories—economic, social, and environmental—the company has continued to score well above most of its competitors. While standards of evaluation may not always be as clear and as transparent as one would like, and while there is ongoing controversy over both the interests behind, and the overall effectiveness of sustainability efforts on the part of for-profit corporations, there is little doubt that British Telecom has for some time now been in the vanguard of social and environmental responsibility initiatives.

See Also: Corporate Social Responsibility; Dow Jones Sustainability Index; Green Technology.

Further Readings

British Telecom. "BT's Sustainability Report 2008." London: British Telecommunications, 2009. http://www.btplc.com/Societyandenvironment/Ourapproach/Sustainabilityreport/index.aspx (Accessed December 2009).

Epstein, Marc J. *Making Sustainability Work: Best Practices in Managing and Measuring Corporate Social, Environmental and Economic Impacts.* San Francisco, CA: Berrett-Koehler Publishers, 2008.

Solymar, Laszlo. *Getting the Message (A History of Communications)*. Oxford, UK: Oxford University Press, 1999.

Witte, Jan Martin, Charlotte Streck and Thorsten Benner, eds. *Progresses or Peril? Partnerships and Networks in Global Environmental Governance. The Post-Johannesburg Agenda*. Berlin, Germany: Global Public Policy Institute, 2003.

Belén García-Olmedo
University of Granada

Brownfield Redevelopment

Unlike "greenfield" development, which encroaches on agricultural or greenbelt land, so-called brownfield redevelopment reuses previously developed land that is often derelict, underutilized, or neglected. This can improve the aesthetic, social, environmental, and economic value of the land, so it is consistent with the broad principles of sustainable development and the ambition of achieving intergenerational equity. However, brownfield land that was previously used for industrial activity is commonly contaminated, necessitating remediation to render the land suitable for the new use. Depending on the type or levels of contamination and the sensitivity of the proposed development to these, such remediation can be costly and time consuming. From an environmental perspective, this can be highly energy intensive, generate waste for disposal elsewhere, or be disruptive to the existing soil structure, groundwater, or ecosystems.

Brownfield redevelopment finds uses for derelict, underutilized, or neglected land, such as this ground-mounted solar photovoltaic array situated on the site of a former landfill at Fort Carson, Colorado.

Source: Western Area Power Administration

In 1998, the government of the United Kingdom announced that by 2008 over 60 percent of new housing should be on brownfield land through residential development or conversion of existing buildings. This represented a small but significant increase in a figure that had previously been just above 50 percent. In its first annual report in 2001, the National Land Use Database of Previously Developed Land indicated that an estimated 66,000 hectares (ha) of brownfield land was available in the UK, of which 36,000 ha was vacant or derelict. Vacant land is defined as land where there are no buildings that could be redeveloped without treatment. Derelict land includes land (and any buildings) that has become damaged by industrial or other development and is currently beyond beneficial use without treatment, such as demolishing buildings, clearing foundations, leveling, or removing contamination.

By 2007, the areas of brownfield and vacant or derelict land in the UK had fallen to 62,130 ha and 33,600 ha, respectively, with 77 percent of new housing from brownfield redevelopment. In the London area for example, residential development on previously developed land rose to 97 percent in the same period. This rise has been accompanied by an increase in the average density of housing, from 25/ha in 2001 to 44/ha in 2007. At this density, the available brownfield land area is equivalent to the area required by housing demand for over a decade.

By definition, in the UK a brownfield is previously developed land that is or was occupied by a permanent structure (excluding agricultural or forestry buildings), including the curtilage of the developed land and any associated fixed surface infrastructure. This definition is independent of the nature of the previous land use, whether residential, industrial, or potentially contaminative. As a consequence of this definition, the development of higher-density housing or flats on a former house and garden site is treated equally favorably under the planning regime as the remediation and redevelopment of a contaminated industrial site. This has led to an increasing urbanization of popular suburban areas under pressure of development and a loss of urban wildlife habitats, rather than the reuse and remediation of derelict areas that was intended.

If left undeveloped, a brownfield site can provide a distinctive type of ecosystem, typified by partial cover of pioneer plant species due to extreme soil compositions or physical conditions. Such open mosaic habitats on previously developed land have recently been included in the new list of UK Biodiversity Action Plan priority habitats and species.

In the United States, a brownfield is a property, the expansion, redevelopment, or reuse of which may be complicated by the presence or potential presence of a hazardous substance, pollutant, or contaminant. Typical examples include former industrial or commercial land affected by mine scarring, low-hazard petroleum products, or other contaminants. As such, the definition specifically excludes land that is more severely contaminated and has high concentrations of hazardous waste or pollution, such as those covered by the "Superfund" legislation and the National Priorities List. It also excludes closed landfills and government sites. It is estimated that there are more than 450,000 such brownfields in the United States. So-called mothballed brownfields are properties where the owners are unwilling to transfer the brownfield or put it back into productive reuse. In contrast, cleanup and reinvestment in these properties can increase local taxation revenues, facilitate job growth, utilize existing infrastructure, take development pressures off undeveloped open land, and both improve and protect the environment.

For brownfield land to be considered contaminated land, it requires there to be a significant risk of harm occurring to the site user, or a negative impact on the site development or the environment if no mitigation measures are put in place. It is common to analyze such risks using a pollution linkage model. This requires identification of all potential sources of contaminants or pollution agents on the site. These are compared to the likely receptors on site or introduced to the site under redevelopment. These typically include site users, property, ecosystems, surface water, and groundwater. It is then possible to identify and assess all potential pathways in the new development that could link sources to receptors. This should include any pollution linkages to off-site receptors through dispersion by migration of contaminants in surface water, groundwater, ground gases, or windborne dust. If any significant pollution linkages are identified, then remediation measures can be planned to remove the pollution source, block the pathway, or remove the receptor (e.g., by exclusion or a change in use).

Redeveloping brownfield land for new industrial or commercial land has the advantage that the land may already be suitable for use. Site development with engineered platforms, concrete floors, or hard-standing will often provide a barrier to soil contamination without recourse to off-site disposal or treatment. The shorter exposure period of adult workers means that higher concentrations of contaminants are tolerable than for residential developments, where children playing in residential gardens are more vulnerable receptors.

For more seriously contaminated sites, the cost of remediation may exceed the redevelopment value. So-called hardcore sites are those where the cost or difficulty of remediation has prevented redevelopment. In regions of widespread industrial decline, or in times of economic recession, it is possible that no economically viable redevelopment option can be found for some brownfield sites. Government funds may be required to render the land suitable for use as public open space for recreation, green corridors for wildlife, or other "soft" end uses.

Efforts to reduce the use of fossil fuels and greenhouse gas emissions during remediation have focused attention on reusing brownfield land for renewable energy production. The initial investment in solar or wind power can be used to offset power use during "green remediation," one of several ways in which sustainable environmental practices can be incorporated. Thereafter, the same equipment can supply electricity to the grid. A former industrial site will typically already have grid connections, saving one of the major establishment costs. Brownfield lands from former mining, military, or landfill sites provide large, isolated locations suitable for installation of wind farms (e.g., Klettwitz, Germany) or solar photovoltaic arrays (e.g., Fort Carson, United States). Using nonagricultural (brownfield) land for biofuel or energy-crop production has the added ethical advantage that it does not impact global food production.

See Also: National Priorities List; Quantitative Risk Assessment; Restoration; Superfund; Sustainability; Sustainable Development; Triple Bottom Line.

Further Readings

Dixon Tim, Mike Raco, Philip Catney and David Lerner. *Sustainable Brownfield Regeneration: Liveable Places From Problem Spaces*. Hoboken, NJ: Wiley-Blackwell, 2007.
National Land Use Database. http://www.nlud.org.uk (Accessed April 2009).
Syms, Paul. *Releasing Brownfields*. London: RICS Foundation, 2001.
U.S. Environmental Protection Agency. "Brownfield and Land Revitalization." http://www .epa.gov/swerosps/bf/index.html (Accessed April 2009).
U.S. Environmental Protection Agency. "Green Remediation." http://cluin.org/green remediation (Accessed April 2009).

Richard Alastair Lord
University of Teesside

CARBON FOOTPRINT

There has been a rising awareness of the social and ethical responsibility of business among many stakeholders such as consumers, government, and environmentalists. This awareness and laws governing pollution emissions, particularly from greenhouse gases (GHGs), for example, carbon dioxide (CO_2), has prompted many companies to reexamine both their methods of production as well as the negative impacts of their products on the environment. The carbon footprint is one such method for examining the environmental impact of industry. Generally, the carbon footprint refers to the total CO_2 and other GHGs that are released into the atmosphere when products are created, consumed, transported, or stored. For a business, the carbon footprint represents the effect an organization has on the climate based on the total amount of GHGs produced, measured in equivalent units of CO_2.

Carbon dioxide emissions are included in the calculation of a company's carbon footprint. Here is a coal-fired power plant in North Rhine-Westphalia, Germany.

Source: iStockphoto

Calculating the Carbon Footprint

Since many companies calculate the carbon footprint of their products differently, it is difficult to directly compare products on their relative impacts on global warming. For example, for some products, how they are used may have a greater impact on global warming than how they are made, although it may be the reverse situation for other products. And yet another class of products may have global warming impacts that are inherent both in their manufacture and in their continued use. Thus, manufacturing a pair of leather shoes may create more carbon emissions, but their continued use may offset that impact. On the other hand, a car would create CO_2 not only in its manufacture, but in its continued use as well. Similarly, fruits and vegetables may create carbon emissions due to refrigeration and transportation, but not as much in their manufacture or consumption.

Much of the confusion about calculating carbon footprints come from ambiguities in the definition of "carbon footprint" itself. While there is some agreement about the inclusion of carbon dioxide emissions, there is variation in whether other GHGs, such as methane (CH_4), should also be included in the definition. A more difficult issue is drawing the boundary on the life cycle impacts. Should only direct emissions be included and not indirect emissions? Should all life cycle impacts be included? Should the emissions be restricted to greenhouse warming potential? Should they exclude CO_2 from soils and other nonfuel sources?

Ultimately, the purpose of measuring the carbon footprint is to understand the consequences of different methods of manufacturing, storage, transportation, and consumption. The idea is to then use this knowledge to devise alternate consumption, production, distribution, and storage that will have less total carbon emissions than before.

One hundred and seventy nations formed an international agreement known as the Kyoto Protocol. Members who ratified the Kyoto Protocol were obliged to attempt to observe the targets and timetable to reduce their GHG emissions. Countries that experience shortfalls in their Kyoto emissions requirements may engage in emissions trading to purchase certified emissions reductions (CERs) or emission reduction units (ERUs). This serves to create a market value for emissions, the buying and selling of carbon credits, and also encourages greater vigilance over total national carbon emissions per country.

Consumers and Carbon Trading

At the individual level, personal carbon trading has also become a new activity for the environmentally conscious consumer. Various carbon offset programs are available as well as sites and formulas for calculating the CO_2 emissions that daily activity can generate, such as flying, driving, washing laundry, and air conditioning. People may choose activities that have lesser carbon footprints or purchase carbon credits to offset the emissions they may generate. If more products display carbon footprint data on their packaging, it is possible that consumers will compare and choose goods that reflect lower carbon emissions.

Similar to this trend, a parallel may be seen in the rising discontent over the use of fossil fuels, nonrenewable energy sources, and the disturbance and irrevocable damage caused by the emission of CO_2 and other GHGs.

Carbon emissions and other environmental sustainability indicators are increasingly being made available to the public. In 2009, over 1,200 companies worldwide supplied sustainability reports, including information on carbon emission by weight, to the Global Reporting Initiative (GRI), a network-based organization that provides a standard framework for sustainability reporting. The number of businesses reporting in 2009 was nearly double that of 2007. By region, European-based businesses comprise nearly half of all GRI sustainability reports and by country, Spain, the United States, and Brazil are the reporting leaders.

The World Resources Institute (WRI) and the World Business Council for Sustainable Development (WBCSD) have recently created two new standards—the Product Life Cycle and Reporting Standard and the Scope 3 (Corporate Value Chain) Accounting and Reporting Standard—for their GHG Protocol Initiative that outlines methods to account for GHG emissions associated with products over the course of their life cycle, from raw material acquisition to production and retailing to user consumption. Beginning in 2010, these new standards will require corporations to report their carbon footprint based not only on point-specific operations, but to examine their entire supply chain and include outsourced activities and product distribution.

Several companies such as Google, Yahoo!, PepsiCo, Timberland, Coca-Cola, BP, and Dell have also promised to take steps toward becoming carbon neutral. PepsiCo plans on being among the first to provide its consumers with an absolute number for its products' carbon footprints. It is expected that this will set a trend that other companies will soon be compelled to follow as well. However, it is not clear whether consumers will understand or even care about what this means. Furthermore, its understanding is compounded by the plethora of methods to calculate it. So, although it will give some indicators to the consumer interested in knowing about a product's carbon emissions, it may be difficult for companies to leverage this information as a marketing tool against competitors.

It is likely that the definition of carbon footprint as well as its measurement will evolve and arrive at a standard metric that is easily quantifiable and widely understood. Such a development will mature the field of carbon emissions considerably. Direct comparisons of carbon emissions between products, activities, sourcing materials, distribution, and logistics will be possible, such as the GRI, WRI and WBCSD emissions reporting guidelines. Robust measurement will lead to precise evaluation of greenhouse impacts for individuals, activities, and products. Comparative rankings of countries on their total carbon emissions will also be enabled. Industry- and firm-specific carbon emission data will allow scientists and researchers to develop alternate and more effective procedures and products. Policy makers will also be empowered by this information to create institutional incentive structures to check, censure, and reward companies on their carbon emissions. In sum, the refinements in definition and measurement will have a positive impact on reducing carbon footprints worldwide.

See Also: Carbon Neutral; Carbon Sequestration; Carbon Trading; Ecological Footprint; Emissions Trading.

Further Readings

Ball, Jeffrey. "Six Products, Six Carbon Footprints." *Wall Street Journal* (March 1, 2009) http://online.wsj.com/article/SB122304950601802565.html (Accessed January 2010).

Global Reporting Initiative (GRI). http://www.globalreporting.org (Accessed January 2010).

Hoffman, Andrew J. *Carbon Strategies: How Leading Companies Are Reducing Their Climate Change Footprint.* Ann Arbor: University of Michigan Press, 2007.

Martin, Andrew. "How Green Is My Orange?" *New York Times* (January 21, 2009). http://www.nytimes.com/2009/01/22/business/22pepsi.html (Accessed January 2010).

Stern, Nicholas. *The Economics of Climate Change: The Stern Review.* Cambridge, UK: Cambridge University Press, 2007.

Wiedmann, Thomas and Jan Minx. "A Definition of 'Carbon Footprint.'" In *Ecological Economics Research Trend,* C. C. Pertsova, ed. Hauppauge, NY: Nova Science Publishers, 2008.

World Resources Institute (WRI). "Sixty Corporations Begin Measuring Emissions From Products and Supply Chains." http://www.wri.org/press/2010/01/sixty-corporations-begin -measuring-emissions-products-and-supply-chains (Accessed January 2010).

Manjula S. Salimath
University of North Texas

CARBON NEUTRAL

Becoming carbon neutral, sometimes known as CO_2 neutral, most commonly refers to achieving net zero greenhouse gas (GHG) emissions (including all greenhouse gases weighted according to their CO_2-equivalent impact). Carbon neutrality can be achieved by reducing and avoiding GHG emissions at the point of origin/production, or by balancing greenhouse gas emissions through the purchase of an equivalent amount of carbon offsets. This combination of activities allows companies, charities, governments, nongovernmental organizations (NGOs), and individuals to reduce their net climate change impact to nil. To emphasize reduction of GHG emissions is generally recognized as a "best practice," and to offset, if necessary, residual emissions.

Carbon neutrality is achieved by reducing carbon emissions and balancing greenhouse gas emissions through the purchase and cancellation of an equivalent amount of carbon offsets. The English village of Ashton Hayes aspires to become England's first carbon-neutral village.

Source: goingcarbonneutral.co.uk

In theory, the way toward carbon neutrality follows three sequential steps. First, there is a need for a reliable and transparent measure of GHG emissions. Homeowners, for instance, can use one of the several available online calculators to produce reasonably precise values. For businesses or public services, a much more detailed carbon footprint assessment is required, for example, using the World Resources Institute (WRI) GHG Protocol or the French Environment and Energy Management Agency (ADEME) program titled Bilan Carbone. The second step involves implementing measures to reduce greenhouse gas emissions, and the third step consists of acquiring carbon offsets (also called carbon credits) generated by others to balance out one's remaining emissions.

Carbon offsets are generated through a "baseline and credit" scheme. Certain companies specialize in providing these offsets, such as co2logic, JPMorgan's ClimateCare, Klimaatneutraal, or TerraPass. Under such schemes, project developers, often in the developing and emerging world, are encouraged to reduce their emissions below a baseline to generate emission reduction credits that can then be sold to entities with mandatory GHG targets or people with "carbon neutral" objectives. The baseline scenario is usually defined as the business as usual scenario, for example, what would have occurred if the climate-friendly project had not been implemented.

Offsets come primarily from four types of projects: renewable energy, energy efficiency, forestry, and waste management. A feature of projects used for offsetting to reach carbon neutrality is that they are often small scale and frequently emanate from voluntary efforts that follow standards such as the Voluntary Carbon Standard or the Gold Standard. Increasingly, however, corporations engage in larger-scale efforts to gain offset credits, such as large tree-planting initiatives. The Kyoto Protocol provides an official offset program, or cap-and-trade system, for nations to "cap" or reduce their GHG emissions. While nation-states are the actual signatories to the Kyoto protocol, corporations can participate in this growing market, broadly overseen by the United Nations Framework Convention on Climate Change. Offsets are now traded on major stock exchanges and organized by institutional frameworks such as the European Union Emission Trading Scheme (EU ETS). Examples of projects that have been used to offset emissions include replacement of inefficient diesel motors with a gas motor in Africa, building of wind farms in India, or support of reforestation programs in Indonesia. More than one-third of voluntary offset credits that are essentially used to help companies reach carbon neutrality come from renewable energy projects, and another third come from forestry projects.

While the term *carbon neutral* was the *New Oxford American Dictionary*'s "Word of the Year" in 2006, debates about a clear and operational definition remain. In addition to the operational problem of what precise pollutants are included in carbon neutral definitions, governmental regulations striving for carbon neutrality, for instance, allow those companies that did not meet set targets to buy pollution credits from companies that overshoot targets. Such trades, however, contain no considerations for what kinds of technologies are used in which industries with what kind of broader consequences for the environment and surrounding communities. There are also ongoing debates about exactly which parts of the production process should be included in carbon assessments: is it only production per se, or should it include things like storage, shipping, and handling? And what about the carbon footprint of materials and products purchased to enable the production process in the first place?

The carbon-offsetting sector is arguably facing some of the same dilemmas as organic food suppliers or fair trade companies. On the one hand, carbon offset providers wish to deliver quality products that give strong guarantees and promote sustainable development, but on the other hand, their costs must remain reasonable. A single term—whether organic, fair trade, or carbon neutral—can cover varying degrees of efforts and is consequently open to abuse. In the case of organic food, given the interests at stake, the use of the word has been subject to regulation in most countries (since 1991 in the European Union, 2000 in Japan, and 2002 in the United States), whereas fair trade certification remains in the hands of the voluntary sector. Despite a recent increase in transparency in regard to what has been included in the neutrality scope, the lack of clear definitions and boundaries in the use of the term *carbon neutral* can lead to confusion among companies, advertisers, and regulators. In 2009, the UK Department of Energy and Climate Change (DECC) launched a public consultation project in order to investigate an accepted definition and discuss how carbon neutrality is applied. In the future, some companies that currently claim they are carbon neutral may need to adjust their practices to fit new emerging definitions. What is clear is that there is a growing demand for "carbon neutrality" among consumers—even without clear definition. Companies have responded with a variety of claims, such as Eurostar's "all journeys are carbon neutral" or the Flying Group's "carbon neutral flying." Even municipalities now advertise their carbon footprint, such as Ashton Hayes's claim to be "England's first carbon neutral village." To date, Norway remains the only country to advertise "carbon neutrality by 2030." Far more difficult is to ascertain the actual reality behind such claims.

Besides the risk of misrepresenting of the term for greenwashing motives, one of the recurrent criticisms of carbon neutrality achieved through offsetting is that it does not reduce emissions at the source. Some critics believe that such mechanisms distract businesses and citizens from the main objective: reducing emissions here and now. These critics believe that companies or citizens that offset will continue to act as before, simply paying a few more dollars, allowing to soothe their environmental conscience by off-shoring their pollutant-generating activities to others (often in the poorer and underdeveloped world). For them, offsetting is a new form of indulgence and avoids structural and behavioral changes in the fight against global warming.

However, for its proponents voluntarily seeking to reach carbon neutrality through use, offsets play a triple role and should ultimately lead to emissions reductions. First, offsetting is a tool to raise awareness and educate individuals and companies about their climate impact. In order to become carbon neutral, individuals and businesses measure their emissions and become aware of their own climate impact and are, therefore, better prepared to act on it. Second, by putting a price on pollution, companies are internalizing their emissions costs, which encourages them to reduce them. For instance, if one takes into account the offsetting cost for a short-haul flight, it is possible that an alternative, cleaner choice such as high-speed trains becomes economically more attractive. Finally, some companies, realizing the growing costs involved in offsetting their toxic emissions, may decide to invest directly in emission reductions at source—by budgeting the return on investment for a more efficient boiler, for instance.

In the end, it should be noted that capital gained through offsetting generally goes toward emission reduction projects, often promoting sustainable development in developing countries. While far from ideal, in short, proponents of the articulation and implementation of clear carbon neutral objectives see the goal of carbon neutrality as an important practical tool in the fight against climate change.

See Also: Carbon Footprint; Carbon Trading; Corporate Social Responsibility; Ecological Footprint; Greenwashing.

Further Readings

Bayon, R., et al. *Voluntary Carbon Markets: An International Business Guide to What They Are and How They Work*. London: Earthscan, 2007.

Brohé, Arnaud, et al. *Carbon Markets: An International Business Guide*. London: Earthscan, 2009.

Kirby, Alex. *Kick the Habit: A UN Guide to Climate Neutrality*. New York: United Nations Environment Programme, 2008.

Smith, Kevin. "The Carbon Neutral Myth." Amsterdam, the Netherlands: Carbon Trade Watch/Transnational Institute, 2007. http://www.carbontradewatch.org/pubs/carbon_neutral_myth.pdf (Accessed January 2010).

Arnaud Brohé
Université Libre de Bruxelles

CARBON SEQUESTRATION

Carbon sequestration is the biological, physical, and chemical capture and long-term storage of carbon, usually from carbon dioxide (CO_2), so as to reduce greenhouse gas emissions and prevent those that are produced from contributing to climate change.

CO_2 concentration in the atmosphere is increasing primarily due to the combustion of fossil fuels—in particular, coal—and also oil and natural gas. CO_2 contributes to global warming, and in 1997, most industrialized countries signed the Kyoto Protocol, agreeing to reduce their emissions of greenhouse gases (primarily CO_2). However, most of those nations are not meeting their agreed reductions. Carbon sequestration has the potential to slow down the rate of increase of CO_2 levels or even reduce levels of CO_2 in the atmosphere.

As part of their Kyoto Protocol commitments, the European Union introduced an emissions trading scheme that involved allocating greenhouse gas allowances to individual companies and allowing them to use carbon trading and offsets to cover additional emissions. Offsets involve paying others for greenhouse gas reductions and carbon sequestration. Other nations are now also proposing to introduce emissions trading, including the United States and Australia.

Companies that do not wish to pay for extra greenhouse gas allowances, or that want to reduce their greenhouse gas emissions as a way of greening their business, may choose to capture their CO_2 emissions or pay for carbon sequestration elsewhere. Also, a number of companies, such as the Carbon Neutral Company, now sell the opportunity to be carbon neutral to people taking airplane trips, driving cars, and engaging in other greenhouse gas–generating activities. The money is often used to fund carbon sequestration, in part, by planting trees.

Nonpoint Source Sequestration

On land, much carbon is sequestered in living plants, in particular, forests and grassland systems via the process of photosynthesis. Linked to this, soil is also a major store of carbon in organic matter accumulated via plant litter and other biomass. Over the past 50 years, conversion of grasslands to croplands and deforestation with associated soil disturbance has resulted in the conversion of substantial soil organic carbon to atmospheric CO_2.

Increasing grassland or forests could transfer CO_2 from the atmosphere into biomass. Peat bogs are also a major store of primarily plant-derived carbon. The creation of new peat bogs or enhancing existing ones could also aid in transferring carbon from the atmosphere. However, it is difficult to know how much carbon is sequestered by grasslands, forests, or peat bogs, and the sequestration is often not permanent. Forests, in particular, are vulnerable to fires, disease, and illegal logging, and after a number of decades, can no longer absorb carbon. Tree plantations also reduce soil fertility, use up valuable water resources in dry areas, and increase erosion and compaction of the soil.

Much carbon is sequestered in the oceans in phytoplankton via the process of photosynthesis. It has been proposed that increasing phytoplankton growth—by adding iron to the oceans, by nitrogen fertilization of oceans, or by transferring nutrient-rich lower ocean layers to the surface—could remove substantial carbon from the atmosphere into living phytoplankton. When phytoplankton die, they sink to the ocean bottom, thus locking away the carbon they absorbed. However, the potential of this approach to capturing CO_2 remains unknown, and its effects on marine ecosystems may be detrimental as the nutrient balance of the ocean could be upset.

CO_2 absorbed into the ocean from the atmosphere reacts with water to form a range of products, primarily carbonic acid, and hence bicarbonate and carbonate ions. The relative concentrations of CO_2, bicarbonate ions, and carbonate ions determine the acidity level of the oceans. Uptake of anthropogenic CO_2 reduces the carbonate ion levels and the buffering capacity of oceans and, hence, ocean acidity increases. It has been proposed that adding crushed limestone or volcanic rock to oceans will increase their buffering capacity and allow them to absorb more CO_2 from the atmosphere. Alternatively, it has been suggested that hydrochloric acid could be removed from the ocean by electrolysis and then neutralized through reactions with silicate.

Point Source Sequestration

Carbon sequestration of CO_2 from industry can be achieved biologically or nonbiologically.

Pyrolysis of biomass, that is, biological material from wood, waste, and alcohol fuels, can be used to create charcoal and this "biochar" can be landfilled or added to soils. There is strong evidence that the addition of biochar to agricultural soils can increase their nutrient retention and hence decrease the requirements for crops fertilization. There is also evidence that biochar can reduce the emissions of nitrous oxide (another greenhouse gas) from croplands. Initiatives have been established in several countries to assess the potential of biochar in relation to carbon sequestration.

Nonbiological capture of CO_2 is "easiest" at large point sources, such as energy (power plants), and at sources of cement, lime, iron, and steel production, although most industries are responsible for CO_2 emissions associated with energy use. Power generation accounts for about one-quarter of global CO_2 emissions and around 40 percent of CO_2 emissions in the United States. Theoretically, CO_2 can be captured directly from the atmosphere—termed *carbon dioxide scrubbing*—but this is technically difficult, energy intensive, and expensive.

After capture, the CO_2 must be transported to process or storage sites; generally, this is done by pipeline. Captured CO_2 could be converted into hydrocarbons, and then stored or used as fuel or to make plastics. Also, the possibility of sequestering carbon on land and in the oceans has been examined. Carbon could be sequestered as gaseous CO_2 in various deep geological formations, such as exhausted oil and gas fields, or in saline aquifers. Currently, CO_2 is transported to oil production fields, where it is injected into older fields to increase oil removal.

Carbon could also be sequestered as liquid CO_2 in deep ocean masses below at least 1,000 meters deep, or as solid storage by using a reaction of CO_2 with metal oxides to produce stable compounds.

The long-term storage of CO_2 is a relatively untried concept, and trapping mechanisms would need to be developed to stop the CO_2 from escaping into the atmosphere.

See Also: Carbon Footprint; Carbon Neutral; Carbon Trading; Clean Fuels; Clean Production; Clean Technology; Ecological Economics; Ecological Footprint; Emissions Trading; Environmental Accounting; Pollution Offsets; Pollution Prevention.

Further Readings

Carbon Monitoring for Action (CARMA). http://carma.org/ (Accessed May 2009).

Intergovernmental Panel on Climate Change (IPCC). *Carbon Dioxide Capture and Storage.* Cambridge, UK: Cambridge University Press, 2005.

Intergovernmental Panel on Climate Change. http://www.ipcc.ch (Accessed May 2009).

International Energy Agency (IEA). "CO$_2$ Capture and Storage." http://www.co2capture andstorage.info (Accessed January 2010).

Wilson, Elizabeth J. and David Gerard, eds. *Carbon Capture and Sequestration: Integrating Technology, Monitoring and Regulation.* Ames, IA: Blackwell Publishing, 2007.

Mitchell Andrews
Derek Watson
University of Sunderland

CARBON TRADING

Carbon trading is an administrative approach to controlling greenhouse gas emissions. It aims to provide economic incentives to companies to reduce emissions. It is also referred to as emissions trading and cap-and-trade.

Carbon trading has received mixed responses from the public, as well as the policy community, and there is a certain degree of skepticism that an economic tool will be able to transform heavy polluters into environmental champions.

In practice, a central authority (usually a governmental body, but some voluntary schemes are managed by private entities) sets a limit or cap on the amount of greenhouse gas emissions (expressed as a carbon dioxide [CO$_2$] equivalent) that can be emitted. Companies covered by a cap-and-trade carbon trading scheme are issued emission allowances and are required to hold an equivalent number of allowances (or credits) that represent the right to emit a specific amount of pollutant. In practice, the unit is one metric ton of CO$_2$ equivalents in United Nations and European Union schemes, and one short ton in the U.S. scheme Regional Greenhouse Gas Initiative (RGGI).

The total amount of allowances and credits cannot exceed the cap, limiting total emissions to that level. Entities that emit more than their allocated total of allowances must buy allowances or credits from those who emit less than the amount they were allocated. In effect, the buyer is paying for emitting CO$_2$, while the seller is being rewarded for having reduced emissions by more than was needed. Thus, in theory, companies that can reduce their emissions most affordably will do so, creating a virtuous dynamic where actors are rewarded for emission reductions.

In particular, five essential elements must be thoroughly defined in order for a carbon trading scheme to be environmentally, economically, and socially effective:

- Defining a scope (a cap and a commitment period)
- Allocating allowances
- Managing the price volatility
- Monitoring, reporting, and tracking allowances on a registry
- Reconciling emissions with allowances and setting penalties for noncompliance

The definition of the scope is based on several parameters, including geographical coverage, temporal range, and the gases covered. Demand for allowances will depend on the severity of the cap, but also on the level of actual emissions from companies included in the scheme. If the reduction target is small, demand for allowances will be weak. Similarly, if companies are able to significantly reduce their emissions, thus remaining within their cap, then demand for allowances will again be weak and prices will remain low to moderate. This can occur either because of the use of mitigation technologies, such as improving energy efficiency, or because of a fall in production during an economic downturn.

The commitment period is the time period for attaining emissions reductions. If the benefits of emissions trading are to be realized, the system must balance predictability in its shape and rules, and have the flexibility to take advantage of changing circumstances. A long commitment period, with banking and borrowing of emissions credits between periods, can provide greater certainty and reduce policy risk.

The creation of a new emission market requires property rights to be identified and allocated where there previously were none. There are two main allocation approaches: selling the rights to pollute, that is, allowances; or giving them away. Free allocation involves giving pollution rights away free of charge under some predefined rule. The most common method of allocation is "grandfathering," where allowances are allocated on the basis of prior use. This allocation method is usually strongly advocated by polluters as it recognizes their implicit right to use the environment as they always have, albeit now under the constraint of a cap. In a free allocation process, the allocation is political and is, therefore, influenced by various forms of lobbying and can be very laborious. It also often results in overallocation.

Alternatively, if the government elects to sell or auction permits, it assumes that polluters had no prior right to the environment and that the atmosphere is a commons effectively owned by all. Under this approach, the companies covered by the scheme face an up-front cost of participation. In practice, governments sometimes develop a hybrid allocation method with a mix of free allocation and auctioning.

Carbon trading can create significant price variability for the involved parties. Such volatility can potentially pose a threat to carbon-constrained economies. Various mechanisms can be used to help control volatility. The first option is to allow banking of allowances for future use. This allows governments to encourage companies to further reduce their emissions now by allowing them to establish a reserve of allowances for the future and can limit price volatility between trading periods and smooth prices.

Setting price floors and/or ceilings is another method that can be used. These would aim to provide a mechanism of "safety valves" to reduce the risk to investments in emissions reductions. The price floor would ensure the regulator against the emissions market collapsing, either due to an overallocation of permits or a fall in demand for permits. The price ceiling would ensure the covered entities against extremely high costs of abatement; however, this would need to be weighed against the loss in environmental integrity induced by the addition of permits to the system.

A third method used to limit price volatility is to link carbon trading schemes to baseline and credits projects outside the capped system. With a baseline and credits project, an investor can generate additional emission credits by investing in emissions reductions in other sectors or areas. These credits can then be used for compliance purposes in a carbon trading scheme and are referred to as offsets.

The monitoring and reporting of emissions is the next critical element. The definition of clear rules and standardized methods for calculating emissions are a prerequisite for the credibility of any emissions trading system. It is important that the measurement methods are reliable and consistent so that a ton of CO_2 means the same thing across different sectors and nations.

Reliable registries are also needed to ensure that emissions allowances can be traced, thereby preserving the environmental integrity of the system. At the end of the accounting period, reconciliation between actual emissions and emissions allowances held by the participants is performed using the data booked in the registry.

Finally, in order to ensure environmental integrity of the system, the regulator must set enforcement rules. A system of fines discourages polluting entities from emitting more greenhouse gases than their allowances cover. However, fines alone are not always enough to ensure environmental integrity. When demand for allowances and carbon prices are high, the polluter may choose to pay the fine, rather than attempt to buy emissions permits. To avoid this pitfall, governments may declare that the payment of a penalty does not release a company from the obligation to reduce emissions. Therefore, the company in default must also reduce their emissions by the additional default amount during the next compliance period.

The Kyoto Protocol established the principle of trading carbon emissions between countries to achieve cost advantages in the reduction of greenhouse gas emissions. However, the largest carbon trading scheme established to date is the European Union's Emissions Trading Scheme (EU ETS), which currently covers more than 10,000 installations in the energy and industrial sectors. These are collectively responsible for nearly half of the EU's CO_2 emissions and 40 percent of the EU's total greenhouse gas emissions. The first trading period (2005–07) put the necessary infrastructure in place. The environmental benefits were doubtful, however, because an overallocation of allowances by most member states meant that the price for allowances was very low, and companies had little incentive to reduce their emissions.

While the United States pioneered emissions trading in its regulation of sulfur dioxide dating from around 1990, the development of a U.S. national carbon market has been slow to emerge. However, recent federal inaction has not meant the complete stalling of the country's engagement with CO_2 emissions trading. Over the last decade, lack of progress at a national level with emissions trading has often masked action being taken at a state level. In recent years, several regional cap-and-trade schemes have emerged. These include the Regional Greenhouse Gas Initiative (RGGI), the Western Climate Initiative, including many Canadian provinces, and the Midwestern Greenhouse Gas Accord. Together, these will encompass most of North America's emissions when fully implemented. A new economy-wide national emissions trading system has now been proposed, with the long-term goal of reducing national emissions by 80 percent by 2050.

See Also: Carbon Neutral; Corporate Social Responsibility; Emissions Trading; Environmental Economics.

Further Readings

Brohé, Arnaud, et al. *Carbon Markets: An International Business Guide.* London/Sterling, VA: Earthscan, 2009.
Stern, Nicholas. *The Stern Review: The Economics of Climate Change.* Cambridge, UK: Cambridge University Press, 2009.

World Bank. "State and Trends of the Carbon Market Report." Washington, D.C.: World
    Bank, 2009. http://www.climatechallengeindia.org/Download-document/116-State-and
    -trends-of-the-carbon-market (Accessed January 2010).

Arnaud Brohé
Université Libre de Bruxelles

CAUSE-RELATED MARKETING

Cause-related marketing (CRM) is defined as the public association of a for-profit company
with a nonprofit organization, intended to raise mutual awareness and benefit. This may
involve donating a percentage of revenues to a specific cause based on the revenue from a
specific time period. McDonald's earmarking $1 for the Muscular Dystrophy Association
from the sale of every large order of fries is a good example of this practice. Companies use
CRM in order to enhance their corporate reputation, raise brand awareness, increase customer loyalty, build sales, and over-
all improve their brand image. CRM
is considered to be distinct from
corporate philanthropy, since the
former term is a marketing relationship and not a donation—the money
is not an outright gift to the not-for-
profit organization and is not tax
deductible.

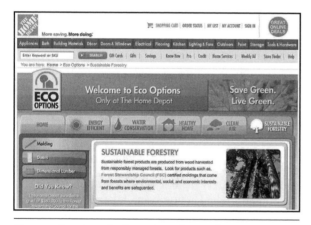

Cause-related marketing pairs a for-profit company
with a nonprofit organization and intends to raise
awareness for mutual benefit. Here, a screen grab from
Home Depot's corporate site touts the company's
ecological efforts and support of the Forest Stewardship
Council's work toward sustainable forestry.

Source: HomeDepot.com

History

The term was first used by American
Express in 1983 to describe its campaign to raise money for the restoration of the Statue of Liberty. The
company donated 1 cent to the
Statue of Liberty every time someone
used their credit card, and $1 for
every new card issued. Over $2 mil-
lion was raised during the first four
months, and, very soon, the number
of new cardholders grew by 45 per-
cent and card usage increased by 28 percent. CRM has continued to grow rapidly in the past
decade. In recent years, the term has encompassed other cooperative activities involving business and charitable causes, harnessing the power of "doing well by doing good."

In the United States, the birth of CRM can be traced back to one of the founding
fathers, Benjamin Franklin, who in 1747 publicized a "scheme of the Philadelphia
Lottery" to raise money for the defense of Philadelphia. Over 30,000 tickets were
printed, and the money raised was used mostly for the creation of a militia and the

construction of a fort, called the battery, on the Pennsylvania side of the Delaware River, just south of Philadelphia. Later, in an effort to raise money for the first public hospital in the United States, The Pennsylvania Hospital, in 1751, he came up with the new idea of combining public money with private donations, thus creating the first matching grant.

Moving ahead to the late 20th century, another notable CRM campaign occurred in 1976 through a partnership between the Marriott Corporation and the March of Dimes. Marriott's goal was to create a cost-effective media coverage and public relations campaign for its 200-acre family entertainment center, Marriott's Great America, in Santa Clara, California. The objective of March of Dimes, on the other hand, was to increase fund-raising by increasing the collection of pledges. The promotion was conducted in 67 cities throughout the western United States and was deemed to be a huge success for both organizations.

Another example of a CRM campaign was created by Rosica, Mulhern and Associates on behalf of Famous Amos cookies in 1979. The chairman and founder of the company, Wally Amos, spoke on behalf of the Literacy Volunteers of America, and gave back a percentage of the sales of his products to this cause. This cause-marketing tie-in was a "win–win" situation for both Famous Amos cookies and for maintaining visibility of the literacy programs.

Firms embrace CRM programs in order to diversify and enhance their funding bases. This can take many forms, including product-, service-, or transaction-specific promotions of a common message, product licensing, endorsements, and certifications, local partnerships, and employee service programs among others. Other examples of notable CRM campaigns include the following:

- In 1984, Sears, Roebuck and Company hired pop star Phil Collins to appear in its print advertisements and sponsored his 40-city concert tour, with the goal of raising more than $1 million for services to help the homeless in the United States.
- The Avon Walk for Breast Cancer program was launched in the United States in 1993 and raises funds through many programs including the sales of special crusade "Pink Ribbon" products by Avon's nearly 600,000 independent sales representatives via concerts, walks, races, and other special events around the world.
- In 2005, rock star Bono started "RED," a campaign that combined consumerism and altruism, whereby firms paid RED a licensing fee to label one or more of their products. They then paid a portion of the sales of their products to the Global Fund, a public-private charity set up to fight AIDS, malaria, and tuberculosis in Africa. Apple Computer, The Gap, Giorgio Armani, and Motorola were some of the companies that participated in the program.
- In 2004, 3M introduced Post-it Super Sticky Notes imprinted with pink ribbons—a portion of the sales revenue was donated to cancer research and treatment.
- Ethos Water (now owned by Starbuck's) gives a nickel for each bottle sold to provide clean water, and Snapple provides a small percentage of its SNAP 2.0 bottled water sales to build playgrounds in poor communities.
- Yoplait's "Save Lids to Save Lives" campaign supports the Susan G. Komen Foundation for the Cure of Breast Cancer. The company packages specific products with a pink lid that customers turn in, and in turn, Yoplait donates 10 cents for each lid.
- Ben & Jerry's has undertaken several social initiatives, tying its products to human well-being as well as ecological protection for the environment. Before selling the company to Unilever, the original owners wanted assurance that such practices would be continued in the future.

- The American Heart Association has given a stamp of approval to Cheerios, the popular breakfast cereal, and about a dozen other products to use its "Heart Check" icon. This signifies that the product meets the association's low-fat, low-cholesterol standards.
- Dannon, the yogurt producer, launched a line of products for children, called Dannon Danimals, featuring containers decorated with wild animals. The National Wildlife Federation received 1.5 percent of the price of each container sold.

Research on Cause-Related Marketing

According to several reports, CRM activities by U.S. businesses have risen dramatically over the past decade. IEG, Inc., reported that expenditures in CRM programs were $1.11 billion (2005), $1.34 billion (2006), $1.44 billion (2007), $1.52 billion (2008), and projected to grow to $1.57 billion in 2009, despite the recession.

A recent study conducted by Cone Millennial Cause found that 89 percent of young individuals (aged 13–25) in the United States would switch brands to a comparable product (and price) if the latter brand was associated with a "good cause." It also found that a significant number would prefer to work for a company that was considered "socially responsible" and a "good corporate citizen." There are both risks and rewards associated with CRM programs for companies. Some rewards include low-cost exposure, increased ability to win customer support, and the ability to cut through the advertising and promotion clutter. Yet, there is also a potential backlash if an initiative is perceived as an abused marketing tool, and firms are also likely to face financial risks for causes with little or no synergy.

There are several lessons that managers can use in developing CRM programs. First, it is important to note that not all customers react to CRM programs the same way. Marketers should do their homework and assess their target audiences' interests and match the CRM program with the needs of their customers. Second, in the short run, some campaigns may simply impact attitudes, not behavior. Managers may have to be patient to see the impact of the cause on the wallets of their customers. Third, not all causes may be a good fit for every company. To find the right match requires considerable amount of research and planning. Fourth, the not-for-profit association and the cause are likely to benefit more than the company sponsoring the CRM program. Finally, it is important to assess the return on investment (ROI) of any CRM program. It is important to demonstrate how every campaign impacts not only the business bottom line, but also the community or cause that was supported.

Online Cause-Related Marketing

In recent years, companies have leveraged the power of the Internet to develop CRM programs. For example, online auctions have helped them create programs that help sellers and firms donate a percentage of their sales to a not-for-profit organization. Businesses and nonprofit organizations can also use the program for cause marketing and nonprofit fundraising programs. In many cases, firms have integrated online marketing channels along with offline channels like printed media, to create effective campaigns.

With consumers around the world spending over 5 billion minutes using Facebook, posting over 2 million tweets and 4 billion photos on Flickr every day (as of August 2009), there is a tremendous potential for developing CRM strategies utilizing the power of these networks. For example, Charity Water used Twitter and Facebook to encourage people with birthdays that month to solicit donations instead of traditional birthday gifts—the campaign has raised almost $1 million in less than a year. Similarly, the Joyful Heart Foundation devised a one-day virtual event on their Facebook page. The goal was to raise awareness of the organization—fans were asked to share a positive message about the foundation's mission. This low-cost effective strategy tripled web traffic to the foundation's website.

Problems With Cause-Related Marketing Campaigns

Despite overall success of many programs, CRM campaigns have had some shortcomings. Often, there is a lack of transparency in how much is made and how much is spent on the causes. For example, in the "RED" initiative begun by Bono, neither the organization nor the firms marketing through them disclosed revenues or total contributions by company or product. Sometimes, CRM is associated with price increases. For example, in April 2008, Gap offered a cause-related shirt (Gap Red T-shirt) for $28, although the average Gap T-shirt was priced at $16.50, suggesting a direct linkage between price and CRM strategy. Interestingly, however, the Gap Red T-shirt was a bestseller for the company in 2007.

Other criticisms include that it undermines traditional corporate philanthropy, which typically involves noncontroversial giving. CRM programs are often aligned with contentious programs, hence raising the issue of ethics of association with partners of questionable reputations or conflicting interests. For example, Home Depot's alliance with the National Wildlife Federation, which involved offering products, information, and expertise to help consumers transform backyards and outdoor areas into wildlife and environmentally friendly areas, raised ethical issues because of conflicting interests between the two organizations. Last but not least, CRM campaigns can be used to cover up otherwise socially and environmentally irresponsible corporate practices, and, regardless of the actual result, are frequently initiated for reasons that have little or nothing to do with the underlying cause. In the end, measuring the net overall results of CRM campaigns on the health of communities, the environment, or specifically targeted groups remains exceedingly difficult, despite demonstrable profits on the part of corporations and beneficiaries directly involved.

See Also: Best Management Practices; Body Shop, The; Bottom of the Pyramid; Corporate Social Responsibility; Environmental Marketing; Integrated Bottom Line; Newman's Own Organics; Social Entrepreneurship; Socially Responsible Investing; Social Marketing; Social Return on Investment; Stonyfield Farm.

Further Readings

Adkins, Sue. *Cause-Related Marketing: Who Cares Wins.* Oxford, UK: Butterworth-Heinemann, 1999.

Baghi, I., E. Rubaltelli and M. Tedeschi. "A Strategy to Communicate Corporate Social Responsibility: Cause-Related Marketing and Its Dark Side." *Corporate Social Responsibility and Environmental Management*, 16/1:15–21 (2009).

Barone, M., A. Norman and A. Miyazaki. "Consumer Response to Retailer Use of Cause-Related Marketing: Is More Fit Better?" *Journal of Retailing*, 83/4:437–45 (2007).

Berglind, Matthew and Cheryl Nakata. "Cause-Related Marketing: More Buck Than Bang?" *Business Horizons*, 48/5:443–53 (2005).

Chiagouris, Larry and Ray Ipshita. "Saving the World With Cause-Related Marketing." *Marketing Management*, 16/4:48–51 (2007).

Cunningham, Peggy. "Sleeping With the Devil? Exploring Ethical Concerns Associated With Cause-Related Marketing." *New Directions for Philanthropic Fundraising*, 18:55–76 (Winter 1997).

Gupta, Shruti and Julie Pirsch. "A Taxonomy of Cause-Related Marketing Research: Current Findings and Future Research Directions." *Journal of Nonprofit & Public Sector Marketing*, 15/1–2:25–43 (2006).

Olsen, G. Douglas, John W. Pracejus and Norman R. Brown. "When Profit Equals Price: Consumer Confusion About Donation Amounts in Cause-Related Marketing." *Journal of Public Policy & Marketing*, 22/2:170–80 (2003).

Ptacek, Joseph J. and Gina Salazar. "Enlightened Self-Interest: Selling Business on the Benefits of Cause-Related Marketing." *Nonprofit World*, 15:9–13 (July–August 1997).

Wu, S. and J. Hung. "A Performance Evaluation Model of CRM on Non-Profit Organizations." *Total Quality Management & Business Excellence*, 19/4:321–30 (2008).

Wymer, Walter W., Jr. and Sridhar Samu, eds. *Nonprofit and Business Sector Collaboration: Social Enterprises, Cause-Related Marketing, Sponsorships, and Other Corporate Non-Profit Dealings*. Binghamton, NY: Business Books, 2003.

Abhijit Roy
University of Scranton

CERES PRINCIPLES

The Ceres Principles are a set of 10 core principles concerning corporate environmental conduct, intended for public adoption by companies in acknowledgment of their commitment to environmental stewardship. Ceres, an acronym for the Coalition for Environmentally Responsible Economics, was formed in 1989, and includes a diverse group of U.S. environmental organizations, socially responsible investors, public pension funds, labor unions, religious organizations, and other concerned groups. The founding goal of the group was to encourage corporations to commit to environmentally responsible and sustainable forms of business practice.

By bringing together environmental groups, investors, and other significant stakeholders, the mission of Ceres was to use the collective power of a diverse array of stakeholders to persuade the business community to recognize environmental values and adopt more environmentally responsible business practices. The coalition was founded in the same year as the notorious *Exxon Valdez* oil spill, an event that brought intense public focus upon the potentially devastating environmental impact of corporate conduct, and hence awareness about the growing need for corporations to incorporate environmental principles into their practices. As a result, Ceres drafted a set of principles, first termed the *Valdez principles*, in acknowledgment of the environmental significance of that event, designed to serve as part of the environmental mission statement of corporations that voluntarily acknowledge their commitment to environmentally responsible business practices. Introduced in fall 1989, the 10 Ceres Principles specify that corporations commit to the following:

- Protection of the biosphere
- Sustainable use of natural resources

- Reduction and disposal of wastes
- Energy conversation
- Risk reduction
- Safe products and services
- Environmental restoration
- Informing the public
- Management commitment
- Audits and reports

In the case of each principle, Ceres provides a more specific summary of what the principle entails for those corporations that adopt the principles. For instance, Ceres stipulates that the principle of energy conservation requires that a company agrees to improve the energy efficiency of its operations as well as of the products it sells and to avail itself of sustainable energy sources. Likewise, the principle of audits and reports mandates that companies engage in an annual self-evaluation of their progress in implementing the principles and undergo routine environmental audits. Companies that commit to the Ceres Principles agree to go beyond mere legal compliance in fulfilling the mandates established in these principles, and the principles provide a benchmark for investors and other stakeholders to gauge the environmental commitment and performance of companies. Ceres provides corporations committed to the Ceres Principles with access to various services through the Ceres organization, including expert advice in areas such as energy and policy analysis, as well as guidance on sustainable reporting and access to investor networks.

After formulating the Ceres Principles, Ceres worked to leverage its market and social capital to encourage companies to adopt the principles. Initially, the Ceres Principles were adopted largely by companies, such as The Body Shop and Ben & Jerry's, which already were identified with a strong commitment to green business practices. However, in 1993, the impact of the Ceres Principles upon the wider corporate world began to grow as Sunoco became the first Fortune 500 company to endorse the principles. Since that time, numerous well-known corporations, such as Nike, Levi Strauss & Co., American Airlines, eBay, and General Mills, have endorsed the principles. As of 2009, well over 50 companies have endorsed the Ceres Principles, including 13 Fortune 500 companies. A complete list of companies endorsing the Ceres Principles is available at the Ceres website. Companies endorsing the Ceres Principles come from diverse sectors of industry, including banking, energy and natural resources, manufacturing, the airline and automobile industries, and healthcare. Ceres works with corporations endorsing the Ceres Principles to establish a plan to incorporate the principles into the mission and operations of the company and requires that companies publicly report on their ongoing performance and initiatives in relation to their commitment to the principles.

The development and propagation of the Ceres Principles within the corporate world represents a significant achievement in the green business movement, as the principles were the first substantial environmental standards to be adopted by major corporations that were produced by a non-industry entity. As an organization, Ceres was noteworthy in recognizing the power of utilizing the resources of investor groups and shareholders such as pension funds to influence corporate behavior. Further, in providing a clear-cut, unitary, and global set of standards to serve as the fundamental starting point for developing and fostering environmental and sustainability efforts within the corporate world, the Ceres Principles have proved to be a key element in integrating an environmental ethic into the business community. Though the adoption of the Ceres Principles by a growing number of businesses is an important step toward making corporations more environmentally sensitive, no code in itself is sufficient to guarantee responsible behavior. In this regard, Ceres has worked diligently to pursue complementary means of cultivating environmental responsiveness in the corporate world. Such efforts include developing the Global Reporting Initiative (GRI) for corporate reporting on environmental, social, and economic performance, sponsoring various efforts to address global climate change, and directing a large group of institutional investors through the Investor Network on Climate Risk (INCR). In this regard, the Ceres Principles represent one prong in a multifaceted approach to creating corporate cultures committed to environmental and sustainable business practices.

See Also: Corporate Social Responsibility; Environmental Audit; Global Reporting Initiative; Socially Responsible Investing; Stewardship; Triple Bottom Line; Voluntary Standards.

Further Readings

Buren, Adriane Van. "Shareholders Seek to Color the World Green." *Business and Society Review*, 90 (Summer 1994).
Cahill, Lawrence B. and Raymond W. Kane. "Corporate Environmental Performance Expectations in the 1990s: More Than Just Compliance." *Environmental Quality Management*, 3/4 (1994).
Ceres. "Ceres Principles." http://www.ceres.org/page.aspx?pid=705 (Accessed April 2009).

Daniel E. Palmer
Kent State University, Trumbull

CERTIFICATION

Certification refers to the verification of a standardized set of characteristics possessed by a product, a production process, or an organization or company. In the realm of green business, the concept of certification is especially important as consumers increasingly demand products and services that are more environmentally friendly and that have humane and just conditions of production. Thus more and more businesses look to third-party certifications often offered by nongovernmental organizations to provide a way to communicate to consumers that their product or service meets certain criteria in a transparent and traceable way. Certified products or business operations then receive the logo or ecolabel from the regulating body, thus enabling environmentally conscious consumers to make informed buying decisions and minimizing the influences of deceptive green marketing practices (i.e., greenwashing). Certification has been used extensively for products and policies related to agrifood commodities, energy efficiency, building construction, and forest stewardship.

Why Certify?

The last few decades have seen the growth of increased consumer awareness and concern for both environmental and social causes. This has been catalyzed by the globalization of information through revolutionary changes in media technology, in which a person in rural Kansas can quickly and easily gain information about deforestation in the Amazon or indigenous protests in Guatemala, for example, simply by opening an Internet browser. This access to information leaves an uneasy dilemma in which the virtual witness is left with limited options to respond and interact with the events that unfold on their screen. The most obvious is to use the tool that is most readily available: purchase power. The consumer can vote with his or her money, a solution that also serves the dual purpose of further raising awareness of the role that things, be they commodities or finished goods, have in linking us to places that are otherwise inaccessible to us in the day-to-day.

But how do consumers know how to vote with their money? One way is through certifications. Early certification initiatives focused on ensuring food safety and quality and product standardization. These included certification systems developed by the International Standardization Organization (ISO), an international network of state- and non-state-standardization agencies. More recently, however, the governance of the value chain is shifting from quality-oriented standards imposed by public institutions to process-oriented standards that are part of third-party voluntary management systems. Thus, the Codex Alimentarius standards utilized by the World Trade Organization (WTO) to govern food health and safety are also accompanied by U.S. Department of Agriculture (USDA) Organic Standards that transform organic agriculture into a legal designation, as well as third-party certifications like Fair Trade that govern social relations, environmental practices, and trade processes in the value chain.

The growth of the number and prevalence of certifications in commodities is indicative of a wider demand in the global agrifood system for product traceability to ensure product safety; however, this growth is not limited to the agrifood industry. Product traceability has become an important impetus for environmental sustainability issues, such as forest stewardship and energy conservation.

Traceability is a set of collection, documentation, and application of information practices that guarantee the origin and life history of a product to the consumer; certification is the verification of traceability. Many different kinds of information about products can be transmitted—information about time, place, material linkages with other products or components, and people involved with the product at different points of production or processing, as well as quality indicators. The goal of any traceability certification system depends on the wider political context in which the product or business practice itself moves. A basic traceability system will be implemented within a company to achieve simple goals such as supply-side management, quality control, and basic safety or hygiene (in the case of agrifood). Much of the recent development in traceability systems has focused on the environmental impacts of doing business from product inception to completion as well as from day-to-day operations.

Parallel to government efforts in both North America and the European Union, the private sector has also developed responses to consumer demands for environmental-related information about products and practices. These have taken the form of certification and verification systems that often broaden the types of information that is traced, from quality and safety to information about labor conditions, environmental conditions, sustainability practices, market prices, and organizational structure at both the producer and consuming end. Forest stewardship is one such product that has received much international attention regarding traceability. Independent third-party organizations, such as the nonprofit Forest Stewardship Council (FSC), provide certification that the timber harvested and its subsequent derived products (e.g., paper, furniture) originate from sustainable forests and plantations and not from locations susceptible to irrevocable environmental degradation. The FSC label has gained widespread international acceptance, with FSC programs currently represented in over 50 countries. Many major corporations (e.g., Coca-Cola, Dell), government agencies, and nongovernmental organizations have committed to purchasing FSC-certified wood and paper products as part of their sustainability programs. For instance, the head green building certification program LEED (Leadership in Energy and Environmental Design) requires that at least half of the wood used in construction be FSC certified.

Launched in 1998 by the U.S. Green Building Council (USGBC), the LEED program is a third-party verification system that provides a set of building and design standards for environmentally sustainable construction and building maintenance. LEED-certified buildings are aimed at increasing energy and water conservation, reducing short-term and long-term greenhouse gas emissions, reducing construction material waste, and lowering daily operating expenses. Both residential and commercial buildings can be LEED certified, and certification is not limited to new construction; there are specialized LEED certification guidelines for existing buildings as well. LEED certification has influenced over 10,000 building projects in the United States since its inception, and has extended to more than 30 countries.

Along the same fundamental principles as LEED, the Energy Star program seeks to improve environmental sustainability by focusing on energy efficiency. Begun in 1992 as a joint venture between the U.S. Environmental Protection Agency (EPA) and U.S. Department of Energy, the Energy Star program is a voluntary certification program to identify products that are more energy efficient in an effort to reduce greenhouse gas emissions. Initially, Energy Star certification was focused on the burgeoning electronics industry (especially computers), but soon expanded into several business sectors, including office equipment, commercial residential heating and cooling, appliances, and building construction and maintenance. Today, the Energy Star ecolabel is well recognized. The program has initiated important changes in business operations and product development, focusing on goods and services that deliver high performance with lower energy inputs and long-term financial gains.

The Politics of Certification

Certifications have had a positive impact on the market, not only because of their higher environmental or social standards of production, but also because they have demonstrated to the mainstream market that commodity supply chains can be organized in a sustainable way. This has led to the mainstreaming of market certifications, increased demand for certified products, and to internal initiatives within companies to improve sustainable supply chain management, even if perfunctorily as a marketing strategy. However, "alternative" commodity certifications, even those with roots in anti-neoliberal ideologies, cannot be seen strictly as alternatives to the dominant neoliberal market system because they are set firmly within the market system; instead, they must be seen as "resilience" within and a "reworking" of neoliberalism, more nuanced distinctions. Many development alternatives realized by nongovernmental organizations (NGOs) have usually been proposed in conjunction with political agendas of social movements. Alternative commodity certifications are no exception. The result has been the evolution of various market certifications that sometimes have overlapping goals and guiding principles, but often represent very different concerns and models for approaching solutions to those concerns, as well as very different politics.

Commodity certifications, from the consumer point of view, offer an easy and accessible way of "caring at a distance." In this sense, they are increasingly talked about (and critiqued) as global governance instruments, in which first world consumers "govern" the environmental, social, and economic conditions of production of the commodities they consume, via governance actors such as NGOs and the business community; however, the framework of governance provides little insight into consumer motivations for reflecting on and engaging with the morality of personal actions like shopping.

Among NGOs and state development agencies working in the rural third world, commodity certifications are often seen as instruments of rural development and conservation, which further complicates the politics surrounding a seemingly simple process of certification. Within the realm of debates on agricultural and timber production, energy efficiency and rural development, discussions are increasingly focused on market-led rural governance and conservation strategies that take the form of voluntary commodity certifications. Embedded within these strategies are discourses and practices aimed at creating and strengthening a global social movement around local control of resources, environmental conservation, and social justice. Certification schemes are, in effect, voluntary regulatory mechanisms with complex, multistakeholder governance structures focusing on varying ethical, social, environmental, and economic priorities. This development instrument approach to certifications must be problematized in a similar way as the governance approach as the same questions arise of who determines or defines rural development for whom.

The certification of coffee offers a poignant example of the complex politics and challenges surrounding agrifood certification in general. Before 1989, coffee prices were regulated by the International Coffee Organization (ICO) through a quota system that kept prices within a range of a defined indicator price. The breakup of the ICO in 1989, which coincided with the end of the Cold War, caused prices to plummet by half between January and August of that year. Since then, market coffee prices have passed through cycles of elevation and decline, determined by cyclical supply and demand that have been exasperated by development aid programs encouraging production in high-price periods resulting in increased supply, in effect increasing supply and sending prices plummeting. The most recent crisis, between 1999 and 2004, was especially severe and resulted in international NGOs like Oxfam responding with campaigns to "Make Trade Fair"; this call reverberated throughout Europe and North America and student organizations, religious organizations, and businesses responded by promoting the burgeoning Fair Trade certification initiative as a strategy to ensure stable livelihoods to small-scale coffee farmers in the developing world and to resist the oligopoly of transnational corporations over the coffee industry. The "coffee crisis" thus played a role in exponentially increasing the demand for, and consumption of, Fair Trade–certified coffee.

Certification is not only a consumer issue, but one that is also problematic from the producer point of view. Producers of certified commodities are faced with demands to change and document production practices in order to meet the demands of consumers demanding green products, and often their livelihoods depend on meeting the often-confusing sets of certification standards and the accompanying costs of certification. Although large-scale producers and companies might already have the capital and capacity to navigate certification processes and to negotiate the standards and costs, small-scale producers are faced with financial and knowledge barriers to achieving certification and receiving its benefits.

A Certified World?

Can certification be an effective green business model? Can it transform systems of production, distribution, and consumption of commodities into "sustainable" structures? There are essentially two sides to this debate. The first side says that yes, certification, accompanied by advances in technology that enable transparency and traceability in the life of a commodity, will force all producers, distributors, and retailers to clean up their practices and become green. In other words, the right choices will be clear to all, so those who do not choose the path of sustainability will lose their market. The other side of the argument, however, holds that certification does make a difference by offering incentives to producers and businesses to go sustainable, but that it can only go so far as long as consumers themselves are not assuming the responsibility of questioning and changing their habits of consumption. Stated another way, if the goal is truly to achieve a green business, placing the burden of responsibility on producers and businesses is not only unjust but ineffective in creating real changes. Consumers, then, must change their consumption habits by focusing on reducing consumption, placing preference on the consumption of local commodities and products, and recycling and reusing. Certification can then be combined with changes in consumption habits.

See Also: Ecolabels; Fair Trade; International Organization for Standardization; Organic; Responsible Sourcing; Sustainability; Transparency; Voluntary Standards.

Further Readings

Forest Stewardship Council. "About FSC." http://www.fsc.org/about-fsc.html (Accessed January 2010).

Goodman, David and Michael J. Watts, eds. *Globalising Food: Agrarian Questions and Global Restructuring*. London: Routledge, 1997.

Hughes, Alex and Suzanne Reimer. *Geographies of Commodity Chains*. London: Routledge, 2004.

Raynolds, L. T. "Re-Embedding Global Agriculture: The International Organic and Fair Trade Movements." *Agriculture and Human Values,* 17:297–309 (2000).

U.S. Green Building Council (USGBC). "An Introduction to LEED." http://www.usgbc.org/DisplayPage.aspx?CMSPageID=1988 (Accessed January 2010).

Heather R. Putnam
University of Kansas

Jill Coleman
Ball State University

Clean Fuels

Clean fuels as a concept lacks clear definition, though the term generally refers to blends or substitutes for gasoline fuels, including compressed natural gas, methanol, ethanol, liquefied gas, and even electric power from renewable sources. Clean fuels as a possible alternative to straight fossil fuels have primarily emerged as a lower pollution alternative in the transportation sector. The U.S. federal tax code, for instance, defines "clean fuels" that are potentially eligible for rebates or tax breaks for motor vehicles designed to use one of the following fuels:

- Natural gas
- Liquefied natural gas (LNG)
- Liquefied petroleum gas (LPG)
- Hydrogen
- Electricity (e.g., some gasoline/electric hybrids)
- Any other fuel that is at least 85 percent alcohol or ether (e.g., E85)

Many heavy diesel engines have been converted for use in urban areas to reduce emissions, such as these Union Pacific clean-diesel locomotives at the Port of Houston, Texas.

Source: Eric Vance/U.S. Environmental Protection Agency

To this day, the American economy is powered primarily by fossil fuels—from energy to production and transportation. Power and energy is essentially produced through the combustion of fossil fuels such as oil, gasoline, or coal. Oil consumption alone exceeds 20 million barrels each day in the American economy. This level of consumption, in turn, is creating a variety of ever-more-pressing problems: depletion of resources; dependence on foreign providers; increasing costs; pollution and environmental degradation. Burning of fossil fuels is recognized as the single largest contributor to the release of greenhouse gas emissions such as carbon and methane, leading to potentially catastrophic problems such as global warming and related climate changes ranging from an increase in natural disasters to aridity and rising sea levels. With fossil-generated power plants leading the way, the U.S. economy emits over 3.2 billion tons of carbon dioxide. Almost 84 percent of U.S. energy production still comes from nonrenewable, fossil sources (9 percent is nuclear, and only roughly 7 percent is renewable, which includes solar, hydroelectric, biomass, wind, and geothermal).

While there is ongoing debate about short-, medium-, and long-term political, economic, and environmental costs of this volume of fossil fuel consumption, there no longer is any doubt that, if consumption levels are to continue on their current trajectory, it would not only mean eventual depletion of fossil resources, but also ever more global competition over scarce resources, and translate into irreparable harm to the environment.

The prospect of "clean" alternative fuels, in short, entails great hope for an alternative that is less toxic and more readily available. The biggest promise that is fueling the "clean fuels" ascendancy is that they are seen as a substitute for fossil fuels that would prevent us from having to make any radical changes to modern energy generation, production, and transportation technologies (put biofuels rather than fossil fuels into the tank, but don't change the basic technology of the vehicle, much less the way we get from point A to point B). Initial clean fuel prospects such as ethanol turned out to be very disappointing. Ethanol, it was realized after years of investment, requires more energy to produce than it yields, and has a range of adverse effects on availability and prices of feedstock and food. But particularly with rising fuel prices, research and investment capital is continuously spent on the discovery and development of new and better fuels that could provide more viable alternatives to fossil fuels, such as liquid propane, biodiesel, hydrogen, and a variety of fuel-cell and battery technologies.

Gas

LPG, also known in some markets, such as the United States, as propane, is in fact a mixture of butane and propane. It is a by-product of natural gas production and also of the oil refining process and therefore widely available wherever oil refineries operate. Nevertheless, its popularity has been limited to a few countries, notably the Netherlands, Italy, and more recently Australia, while in some locations, like Japan, Hong Kong, and South Korea, LPG has been popular for taxi use. The emissions advantage of LPG has been known for many years, yet it was not until the late 1980s that it began to be actively promoted as a cleaner alternative fuel. LPG is considered particularly attractive for commercial vehicles that operate in an urban environment. Thus a number of buses and local authority vehicles have been converted.

Natural gas, or methane, has a lower energy density and therefore needs a larger additional tank, although it has a slight emissions advantage over LPG. Due to its simple molecular structure—especially compared with gasoline—it burns cleanly and is readily available in many parts of the world, with a more equal distribution than oil. It has proven popular with fleet users in countries such as Canada, Australia, Sweden, the United Kingdom, Germany, and many others. It comes in two forms, either as compressed natural gas (CNG), or in the alternative LNG, which requires storage at very low temperatures, although more can be carried in the same volume than with CNG. LNG availability is more limited as this technique is used mainly for shipping natural gas and for storing longer-term reserves, while CNG is readily available in many countries from the domestic distribution infrastructure that supplies natural gas for heating and cooking. Heavy diesel engines are relatively easy to convert to run on LPG or CNG, although they have to be turned into spark-ignition engines, while an additional compressed gas tank is also required. Gasoline engines, as used in most passenger cars, are cheaper to convert.

Alcohol

Alcohol fuels, ethanol and methanol, have been used in automotive applications for a long time, particularly as high-octane fuels for racing cars. Their suitability in motor racing is due to the fact that they allow higher-compression engines to be used. Even though they are also hydrocarbon fuels, they produce somewhat lower emissions. Ethanol can be produced from the fermentation of a range of crops. In the mid-1970s, the Brazilian government launched the "Proalcool" program as an import substitution project. In the wake of the oil crisis of 1973–74, Brazil felt it spent too much on importing oil to run its cars, and a means was devised to substitute this with ethanol produced from sugarcane. Brazil has been a major sugar producer since the 18th century and the industry has a powerful lobby in Brasilia. With volatile world sugar markets, the industry also felt some diversification would be helpful and it was very supportive of the program. Although the Proalcool scheme suffered a decline in the 1990s, environmental concerns combined with the availability of new flex fuel vehicle technologies have prompted a revival in recent years, with Brazil now exporting ethanol to many other countries. Flex fuel technology was first developed in the United States, but Brazil was able to develop its own intellectual property rights in this area. This technology allows the user to fill the fuel tank with any combination of gasoline and ethanol, and the engine management system automatically adjusts the performance according to the fuel mix it encounters. This reduces the risk to the user, because a price increase or supply constraint of one fuel need not stop the car. Another benefit of ethanol is that it provides higher octane ratings, and hence better performance than gasoline.

In the United States, a large-scale conversion of corn crops to ethanol started in the 1980s, followed by growing interest in developing countries and Europe. In practical terms, there are limitations to this approach as vast areas of dedicated crop cultivation would be required to run a significant proportion of the world's cars on this fuel, although where surpluses of crops rich in sugar exist—such as in Brazil—it may be feasible locally. The biofuels industry is more optimistic about the potential of so-called second-generation biofuels, produced from nonfood crops and waste biomass. There are also problems with the ethanol supply infrastructure, as—unlike gasoline—ethanol tends to absorb water, which does not always enhance its performance as a fuel. With gasoline, any water in the system will sink to the bottom and can easily be separated. Ethanol needs special provisions to deal with water, which adds to the cost of infrastructure.

Methanol, the other alcohol fuel, is more dangerous to handle than ethanol, or even gasoline, and requires a completely different fuel delivery system as it corrodes most existing fuel system materials. Nevertheless, it enjoys some popularity as an alternative fuel. In practice, it is usually mixed with gasoline in order to control its effects somewhat and make cold starting easier. M85 (85 percent methanol, 15 percent gasoline) is produced this way. M85 became a popular alternative fuel in parts of the United States from the early 1990s onward. More recently, methanol has come to be regarded as a useful source of hydrogen for feeding fuel cells. In this application, it may prove more useful than as a direct fuel for internal combustion engines.

Diesel Alternatives

There have also been several attempts to develop alternative fuels for diesel engines. Many heavy diesel engines have now been converted to run on gaseous fuels for use in urban areas in order to reduce emissions. These, however, need to be converted from compression ignition to spark ignition. By the late 1990s, one of the most promising diesel alternatives was dimethyl esther (DME), which is usually derived from natural gas, but can be made from a range of feedstocks including biomass. While engines require modifications to the injection system, emissions are much improved. Much of the development work on DME is carried out in Scandinavia, where its potential as a vehicle fuel was first noted. Other alternatives to diesel oil are the so-called biodiesels. These are derived from biomass from oil-rich plants, such as rapeseed oil, sunflowers, or soy and can be used with little or no modification. Volkswagen was one of the first to make all of its diesel engines capable of running on biodiesel. The European Union has a legal requirement to make all vehicle fuels contain at least 5.75 percent biodiesel or bioethanol by 2010, with the proportion rising thereafter. In North America many private enthusiasts run older-generation diesel cars on home-grown biodiesel. It is relatively easy to produce at home and can also be derived from fresh or used cooking oil at fairly low cost.

It is also possible to turn natural gas into a liquid fuel, using gas-to-liquid (GTL) technology. The technology is used to produce a particularly clean form of diesel fuel. Shell is a major producer and has refineries operating this process in Malaysia and Qatar. Shell's GTL diesel has been used by Audi in its diesel-powered racing cars, which have won the prestigious Le Mans 24-hour race. It can be blended with conventional diesel in order to clean it up, allowing it to meet cleaner fuel standards. With natural gas reserves considered to be greater than those for oil, many countries are expecting to be able to supply natural gas for conversion to GTL liquid fuel for many years to come

Hydrogen (H$_2$)

On the face of it, hydrogen appears to offer the ideal solution. It is not a fuel as such, but an energy carrier in that energy needs to be used to extract it from either water or hydrocarbon fuels, including biomass. However, it is suited for burning in internal combustion engines with relatively minor modifications, and as the combustion process involves a reaction with oxygen, its emissions are essentially water. In practice, very low levels of hydrocarbons are also emitted because of the lubricating oil that is still required by the engine. Hydrogen can be burned in existing internal combustion engines with some modifications, so a wholesale move away from existing engine technology and production facilities would not be required. It is for this reason that the vehicle manufacturers have been broadly supportive of this fuel. Mazda has reported that their Wankel rotary engines are particularly suited to running on hydrogen.

However, there are some problems, mainly centering around hydrogen production and storage. Hydrogen does not occur naturally in its pure form on our planet. It is normally bound with oxygen or carbon in some form and is usually produced from water or some hydrocarbon fuel such as methanol. This process can be quite energy intensive and therefore raises the question of what energy source to use to make hydrogen, a process that can itself be polluting. Currently, only about 4–5 percent of hydrogen worldwide is produced from renewable energy sources and can therefore be considered "sustainable." These include hydro and geothermal, while most hydrogen is produced using fossil energy sources.

Hydrogen also presents storage problems. Existing storage solutions such as compressed hydrogen tanks or metal hydride are quite bulky and hydrogen tends to escape through evaporation over a relatively short period of time. By the late 1990s, thinking therefore moved more toward generating hydrogen on board the vehicle from a hydrocarbon fuel such as methanol or even gasoline; the latter option was promoted by Chrysler, in particular. However, as most hydrogen is derived from fossil fuels, it does not solve the problem of our overreliance on scarce oil reserves, unless the hydrogen is derived from water using renewable energy, as is done in Iceland. Where hydrogen may have a role to play is in fuel cells. Some hydrogen fuel cell vehicles are already in use on an experimental basis, but more will gradually appear during the early 21st century, although the hydrogen generation problem has yet to be solved.

Many other alternatives have been tried in the search for cleaner and more sustainable automotive fuels. However, there are also other relevant trends. In our search for new fuel feedstock technologies, current thinking aims to derive gasoline and diesel from other fossil carbon sources such as coal, tar sands, and oil shale. Coal-to-liquid technology was originally developed by Fischer and Tropsch (and is hence known as the Fischer-Tropsch process) in Germany in the 1930s, then adopted by South Africa during the apartheid-era blockades, and is now being reconsidered as a solution to ensuring future fuel supplies. These are far removed from being clean fuels, as they would generate considerable additional carbon dioxide emissions compared even with our current fuels. However, they offer the possibility of carbon capture and storage, whereby the carbon is separated out at the processing stage and buried or pumped underground for long-term storage. Using such technologies—currently still in their infancy—these potentially much dirtier fuels can become much cleaner. If such technologies—which in themselves require very high capital investments—can be developed at an appropriate cost, many new regions can become major sources of cleaner carbon fuels. China, for example, with its rapidly motorizing economy has very large coal reserves. India is in a similar position, while the United States and Australia could also become significant sources of coal-derived fuels.

Alternative Powertrain

In addition to alternative fuels for internal combustion engines, alternatives have also extended to alternatives for the internal combustion engine itself. This is generally considered a less attractive option by the motor industry as it means scrapping the existing engine production facilities and losing the expertise in conventional internal combustion technology built up over more than a century. Among alternative powertrains, battery electric vehicles have a long history, and recent advances in battery technology are making these increasingly viable. The gasoline-electric hybrid, popularized by the Toyota Prius and Honda Insight, combines conventional with alternative powertrain, making it more appealing to carmakers. It has met with considerable success, and diesel-electric hybrid powertrains are well established for trains and are also already in use for some commercial vehicles, while they are under development for cars in Europe. Plug-in hybrids form a compromise between battery electrics and existing hybrids. They do not suffer the limited range of battery-electric vehicles, while their ability to be charged from the grid allows cleaner energy input where sustainable electricity generation is used. Plug-in hybrids use smaller internal combustion engines only for electricity generation and entered the market in limited numbers from 2009 onward. In the longer term, many still expect considerable promise from the hydrogen fuel cell. Around 1,000 experimental fuel cell cars were in use by 2007, and Honda first offered its FCX Clarity fuel cell car for lease in 2008.

Alternative Clean Fuels Energy Sources

While significantly more public focus has been on alternative clean fuels in the transportation sector, the energy-producing industry is actually a larger polluter in the United States than transportation, and thus is the sector that holds more promise to make a significant dent in the reduction of greenhouse gas emissions through the use of clean fuels. Currently, only about 4 percent of energy production relies on clean fuels, of which the largest portion goes to biomass power plants. It is important to note, however, that both the number and variety of clean fuels system, especially for smaller uses, is rapidly increasing in both functionality and affordability. This includes cogeneration systems, hydrogen fuel cells, hydrogen fuel system, and biofuel systems, and several more exotic versions based on algae and bacteria.

In conclusion, public debate centered on fossil fuel alternatives ordinarily focuses on solar, wind, and nuclear power. Most observers agree, however, that in both the transportation and energy-generation sectors, there is an increasing need and growing opportunities for a variety of uses of clean fuels. While clean fuels are not likely to ever reach the scale of usage currently held by fossil fuels, they are very likely to play a significant role in the transition from fossil to renewable energy sources.

See Also: Bio-Based Material; Biofuels; Cogeneration; Green Technology; Toyota.

Further Readings

Mondt, R. *Cleaner Cars: The History and Technology of Emission Control Since the 1960s.* Warrendale, PA: Society of Automotive Engineers, 2000.
Nieuwenhuis, P. and P. Wells. *The Automotive Industry and the Environment: A Technical, Business and Social Future.* Boca Raton, FL: CRC Press, 2003.
Roberts, P. *The End of Oil: The Decline of the Petroleum Economy and the Rise of a New Energy Order.* London: Bloomsbury, 2004.
Romm, J. *The Hype About Hydrogen: Fact and Fiction in the Race to Save the Climate.* Washington, D.C.: Island Press, 2004.

Paul Nieuwenhuis
Cardiff University

CLEAN PRODUCTION

Clean production is a philosophical and analytical approach to production which considers all the effects a product may have on the environment and on human health during all phases of its life cycle, with particular emphasis on the reduction of immediate and long-term harm. A primary focus of clean production is the reduction or elimination of releasing hazardous substances into the environment, preferably through upstream methods emphasizing product and process redesign, as well as reuse and recycling of waste materials in the manufacturing process as opposed to downstream methods, which move the waste elsewhere. Because of the difficulties in achieving truly clean production, some prefer the term *cleaner production*, which denotes improvement over standard approaches while also emphasizing that any manufacturing process is likely to have at least some detrimental effect on the environment. Many manufacturers have moved toward adopting clean production methods not only because of external pressures brought by consumers, governments, and other entities, but also because it may provide them with a competitive advantage in terms of both efficiency and public perception.

The international organization Clean Production Action outlines four principles necessary for clean production. The Precautionary Principle states that it is best to avoid releasing substances known to cause damage to the environment, and that in the case of scientific uncertainty, the burden of proof should lie on those advocating a process to show that it is safe or that there are no safer alternatives available, rather than placing the burden on regulators or community members to prove that a substance does cause harm. The Preventive Principle states that it is preferable to prevent environmental damage, for instance by replacing hazardous chemicals used in the production process with those that are safer, than to try to fix the damage after the fact. If hazardous chemicals must be used, then attention should be devoted to preventing accidents, spills, and other releases of toxins into the environment. The Public Participation Principle states that the public must be informed about the types and quantity of materials used in production or which may be released into the environment, including any hazardous chemicals. The Holistic Principle states that clean production requires looking at the entire life cycle of a product, including the ultimate disposal of hazardous waste materials.

Clean production is an important aspect of sustainability because it requires looking at the effects of production at a systems level and across political and geographic boundaries in order to consider the total effect on the environment. This is in contrast to many current environmental regulations that are focused on end-of-pipe requirements for specific types of wastes, for instance those containing chemicals listed in a toxics inventory. Clean production advocates point out that in a manufacturing process, it is possible to reduce emissions of listed chemicals yet increase emissions of others that are hazardous but not included on the inventory, so that overall environmental harm is increased. They also note that current regulations often allow displacing waste, which may reduce harm in one geographical area while increasing it in another. For instance, if a plant is allowed to discharge toxic chemicals into a sewer system, they may be treated at wastewater facilities, which concentrate them into sludge, but that sludge must still be disposed of somewhere, often in a landfill or through incineration, which disperses them into the ground or air, but does not get rid of them. In contrast, clean production strategies focus on reducing the amount of toxic chemicals used in a process (for instance, by substituting less toxic chemicals) as well as ways to reuse or recycle them, thus reducing the total amount released into the environment.

Although toxic chemicals have been a strong focus of clean production, there are many other concerns, including reducing all types of waste through redesign on the manufacturing process, as well as increased reuse and recycling of materials. Clean production also encompasses many other issues that are important to sustainability, including energy efficiency, reduced use of raw materials, including water, an overall reduction of consumption while retaining quality of life, and protection of biological and social diversity.

Operational Pathways

There are two primary pathways to make production cleaner: reducing the amount of material flowing through an industrial process (i.e., improving material efficiency), and substituting less hazardous materials, products, and services for those that carry more risk.

In a production process, improving material efficiency can take many forms, including improved materials handling, which avoids waste as well as leaks and spills, closing the materials loop by reusing or recycling materials, and redesigning processes so that they use less energy and fewer materials. Looking at product cycles, material efficiency can be improved by means such as reusing, restoring, and recycling both raw materials and manufactured products. In terms of consumption patterns, material efficiency can be increased by designing products to last longer, finding new uses for existing products, and educating people to value products that last and can be reused over those that are disposable.

Improving material efficiency can result in economic savings for a company by allowing them to produce the same amount of goods with less material input and less waste as well as improving their public image and reducing risk, and thus is attractive even to companies not particularly interested in marketing themselves as being environmentally conscious. Some improvements to material efficiency can be adopted relatively quickly and easily, for instance, establishing programs to recycle glass and paper products. However, others require redesigning products, making changes in the manufacturing process, or influencing the attitudes of consumers, all of which require a greater investment in time and effort.

It must be noted that it is not sensible to claim that the material intensity of an economy can be reduced to zero: there are limitations to how efficient any manufacturing process can be, and there are practical and economic constraints as well. It should also be noted that evaluating material efficiency must be done at a systems level; for instance, it is possible that reducing materials throughput at one stage in a process may lead to greater material demand at another stage.

Substitution is the second major strategy that can make a production process cleaner. It is most often applied with materials known to be particularly hazardous to humans, including naturally occurring substances such as mercury and arsenic, as well as synthetic compounds such as polychlorinated biphenyls (PCBs) and dioxin. Some of these substances are toxic to humans at very low dosages, while others pose problems because they are extremely persistent in the environment (i.e., they biodegrade very slowly, if at all). With man-made compounds in particular, the risk to humans and the environment is not known, and thus from a clean production standpoint, it is preferable to avoid their use entirely. For instance, chlorofluorocarbons (CFCs) were once considered safe because they were nonreactive in the lower atmosphere, yet over the years they accumulated to such a degree that they reached the upper atmosphere, and are now considered partly responsible for the reduction of the ozone layer. Although recycling can reduce the amount of such substances released into the environment, it is never 100 percent efficient, and it also presents the possibility of accidental release.

The two strategies are not independent, and most companies attempting to make their production process cleaner will use aspects of both, depending on their particular circumstances, existing regulations, economic pressures, and so on. However, standard recommendations for implementing clean production will include reducing the flow of raw materials through the system; reducing use and disposal of toxic chemicals and persistent synthetic materials; moving toward sustainable use of renewable resources, such as wood and water; creating long-lasting rather than disposable products; and implementing life cycle analysis with regard to products and their manufacture.

Implementing clean production strategies at a place of business requires buy-in from the highest levels of management to the lowest level of employee. Because clean production is a new way of approaching industrial processes, it may require education at all levels, as well as initial investments of cash whose benefits may not be immediately apparent. Typically, implementing clean production requires a team including people both from within and outside of the company representing different scientific disciplines, as well as areas of expertise within manufacturing. The team will review current practices and identify areas where improvements can be made, which can include anything from the sources and consumption of raw materials, to redesigning products and packaging, to waste reduction and disposal. The team will then conduct feasibility analyses to determine which areas are likely to yield the greatest improvements while allowing the company to continue operating. Once the foci of change are agreed upon, practical goals and milestones must be set and these results communicated to everyone in the organization. Further education and training may be required for people whose job duties have changed, and the team will also need to continue monitoring the company and making changes as necessary until the goals of clean production are reached.

Barriers to Clean Production

Although many embrace clean production as a worthy goal, in reality there are potential problems in implementation that must be considered. The first is that realistic goals must be established: although it might be ideal to specify that zero pollution should occur, this is not practical and can act as a barrier to potential reforms. Similarly, although one can wish for zero production of waste, it is not only impossible, given the laws of thermodynamics, but also may not be economically or socially feasible, leading to a backlash against more reasonable reforms. A third problem is defining which wastes deserve the most scrutiny (an approach taken by the construction of inventories of toxins, for instance). This must be done with the understanding that the effects of many chemicals on humans and the environment are unknown and such lists will inevitably be incomplete.

If one attempts to encourage clean production through laws and regulations, another set of difficulties may arise. Laws and regulations are created by people who may not all agree on the harms of pollution, or the necessity of any environmental concerns, for that matter, or may be interested primarily in protecting the environment closest to them with less concern for people living some distance away, or for the ecosystem as a whole. This explains why poor nations and economically disadvantaged areas within a region or country often bear a disproportionate burden of the by-products of the industrial process, including pollution and waste disposal. A related consideration is that manufacturers and other business concerns can and do influence the legislative process, and may successfully help shape laws and regulations in order to gain an advantage over their competitors.

Enforcement of clean production regulations is more problematic than the traditional end-of-pipe approach, in which regulators must deal with a relatively well-defined process and only two players are involved: regulators and manufacturers. Additionally, approaches tend to be quite limited and focused on only a few substances. In contrast, because clean production deliberately includes consideration of the totality of environmental effects through all stages of product manufacture, use and disposal, and across political boundaries, monitoring clean production is much more complex. For instance, a clean production approach to mercury requires consideration not only of mercury but of the chemicals commonly used in conjunction with it in industrial processes (e.g., chlorine), as well as the behavior of end users of products containing mercury (ranging from mascara to thermometers to lamps). For instance, how should consumers dispose of products containing mercury, and how will these rules be enforced?

Management may not be accustomed to thinking of their processes in the larger context that clean production requires. Instead, they may have been trained to think in terms of short-term goals, such as quarterly productivity, and to only be concerned with meeting existing environmental regulations. Engineers and other technical staff may lack the skills necessary to incorporate the clean production ethos into analysis and design, and may also require further education as well as clear indications of support from management. Company cultures are often resistant to change, and lower-level staff may resent changes in their duties or heightened monitoring (for instance, additional procedures put into place to assure that toxic materials are handled safely and supervision to ensure that the procedures are implemented strictly) if no benefit accrues to them and if they are not educated as to the reasons for the changes.

Although a company may benefit in the long term by adopting clean production strategies, they may also suffer in the short term because of additional expenses or changes in relationships with traditional suppliers. In addition, although they may eventually achieve positive public recognition for "going green" at the beginning stages of the process, the effect may be just the opposite: by drawing attention to processes that require change, they may supply ammunition to competitors or to activists who may accuse them of harming the environment (missing the point that they are reforming their practices while their competitors may not be).

Conclusion

Increasingly, businesses, governments, and private citizens are becoming aware that natural resources are being depleted while pollution is increasing, and that in order to solve these problems, we must change our view of the production process. Clean production offers an approach to analyzing industrial processes with regard to their total effect on the ecosystem, rather than the fragmented approach taken by most current regulations. Implementing clean production requires that businesses look at all effects of their products on the environment, from the amount of raw materials consumed in their production to the issue of waste caused by disposal of the end products. While implementing clean production strategies may require cultural change within a company as well as physical changes to aspects of a production process, many companies have found that implementing clean production strategies are economically beneficial because they reduce waste, increase efficiency, and improve the company's image.

See Also: Best Management Practices; Certification; Corporate Social Responsibility; Environmental Management System; Industrial Metabolism; Industrial Nutrients; Sustainable Development; Triple Bottom Line; Waste Reduction.

Further Readings

Allen, David T. and David Shonnard. *Green Engineering: Environmentally Conscious Design of Chemical Processes*. Upper Saddle River, NJ: Prentice-Hall, 2001.

Cato, Molly Scott. *Green Economics: An Introduction to Theory, Policy and Practice*. London: Earthscan, 2009.

Clapp, Jennifer. *Paths to a Green World: The Political Economy of the Global Environment*. Cambridge, MA: MIT Press, 2005.

Clean Production Action. http://www.cleanproduction.org (Accessed December 2009).

Faber, Daniel. "A More 'Productive' Environmental Justice Politics: Movement Alliances in Massachusetts for Clean Production and Regional Equity." In *Environmental Justice and Environmentalism: The Social Justice Challenge to the Environmental Movement*, Ronald Sandler and Phaedra C. Pezzullo, eds. Cambridge, MA: MIT Press, 2007.

Jackson, Tim, ed. *Clean Production Strategies: Developing Preventive Environmental Management in the Industrial Economy*. Boca Raton, FL: Lewis Publishers, 1993.

Misra, Krishna B. *Clean Production: Environmental and Economic Perspectives*. New York: Springer, 1996.

Organisation for Economic Co-operation and Development (OECD). *Business and the Environment: Policy Incentives and Corporate Responses*. Paris: OECD, 2007.

Thorpe, Beverley. "What Is Clean Production?" CleanProduction.org, June 2009. http://www .cleanproduction.org/Publications.php (Accessed January 2009).

Tucker, N. "Clean Production." In *Green Composites: Polymer Composites and the Environment*, Caroline Baillie, ed. Boca Raton, FL: CRC Press, 2005.

Sarah Boslaugh
Washington University in St. Louis

CLEAN TECHNOLOGY

Clean technologies (CTs) are a diverse range of knowledge-based products, services, and processes that harness renewable materials and energy sources; dramatically reduce wastes and pollutants (cut or eliminate emissions and wastes); encourage recover, reuse, and recycle; reduce the pressure on natural resources and restore the balance of the ecosystem and biosphere; and ultimately help in providing ecologically sustainable development. These technologies are, therefore, viable, cost-effective, environmentally superior, and suitable to the climatic, economical, geographical, ecological, and social conditions of the country. Adoption of CTs and cleaner production strategies is considered to provide a balance between development and environment through economic benefits by way of increased resource efficiency, innovation, and reduced cost for environmental management. It is becoming the cornerstone of corporate, investment, and government strategies to profit in the next decade and to guarantee economic competitiveness for years to come.

CTs are competitive with, if not superior to, their conventional counterparts. Many also offer significant additional benefits, notably their ability to improve the lives of those in both developed and developing countries. Commercializing CTs is a profitable enterprise that is moving steadily into mainstream business. When industry giants such as GE, Toyota, and Sharp, and investment firms such as Goldman Sachs are making multibillion-dollar investments in CT, it makes a statement and is noticed. Developing CTs is no longer a social issue championed by environmentalists; it is a profitable enterprise moving solidly into the business mainstream. In fact, as the economy faces unprecedented challenges from high energy prices, resource shortages, and global environmental and security threats, CTs are ideally designed to provide superior performance at a lower cost while creating significantly less waste than conventional offerings, thus acting as the next engine of economic growth.

Historical Overview

CTs are recognized as an effective tool to achieve the major objectives of sustainable industrialization. Since the 1990s, interest in these technologies has increased with two trends: a decline in the relative cost of these technologies and a growing understanding of the link between industrial design used in the 19th century and early 20th century, such as fossil fuel power plants, the internal combustion engine, and chemical manufacturing, and an emerging understanding of human-caused impact on Earth systems resulting from their use.

Cleantech was popularized in large part through the work of Nick Parker and Keith Raab, founders of the Cleantech Venture Network (now Cleantech Group) from 2002 onward, beginning as a term to describe the "green and clean" technologies, including solar, biofuels, fuel cells, water remediation, and renewable power generation, that venture capital investors were turning to in increasing numbers as the next trend in technology investing after the collapse of the tech boom in 2001. Since then, the term has come into wide use in the media, broader investment community, and many of the underlying industries that make up the umbrella sector and spawned numerous conferences, websites, magazines, indices, newsletters, and companies, growing into the third largest venture capital investment sector behind IT and biotech.

The sector and the term came into their own in the 2005 and 2006 time frame, when mainstream institutional investors, led by CalPERS and CalSTRS, began allocating investment into venture funds in the environmental, alternative, and renewable energy sectors and adopted cleantech as a term of choice for the description of that asset class, lending credibility to the sector. But possibly most significant, the 2004 to 2006 time frame saw the emergence of financial and capital market successes in the solar, wind, and ethanol industries that make up large portions of the various cleantech-related stock indices, driven by changes in policy incentives and fuels standards in the United States and Europe. Other factors attributed as major drivers in that time frame include rising energy and commodity prices, increased consumer awareness of sustainability issues, and the start of the Kyoto Protocol–based carbon trading mechanisms. The combination of these events began to attract significant amounts of capital and awareness to the sector.

Categories of Clean Technology

Principal categories of CT include the following:

- Energy generation and renewable energy technologies: solar (including thin-film PV, solar, hydrogen), wind, hydel, OTECC, biofuels, geothermal, combined heat and power plants, others
- Energy storage: fuel cells, advanced batteries, hybrid systems
- Energy distribution/smart grid: homes and businesses will no longer be just energy consumers but also energy producers
- Materials: recyclable, biodegradable/nano, bio, chemical, others
- Water: reverse-osmosis water desalination, water conservation, water treatment, wastewater treatment
- Transportation: fuel-efficient/less polluting vehicles, nonmotorized transport
- Agro-ecology and natural farming
- Eco-aquaculture
- Eco-technologies
- Industrial ecology
- Green buildings

Potential industrial sectors for clean technology applications include pulp and paper, dye, electroplating, textile, dairy, paints, pharmaceuticals, rubber, steel, pesticide/fertilizer, packaging material, tannery, screen printing, plastic components, solar power, wind power, fuel cell technologies, small-scale hydropower, and geothermal power.

Clean Development Mechanism

The Clean Development Mechanism (CDM) is a market-based instrument under the Kyoto Protocol of the United Nations Framework Convention on Climate Change under which developed (Annex I) countries can implement greenhouse gas (GHG) mitigation projects in developing (Non-Annex I) countries to meet a part of their emission reduction commitment. Joint Implementation (JI) has a similar definition, but it will be operational within Annex I countries.

The benefits of CDM to developing countries are as follows:

- New source of foreign investments
- Transfer of technology and expertise

- Employment generation
- Infrastructure development
- Reduction in imported energy demand

The benefits of CDM to developed countries are as follows:

- Reduction in emission mitigation costs, to support project activities that reduce GHG emissions in the developing countries in return for Certified Emission Reductions (CERs)/ Carbon Credits
- More flexibility for meeting their commitment
- Market for new and advanced technologies
- New investment opportunities

Potential CDM projects include biomass cogeneration/power generation, biogas cogeneration/power generation, methane capture projects, wind, geothermal, hydropower, fuel switching, ethanol, biodiesel, efficiency and capacity upgrades.

Barriers for Clean Technology

Despite recognizing the potential benefits of addressing environmental concerns, small and medium-sized enterprises in particular struggle to take the practical measures necessary to reduce their impact on the environment. This is often due to businesses not being aware of their legislative and environmental obligations or simply not having sufficient resources to implement change. Lack of capacity in development of financial institutions for appraisal of proposals for switching existing production facilities to CTs is another barrier for its adoption. The lack of investment in research and development efforts aimed at developing commercially viable clean technologies is also an obstacle.

Future Directions

The potential of new technologies combined with innovative public policy and strategic investment to stimulate the growth of new markets for environmentally sound products and services should be explored fully while also reinvigorating slowing markets through the widening application of new technologies across the entire economy. While market-based mechanisms are part of the answer of a short- to medium-term solution in the area of clean technology transfer, significant help is needed at a very fundamental level to build sound legal and governance capacity and to create a new sustainable growth paradigm. Government must extend tax credits and create long-term incentives for renewable energy and CT. The Renewable Energy Law providing government support in the form of tax breaks, targeted loan subsidies and special funding for wind, solar, water, and biomass should be enforced. Likewise, public policy plays a critical role in encouraging the adoption of new technology that serves to jump-start market demand and reduce entry costs. Use of revenue-enhancing fiscal instruments to promote shifts to clean technologies in both existing and new units should be considered.

A viable environment for continued innovation also requires innovative public policy that is forward thinking, collaborative with the private sector, and globally oriented. Combining known methods with new technical know-how can enhance and protect the innovation. By accessing and mobilizing appropriate resources, new opportunities can be created or activated. Technology needs assessment, joint research and development programs, and a healthy technology-transfer environment should be developed for adoption of clean technologies. University partnerships and research should be collaborated on to generate a pipeline of innovation and talent. Further, entrepreneurs should continually review their business ideas in the light of experience and alter them as they learn more about market needs and the resources at their disposal.

Basing technology diffusion solely on transnational firms or north–south transfers to established companies misses the important role of indigenous firms that arguably are more aware of local needs and responses, and are better able to implement technologies concomitant with the demand and economic potential of their surroundings. Therefore, active participation from all the local stakeholders is an important part of a successful CT program.

In 2008, CT venture investments in North America, Europe, China, and India totaled a record $8.4 billion—up 38 percent from $6.1 billion in 2007—the seventh consecutive year of growth for CT venture investing, making it the fastest-growing area of venture capital investment. According to the published research, the top CT sectors in 2008 were solar, biofuels, transportation, and wind. Solar accounted for almost 40 percent of total CT investment dollars in 2008, followed by biofuels at 11 percent.

The 2009 United Nations Climate Change Conference in Copenhagen is expected to create a framework whereby limits would eventually be placed on GHG emissions. Many proponents of the cleantech industry hope an agreement is established there to replace the Kyoto Protocol. As this treaty is expected, scholars have suggested a profound and inevitable shift from "business as usual."

The market for green technology in developed countries is rapidly growing, but green technology in developing countries has not been greatly developed. Creating a global exchange network should involve firms already using green technology, venture capitalists, and those in developing countries who are eager to expand green technology and are knowledgeable in local laws, regulations, and customs.

CT is still in an early stage of development. For instance, it took coal nearly 100 years to bypass traditional energy sources (such as the burning of wood) as the world's primary energy source. It then took nearly 100 years for oil to surpass coal usage. Natural gas has been in development for more than 100 years, and represents about 20 percent of global primary energy use. Similarly, it will take new renewables, such as wind, solar, and biofuels, 10, 20, 30 years or more to catch up with coal, oil, and natural gas. This is one of the reasons why long-term thinking is so crucial.

The cleantech revolution has already been in the making for 30 to 50 years; the first conversion of sunlight to electricity in a solar PV cell, for example, took place at Bell Labs in 1954. Moving forward, CT will exhibit both disruptive sudden advances and more deliberate, incremental change. In a world of increasingly constrained natural resources, it would be hard to imagine a sector that offers more promising long-term returns and rewards.

See Also: Appropriate Technology; Best Available Control Technology; Best Management Practices; Clean Fuels; Closed-Loop Supply Chain; Cradle-to-Cradle; Extended Producer Responsibility; Extended Product Responsibility; Green Technology; Life Cycle Analysis; Waste Reduction.

Further Readings

Asplund, Richard W. *Profiting From Clean Energy: A Complete Guide to Trading Green in Solar, Wind, Ethanol, Fuel Cell, Carbon Credit Industries, and More.* Hoboken, NJ: John Wiley & Sons, 2008.

Makower, Joel and Cara Pike. *Strategies for the Green Economy: Opportunities and Challenges in the New World of Business.* New York: McGraw-Hill, 2008.

Pernick, R. and C. Wilder. *The Clean Tech Revolution: Discover the Top Trends, Technologies, and Companies to Watch.* New York: HarperCollins, 2008.

Schaeffer, J. *Real Goods Solar Living Sourcebook: Special 30th Anniversary Edition; The Complete Guide to Renewable Energy Technologies and Sustainable Living.* Gabriola Island, British Columbia, Canada: New Society Publishers, 2007.

Gopalsamy Poyyamoli
Rasmi Patnaik
Pondicherry University

CLOSED-LOOP SUPPLY CHAIN

A closed-loop supply chain is ideally one that completely reuses, recycles, or composts all materials used to manufacture a product, as well as the end product itself at the end of its lifetime. The aim is to maximize use of natural resources and minimize environmentally damaging wastes.

Natural ecosystems tend to be closed-loop supply systems. However, for consumer goods, a system of total closure might only be achievable for products made entirely from natural components like organically grown, unpackaged farm produce, and perhaps some timber products. Manufactured goods that use synthetic materials are likely to have inputs and outputs that cannot be recaptured and therefore a closed loop will only be partial, involving some, but not all, of their materials.

To achieve a partial closed loop, manufacturers may adjust their manufacturing processes in order to recycle factory wastes for reuse in the factory or, more ambitiously, they may implement a policy which recaptures used products from consumers. The components of recaptured product can then be broken down and fed back into the manufacturing system, thereby closing the loop.

Closed-loop supply chains are an alternative to the more usual forward supply chains where the customer is the end user and the product is disposed of, often in municipal waste streams. Much of this waste stream currently goes to landfill, but it contains hazardous materials and poses environmental risks. Closing a supply chain is almost always a move toward better sustainability, and the more components of a product that are captured within the closed loop, the greener the initiative is.

Recapturing the product from the consumer usually necessitates a policy of Extended Producer Responsibility (EPR) or product stewardship. EPR is based on the "polluter pays" principle, but goes beyond a manufacturer's responsibility for pollution from product manufacture to making the manufacturer responsible for the environmental impact of a product for its entire life cycle: from manufacture to consumer

Corporations such as Hewlett-Packard attempt a closed-loop supply chain by encouraging consumers to send back their used printer cartridges for recycling.

Source: Hewlett-Packard

disposal, how the product should be returned and repaired, recycled, or reused, or if its components should be fed back into the manufacturing process.

Product stewardship is a related idea in that it is concerned with the environmental impacts of the product throughout its life cycle. However, product stewardship shares responsibility between all those involved in a product's life cycle—including designers, suppliers, manufacturers, distributors, retailers, and consumers—rather than shifting responsibility to the manufacturer.

Normally, government authorities take responsibility for disposal of products, and therefore the burden is paid for with taxpayer dollars. However, product design and manufacturing decisions can determine how environmentally damaging a product will be when used and disposed of, and how easily it can be recycled. Because governments have taken responsibility for waste management, manufacturers have created an excess of disposable products and packaging without thought to the environmental and other costs associated with them. Manufactured goods now make up more than three-quarters of municipal waste.

Legislative Requirements

EPR was adopted during the 1990s by various Organisation for Economic Co-operation and Development (OECD) countries. The Swedish Eco-Cycle legislation embraces the concept. In Germany, the Netherlands, Austria, Switzerland, and France, manufacturers have legal responsibility for taking back packaging and recycling their products and there is "end-of-life legislation and voluntary agreements concerning a number of complex products," such as cars and batteries.

The EU's directive on waste electrical and electronic equipment (WEEE) also requires that manufacturers take back used consumer products. In 2001, the EU environment ministers proposed extending the polluter pays principle to cover disposal of products at the end of their useful life. In this case, they defined the polluter not as the consumer, but as the manufacturers of the electrical and electronic equipment. They reasoned that manufacturers should be responsible for the disposal and recycling of these products after consumers were finished with them.

The 2003 WEEE directive aimed at "as a first priority, the prevention of waste electrical and electronic equipment, and in addition, the reuse, recycling and other forms of recovery of such wastes so as to reduce the disposal of waste." It was therefore intended to encourage manufacturers to design products to enhance their potential for reuse, recovery, and recycling.

The equipment covered by the directive includes household appliances, information technology and telecommunications equipment, consumer and lighting equipment, electrical and electronic tools, toys, leisure and sporting equipment, medical equipment, monitoring and control instruments, and automatic dispensers. The final treatment of the collected equipment should use the "best available treatment, recovery and recycling techniques."

Design

Closed-loop supply chains require that design, manufacturing, and packaging decisions are made with a view to recapturing the end products for input back into the manufacturing process.

What is termed *cradle-to-cradle* design is the key to achieving the most sustainable outcome for manufactured goods. The kind of design mind-set required often involves referring to nature for inspiration, where one organism's waste is food for another and nutrients flow indefinitely in cradle-to-cradle cycles of birth, decay, and rebirth. In nature, waste equals food. When engineers and designers come to understand these regenerative systems, they should be able to recognize that all materials can be designed as nutrients that flow and then recycle through manufacturing systems. The biological metabolism in ecosystems can be mirrored in the technological systems they design so there is a closed loop in which valuable resources move in cycles of production, use, recovery, and remanufacture.

To achieve this cradle-to-cradle supply framework, designers and engineers must assess the final outcome of a manufacturing process and only select safe materials that will not only optimize products and services, but also allow a closed loop in the material flows so that both the manufacturing process and the product are inherently benign and sustainable. Any materials that are to be left outside of the loop should be designed, where possible, to biodegrade, such as textiles and packaging made from natural fibers. Synthetic materials designed for use within the closed loop, such as carpet yarns made from synthetics, can be repeatedly recycled to provide high-quality, high-tech ingredients for successive generations of synthetic products.

Logistics

Closed-loop supply chains require used consumer products to be collected before they can be reintegrated into the manufacturing process. Collected products must then be sorted and distributed for reprocessing, which may involve cleaning, disassembly, shredding, or other treatment. The product may then be repaired, reused, recycled, remanufactured, or the components reused. New products created from this process need to be marketed.

Retrieval of products from consumers may involve a variety of take-back systems, such as the beverage container deposit system, where consumers receive a small monetary incentive for taking back a product. This cost is paid for as part of the purchase price of the product. The 2005 EU Directive requires that consumers should be able to return their used equipment free of charge, and that governments ensure collection facilities are made available. IBM charges its U.S. customers for taking back used equipment. The equipment is sorted for remarketing, refurbishment, recycling of components, or harvesting of spare parts.

Because it is often too expensive for individual companies to finance take-back schemes, there is a tendency for such schemes to be a cooperative effort across industries. For example, battery manufacturers are joining to facilitate the collection of used rechargeable batteries for recycling. In the United States, carpet manufacturers have agreed to make a joint effort to reuse and recycle carpets that consumers no longer want.

Reverse logistics is the name given to the overall operations related to the reuse of products and materials and the development of a closed-loop supply chain. Reverse logistics involves planning and controlling an efficient and cost-effective flow of raw materials and finished goods from consumers back to the point of origin at the factory, for the purpose of recapturing value or proper disposal.

To date, most closed-loop supply chains amount to little more than recycling schemes for packaging and containers: recycling of bottles and cans is well known; SC Johnson recycles aerosol cans from its products; and Xerox recycles toner cartridges from photocopying machines.

Often it is only parts of the returned product that are reused, for example the Indesit Company uses recycled plastic from recovered refrigerator waste to make the back panel for washing machines. In some cases, materials are ground up and used for something altogether different. Nike's Reuse-a-Shoe program enables consumers to have their old sports shoes ground up to make synthetic athletic surfaces.

Nevertheless, closed-loop supply chains are likely to increase in number and scope with the expansion of green business, because they are not only more resource efficient and less waste producing, but also because they can add to a firm's profitability.

See Also: Cradle-to-Cradle; Extended Producer Responsibility; Extended Product Responsibility; Industrial Ecology; Supply Chain Management.

Further Readings

De Brito, M., R. Dekker and S. D. P. Flapper. *Reverse Logistics—A Review of Case Studies.* ERIM Report Series. Rotterdam, the Netherlands: Research in Management, 2003.
Fleischmann, M., et al. "Integrating Closed-Loop Supply Chains and Spare Parts Management at IBM." *Interfaces*, 33/6: 44–56 (2003).
Guide, V. D. R., Jr., V. Jayaraman, R. Srivastara and W. C. Benton. "Supply Chain Management for Recoverable Manufacturing Systems." *Interfaces*, 30/3:125–42 (2000).
Srivastava, S. R. and R. K. Srivastava. "Managing Product Returns for Reverse Logistics." *International Journal of Physical Distribution and Logistics Management*, 36/7:524–46 (2006).

Sharon Beder
University of Wollongong

Cogeneration

All thermal power plants, whether they are based on fossil fuel or burn coal, natural gas, or petroleum, convert less than half of their energy into electricity. The rest of the energy is generally lost as excess heat, and is either vented into the atmosphere through cooling towers or dissipated into the environment through a "heat sink," such as a large lake, river, or ocean. Cogeneration, or combined heat and power (CHP), takes the heat ordinarily generated as a waste product from thermal generation and uses it to provide heat for dwellings and businesses, as well as low-grade process heat. More generally, cogeneration refers to the simultaneous production of two forms of energy from a single fuel source. A simple example for cogeneration is the double function of an automobile engine during cold winter days: the engine powers the car and, at the same time, uses its excess heat to warm the interior of the vehicle. This simple example can also illustrate some of the biggest obstacles to cogeneration, for it requires a close physical proximity between power generation and the potential use of excess heat. Transporting heat over longer distances—either as steam or as hot water—inevitably translates into exponentially increasing efficiency losses.

Some of the principles of cogeneration/CHP systems are as follows:

- They use the "waste" heat as a by-product of electricity production to provide warmth for buildings, and provide process heat for industrial uses.
- They produce electricity and heat near to where it is being used, avoiding the inefficiencies that arise as a result of distribution losses.

At present, in a centralized power distribution, there is inherent waste—power is lost as a result of resistance in the production, transmission, and distribution network. At the point of production, thermal power stations, through nuclear fission or combustion of fuel or burning of coal or gas, transform thermal energy into mechanical power in order to generate electricity, in the process producing a large amount of excess heat. As the extra heat is routinely released into the atmosphere, conventional thermal power plants represent an extremely inefficient use of resources, as the heat is "wasted" and not put to useful work. Thermal power plants that "cogenerate" both energy and heat and in turn put both to use, on the other hand, can greatly increase the energy efficiency, often operating at 50 to 70 percent higher efficiency rates than single-generation facilities.

Practical Applications

There are a great many practical applications of cogeneration or CHP, as otherwise wasted heat can be used to provide heat for warming, provide electricity in buildings in which a unit operates, channel heat into absorption chillers to generate cold air, or supply hot air for drying processes of smaller businesses. Cogeneration can be used in large power plants, but also in a great variety of smaller applications, ranging from the landfill use of biogas, to hot water generation from independent power plants in hospitals or businesses, to natural gas–operated electricity-producing cogeneration modules in greenhouses. Indeed, most cogeneration plants in the United States and Europe today are relatively small facilities operated by small to mid-sized companies, hospitals, universities, and the military. Particularly with rising fuel prices, cogeneration facilities benefit a company's bottom line, and have the added benefit that, by using waste heat, they prevent the additional burning of toxic pollutant-spewing fuels.

While cogeneration still makes up a relatively small segment of the American energy market, it is steadily growing in significance. Industry analysts estimate that cogeneration now produces almost 10 percent of our nation's electricity, saves customers up to 40 percent on their energy expenses, and provides even greater benefits to the environment. Cogeneration energy production in Europe has reached about 12 percent of the market as of 2009, with a large variance of between 2 percent for Poland and up to 60 percent for Denmark. Observers suggest, however, that the remaining difference between Europe and the United States is mostly due to higher fuel prices in Europe.

There are a large number of very successful examples of cogeneration in the United States. According to reports, the Sunnyvale corporation operates a gas-powered cogeneration plant in its California-based computer networking plant, and estimates that it saves up to $300,000 in energy costs each year. An Illinois-based company called Epcor USA Ventures has started to operate several cogeneration power plants in San Diego to provide electrical energy to the U.S. Marine Corps and Navy bases there, while at the same time using excess heat to drive steam generators hooked into the bases' centralized heating and cooling systems. In San Francisco's financial district, the 530,000 square foot commercial office building Transamerica Pyramid is operating a gas-fired reciprocating generator that provides most of the electrical as well as the hot water, heating, and cooling needs of the building, reducing greenhouse gas emissions by up to 40 percent in the process.

Technologies Used in Cogeneration/CHP

For larger cogeneration/CHP installations, Combined Cycle Generating Plants can be used, where the gas first passes through a gas turbine, with the waste heat from that being used to raise steam to power a steam turbine, and the water to be heated providing a condenser for the steam turbine. Gas turbines and steam turbines can also be used in isolation.

Medium-sized installations are more likely to use internal combustion engines. These turn the shaft of a generator, producing electricity, while the engine coolant heats the heat main. These are often fitted with fuel delivery equipment capable of working with gas. Some installations, however, are designed specifically to run on a range of fuels.

One technology that is rapidly gaining market share is the fuel cell. High-temperature fuel cells electrochemically produce energy, reforming gas into hydrogen on-site. They produce a very clean exhaust, and, because they have no moving parts, offer many advantages over reciprocating or rotating engines.

Some manufacturers have been looking to produce cogeneration/CHP units on the domestic scale. These units would take the place of a typical boiler in a house, and would produce electricity and heat for a single dwelling. These are likely to use internal combustion engines, fuel cells, or Stirling engine technology.

Tri-/Quad-/Poly-Generation

In addition to the basic technology of combined heat and power, there are a few variations that can be implemented with careful technology selection to provide greater amenity.

Absorption chillers are refrigeration systems that use heat as the energy source to drive a cooling cycle which can be used for air conditioning or for storing refrigerated goods. As the absorption chiller is driven by heat, it can often be powered by heat the cogeneration/CHP unit produces in the summer months that is otherwise "surplus to requirement," helping to balance out the annual demand profile. These are often used with cogeneration/CHP units, and the technology combination is termed *tri-generation*.

When using fuel cells as the technology for converting primary fuel into heat and electricity, high-quality water is produced as a by-product. This water can be used in other applications, for example, for replacing water that is lost from a heated swimming pool. This is termed quad-generation. The term *quad-generation* is also sometimes used in cases where a CHP unit integrates with a supplemental solar heating/cooling system.

Overall, co- or multiple-generation applications will undoubtedly grow in significance, as they not only provide more efficiency and substantial savings, but also significantly help cut down on greenhouse emissions. New cogeneration technologies and applications are continuously developed and will further contribute to both economic viability and the environmental urgency for increased use.

See Also: Appropriate Technology; Carbon Footprint; Distributed Energy; Ecoefficiency.

Further Readings

Elliott, R. N. and M. Spurr. "Combined Heat and Power: Capturing Wasted Energy." (1999). http://www.osti.gov (Accessed April 2009).
Havelsky, V. "Energetic Efficiency of Cogeneration Systems for Combined Heat, Cold and Power Production." *International Journal of Refrigeration* (1999).
Kolanowski, Bernard F. *Small-Scale Cogeneration Handbook.* Lilburn, GA: Fairmont Press, 2008.
Pehnt, Martin. *Micro Cogeneration: Towards Decentralized Energy Systems.* New York: Springer, 2005.
Sirchis, J. *Combined Production of Heat and Power.* London: Taylor and Francis, 2007.

Gavin D. J. Harper
Cardiff University

Compliance

Compliance refers to the goal of businesses that their operations and personnel comply with all relevant laws and regulations. Complying with environmental laws and regulations is the minimum that any firm should be doing. Green businesses seek to go beyond regulatory compliance to protect the environment.

Regulations are mandatory requirements that can apply to individuals, businesses, state or local governments, nonprofit institutions, or others. Governments may require that certain technologies be used in factories, for example, to reduce pollution. They may specify production processes to be used or pollution controls to be put in place. Another approach is to set standards for emissions and products coming from a factory, and to allow businesses to meet these standards however they like. A related approach is to set standards to be met in the surrounding environment, such as air and water quality standards. Sometimes, air or water quality standards are set; but it is the emissions that are monitored, and the concentration in the surrounding air or water is calculated by assuming that the waste stream dilutes. Standards may also be set in terms of concentrations of wastes in people, animals, and plants exposed to the waste streams. These might be residue levels in the body or tissues, or might be some measure in the air or water where the communities are situated, breed, or feed.

In the United States, more than 100 independent federal ecological statutes and a variety of state-level regulations have been enacted to promote health and reduce dangerous chemical substance levels in the environment. Acts such as the Clean Air Act, Clean Water Act, Emergency Planning, and Community Right-to-Know Act, Pollution Prevention Act, and the Resource Conservation and Recovery Act represent examples of these government interventions.

Enforcing Compliance

Enforcement deters those who might otherwise profit from violating the law and creates a fair playing ground for compliant companies. To be effective, regulations need full political support so that regulatory agencies have the financial and human resources to monitor and enforce standards properly.

There are a number of options available to the regulator to enforce standards and regulations if they are not being met by a firm. Standards can be enforced, for example, by licensing a plant or product. The license sets the standards that the plant or product has to meet. Performance is then monitored to see if the firm complies with the license conditions; if they do not, a penalty is imposed.

Regulatory authorities may try a soft approach of persuasion and warnings. Regulators use other tools to encourage businesses to comply with their requirements, such as partnerships, educational programs, and grants. Alternatively, they may take a tougher approach and take noncompliant firms to court to have them fined, claim any bonds that the firm has put up, or even withdraw the firm's license to operate.

Commitment to compliance is best generated when there are substantial sanctions for noncompliance with a clear legal mandate. To impose compliance in the United States, regulatory agencies, such as the U.S. Environmental Protection Agency (EPA), use a variety of tools and approaches. The EPA describes its role as creating regulations that set specific requirements about what is legal and what is not, and then tries to ensure that these laws are adhered to. For example, a regulation issued by the EPA to implement the Clean Air Act might explain the levels of a pollutant that are considered safe (e.g., sulfur dioxide). The EPA continually works to help businesses comply with the law and to enforce the law when they do not comply.

The EPA's civil, cleanup, and criminal enforcement programs work with the Department of Justice, and state and tribal governments to take legal actions in both federal and state courts to bring polluters into compliance with federal environmental laws. The agency emphasizes actions that reduce significant risks to human health or the environment, and consults extensively with states and other stakeholders in determining risk-based priorities.

In Australia and the United Kingdom, regulatory authorities tend to take a less legislative approach to regulation, preferring to work with noncompliant companies, encouraging them to improve their practices rather than take them to court or fine them.

Achieving Compliance

Large companies that are concerned with the financial and reputational risks of noncompliance tend to implement company-wide and training systems to ensure compliance with the many rules and regulations they have to adhere to. It is not productive to write e-mails, make announcements, and form policies if employees believe that they still have the ability to make compliance decisions based upon situational variables. Organizations must create internally driven and designed programs that train for compliance as a part of employee organizational tasks. Environmental managements systems (EMS) provide a framework for this and guidelines are provided by ISO 14001.

Companies may also commission environmental audits by outside consultants or internal staff. Such audits will include a review of company policies, protocols, documents and records, as well as interviews with staff and site inspections. Such audits may extend to suppliers and others in the supply chain.

In the United States, the EPA authorizes reduced penalties for firms that voluntarily undertake compliance audits, and then correct and report any discovered violations to the agency. Self-auditing is beneficial because it permits self-policing to occur. However, researchers have found that it is more likely to be socially beneficial when the damages caused by such violations are large and current audit policy explicitly excludes violations being punished in this way. The program is currently designed to address small violations, which actually increase social costs, rendering the program potentially less effective.

The Sarbanes-Oxley Act of 2002 (SOX) was a response in the United States to the much publicized fraudulent activities of companies like Enron. It was an amendment to the Federal Sentencing Guidelines for Organizations, and helps organizations create a structured and systematic approach toward achieving compliance and ethics standards within the firm. The guidelines read as follows:

1. The organization must establish standards and procedures to prevent and detect criminal conduct.

2. High-level personnel of the organization shall ensure that the organization has an effective compliance and ethics program.

3. The organization must exercise due diligence and not place individuals with questionable backgrounds into positions of substantial authority.

4. The organization must communicate the standards and procedures of its compliance and ethics program to all of its employees and, as appropriate, to the organization's agents. It must also conduct effective ethics training programs.

5. The organization must ensure its compliance and ethics program is followed, including monitoring and auditing to detect criminal conduct. The organization must also periodically evaluate the effectiveness of its program and must include and publicize an anonymous or confidential system for the reporting of suspected wrongdoing.

6. The organization's compliance and ethics program shall be promoted and enforced consistently throughout the organization.

7. After criminal conduct has been detected, the organization shall take reasonable steps to respond appropriately to the criminal conduct.

The Limits of Compliance

Regulations are often resisted by industry and business, who fear that the costs of remedying breaches might lead to cutbacks in their operations or reduce their profitability. Industry groups do not like infringement on management autonomy; for this reason, they favor self-regulation by industry associations, voluntary codes of practice, motivational and environmental management training, and financial incentives for firms that do the right thing rather than regulatory measures.

The push–pull relationship between regulators, environmentalists, and special interest groups continues to challenge the effectiveness of compliance. For example, in 2006, the EPA proposed to reduce allowable levels of particulate matter (PM) in the air, but not by as much as its Clean Air Scientific Advisory Committee (CASAC) recommended, the latter group claiming that the rule would not be sufficiently protective.

Regulations can be general or specific, but only those with narrowly focused prescribed requirements have a record of being followed. Ironically, when they are more specific, it may be easier for those being regulated to build a case that supports alternative vantage points, which may not favor the purpose of the regulation.

A government may require that a firm use the "best available technology" (BAT), which requires them to install the best commercially available and proven techniques. However, more often governments require the "best practicable technology" (BPT), which requires firms to install the best currently used technology that they can readily afford. Environmental standards are usually based on what can be economically achieved using existing technologies. Therefore, environmental reform is almost always limited to what can be achieved by the enforced adoption of readily available technologies that are not too expensive. This may have been sufficient in the past, but the continued degradation of the environment at both a local and a global level suggests that a new approach is necessary for the future. Companies need to treat such standards as a bare minimum.

The other approach is to regulate by setting effluent or emissions standards and allowing polluters to meet those standards in any way they see fit. Within this approach, either uniform or ambient emission standards can be set. Uniform emission standards can be set for the waste streams of all industries, wherever they are located, and whatever their financial position. A uniform standard is more equitable and simplest to administer but is thought to impose "unnecessary" costs on government and industry. Ambient emission standards, however, are standards that vary according to the existing environmental conditions in the local area. The existing environmental conditions may include biological properties of the area, the uses to which the waters are put, and the actual despoiling that has already been suffered, a degraded area warranting less protection than a pristine one.

The translation from air or water quality standards to effluent or emissions standards is not a simple matter of mathematics or analysis. A number of assumptions and value judgments need to be made, and what often ends up happening is that effluent standards tend to be based on "good practice" rather than being directly related to water quality criteria. Thus, the two approaches to standard setting are often very similar.

Often, standards are not as stringent as they need to be to protect the environment as a result of regulatory capture, where regulatory authorities act in the interests of those they are regulating rather than in the public interest. Moreover, consideration of the relative assimilative capacities of different waterways often allows polluters to save money and use less than the best practicable technology.

A concern with regulations in general is that the rules are based upon prior knowledge. As researchers continue to pursue new knowledge, additional information is uncovered that can render rules obsolete and/or ineffective. For example, since the early 1970s, water-cooled power plants have installed fish protection technologies with limited guidance from the EPA. In 2004, the EPA published rules for implementing the Clean Water Act at existing plants. Researchers examined possible sources of injury and stress that fish encounter as they move through the cooling water intake system, and previous assumptions that influenced regulation parameters were challenged. Such findings demonstrate how existing regulations may not be based upon complete and accurate assumptions, and how existing rules can become outdated. As new technologies and knowledge emerge, rule adherence can be less than effective.

One central feature of compliance is that regulations often require firms and communities to self-identify, monitor, report, or complete requirements on their own recognizance. For example in the United States, state- and federal-issued general permits for stormwater discharges associated with industrial activities came into effect back in the early 1990s. Compliance was not achieved until 2004.

Self-monitoring is favored by industry because it reduces outside interference. Regulatory authorities often find self-monitoring necessary because they do not have the resources to provide sufficient inspectors, and are reluctant to levy this cost from the people they are regulating (although environmentalists argue that this is what they should do). Occasionally, outsiders, such as environmental groups and the media, identify a firm that is not meeting its license conditions.

Because standard setting is a political process, it will vary from government to government. Some governments may prefer to keep environmental standards low and flexible in order to attract industry. There is the added complication that standards may be different in each state. What is more, compliance can vary in its effectiveness among states and different urbanized regions. One U.S. study showed how Texas and Oklahoma attained higher compliance rates than California.

Taking this information together, we see how the term *beyond compliance* emerged, which describes actions taken that are not required by law but assumed to be relevant, important, and adopted voluntarily. Given the variety of challenges present when trying to initiate change through compliance, laws and regulations are just not enough.

Going Beyond Compliance

Most companies still rely on the course set by their national governments: following the adoption of the Kyoto Protocol, and waiting for the implementation of compliance mandates before they take action. In other words, they do not go beyond compliance. To address green business objectives, companies must not only abide by environmental regulations to avoid punishment and negative consequences, they must also advance the ethical intent behind them. The firm must go beyond compliance to meet the general practices and targeted resource conservation and pollution prevention measures.

Green businesses realize the need to do more than what governments ask, especially when stakeholder forces are beginning to insist upon more rigorous green product development and creation. Resource efficiency is good business, but when companies face mandates from their customers and market forces, we are likely to see more demonstrative change. If customers, end users, employees, or other external stakeholders and influencers begin to set more stringent standards, then industry will move to address this demand. It remains a challenge for customers to prefer green products, but it is an important responsibility on the part of industry to surmount real and/or perceived price and quality issues. Customers can influence standards, and mega-retailers like Wal-Mart can demand their own controls and phaseouts (e.g., phthalates in plastic toys). Because the supply chain framework supports our global economy, when a food or chemical scare runs through a business, it affects manufacturers and retailers alike.

The EPA models how to go beyond compliance by providing voluntary programs to help firms procure green power as a way to reduce the environmental impacts associated with electricity use. By partnering with the EPA, hundreds of organizations opt to purchase billions of kilowatt-hours of green power annually. One of the leaders of this initiative was the New Belgium Brewing Company, which became the first brewery in the United States to use 100 percent wind power to meet its electricity needs.

Partnering with the EPA can help organizations lower their transaction costs of buying green power, reduce their carbon footprint, and communicate green leadership to key stakeholders. The choice to buy green power is claimed as one of the easiest and most effective ways to improve a firm's environmental performance.

When companies establish their own standards, they can mirror, partner with, and/or exceed government program expectations. For example, Andrew Winston describes how Verizon chose to go beyond its compliance regulations by setting energy performance standards for their suppliers. Home Depot examined the environmental regulations regarding air pollutants for diesel trucks. Although recent 2007 legislation was to be phased in and carried "grandfather" clauses (the manufacturer has a specific period of time before all trucks must meet the new standard), it informed its suppliers and distribution companies that it would impose a changeover to its infrastructure sooner than the government required.

To effectively address the agenda for a green society, organizations need to direct and guide employee behaviors by imposing measurable "green" objectives that are specific and incorporated into everyday decision-making efforts. They must also educate employees to go beyond compliance, understanding the intentions behind the rules to ensure that the green business strategy is realized through consistently applied decisions and actions.

The most successful organizations have already begun to combine their compliance and ethics efforts and have even reshaped compliance as a means to build value through corporate culture rather than legalistic thinking that simply affirms a myopic "check the box" mind-set. Compliance and ethics must be a part of the climate and culture of the firm, which is driven from the top down and implemented at every level of the organization.

See Also: Carbon Trading; Environmental Audit; Environmental Management System; Leadership in Green Business; Sustainability.

Further Readings

Dizard, John. "Hard Battle Ahead on Market Solution for Pollution." *Financial Times* (January 20, 2009).

Friesen, Lana. "The Social Welfare Implications of Industry Self-Auditing." *Journal of Environmental Economics and Management,* 51/3 (2006).

Michaelson, Christopher. "Compliance and the Illusion of Ethical Progress." *Journal of Business Ethics,* 66/2 (2006).

Potoski, M. and A. Prakash. "Green Clubs and Voluntary Governance: ISO 14001 and Firms' Regulatory Compliance." *American Journal of Political Science,* 49/2 (2007).

Renner, Rebecca and Kris Christen. "EPA's Science Advisers Disappointed by Proposed PM Standards." *Environmental Science & Technology,* 40/7 (2006).

Sekerka, Leslie E. and Roxanne Zolin. "Rule Bending: Can Prudential Judgment Affect Rule Compliance and Values in the Workplace?" *Public Integrity,* 9/3 (2007).

Taft, E. P., J. L. Black and L. R. Tuttle. "Clean Water Act Compliance: Fish and Cooling Water Intakes: Debunking the Myths." *Alden Research Laboratory Inc. and Electric Power Research Institute,* 151/3 (2007).

U.S. Environmental Protection Agency (EPA). "Green Power Partnership." http://epa.gov/grnpower/partners/partners/newbelgiumbrewingcompany.htm (Accessed May 2009).

Verschoor, Curtis C. "Interactions Between Compliance and Ethics." *Strategic Finance,* 87/12 (2006).

Weiss, Joseph W. *Business Ethics: A Stakeholder and Issues Management Approach With Cases.* Mason, OH: South-Western Cengage Learning, 2009.

Winston, Andrew. "Green Business and Compliance—The Government Is the Least of Your Worries." *Harvard Business Online* (September 15, 2008). http://www.huffingtonpost .com/andrew-winston/green-business-and-compli_b_123496.html (Accessed May 2009).

Leslie E. Sekerka
Menlo College

CONSERVATION

Conservation can broadly be defined as the protection of biological diversity and natural resources from human-induced change. The concept of conservation has a range of meanings to different people. Over the course of history, conservation ideals have not only been used to save animals and plants from extinction, but also to promote ideas about land and agriculture, to colonize peoples, to promote nation building, to fight against social and political injustices and, most importantly, to provide sustainable resources for green business.

Conservation is a process; it attempts to regulate and protect biodiversity and natural resources for the longer term. Usually, it is a response to anthropogenic threats—originating from humans—of one kind or another, including industrial, urban, or agricultural developments, human encroachment, or general unsustainable human practices. Its object is mostly nonhuman, although it is now commonly accepted that "cultural heritage" should be considered equally worthy of conservation. Nonetheless, much of contemporary conservation efforts focus on the natural environment, especially biological diversity, in terms of plant and animal species and natural resources, such as water, forests, soil, air (quality), mountains, wetlands, but also landscapes, ecosystems, and so forth.

One branch of conservation aims to regulate and preserve biodiversity and natural resources from agricultural development. Here, conservationists Bobbi McDermott and Neal Hoy review a conservation plan in a field of leaf lettuce in Arizona.

Source: Jeff Vanuga/U.S. Department of Agriculture Natural Resources Conservation Service

Conservation is not only about "nature." It is as much—if not more—about people and the interaction between people and nature. It is often assumed that humans have a dominant and therefore responsible role; dominant because humans are "allowed" to use biodiversity and natural resources, but they must do so in a responsible manner. This usually entails limiting the impact of humans on nature and ensuring that nature has a place to exist without too much negative human interference.

Human Reliance on Ecosystems

The human reliance on ecosystems is one of the major arguments underlying the conservation imperative. Humans benefit from a vast number of natural processes that are provided by natural ecosystems. These benefits are collectively known as ecosystem services, and include basic necessities like clean drinking water and natural processes such as the decomposition of wastes. Conservation of specific ecosystems may maintain a vast range of ecosystem services for a locality, including moderation of weather extremes; dispersal of seeds; mitigation of drought and floods; cycling of nutrients; protection of streams, rivers channels, and coastal shores from erosion; control of agricultural pests; purification of air and water; and pollination of crops.

The concept of use is crucial in differentiating between conservation and preservation. Conservation usually allows some use of natural resources or biodiversity, while preservation generally allows no use at all. This use can be consumptive and extractive, or nonconsumptive and nonextractive. Examples of the former include harvesting plants for medicinal use or selective logging as a part of sustainable forest management. Examples of the latter include taking photos of animals in their natural habitat.

The idea of use in conservation is evidently important when thinking about "green business." After all, natural resources—such as oil, gas, water, minerals, and trees—continue to form the basis on which contemporary capitalist society rests. As such, nature indeed provides the "resources" that lead to all types of benefits that humans enjoy, and—so the argument goes—in order to guarantee these benefits and the economic system behind it, humans need to conserve these "resources." Talking about "natural resources" is therefore another indication that we are not talking about preservation, which connotes more with the term wilderness. Yet, the boundaries between conservation and preservation are not firm, neither in theory nor in practice.

Historically (and in the Western hemisphere), the concept of conservation gained prominence in the 18th and 19th centuries. Before and during those times—and continuing today—many people believed that nature needed to be "conquered" and the land made productive for human benefits. As a result, flora and fauna all over the world were rapidly harvested, cut, killed, or cleared with the objectives of securing land for agriculture, using resources for subsistence, trade, and industrial development and for purposes of recreation. Much of this trend came forth from the mercantilist and capitalist expansions from Europe from approximately 1600–1920. Some of the main environmental consequences include the following:

- Massive slaughter of animals, especially for trade in fur and hides (sables, beavers, otters, bears, etc.) and meat, but also to protect agricultural land
- The development of a world system of cash-crop production that led to the subjugation of natural environments for monocropping, most notably sugar, but also tobacco and other crops
- Massive clearing of forests

These developments had a tremendous impact on the natural environment and on the relationship between people and nature. Some of the main environmental impacts were land and soil degradation, loss of biodiversity, loss of environmental and ecosystem regulation capacities, pollution, loss of "wild" habitat, and so forth. In response, countermovements and ideas arose during the 18th and 19th centuries, most notably in Great Britain and in the United States, but also in areas of mainland Europe and other parts of the world. The idea was to "save" nature from its destruction and restore a balance between humans and nature. While there was no coherent movement, two distinct tendencies within this trend were visible.

The first was a romanticization of nature and wilderness, especially in Britain and the United States, but also later in colonial Africa. Wildlife and biodiversity were revered as having an almost-mythical status because of their aesthetic value and grandeur. The second tendency was based in a belief in "progress" and scientific principles, and recognized the use and economic values of ecosystems. This latter tendency was most notable in mainland Europe, for example, in Germany, where ideas about forestry increasingly came together around the need for a continued timber supply for economic development and for the regulation of wetlands and water tables.

There are numerous conservation schemes under way around the world designed to maintain ecosystem services that involve green businesses either directly or indirectly. A good example involves the tourism industry based on the Great Barrier Reef in Queensland, Australia. Collectively, the industry has successfully lobbied the Australian government to expand the network of marine sanctuaries within the reef in order to protect and increase the ecosystem's ability to sustain the industry into the future.

Resource Conservation

As well as referring to ecosystem resources, conservation can also refer to industrial inputs that interact with nature either at their source, such as water, or in the disposal of their waste products, such as energy and minerals. When businesses are attempting to become greener, conservation of water and energy resources are often the first measures they undertake.

Water conservation presents many businesses with immediate opportunities to save money and protect the environment at the same time. In many parts of the world, demand for water is outpacing supply. For instance, both PepsiCo and the Coca-Cola Company have found a need to close bottling plants in India because of competition for water resources from farmers and urban centers. Increasingly, hydropower companies are finding that recurring drought conditions force them to temporarily close off generation for extended periods. There are now many green businesses providing products and services to assist large corporations to conserve their water supplies. Oil companies, mines, utilities, and beverage companies are just some of the industries already benefiting from this assistance.

Energy conservation often means engaging with simple ideas like insisting that employees turn off their computers at night, installing more energy-efficient lights, and adjusting the thermostat so that less heating is required in winter and less cooling in summer. But even these simple ideas can often save large corporations millions of dollars a year. There are now many green business opportunities to be found in drawing up energy conservation plans for companies and working out methods for them that use less energy.

Commons Management

A lot of the difficulties associated with conservation arise from the problem of commons management. Although the idea of "the commons" has in recent years extended beyond the natural environment to include cultural products like software, traditionally the commons have been defined as collectively owned natural assets like forests, the atmosphere, fisheries, and grazing land. The management and conservation of these commonly held assets has often been particularly difficult due to the variety of users and the level of demand for the resources held by them.

Much of the underlying logic of commons management difficulties was summed up in a landmark 1968 article by Garrett Hardin, titled "The Tragedy of the Commons." This article argued that multiple users, acting independently, and only guided by their own self-interest, will ultimately deplete a limited, shared resource despite the fact that this is against everyone's long-term interest. Hardin used a medieval grazing common as his prime example, where all the individual herders had a personal interest in increasing the size of their own herds. When one herder added a cow, although he benefited personally, the grazing land was slightly degraded as a result and all the herders suffered. But the herder with the extra cow benefited more than he lost, and this encouraged the other herders to follow his example. The end result was overgrazing and severely depleted resources.

To counter this problem, a number of commons management practices have been adopted in differing situations with varying results. One of the most common management practices is to remove the resource from the public, often in the form of a national park or nature reserve.

During the mid-20th century, protectionist and coercive conservation policies, known under the banner of "fortress conservation," became dominant. The core elements of fortress conservation policies consisted of establishing protected areas, excluding people and limiting or forbidding their rights for consumptive use, and using strict enforcement of these rules through a "fences and fines" approach. Often, this even included relocating communities from areas they had lived in for generations, creating many protected areas with adjacent "border communities" living in abject poverty, most notably again in (southern and eastern) Africa, but also in other parts of the world.

Due to processes of decolonization and a more emancipatory international climate in the 1960s and 1970s, conservation ideas started to change considerably. Most importantly, there was an increasing recognition of the enormous human costs involved in establishing and maintaining policies of fortress conservation. Hence, "community-based" approaches to natural resource management became the dominant paradigm globally. Community-based or socially inclusive approaches are known under a myriad of terms and acronyms, such as Community-Based Conservation (CBC) and Community Conservation (CC). Here, we will use Community-Based Natural Resource Management (CBNRM) to denote the paradigm in a broad sense.

According to much of the literature, community approaches to conservation must adhere to three basic principles: giving local people ownership over resources, allowing them to participate in land use decisions, and ensuring they benefit from conservation.

However, the degree to which people are allowed to participate in management of, have ownership over, and benefit from natural resources depends on a wide variety of social and political issues. It involves a redistribution of power and authority, often from more central state levels to local levels, which involves competing interests and strategic compromise. However, these processes of decentralization and devolution of power and/or authority are complex and problematic, and successful outcomes have turned out to be rare. Even one of the most widely acclaimed success stories in CBNRM—the Communal Areas Management Program for Indigenous Resources or CAMPFIRE in Zimbabwe, which in its conceptualization recognized that CBNRM is about issues of power, politics, and governance—largely failed to equally distribute the accrued benefits within the involved communities and fell short of making a positive impact on the majority of local people's livelihoods.

Critics argue that CBNRM is fundamentally flawed by not acknowledging that development and conservation are inherently contradictory and that their integration in policy actually makes both worse off. They (re)emphasize the value of top-down conservation and of protected areas as the last "safe havens" for biodiversity.

Legislation and Regulation

Community-based conservation is only one among a wide variety of hybrids of different conservation models. State-based legislation is an even more common way of managing common resources. When demand for a specific resource is particularly insistent, licenses might be issued to ration access to it, as with fishing licenses and water rights permits.

Standards may be set in terms of concentrations of wastes in people, animals, and plants exposed to the waste streams. These might be residue levels in the body or tissues, or might be some measure in the air or water where the communities are situated, breed, or feed. Standards are enforced by licensing a plant or product. The license sets out the standards that the plant or product has to meet. Performance is then monitored to see if the firm complies with the license conditions; if they do not, some sort of penalty is imposed. Often the monitoring is done by the firm itself, with occasional checks by the regulator.

Legislation and regulation has not been particularly successful at protecting common resources because those regulated have the resources to influence regulations, and because of the prevailing belief that such regulations diminish economic growth. Consequently, governments seldom have the political will to properly conserve the commons in the face of economic and industry pressures.

Environmental impact assessments are often required by government authorities so that the impacts on ecosystems can be assessed and mitigated before developments are allowed to proceed. However, because these assessments are usually commissioned or undertaken by project proponents, they tend to underplay any loss of conservation.

In this context, green businesses need to go beyond regulatory compliance if they wish to conserve ecosystems, resources, and shared commons, including water and air.

Neoliberal Conservation

Increasingly, under neoliberal policies, there is a tendency to solve the tragedy of the commons by simply bringing common ownership to an end, and privatizing the commons. The argument behind this type of privatization is that if the resource is moved into private hands, the individual owner will have more incentive to conserve it than if it were left in public ownership. This thinking has often proved to be unsound, because individual owners are likely to concentrate on extracting a single resource that is of particular value to them at the expense of the ecosystem services that might be lost in the process.

One of the main features of neoliberal conservation is that it tries to achieve conservation objectives through market means or principles of ecological economics. Basically, this comes down to likening the human–nature relationship to a marketplace and creating incentive structures that take conservation into account. A practical example is the currently popular model of "payments for environmental services," whereby the use of particular environmental services should be conserved through payment incentives.

Proponents of neoliberal conservation say that within a capitalist world, it makes sense to take the environment into the orbit of the market. Critics state that this move is akin to using the same tactics that have led us to the problem of environmental degradation in the first place as a remedy to tackle that same problem.

Global Conservation: Actors, Policies, and Places

A great number of actors are involved in the conservation of various components of bio-diversity and natural resources worldwide, ranging from states, local communities, and private actors, to intergovernmental organizations, nongovernmental organizations (NGOs), and conservation societies, to green businesses.

States are generally regarded as the main actors in conservation. They sign international treaties, develop and implement national conservation legislation, and fund most conservation activities. However, with the advent of CBNRM, some say that the role of the state has decreased in favor of local communities, nongovernmental, and private actors. Others argue that while it is true that the state has increasingly shared conservation authority with other actors, they remain one of the, if not the most important actor in conservation.

The conservation actors most known by the general public are arguably some of the major conservation NGOs, such as Conservation International (CI), the World Wildlife Fund, and the Nature Conservancy, to mention just a few. Other, lesser known but equally important conservation actors range from the United Nations Environment Programme and the International Union for the Conservation of Nature (IUCN) to the Global Environment Facility (GEF), an international funding facility that aims to help developing countries meet their goals under some of the main global conservation treaties.

One of the most well known and important of these events was the 1993 Convention on Biological Diversity (CBD), which aimed to conserve biological diversity, to use biological diversity in a sustainable fashion, and to share the benefits of biological diversity in a sustainable fashion. Again, one can see in these objectives the importance of the concept of use, and the idea that humans should be able to benefit from biological "resources." These elements are recurrent in many global conservation policies, other examples of which include the Convention on Wetlands of International Importance (in Ramsar, Iran, 1971) and the Convention on International Trade in Endangered Species of Wild Fauna and Flora (in Washington, D.C., 1973).

Business also plays a decisive role in conservation. They can function either as powerful actors seeking to undermine and water down regulatory activities of the state, or as socially responsible green businesses seeking to reduce their own resource use, wastes, and carbon footprint or to contribute to the conservation of ecosystems that are under threat.

See Also: Compliance; Ecosystem Services; Environmental Services; Resource Management.

Further Readings

Adams, W. M. *Against Extinction: The Story of Conservation*. London: Earthscan, 2004.

Hackle, J. D. "Community Conservation and the Future of Africa's Wildlife." *Conservation Biology*, 13/4:726–34 (1999).

Hardin, Garrett. "The Tragedy of the Commons." Science, 162:1243–48 (1968). http://dieoff .org/page95.htm (Accessed January 2010).

Heynen, N., S. Prudham, J. McCarthy and P. Robbins. *Neoliberal Environments: False Promises and Unnatural Consequences*. London: Routledge, 2007.

Hulme, D. and M. Murphree, eds. *African Wildlife and Livelihoods: The Promise and Performance of Community Conservation*. Oxford, UK: James Currey, 2001.

Hutton, J., W. M. Adams and J. C. Murombedzi. "Back to the Barriers? Changing Narratives in Biodiversity Conservation." *Forum for Development Studies*, 2:341–70 (2005).

Oates, J. F. *Myth and Reality in the Rain Forest: How Conservation Strategies Are Failing in West Africa*. Berkeley: University of California Press, 1999.

Sarkar, S. "Wilderness Preservation and Biodiversity Conservation—Keeping Divergent Goals Distinct." *BioScience*, 49/5:405–12 (1999).

Bram Büscher
Institute of Social Studies, the Netherlands
University of Johannesburg

Stacey Büscher
HIVOS, the Netherlands

COOPETITION

Coopetition occurs when competing firms cooperate in some aspects of their production and marketing in order to maximize their profits. Coopetition can be used as a green business strategy to increase the sustainability of competing businesses.

The term *coopetition*, created by combining the words *cooperation* and *competition*, was coined by Ray Noorda, founder of the networking software company Novell. It was popularized by Adam Brandenburger from Harvard Business School and Barry Nalebuff from Yale School of Management. Coopetition describes a fusion of these antithetical concepts, which, although it may seem paradoxical, already has a considerable history of business success.

Noteworthy examples of coopetition success involve arrangements between competing airlines in what are called "code sharing" agreements, which allow partner airlines to market seats on each others' flights, and arrangements between motor vehicle manufacturers to share some common components in competing automobile models. A good example of complementarities in coopetiton is computer hardware and software. Faster hardware prompts customers to upgrade to more powerful software, and vice versa. The conventional wisdom of "winner take all" philosophy is gradually giving way to a realization that, in networked globalized economies, businesses must both cooperate and compete in order to create maximum value in the marketplace.

Coopetition holds promising potential for green business. In a landmark article published in *Scientific American* in 1989, R. Frosch and N. Gallopoulos theorized that industrial systems could be made to function like natural ecosystems with the wastes of one manufacturer being used as a resource for another, thereby avoiding the costs of waste disposal for the one, while providing raw materials for the other. Apart from the economic advantages for partners in this kind of industrial ecology, there is the added advantage of less environmental costs due to less overall resource use and less waste disposal. For these arrangements to work properly, however, a philosophy of coopetition is often required to overcome the culture of suspicion inherent in traditional competitive business practices.

This idea of cooperation between complementary firms can be applied at every stage of the supply chain, and can be facilitated by co-location of manufacturing operations of complementary firms, that is the operation of one company on the premises of another company. This has the added environmental advantage of not having to transport materials long distances.

In other cases, it is often in the interests of competing companies that rely on the same resources, particularly natural resources, to cooperate to ensure those resources are managed sustainably. Competing companies may also cooperate in facilitating the collection of postconsumer waste products for reuse and recycling, which may be too expensive for individual companies. For example, in the United States, carpet and fiber manufacturers have agreed to increase the reuse and recycling of waste carpets.

Perhaps the most important aspect of coopetition is the potential it has to drive the greening of a company's network of stakeholders, customers, suppliers, and complementors. Under normal competitive business conditions, there is usually a prevailing imperative placed on suppliers to provide the cheapest product. This can act as a deterrent for the suppliers to improve the sustainability of their inputs if greening them is likely to raise costs. However, coopetition introduces an opportunity either for competing suppliers, or for the companies in receipt of the inputs, or both, to put aside competitive pressures and cooperate by agreeing on the need for greener inputs and outputs. When one company asks for more sustainable practices or products from suppliers, it can easily be ignored, but when several companies demand more sustainable practices from suppliers, suppliers have little choice but to change their practices.

Similarly, customers can be primarily driven by cost considerations, and when sustainable products are more expensive, they are often at a disadvantage. However, manufacturers can cooperate to ensure that competing products all reach a minimum level of sustainability so that sustainability is no longer accompanied by a price disadvantage. For example, Swiss retailers Coop and Remei partnered together to produce and market clothing made from organic cotton.

In the past, this kind of green shift has usually required government regulation to facilitate it, but in recent years, businesses have been learning to take advantage of industry self-regulation and, through coopetition agreements, are taking greening initiatives themselves. To this end, competing companies may cooperate by forming coalitions, such as the Electronic Industry Citizenship Coalition (EICC), which has put together a code of conduct which includes guidelines for corporate social responsibility, as well as tools for helping member companies to uphold the code. The advantage of this sort of cooperation is that corporate social responsibility often comes at a price, and if competing companies all agree to undertake certain environmental practices, it is less likely that a market disadvantage will be attached to them.

Coopetition can also encourage sustainable innovation. By collaborating in research partnerships and joint ventures, firms are able to combine financial resources, facilities, and expertise to create innovative technologies and solutions. Business networks, partnerships, and consortia have been a major factor in contributing to technological innovation in the United States. Such alliances increased markedly in the United States during the 1990s, while they declined in Europe and Japan.

The three big American automobile manufacturers have had a seven-year research collaboration that involves sharing resources while the technology is in the precompetitive stage. Outcomes of their partnership have included plug-in hybrid cars, including the Chevy Volt, and Ford's EcoBoost direct injection technology, aimed at using less petrol and reducing carbon dioxide emissions. The companies are collaborating because they claim that the resources of individual companies are not enough to make the technological breakthroughs necessary for a sustainable transport future.

The advantage of coopetition for sustainable innovation is particularly true when competing businesses partner to take advantage of the many government subsidies that are becoming available for new green investment. Parameters for receiving green investment subsidies often require innovations of one kind or another, and by pooling design expertise, partners can minimize the risk and expense of innovative green ventures.

See Also: Best Management Practices; Closed-Loop Supply Chain; Integrated Bottom Line; Responsible Sourcing; Value Chain.

Further Readings

Bonel, Elena and Elena Rocco. "Co-Opeting to Survive: Surviving Coopetition." *International Studies of Management & Organization,* 37/2:70–96 (2007).
Brandenburger, Adam M. and Barry Nalebuff. *Co-Opetition.* New York: Doubleday, 1996.
Eriksson, Erik. "Achieving Suitable Coopetition in Buyer-Supplier Relationships: The Case of AstraZeneca." *Journal of Business to Business Marketing,* 15/4:425–40 (2008).
Frosch, R. A. and N. E. Gallopoulos. "Strategies for Manufacturing." *Scientific American,* 261/3:144–52 (1989).
Luo, Xueming, Rebecca J. Slotegraaf and Xing Pan. "Cross-Functional 'Coopetition': The Simultaneous Role of Cooperation and Competition Within Firms." *Journal of Marketing,* 70/2:67–80 (2006).
Luo, Yadong. "A Coopetition Perspective of Global Competition." *Journal of World Business,* 42/2:129–40 (2007).
Padula, Giovanna and Giovanni Battista Dagnino. "Untangling the Rise of Co-opetition: The Intrusion of Competition in a Cooperative Game Structure." *International Studies of Management & Organization,* 37/2:32–52 (2007).
Ritala, P., P. Hurmelinna-Laukkanen and K. Blomqvist. "Tug of War in Innovation— Coopetitive Service Development." *International Journal of Services Technology and Management,* 12/3:255–70 (2009).
Ross, William T., Jr. and Diana C. Robertson. "Compound Relationships Between Firms." *Journal of Marketing,* 71/3:108–20 (2007).
Walley, Keith. "Coopetition: An Introduction to the Subject and an Agenda for Research." *Review of International Studies of Management & Organization,* 37/2:11–31 (2007).

Sharon Beder
University of Wollongong

CORE COMPETENCIES

Core competencies are fundamental ways in which a firm excels. In nature, a cheetah's core competency would be its ability to outrun its prey, an eagle's would be its sharp eyesight, and the core competency of a cactus would be its ability to survive for long periods without water. Green businesses that can make sustainability their core competency will have a competitive advantage and can expand their market share and profitability.

The core competencies of a business are the factors that it considers central to the way it functions: factors that can be leveraged to multiple products and markets, that are not easily imitated by competitors, and that benefit the consumer. The net effect is competitive advantage in the market. The core competency represents the company's added value compared to other companies in the field. Core competence is one of the most important concepts driving the globalized economy. It is the key idea behind the current wave of outsourcing, as corporations concentrate their efforts on things they do well, and outsource as much of the rest as they can.

There is a wide variety in the types of factors that can be considered core competencies. Core competencies can be proprietary methods or products, as in the case of Microsoft or Google. Toyota has built core competencies in its methods, including supply chain management, its assembly line, and a core competency in its products. It has become an industry leader in efficient mobility, including the production of hybrid cars, which are more environmentally friendly than petrol-only equivalents.

Strong brand identity can be a core competency, as in the case of Apple computers and the Coca-Cola Company. For green businesses, having their brands identified as "green" can comprise a core competency. Planet Ark is an Australian brand that has sought to achieve this. The company now considers its brand to be its major intellectual asset. It protects its brand by offering only products that meet specific environmental standards and promotes it by association with Plant Ark Environmental Foundation's nonprofit activities.

Other core competencies can include unique corporate structures (such as the franchise system of many fast food chains), or internal programs that yield more highly skilled or productive labor (training, benefits to attract high-quality hires, on-site child care, flextime, incentives for innovation, or further training). Green businesses could develop core competencies among their staff in terms of sustainable practices through development of core competencies in employee training programs, hiring practices, and incentive schemes.

Identifying core competencies requires identifying what is called "core business," an idealized model of the business's primary activity. Businesses tend to be engaged in multiple business activities that may or may not interrelate. It is the core business, the "true" business, that is relevant to a firm's core competencies. Firms with environmentally damaging core businesses, such as coal-fired power stations, are unlikely to develop core competencies related to sustainability without first changing its core business.

The company's core competency may not at first glance be what it regards as its core business. A retail chain's core competency may be not the selling or selection of the goods, but the transportation network responsible for bringing them to market. A restaurant chain's core competency may be its management training program, as is the case with McDonald's. For a green business, a core competency could be the system of reverse logistics with which it operates a closed-loop supply chain, ensuring the recapture of used products for reintegration in the manufacturing process. Another core competency might be clean technology production methods.

The flipside of core competency is core rigidity. A company's core rigidities are caused by over-relying on their advantages and core competencies. When one is especially skilled in one area, one may try to solve every problem with that skill, even when it amounts to turning left three times in order to make a right. This leads to inefficiency, resource over-consumption, and blind spots—not to mention that it leaves less skilled areas unskilled, instead of taking the opportunity to develop them.

The real problem occurs when a company's core competencies cease to be core competencies—when its advantages are made obsolete by market changes, as competitors become stronger, or as a technological shift makes an old way of doing things undesirable. Those circumstances are when core rigidity can be the ruin of a company. For example, some car companies were slow to change their core competencies from the production of larger, inefficient cars and gas-intensive SUVs, and were stymied as market demand plummeted as a result of rising gas prices, followed by an economic downturn. As nations prepare to tackle the challenge of climate change, companies that cling to core competencies dependent on heavy use of fossil fuels will similarly find their markets diminishing.

Greening Requires New Core Competencies

Some businesses mistakenly think that they can "green" themselves without developing any new core competencies in the area of sustainability. One example is the Ford Motor Company, which, in 2004, won the Green Roofs for Healthy Cities Award of Excellence in the Extensive Industrial Commercial category for redevelopment of the River Rouge truck factory near Dearborn, Michigan. The "green" renovation involved a new assembly plant with a living roof, covered with a succulent groundcover and other plants. The roof was designed to reduce stormwater runoff and absorb carbon dioxide, so that oxygen is emitted and greenhouse gases are reduced. Meanwhile, inside the factory, Ford followed its traditional core competency and continued to manufacture gas-guzzling trucks. Together with the other big two American motor companies, Ford experienced financial difficulties during the global financial crisis.

Recently, apparently learning from its past mistakes, Ford is emulating a wiser competitor, Toyota, and announced that it will position greener motor vehicles among its core competencies. In January 2010, Ford announced it would invest an additional $450 million in Michigan for hybrid and plug-in vehicle production, and that it would make battery system design and development an in-house core competency. Ford has found that making sustainability a core competency can be far more profitable and market building than merely greening a factory roof.

Sustainability as Core Competency

Periodically, the market will change because of external influences, and companies that are able to build a core competency that suits the new shape of the market are more likely to prosper. That was the case when motion pictures added sound, when television expanded the programming that had been developed on radio, and when the Internet changed the way consumers shop. Such changes were technological, but there are other factors that can alter the landscape, such as government intervention. For example, corn subsidies changed the way some agricultural businesses worked, while the Glass-Steagall Act of 1933 forced banks to decide to be either a commercial lending bank or an investment bank.

Changes to environmental regulations, changes resulting from declining environmental amenities, such as water shortages and increasing energy costs, as well as changing environmental concerns in the community can similarly force companies to build new core competencies. The hospitality industry has already encouraged hotels to introduce new positions for employees, such as director of sustainability, or vice president of responsible business, while drawing attention to the environmental impact of water consumption in the hotel business, and the ways that such consumption can be reduced by responsible behavior both on the part of the guests, and improved water management on the part of the hotel. Among the benefits of running a more sustainable hotel are improving public image, strengthening employee morale by emphasizing the positive acts implemented by the company, and the economic gains of reducing resource consumption.

Unilever has found that it must develop core competencies to address problems in developing countries associated with poverty, water scarcity, and the effects of climate change, if it is to stay competitive in coming decades. Roughly 40 percent of the company's sales and the larger part of its growth is now taking place in developing nations. Unilever food products amount to roughly 10 percent of the world's tea harvest and 30 percent of spinach crops. It is also among the world's largest purchasers of fish. As global environmental problems grow deeper and regulations to address these problems grow tighter, Unilever has plans to make green technologies a core competency so that its leadership in packaged foods, soaps, and other goods will not be endangered.

In the 21st century, social pressures, the possibility of national or international regulations, and the ongoing environmental crisis have made green business a practical necessity; however, most companies are not equipped to seamlessly shift into the new paradigm.

In the case of the Internet, being Internet savvy was initially a core competency for those businesses that possessed the skill set before others in their field—both the existing businesses and the new businesses that entered the industry with that competency as part of their brand identity, like Amazon in the bookstore industry or eBay in the auction/consignment/secondhand industry. Eventually, however, Internet competencies became a universal requirement in business rather than a core competency that distinguished some firms from others.

Sustainability offers a similar opportunity. In time, sustainability will become a universal requirement in business. In the meantime, forward-thinking firms can develop it as a core competency and profit from the advantage it gives. Though a sustainable business is not a new concept, a business that is able to operate sustainably not only has added value from the consumer perspective, but it also has the potential to operate more efficiently and more cheaply than nonsustainable companies in the same field.

See Also: Best Management Practices; Closed-Loop Supply Chain; Value Chain.

Further Readings

Cloud, Jaimie. "An Introduction to Education for Sustainability: Core Content, Competencies and Habits of Mind." Paper presented at the Annual Meeting of the North American Association for Environmental Education, October 24, 2005.

Drejer, Anders. *Strategic Management and Core Competencies: Theory and Application.* Santa Barbara, CA: Praeger, 2002.

Hamel, Gary and C. K. Prahalad. *Competing for the Future.* Cambridge, MA: Harvard Business Press, 1996.

Newport, Dave, et al. "Core Competencies for Sustainability Professionals." *Sustainability: The Journal of Record,* 1/4 (August 2008).

Stewart, Richard B. *Reconstructing Climate Policy: Beyond Kyoto.* Washington, D.C.: AEI Press, 2003.

Bill Kte'pi
Independent Scholar

CORPORATE SOCIAL RESPONSIBILITY

"Social responsibility [is the] responsibility of an organization for the impacts of its decisions and activities on society and the environment through transparent and ethical behavior that is consistent with sustainable development and the welfare of society; takes into account the expectations of stakeholders; is in compliance with applicable law and consistent with international norms of behavior; and is integrated throughout the organization." —Working definition, ISO 26000 Working Group on Social Responsibility, Sydney, February 2007.

Corporate social responsibility (CSR) is an evolving concept that does not currently have a universally accepted definition. It frequently overlaps with similar terms, such as corporate sustainability, corporate sustainable development, and corporate citizenship. Whether or not or to what extent businesses must implement CSR has been a heated topic of discussion through the years, but was particularly vociferous during the debate about conducting business in South Africa during the apartheid era. At the time, many corporate executives maintained that their responsibility extended only to their shareholders.

Milton Friedman argued their case in his 1970 *New York Times Magazine* article, "The Social Responsibility of Business Is to Increase Its Profits." In Friedman's Libertarian point of view, a business, since it was not a person, had no responsibility other than to fulfill the purpose for which it was created, that is, to maximize the return on shareholder investment. He said that executives and employees owed allegiance to the firms for which they worked and must keep the profit-maximization goal firmly in mind. He equated efforts to improve society through the conduct of business with "pure and unadulterated socialism."

Although there are still those who hold Friedman's view, most business executives, including many who self-identify as Libertarians, now acknowledge that companies do indeed have a social responsibility.

This page from an October 2006 British Petroleum brochure touts its commitment to "sustained and valued relationships" during its exploration and extraction of liquid natural gas.

Source: BP.com

Forty Years in the Making

In the 40 years since the publication of Friedman's article, much has happened to change businesses' stance on CSR from defiant opposition to acceptance, and in some cases advocacy, but two factors were key in this transformation. The most important was the advent of the information age, in particular the development of and open access to the Internet and social media (e.g., mobile phones, smart phones, text messaging, and Twitter). The second was the advanced understanding of global climate change, which altered the way most people think about natural resources and social responsibility. There is a growing public backlash against accepting the destruction of commonly held resources to meet the selfish ends of a few.

At the time when Friedman made his case that business was not a person and therefore not a moral entity subject to social responsibility, the flow of information was tightly controlled by corporations. The general public only had access to news and events through traditional media (e.g., radio, television, and print). Even stockholders had limited access to information about the corporations in which they were invested, having to rely on annual reports. In such an environment, corporate actions were held up to very little public scrutiny and coordinated consumer actions were rare. As the flow of information loosened, following the introduction of home computers in the 1980s and the expansion of the Internet in the 1990s, it not only became harder for corporations to keep their operations private, but it allowed consumers across vast distances to exchange information and act upon it.

Although scientists have been studying global climate change in earnest since the 1950s, many of the most important advancements in our understanding of the process were occurring simultaneous to the development of the information age. The expansion of the Internet was instrumental in moving the issue from the pages of scientific journals into wider public discourse. As scientists connected the dots between local actions and global consequences, people worldwide saw the implications of their investment and purchasing decisions. By the end of the 20th century, the public was organizing international demonstrations and boycotts.

The response within the business community has evolved over time. After businesses' rejection of the notion of CSR in the 1970s, incidents such as the devastating pollution of Love Canal in Niagara Falls, New York, by the Hooker Chemical Company, and the toxic gas release by Union Carbide in Bhopal, India, led to a decade-long discussion about business ethics. While many in the business community held to the belief, and still do, that a corporation's responsibility to the wider community begins and ends at the judicial bar, tragedies such as the Love Canal and Bhopal incidents convinced the public that merely abiding by the law is not always enough.

The public demand for corporate accountability grew stronger through the 1990s and continued into the 2000s, as the practices of multinational corporations were more frequently reported on, but many of these actions were initially limited to groups with focused interests (e.g., child labor practices and animal testing). The widespread understanding of the threats posed by global climate change raised the hue and cry for corporate accountability to a higher pitch, and businesses were forced to take notice.

Who Is a Stakeholder?

In Friedman's world, the only stakeholders a business should worry about were the owners—which in a corporate setting, are shareholders—and the only responsibility of the company was to maximize the return on the owners' investment. But by the 1980s, faced with an informed public, businesses were increasingly being reminded of the old adage "The customer is always right." Without customers, there would be no profit to maximize, and customers now receive their information from many sources. Accordingly, many businesses have broadened their definition of stakeholders to include not just stockholders, but customers, employees, and suppliers as well. For national and multinational corporations, the stakeholder list also includes governmental agencies.

With an increasing focus on crossborder trade, multinational corporations faced increased demand to adopt CSR policies detailing their approaches to human resource management practices, environmental protection, health and safety, and sustainable economical development in their foreign operations. As the urgency to respond to the challenge of global climate change has become apparent, many are convinced the scope of CSR should be expanded to include the international community.

CSR as a Business Strategy

Taken at face value, when a company establishes a CSR policy, it is simply making a social contract with its stakeholders, however they are defined. Typically, a CSR policy will restate the company's commitment to abide by accepted corporate governance practices, ethical standards, and the laws governing the places where it operates. It may include statements on community involvement, and investment and intentions regarding corporate philanthropy and employee volunteering. It may also establish the accountability, transparency, and reporting standards to which it will adhere. Frequently, a CSR policy will voice support for human rights and respect for diverse cultures and disadvantaged peoples.

Companies frequently have other motives for investing in a CSR policy. The action is often taken primarily in response to concerns of existing stakeholders. But even in the absence of these, businesses often see the move as a way to attract new investors or customers. If their business is the first in its industry or region to adopt and publicize their CSR policy, it can distinguish them from the competitive field. For some companies, a well-publicized CSR policy can be used to justify charging a higher price for their product.

There are numerous examples of reverse situations where the lack of a CSR policy, or even worse, being caught in the act of breaking such a contract, has had serious repercussions for manufacturers or retailers. Some corporations are slow learners. It took multiple incidents for Nike to finally take consumer opposition to child labor seriously. However, most companies are aware that mass consumer action can be organized in a matter of moments via social networking platforms such as Twitter.

Social Responsibility and Investors

Market indices have been developed to provide individual and institutional investors, financial advisers, mutual and equity fund managers, venture capitalists, and other similar organizations with information pertaining a wide range of corporate characteristics including governance, human resource management, health and safety, environmental protection, and community development. The intent of these indices is to facilitate socially responsible investment (SRI), such as investing in corporations that follow the principles of CSR.

Initially, investor approaches to SRI were simple and straightforward—they simply discarded holdings in sectors they deemed undesirable. But as SRI developed, the facets of a firm's governance, social, and environmental record were taken into greater account. As stated by Ronald Brown, president of Chilean Fruit Exporters Association (at Global Reporting Initiative [GRI] G3 meeting in October 2006), "We believe in CSR because it is a proposition aligned with our values, but also because it makes business sense. Our commercial partners expect from us sound environmental and social practices. We get and understand the message and are actively promoting CSR among associates. We want to be recognized as a responsible industry, adding value to our products." Organizations like Chilean Fruit Exporters that perform well with regard to CSR can build their reputation, while those that perform poorly can damage their brand, as well as their company value, when exposed. Reputation, or brand equity, is founded on values such as trust, credibility, reliability, quality, consistency, and accountability, and extends throughout the supply chain.

Implementing CSR

Following CSR policies offers distinct advantages to organizations. It enables corporations not only to recruit and develop personnel, but it also helps them to retain staff, particularly when they implement "family-friendly" policies. Enhanced operational efficiencies and cost savings can also be achieved by turning waste into a revenue-generating stream, or by using energy more efficiently.

A firm's first step toward implementing CSR is to get approval of top management including the board of directors and chief executive officer. A logical first step is to gather and examine pertinent information about the organization's products, services, decision-making processes, and activities to determine where it stands with respect to implementation of CSR. Frequently, an organization can complement its present system without much incremental investment by improving existing employee educational advancement programs or by adopting other environmental or occupational health programs.

After doing the assessment, the next step is to develop a strategy. CSR assessment generates a base of information that can be used to develop a plan for moving ahead, allowing the firm to be successful and fulfilling stakeholders' expectations. Such a strategy helps to ensure that a firm builds, maintains, and continually strengthens its identity, market, and relationships. This is best developed with the support of not only the senior management, but of all employees. It is advisable to research current, state-of-the-art CSR instruments, as traditional approaches can hinder achieving optimum performance. Brainstorming sessions with senior managers, employees, key business partners, and others can clarify the firm's core business objectives, methods, and core competencies.

The next logical step, after having set goals, is to implement those commitments. Failing to meet commitments without any satisfactory explanation can do irreparable damage to the organization's image. Successful implementation can best be achieved by developing an integrated decision-making structure.

Reporting and verification are the final steps of implementing CSR. Verification, through independent third-party on-site inspections, gives stakeholders reliable evidence that the corporation is upholding its CSR commitments.

Putting Words Into Action

Thousands of companies worldwide have adopted CSR policies, but not all have translated those policies into actions extending beyond the normal range of their business activities. Those that have done so have typically partnered with nongovernmental organizations (NGOs), whether to work within their own community or in the international arena. Although some criticize these efforts as self-serving, it is difficult to argue with the positive impacts they have made.

The Swedish furniture company IKEA is one such firm. IKEA has partnered with UNICEF in order to act on a commitment contained in the company's Corporate Social

Initiative to actively work to eradicate child labor. IKEA has been working with UNICEF since 2000 on a suite of programs designed to improve the prospects of children in south Asia, particularly in the carpet-weaving regions of India where child labor abuses have been common. UNICEF reports that over the last decade, IKEA was their single largest corporate contributor, having donated $180 million over the last decade.

The Millennium Villages (MV) project offers another example of corporate social initiatives in action. MV is a public–private joint venture involving United Nations (UN) agencies, private foundations, universities, international NGOs and aid agencies, and corporations. The MV project seeks to bring lasting change to poverty-stricken African communities by building the capacity within these communities to develop and sustain healthy local economies. In remote areas, such as those where MV projects are located, the single-most effective tool in such efforts has turned out to be the mobile phone. This is mainly due to the access phones provide to both markets and market information, but it is also due to the fact that technology can be easily, quickly, and economically introduced, compared to scarce local phone lines. The telecommunications firms Ericsson, Motorola, and Sony Ericsson have donated substantial amounts of equipment, training, and money to the project.

Measuring Words Against Deeds

Natural Logic, a consultancy focused on helping companies devise strategies for both corporate and environmental sustainability, looked at CSR reports issued by companies in the technology and telecommunications sector, for example, to determine whether such reports were truly revealing. They found that despite efforts of the GRI, few companies followed a standard reporting format or offered any meaningful metrics by which to measure progress against goals. They also discovered that companies that compiled the most detailed reports were seldom those that claimed the most progress. But despite the deficiencies of the reports and the difficulty of comparing reports of businesses in the same industry, let alone those in different sectors, Natural Logic concluded that the reports were still valuable tools for tracking what is being done and in defining the leading edge.

What motivates a company to adopt a CSR policy is less important than what is accomplished as a result. Optimistically, perhaps, Natural Logic concluded from its survey that the mere act of reporting on sustainability efforts opens up the possibility of improved efforts, as it requires the organization to self-assess. While that may be arguable, the fact that it is in an organization's best interest to formulate and act on CSR seems clear.

See Also: Dow Jones Sustainability Index; Natural Capital; Socially Responsible Investing; Social Return on Investment.

Further Readings

Friedman, Milton. "The Social Responsibility of Business Is to Increase Its Profits." *New York Times Magazine* (September 13, 1970). http://www.colorado.edu/studentgroups/libertarians/issues/friedman-soc-resp-business.html (Accessed January 2010).
Friend, Gil. "What Do Social Responsibility Reports Really Tell Us?" *World Changing*. http://www.worldchanging.com/archives/002369.html (Accessed January 2010).

Hohnen, Paul. "Corporate Social Responsibility: An Implementation Guide for Business,"
    Jason Potts, ed. International Institute for Sustainable Development. http://www.iisd.org/
    pdf/2007/csr_guide.pdf (Accessed November 2009).
Social Investment Forum Foundation. "2005 Report on Socially Responsible Investing Trends
    in the United States." http://www.sbnow.org/doc/SRITrendsReport2005PPPresentation.pdf
    (Accessed January 2010).
SustainAbility Tomorrow's Value. "The Changing Landscape of Liability." http://www
    .sustainability.com/insight/liability-article.asp?id=180 (Accessed November 2009).
United Nations Environment Programme Finance Initiative. "A World of Risk, a World of
    Opportunity?" http://www.unepfi.org/events/2005/roundtable/press (Accessed November
    2009).

Velma I. Grover
United Nations University–Institute
for Water, Environment and Health (UNU–IWEH)

COST-BENEFIT ANALYSIS

Cost-benefit analysis (CBA) is a tool for comparing the costs of an action with its benefits in order to aid decisions and choices between alternative courses of action. CBA is useful to green businesses because it enables them to integrate environmental and economic goals in their operations and ensure that environmental and social costs are taken into consideration in their purchasing or investment decisions.

In order to weigh costs against benefits, CBA usually attempts to put a monetary value on both costs and benefits so that they are expressed in the same units and all the benefits can be added up and compared with the sum total of costs. Costs and benefits include those in the market, those outside the market, long-term impacts of varying probability, and externalities—costs and benefits external to the parties involved in producing, marketing, and consuming a product. While the simplest form of a CBA may simply ask "is the cost of these energy conservation measures worth the savings on our electricity bill," more complicated CBAs look at situations that have multiple costs and benefits, such as the decision of a company to become a green business.

CBAs are especially associated with large public or private sector endeavors, where the costs and benefits are numerous and may not all be apparent without significant investigation. Indeed, CBAs are now a formal requirement of many large-scale projects undertaken by private enterprises, such as those in the mining sector and the building industry. The costs of a road project, for example, would include the cost of labor and materials used in construction, as well as other costs such as the loss of parkland and homes to make way for roads, and the resulting pollution, disruption to neighborhoods, or the loss of peace and quiet. The benefits of such a project might include time saved to motorists, increased predictability of journey times, and increased accessibility of a particular location.

Nevertheless, even small green businesses can utilize CBA to analyze the various costs and benefits of major decisions. For example, the decision to install waste recovery technologies would not only include the costs of the technologies and the savings in disposal costs, but also the environmental benefits of less pollution in the local waterway, the value of resources recovered, the gain in reputational value, and the lower risks of noncompliance with environmental regulations.

Valuing Environmental Costs and Benefits

Obviously, some costs and benefits are not easy to put into monetary terms. But proponents of CBA see it as helping to make the decision-making process more objective and rational. They argue that it is rational to choose a course of action in which the gains outweigh the losses and that, by putting the gains and losses in numeric terms, it is easier to be objective, consistent, and rational in the assessment.

Economists argue that whenever a decision is made, people weigh the pros and cons of that decision, but they often do so unconsciously or intuitively. By undertaking a formal CBA, the values they are attaching to the costs and benefits are made explicit and are recorded for everyone to see. This means that they have to think about those values in a more systematic and reasoned way.

The quantification of environmental costs and benefits for a CBA can be difficult. Direct costs and benefits are the easiest to estimate. These might include estimating the value of production forgone because of environmental damage, or the value of earnings lost through health problems associated with air and water pollution. However, direct monetary costs tend to underestimate the real costs and benefits provided by the environment. For example, improved health resulting from a cleaner and safer environment is worth more than just the medical bills saved. A clean beach is worth more than just the value of having healthier beachgoers.

There is obviously debate about how to quantify environmental costs and benefits. When considering the environmental impact of a course of action, what is the value of preserving wilderness—not just forested areas that play an important role in the fight to control global climate change, but desert areas, prairies, grasslands, and so forth? What is the value of preserving the national parks, or leaving mountaintops undeveloped, or preventing strip-mining in coal country? Getting even the strongest advocates to agree on a quantified value is a difficult proposition.

For most economists, however, the environment can be priced because all these values can be translated into the aggregated preferences of individuals using contingent valuation. Contingent valuation involves the use of surveys in which people are asked how much they are willing to pay to preserve or improve the environment (willingness to pay), or how much monetary compensation a person is willing to accept for loss of environmental amenity (willingness to sell). However, people may inflate or deflate the amounts they are willing to pay or accept. For example, surveys found that U.S. households were willing to spend $100 each to prevent another spill like the *Exxon Valdez*, but when they were asked how much money they would want before they would allow another spill to happen, not only were the sums much higher, but many people said they would not allow it to happen no matter how much they were paid. Contingent valuation might, however, be far more accurate when used to value reputations for green businesses.

Another method of quantifying environmental costs and benefits is hedonic pricing. This method assumes that the value of environmental assets can be found by considering the prices of the closest market substitutes. For example, a lake that is used for fishing, boating, and swimming might be valued by calculating what people spend on private fishing, boating, and swimming facilities. Another market substitute commonly used is property values. The idea is that houses in a polluted area will be worth less than houses in a nonpolluted area, and part of the difference in house prices will reflect the value the market puts on clean air or, alternatively, the cost of pollution.

Proxies are also used in contingent valuation to avoid directly asking people how much the environment or their health is worth. For example, economists in the U.S. state of North Carolina, wishing to put a monetary value on a chronic case of bronchitis, thought that if they asked people directly what they would pay to avoid getting chronic bronchitis, they would get unrealistically high amounts. They instead asked respondents if they would prefer to live in a more expensive area where the risk of getting bronchitis was lower. They were told what the effects of bronchitis were. The surveyors altered the cost of living and the risk of bronchitis until the shopper being questioned would be equally happy living where they were or moving to the new location with the lower risk of bronchitis. They calculated that a case of chronic bronchitis was worth $883,000 to those surveyed.

Opportunity Costs

The costs investigated by CBA include opportunity costs—the costs of lost options. Because the various options available in a decision-making process are often mutually exclusive—where to build a factory, or whether or not to offer a "green" version of a detergent—picking an option eliminates the other options as possibilities. You can use raw materials to make one thing or another; you can allocate funds to one end or to another. The use of an asset makes it unavailable for other uses, either temporarily or permanently. The impact of eliminating options is an opportunity cost.

Opportunity costs can be used to put a value on an area of the environment. For example, the value of preserving a wetland may be estimated by working out what the land would be worth if it were used for agriculture or housing. For each alternative activity, the economist works out what benefits would have been gained that could not be gained in any other way, and then subtracts the costs that would be involved in getting these benefits. So, for the housing alternative, the cost of building the houses and providing services for them would be subtracted from the value of the houses. And if those same houses could just as easily be built somewhere else, the opportunity cost would only consider the additional benefits from building them in the area under consideration. The highest amount of net benefits (after subtracting costs) that one can get from any alternative course of action that has been forgone is the opportunity cost of preserving that area. This indicates the minimum value placed on the area, since the decision to preserve it has meant that those making the decision were willing to forgo at least those benefits, and maybe more.

This method can be used before a decision is made, so that decision makers or the public can decide whether they believe the area is indeed worth what has been worked out as the opportunity cost. If they decide not to preserve the area, environmental losses can be worked out in terms of the amount it would take to restore the environment to its original state after development has occurred—for example, after mining or logging. Environmentalists do not believe some areas can be restored in this way, and therefore would reject this as being a full measure of the environmental loss.

Opportunity cost can only be a partial measure of environmental value. The value of the area for housing may have no relationship whatsoever to the actual ecological or aesthetic or spiritual value of the area it will be destroying. For example, a wetland might be providing a breeding ground for fish and other aquatic organisms, as well as performing a cleansing function, filtering out pollutants that flow through the area.

CBAs and Moral Considerations

CBAs have contributed to the popular imagination's conception of the "hard numbers" side of business and governance as cold blooded, because of such ideas as a corporation electing to commit an environmentally damaging action because the cost of possible cleanup is lower than the profit rewards of the damage; or of corporations risking negligence lawsuits when their collective cost is less than the cost of preventing them (through recall, higher safety standards, or other expenses).

There is at least one well-known example: when design flaws were discovered that compromised the Ford Pinto's safety in the event of a rear-end collision, Ford elected not to recall it, estimating the cost of wrongful death lawsuits to be less than $50 million, and the cost of the recall nearly three times that. The impact in negative publicity after a *Mother Jones* exposé, however, was significant—though since the company enjoyed healthy sales in the following decades, perhaps not so significant as to render their decision impractical. Ford ended up paying less than $10 million in damages (including both compensatory and punitive) after a 1981 Court of Appeals decision, and 10 years later introduced the Ford Explorer, the vehicle that defined the modern SUV market.

Some of that perception of cold-bloodedness comes from the mathematical necessity of assigning a monetary value to costs and benefits that are not monetary in nature. This leads to such seemingly terrible things as insurance companies' assigning a dollar value to a human life (as well as to various permanent injuries), corporations' and governments' assigning dollar values to environmental damage like species endangerment or accelerated climate change, and so forth. The goal here is not to perceive nonfinancial impacts in financial terms, in order to buy them off; but there must be some common denominator to provide a means of comparison.

One of the areas where CBAs are challenged is in the way they aggregate costs and benefits. Are the costs borne disproportionately by one group? Are the benefits enjoyed disproportionately by another group? This can be cause for concern, and it is frequently alleged that many policies designed to be socially optimal in fact disproportionately distribute their benefits among the middle and upper income brackets, while exacting an unfair cost on lower income brackets.

Furthermore, benefits expected to outweigh costs may exist principally in the future while the costs are paid in the present. In most CBAs, future costs and benefits are discounted (reduced) because it is assumed that they are not worth as much to people as present costs and benefits. In terms of environmental costs, the higher the discount rate that is used, the greater is the bias toward the present and against the future.

CBA has historically not considered sustainability as a benefit in and of itself. Thus, all other things being equal, a CBA will not pick a sustainable development project over another project, in contradiction of the general goals of environmental policy.

Predicting Future Costs and Benefits

The decisions informed by CBAs often involve some educated guesswork, and some dependence on probable or plausible outcomes, rather than fully guaranteed ones. Some outcomes are more easily estimated than others. For instance, a state considering a container deposit law has access to a great deal of relevant data from the states that have already adopted such laws and can construct various models to predict the effect—or a range of effects—of such legislation on the state's economy.

Sometimes environmental costs and benefits only become clear in hindsight. Certainly, the costs of industry's dependence on fossil fuels, of deprioritizing automobile fuel efficiency once the oil price shocks of the 1970s had receded from memory, and of the post–World War II agricultural industry's dependence on cheap energy have become clearer in the 21st century than when those choices were made.

Future environmental consequences are very often hidden or unknown; it took generations during which our understanding of environmental impact lagged behind our ability to create that impact, and generations of new discoveries about carcinogens, toxicity, climate change, and the balance of the ecosystem, before environmental cautiousness was fully embraced.

In some cases, it is impossible to assign costs to environmental consequences whose probability is unknown. Where these consequences are likely to be serious and irreversible, it is more appropriate to use the precautionary principle than try to include such costs in a CBA. In other words, it is appropriate to avoid the possibility of these environmental consequences. In this case, CBA can be used to compare options for doing this.

Environmental costs may even become benefits in the changing context of depletion/ pollution and increased public awareness. For example, a business may regard the waste products of a manufacturing process to be a cost; however, a green business may be able to turn those wastes into a benefit through recycling and reuse. Similarly, a waste product may become an asset because of resource depletion that increases the scarcity of components of the waste.

One of the best ways for a green business to develop accurate CBAs is by continually revisiting their previous CBAs and evaluating how closely their predictions and assumptions matched up to what really happened. Some studies have found that transportation infrastructure CBAs, on average, underestimate the costs of construction, for instance, and have a large margin of error in predicting traffic volume and passenger counts. Theories vary for whether there are generalized errors in judgment leading to these specific errors in projections, though it is generally accepted that relying too heavily on past performance as a predictor of future performance will lead to cost underestimates. Some researchers believe there may be a psychological element leading to cost underestimates in CBAs: because a CBA is rarely going to be done except when there is some reason to desire the action the CBA analyzes, there may be a tendency to underestimate negatives and overestimate benefits because of unconscious bias and the desire to move forward. In any event, investigating the weaknesses—and margins of error—of previous CBAs of similar situations can help to minimize these defects.

Role of CBA in Environmental Policy

CBA can be used to assess natural resource use decisions, such as the rate of exploitation of scarce natural resources and the management of wilderness areas. It is also commonly used in the United States to assess the merit of government regulations and policies. During the late 1990s, more international bodies like the European Union introduced legal provisions requiring CBAs of environmental policy.

CBA has been an increasing part of U.S. federal environmental policy since the 1970s, when the apparatus of environmental policy in the form of bodies like the U.S. Environmental Protection Agency (EPA) was first established. When the EPA decided to phase all the lead out of gasoline in the 1980s, it justified this decision on the basis of CBA that calculated medical costs from lead poisoning as well as the costs of brain damage to children, including remedial education and lower future income. It valued the loss of each IQ point to children exposed to lead at $8,346.

Nowadays, the U.S. Congressional Budget Office (CBO) regularly prepares information not only about expected spending, but the financial impact of implementing bills proposed before Congress. For instance, the CBO prepared conservative estimates of the effect on gross national product and employment of the various measures proposed in 2008 and 2009 to deal with the global economic crisis, in addition to its statement on the cost of those measures.

Environmental regulations are designed to ensure that companies are forced to pay for environmental impacts, either by reducing them, paying a charge or fine, or suffering the consequences of noncompliance. Cap-and-trade or emissions trading programs have been designed to add to the cost of businesses that are heavy polluters, and therefore to alter the balance between costs and benefits. However, the price, charge, or fine seldom represents the true environmental cost of the environmental impact, and a green business has to decide whether to include the actual costs to themselves of compliance or noncompliance in any CBA they undertake, or rather to estimate the broader cost to the community of noncompliance and include the benefit to the community of compliance.

Related Analytical Techniques

Cost-effectiveness analysis (CEA) is a decision-making tool that tends to be used in cases where CBA techniques are undesirable or go into unnecessary depth, particularly when the choices analyzed do not amount to "do we do this or not," but "how shall we do this?" For instance, it is used to analyze the various methods by which a legally required action may be enacted, or by which an existing standard of service can be maintained. It is broadly similar to CBA, in that it compares the results of multiple courses of action, but it is specifically concerned with accomplishing a specific task and finding the cheapest method of accomplishing that task.

Social return on investment analysis (SROI) is a tool that emphasizes the nonfinancial impacts of financial decisions. Just as CBAs necessarily assign a monetary value to nonmonetary things, such as environmental damage and human injury, SROI assigns such values to the nonmonetary gains of actions. The goal is to encourage investment in projects that are not only economically profitable, but also socially responsible and beneficial, and to redefine the value of an investment in order to consider and reward social gains. SROI is much talked about, but has had little impact on the investment market, apart from a small sector of socially responsible hedge funds.

In conclusion, CBA is a useful tool for comparing options; however, it should not be the only criterion for making a decision because it is does not adequately capture environmental values, moral considerations, or equity issues, nor can it be a substitute for the precautionary principle. Nevertheless, CBA does allow green businesses to include environmental considerations when they are considering the economic advantages of one option over another, or when they are making a decision about whether to go ahead with an investment.

See Also: Ecological Footprint; Environmental Accounting; Externalities; Industrial Ecology.

Further Readings

Fuguitt, Diana and Shanton J. Wilcox. *Cost-Benefit Analysis for Public Sector Decision Makers.* Westport, CT: Quorum Books, 1999.

Graves, Philip E. *Environmental Economics: A Critique of Benefit-Cost Analysis.* New York: Rowman & Littlefield, 2007.

Johansson, Per-Olov. *Cost-Benefit Analysis of Environmental Change.* Cambridge, UK: Cambridge University Press, 1993.

Pearce, David William. *Economics and Environment: Essays on Ecological Economics and Sustainable Development.* Williston, VT: Edward Elgar, 2000.

Pearce, David William, Giles Atkinson and Susana Mourato. *Cost-Benefit Analysis and the Environment: Recent Developments.* Paris: Organisation for Economic Co-operation and Development (OECD) Publishing, 2006.

Wasike, Wilson S. K., Mike Dobson, Ilyas Baker and Kevin B. Briggs. "Cost-Benefit Analysis and the Environment." *The Environmentalist*, 15/1 (March 1995).

Bill Kte'pi
Independent Scholar

CRADLE-TO-CRADLE

Cradle-to-cradle (C2C) is based on the principle that the whole life cycle of products and services needs to be redesigned so that every waste is a raw material for another round of production, service, or use. It aims to create systems that are both ecoefficient and waste-free, and can be been applied to industrial designs and manufacturing as well as urban environments, buildings, economics, and social systems.

Any production process will produce some waste, and traditionally most of this waste has ended up in landfills. "Cradle-to-grave" is a process by which companies take responsibility for the disposal of the goods they produce (usually as a waste product of the production process), but the waste is not necessarily reused or recycled. Not much thought is given to whether the waste can be used by another industry or company, where it could be an ingredient or material for production of their goods. In the terms of a life cycle analysis, cradle-to-grave describes the entire life of a material until the time it is disposed of. Consequently, the environmental problems associated with disposal, particularly in regards to landfills and waste incineration, are not dealt with by the cradle-to-grave process.

It is therefore essential to view the life cycle in terms of cradle-to-cradle, and to consider the best way to dispose of the end product when no longer required. The term *cradle-to-cradle* was invented by Walter R. Stahel in the 1970s, but it did not become popular until the book *Cradle to Cradle: Remaking the Way We Make Things,* by William McDonough and Michael Braungart, was published in 2002. Stahel stressed the importance of using the holistic approach, as well as the economic, ecological, and social advantages of a closed-loop economy.

McDonough and Braungart argued that the cradle-to-cradle practice requires a complete turnaround in thinking in some cases—just as when we are driving to a certain destination and we realize that we are going on a wrong path, slowing down does not correct the path. We need to return to a certain point to continue our journey. Similarly, we need a completely different model for dealing with waste.

Zero-waste plans focus on reducing waste and reusing products, followed by recycling and composting through methods such as this waste bin in Nevada.

Source: Nevada Division of Environmental Protection/ Bureau of Waste Management

Cradle-to-cradle design often involves referring to nature for inspiration, where one organism's waste is food for another, and nutrients flow indefinitely in cycles of birth, decay, and rebirth. The biological metabolism in ecosystems can be mirrored in the technological systems they design, so that there is a closed loop in which valuable resources move in cycles of production, use, recovery, and remanufacture.

In the cradle-to-cradle model, all materials used in industrial/ manufacturing processes (such as creating metals, fibers, and dyes) fall into two main categories: technical or biological nutrients. Technical nutrients are restricted to nontoxic, nonharmful synthetic materials that do not have a negative impact on the environment and can be used in continuous cycles as the same product without losing their integrity or quality. Biological nutrients consist mainly of organic materials that can be easily disposed of in any natural environment to decompose into the soil.

McDonough and Braungart identify three key principles in cradle-to-cradle design: waste equals food; use current solar income, that is, use renewable energy such as wind and solar; and celebrate diversity. Celebrating diversity leads to "designs that celebrate and support diversity and locality," rather than taking a one-size-fits-all approach.

Inspired by this principle and its implementation in the province of Fryslân in the Netherlands, a number of other Dutch islands have joined a European Union project called Cradle to Cradle Islands, along with North Sea Islands from Germany, Sweden, Norway, Denmark, and England. These Cradle to Cradle Islands are expected to become a living showcase of the principle on a microcosm level. The islands have invited industries worldwide to demonstrate their innovations in waste elimination, optimization of the water cycle, and solar and other renewable sustainable energy sources.

Several companies have also adopted the principle of cradle-to-cradle, and are certified for the same. For most of these companies, the goal is to become "ecoeffective." Cradle-to-cradle certification is provided by McDonough Braungart Design Chemistry (MBDC). According to the company, certification "provides a company with a means to tangibly, credibly measure achievement in environmentally-intelligent design and helps customers purchase and specify products that are pursuing a broader definition of quality." Certification is based on whether the company is using environmentally safe and healthy materials; designing their products to facilitate recycling and composting; using renewable energy and energy efficiency; efficiently using water; and instituting strategies for social responsibility. Once a product meets the necessary criteria, it is given a Basic, Silver, Gold, or Platinum product certification and the product can be branded as "Cradle-to-Cradle Certified." Certified companies include carpet manufacturers, commercial wall covering manufacturers, and chair designers.

The Sierra Club has also adopted zero waste cradle-to-cradle principles that specify the roles that government, companies/industries, and consumers should play in addressing the waste crisis in the United States. The aim of the plan is "to prevent waste by design" rather than to focus on waste management and disposal. In addition to the cradle-to-cradle approach, the other two components of a zero waste cradle-to-cradle plan are land use policies that promote developing reuse, repair, and recycling businesses in business districts; and local government management of organic waste, such as food scraps and yard waste. In a zero waste plan, governments are expected to divert all organic waste to composting plants.

See Also: Bio-Based Material; Clean Production; Closed-Loop Supply Chain; Ecoeffectiveness; Green Design; Waste Reduction.

Further Readings

Fryslân International. "Cradle to Cradle Islands." http://www.c2cislands.org/sjablonen/default/default.asp?objectID=1207 (Accessed January 2010).
McDonough, William and Michael Braungart. *Cradle to Cradle: Remaking the Way We Make Things.* New York: North Point Press, 2002.
Sedona.biz. "Sierra Club Adopts Zero Waste Cradle-to-Cradle Principles for the 21st Century." http://www.sedona.biz/sierra-club-zero-waste0108.htm (Accessed November 2009).

Velma I. Grover
United Nations University–Institute for
Water, Environment and Health (UNU–IWEH)

DEMAND-SIDE MANAGEMENT

Demand-side management (DSM) refers to any policy that seeks to influence the demand for a good or service. DSM is practiced in both the public and private sectors, and can be used to either increase or reduce demand through direct or indirect methods. The use of DSM is most associated with regulated utility companies, most of which adopted tiered-pricing schemes during the 1980s to encourage energy conservation, but DSM is increasingly being used to address a wide range of capacity issues associated with sustainability. Building standards that require installation of dual-pane windows; purchase rebate programs for energy-efficient water heaters; moral suasion campaigns to reduce, reuse, and recycle; gasoline taxes; and turnpike tolls are all examples of DSM in practice.

Prior to the energy crisis of the 1980s, DSM policies were mainly associated with war-time rationing. Since that time, highly industrialized nations have all bumped up against capacity constraints either because demand increased faster than facilities could be built or because natural systems were being overtaxed, and as a result the use of DSM has become an increasingly common tool in the battle for sustainability.

DSM and Electric Utilities

Electric utilities were among the first to use DSM techniques to address capacity issues. Electric systems are expensive both to build and maintain. For much of the time, they operate below capacity, but to ensure reliability they are sized for those brief periods in the day and season when the network load is at its highest (peak). Depending on the geographic area, peak times may occur in summer or winter. Influencing the peak demand has proven an effective way for operators to manage the rate of network expansion and contain construction and operations costs while keeping electricity prices at affordable levels. Through DSM, utilities could avoid building what would be excess capacity for all but a few hours a day.

In their simplest forms, DSM programs are conservation campaigns exhorting consumers to switch off unnecessary lights and appliances, or to lower their thermostats. During the energy crisis of the 1970s when fuel prices were high, such programs proved effective in reining in consumer demand, but when oil prices began to fall, so did consumer resolve to conserve.

Incentive-based DSMs featuring tiered pricing were first introduced in the 1980s. Under these schemes (modified versions of which are still in use), residential and industrial consumers are charged a relatively low rate for the first tier of kilowatt hours they use. If they exceed this base usage, they pay an incrementally higher rate for the next tier and each successive tier. The increment may also increase—for instance, the difference between the first and second tiers may be an additional 1 cent per kWh, but the difference between the second and third tiers may be 3 cents per kWh. The tier rate may also be higher during one time of day as opposed to another, with rates typically being the highest during weekday business hours when demand tends to peak.

Some providers offer incentives for consumers to relinquish control of their thermostats to the utility during peak hours. Others offer incentives for consumers to install alternative energy systems. In some cases, utilities will reward consumers using alternative energy sources that can feed power to the grid by offering them discounted rates when they are using energy from the grid.

Nonpricing mechanisms are also used. Utilities frequently offer free energy audits or equipment inspections to improve system efficiency.

DSM and Water Purveyors

Water purveyors have generally been slower to adopt DSM, but in arid areas where industrial, agricultural, and residential users compete for a limited supply of water, they are increasingly using the technique to encourage conservation. Applications include tiered pricing systems similar to those used in the electric industry, incentive programs to install low-flow showerheads and low-water-use toilets, and, in drought-stricken areas, temporary surcharges and outdoor watering restrictions.

While DSM provides incentive for users to conserve, the same may not necessarily be said for the water supplier. Depending on the structure of the water company, how it obtains its water, and the regulatory system in place, the purveyor may determine that it is in its financial best interest to use any water savings realized from DSM to expand its customer base. This is likely to occur when water companies are for-profit concerns, because profit is only made on water sold, not on water sitting in a reservoir. Furthermore, reduced sales make it harder to cover what are typically substantial fixed costs without seriously cutting into the profit margin. In such situations, companies are motivated to expand, rather than constrain, demand. It was common practice during the construction boom of the 1990s for companies that had DSM in place to issue commitments to serve new developments rather than bank water savings attributable to DSM in aquifers or reservoirs.

Municipal companies have had better track records with DSM. During the serious drought that affected the southeast United States in 2008, the City of Birmingham, Alabama, instituted a water surcharge that resulted in a 17 percent reduction in demand for the duration of the program.

Other Applications of DSM

DSM has many applications in the transportation sector. Designated bus and high occupancy vehicle (HOV) lanes, bridge tolls, parking meters, metered on-ramps, and congestion charges are all examples of DSM. Some have been highly effective, others not at all. HOV lanes do little to encourage people to take the bus or carpool. Metered on-ramps marginally improve traffic flows. Pricing schemes, like the congestion charge in London's central district, are more successful in changing demand, though the changes generally do not last if the measure is abandoned.

The telecommunications industry uses DSM pricing. For landlines, different rates are charged for local and long distance calls, higher rates are normally charged for peak hours than for off-peak times (weekends and late night). Mobile phone companies typically structure their rates so that calls made within their own system are less expensive to the consumer than those made to phones supported by other companies (whether landline or mobile).

The Future for DSM

Utility managers will turn toward DSM more and more in order to deal with uncertainty of future resource supplies and in the quest to reduce greenhouse gas emissions. In the water resources sector, the changing climate and reductions in freshwater supplies will drive adoption of DSM. For private sector companies, DSM will continue to be attractive because it holds the potential to shape consumer demand for more profitable operations.

Regardless of the application, the effectiveness of DSM in addressing sustainability issues rests on finding the right pressure points and discerning how the intrinsic incentives will play out in order to avoid unintended consequences.

See Also: Ecoefficiency; Economic Value Added; Sustainability; Sustainable Development.

Further Readings

Berry, Michael A. and Dennis A. Rondinelli. "Proactive Corporate Environmental Management: A New Industrial Revolution." *The Academy of Management Executive*, 12/2:38–50 (May 1998).

Gellings, C. and W. M. Smith. "Integrating Demand-Side Management Into Utility Planning." *Proceedings of the IEEE*, 77/6:908–18 (1989).

International Energy Agency. "Demand-Side Management." http://www.iea.org/Textbase/techno/iaresults.asp?id_ia=8 (Accessed November 2009).

Maas, T. *What the Experts Think: Understanding Urban Water Demand Management in Canada*. Victoria, British Columbia, Canada: The POLIS Project on Ecological Governance, 2003.

Velma I. Grover
United Nations University–Institute
for Water, Environment and Health (UNU–IWEH)

DEMATERIALIZATION

The concept of dematerialization refers broadly to the reduction of materials used by society over time. Definitions of the term *material use* vary, but generally include some estimate of ecosystem appropriation and waste generation. Dematerialization studies have been done for products, businesses, regions, nations, even the globe. Measures of dematerialization are often expressed as relative, rather than as absolute values. Such relative values, often called "intensity factors," express total material use per unit of economic output. It should be noted that with the steep upward curves typical for economic output and population, declines in relative values may be visible even as absolute materials use continues to climb. This makes the scale and scope of analysis crucial when measuring dematerialization.

Background

The study of dematerialization to date has largely focused on minerals and their use. The work, particularly in the United States, has been largely empirical and was often produced for national security assessments. Scientists at the Wuppertal Institute in Germany were the first to expand on and develop the concept by proposing "material input per service unit." This measures both the direct materials embedded in products, as well as the indirect materials used throughout the stages of the production process. Hidden flows of materials, such as mining waste, are included. This "ecological backpack" approach traces five major categories of materials: (1) abiotic raw materials (e.g., minerals, fossil fuels, and excavation residues); (2) biotic raw materials (e.g., farm and nonfarmed plant biomass); (3) soil movements (e.g., erosion); (4) water use; and (5) air (e.g., combustion by-products). So, in essence, all stages of production, including disposal, are incorporated.

A primary appeal of dematerialization is that, through technological innovation, there is the possibility of decoupling economic growth from material consumption. This decoupling is often expressed as a "Factor X." It is observed either as a slowing of material use as the economy continues to grow (i.e., declining intensity), or theoretically as a complete break where overall material use actually declines in spite of growing economy. Noted energy expert Amory Lovins maintains that it is technically feasible to double economic welfare while halving energy use across 50 different economic activities. Other scientists argue that with lifestyle changes and technological innovation, much higher factors can be achieved.

Most dematerialization efforts hinge on substituting conventional energy sources and materials with "lighter" alternatives, such as using fiber optics instead of copper in telecommunication networks. However, this substitution may not lead to reductions in the use of the replaced material. Although use of fiber optics has displaced demand in telecommunications, overall use of copper has increased for uses like electrical wiring. Other "substitutions" can be much broader, such as the emergence of digital delivery of music, movies, and even books. The complexity lies in assessing whether such change actually leads to overall trends of dematerialization.

Measuring Dematerialization: Scope and Scale

The system boundaries—in essence scope, temporal, and geographic scale—are fundamental in terms of shaping how dematerialization is measured and assessed. For example, measuring the material mass in the product cycle at a given point in time can mislead if large amounts of materials were displaced long ago from mining activity, or if large amounts of end-of-life wastes will be disposed of at a later date. Do we incorporate air and water as appropriated from ecosystems? And if so, how do we measure this? How do we model energy or human work? Both of these can be substitutes for appropriation of materials.

One way to address this challenge is by incorporating "exergy" as a representation of the level of transformation that materials have already undergone from their natural state. The energy embedded in materials is essentially included and can be applied systematically across a large range of materials. But this does not resolve the challenges of scope, particularly for the purpose of relating materials use to particular economic welfare values.

Economic welfare is typically measured as gross domestic product (GDP). The measure is clear enough until one considers how to represent specific economic functions versus materials use at greater levels of detail, as in assessments of trends or structural changes. It is also often difficult to identify equivalent economic functions over time. From the standpoint of dematerialization, similar economic functions are routinely shifted to other economic sectors. In the digital media example, entertainment moves out of theater establishments and into retail (e.g., big screen, high-definition televisions), telecommunications (e.g., information and communications technology, digital broadband), and household energy consumption. Nevertheless, assessing the level of dematerialization achieved by digital media requires making a complex cross-sectoral assessment.

At the level of individual businesses, declines in material intensity have been achieved primarily through material reduction and substitution. The mass of the standard aluminum beverage can, for example, has declined for some time as new alloys are discovered that give adequate strength despite thinner container walls. Companies can often save money and reduce environmental impact through basic material reduction strategies. For example, by reshaping and reducing the cap size, Nestlé Waters, North America, has thinned the plastic in its half-liter water bottle, claiming that it uses 30 percent less plastic than the average half-liter bottle.

Another well-known example of dematerialization in green business is Interface Corporation. It achieved dematerialization by designing carpet in tiles so that carpet could be selectively replaced in high-traffic areas. This eliminates the substantial waste associated with recarpeting an entire floor. New materials and systems have also been implemented to recover materials from used tiles to produce new ones ("closed-loop" manufacturing). In *The Ecology of Commerce*, Paul Hawken explains that the central insight here was reframing the carpeting product as instead the "service" of floor covering. Conceptualizing products as services can lead companies to strategies that create value with less material. In the green business movement, generally, *dematerialization* has become a favored term. In their book, *Natural Capitalism*, Paul Hawken, Amory Lovins, and L. Hunter Lovins outline the many forms dematerialization can take, and introduce green business principles, such as cradle-to-cradle design, biomimicry, just-in-time production, and extended producer responsibility.

Many dematerialization studies are executed at a national level. Some studies generally indicate dematerialization occurs with the shift to a service sector economy. But these dematerialization studies generally restrict their analysis to a particular geographic scale, neglecting to incorporate, for instance, the embodied energy within goods that are manufactured overseas and imported. However, Cleveland and Ruth find no compelling evidence that the U.S. economy has decoupled significantly from material inputs. There is a need for more dematerialization studies at a multinational or global level to more fully assess whether economies are actually dematerializing. Without such evidence, it begs the question, "Is dematerialization in industrialized society dependent on increased materialization elsewhere?"

Trends and the "Rebound Effect"

After evaluating 40 empirical studies of dematerialization, Cleveland and Ruth concluded that dematerialization trends were unclear, due to the varying scale, scope, and rigor of the studies. A decade later, this situation has not markedly improved. Nevertheless, certain factors broadly driving or inhibiting dematerialization can be identified.

The mining and refining of primary ores and fuels has become more efficient. Material substitution through greater use of alloys and composites has encouraged the dematerialization of some products (e.g., lighter aluminum cans). Other strategies include design improvements like miniaturization, greater life spans of certain products, and increased recycling.

In contrast, declining ore grades necessitate greater use of energy or water in the production process (e.g., the substitution of tar sands or oil shale for petroleum deposits). Ironically, the material innovation strategies that initially encouraged dematerialization, with respect to production, may actually inhibit overall dematerialization by shortening the product life or by increasing product complexity in ways that work against recycling (e.g., the proliferation of polymer composites or biopolymers).

Finally, scholars have identified a "rebound effect" (also known as Jevons Paradox), whereby relative dematerialization, such as increased efficiency, actually leads to greater overall consumption. This is often connected to cost, as the material or product becomes cheaper to produce due to greater efficiencies and this, over the long term, induces greater overall consumption. A study led by Robert Ayres found that in every case investigated, increased demand seemed to nullify efficiency gains.

See Also: Best Management Practices; Corporate Social Responsibility; Cradle-to-Cradle; Ecoefficiency; Ecolabels; Ecological Footprint.

Further Readings

Ayres, R. U., L. W. Ayres and B. Warr. "Is the U.S. Economy Dematerializing? Main Indicators and Drivers." In *Economics of Industrial Ecology: Materials, Structural Change, and Spatial Scales*, C. Jeroen, J. M. v. d. Bergh and M. A. Janssen, eds. Boston: MIT Press, 2004.

Cleveland, Cutler J. and Matthias Ruth. "Indicators of Dematerialization and the Materials Intensity of Use." *Journal of Industrial Ecology*, 2/3:15–50, 1999.

Hawken, P. *The Ecology of Commerce: A Declaration of Sustainability*. New York: HarperBusiness, 1994.

Hawken, P., A. Lovins and L. Hunter Lovins. *Natural Capitalism: Creating the Next Industrial Revolution*. Boston: Little, Brown and Company, 1999.

Jeroen, C. J. M. van den Bergh and Marco A. Janssen. "The Interface Between Economics and Industrial Ecology: A Survey." In *Economics of Industrial Ecology: Materials, Structural Change, and Spatial Scales*, C. Jeroen, J. M. v. d. Bergh and M. A. Janssen, eds. Boston: MIT Press, 2004.

Linderhof, V. G. M., J. M. van den Bergh and H. Verbruggen. "Digital Dematerialization: Economic Mechanisms Behind the Net Impact of ICT on Materials Use." In *Climate Change and Sustainable Development: New Challenges for Poverty Reduction*, M. A. M Salih, J. Cramer and L. Box, eds. Cheltenham, UK: Edward Elgar Publishing, 2009.

Sun, J. W. and T. Meristo. "Measurement of Dematerialization/Materialization: A Case Analysis of Energy Saving and Decarbonization in OECD Countries, 1960–95." *Technological Forecasting and Social Change*, 60:275–94 (1999).

Weizsacker, E. von, A. B. Lovins and H. L. Lovins. *Factor Four: Doubling Wealth-Halving Resource Use*. London: Earthscan, 1997.

Robert O. Vos
Josh Newell
University of Southern California

DEPOSIT SYSTEMS

Deposit systems are programs designed to divert reusable, recyclable, or hazardous materials from illegal disposal or from municipal landfills by providing consumers with financial incentives to act responsibly. A deposit is charged at the point of purchase and refunded upon return of the good or its recyclable container to an authorized redemption center. Most commonly, deposit systems have been set up for glass or aluminum beverage containers, but they can be applied to any consumer good. They are currently components of a few cradle-to-cradle (C2C) and cradle-to-grave (C2G) product management schemes, and will likely become commonplace as the public's understanding of life cycle costs increases. Costs associated with the disposal, particularly for hazardous or nonbiodegradable materials, are increasingly considered the responsibility of the producer or consumer.

Multiple benefits are gained from deposit systems. They help limit the nearly intractable local government problem of illegal dumping. For society at large and producers, they provide incentives to reclaim materials for reuse in industrial processes and to redesign production processes, final products, and packaging to exert fewer environmental impacts. Deposit systems are particularly valuable when the cost of cleaning up improper disposal is far greater than that of running a recycling program. When deposit systems falter or fail, it is generally because the inconvenience of participating outweighs the perceived value of returning the product for the refund.

Most deposit systems apply to beverage containers. Under these programs, a deposit on aluminum, glass, or plastic beverage containers is collected when the beverage is sold. Deposits are usually reclaimed from authorized redemption centers or from vendors. Many of the earliest government container deposit programs (CDPs) predated curbside recycling collection and were intended to encourage recycling and extend the usable lifetime of municipal landfills. They were also promoted as a way to reduce littering in general, particularly along highways and in lakes and rivers. When CDPs are government administered, unredeemed deposits are most often earmarked for costs associated with the CDP or other environmental programs. When refunds are channeled through vendors or manufacturers, application of unredeemed deposits is often left to their discretion.

There is no U.S. federal container deposit law. Bills to establish one were unsuccessful introduced in Congress in 2002 and 2007. The first CDP law adopted in the United States was the 1972 Oregon Bottle Bill. Only 10 other states followed suit. Efforts in the remaining 39 states, and later attempts to amend existing laws, have been politically contentious. The beverage container industry, beverage companies, and corporate owners of grocery and convenience stores spend vast amounts lobbying against CDPs. They are frequently joined in opposition by consumer advocate groups. Objections generally revolve around increased costs for the business community and consumers. Local governments have sometimes opposed CDPs on the grounds that they divert aluminum cans from curbside recycling, thus removing an important income stream from their recycling program business models. Others criticize CDPs for inconsistencies in application (e.g., having a deposit for soft drink bottles but not for bottled water), pointing out they can cause consumers to shift away from containers covered by the CDP to those that are not.

CDPs have been effective, but only to a degree. Some studies show that roadside litter initially decreased by between 30–64 percent in states with bottle bills. In these states, approximately 70 percent of containers covered by the CDP were recycled in 2006, while the national average was 33 percent. It can be argued these figures in part reflect the higher propensity to recycle among citizens inclined to vote for a CDP.

Since 2000, the United States has seen significant growth in beverage container packaging, primarily as a result of the rapid growth of bottled water sales. Meanwhile, national beverage container recycling rates declined. A study conducted in 2008 by As You Sow, a nonprofit group promoting corporate responsibility, found that of the 12.1 million tons of disposable beverage containers sold in 2006, only about 2.8 million tons were recycled. Not surprisingly, aluminum, given its high salvage value, was recycled at the highest rate (37.3 percent).

Deposit programs also are in place for other products, such as automobile tires and batteries, which either have a high potential for recycling or represent a potential hazardous waste problem to landfill operations. Redemption of such products is generally conducted through certified scrap dealers, tire or automotive part retailers, or government-run collection centers.

The substantial increase since 1990 in the number of electronic products discarded annually poses multiple serious hazardous waste risks. In response to various aspects of this problem, the European Union places the financial burden for the collection, recovery, treatment, and/or disposal of electronic waste on producers. Generally, such programs are funded by unit sales surcharges. In other instances, the producers run their own programs, accepting and handling only products carrying their label. These need not necessarily be deposit systems, though many offer a financial incentive for consumers to participate by discounting the prices of a subsequent purchase or trade-in.

Quasi-deposit systems have been voluntarily established by companies using cradle-to-cradle or cradle-to-grave product management. Flooring manufacturer Interface, Inc., tracks its environmental performance cradle-to-cradle. Interface leases rather than sells its products, recycles, and reuses components of returned worn or damaged merchandise. Such systems seem the way of the future, as consumers are increasingly concerned with responsible resource use.

See Also: Cradle-to-Cradle; Ecoefficiency; Recycling, Business of; Voluntary Standards; Waste Reduction.

Further Readings

European Union. Directive 2002/96/EC of the European Parliament and of the Council of January 27, 2003 on Waste Electrical and Electronic Equipment (WEEE). http://eur-lex

.europa.eu/LexUriServ/LexUriServ.do?uri=OJ:L:2003:037:0024:0038:EN:PDF (Accessed January 2010).

Galland, Amy. "Waste and Opportunity: U.S. Beverage Container Recycling Scorecard and Report, 2008." San Francisco, CA: As You Sow Foundation, 2008. http://www.asyousow.org/sustainability/bev_survey.shtml (Accessed January 2010).

Gitlitz, Jenny and Pat Franklin. "The 10¢ Incentive to Recycle." Culver City, CA: Container Recycling Institute, 2006. http://www.container-recycling.org/publications/ (Accessed January 2010).

Stavins, Robert N. "Market-Based Environmental Policies: Discussion Paper 98-26." Washington, D.C.: Resources for the Future, 1998. http://www.rff.org/documents/RFF-DP-03-43.pdf (Accessed January 2010).

Bill Kte'pi
Independent Scholar

Design for Environment

Green design, also known as ecodesign, design for the environment, and sustainable design, refers to the practice of designing products for sustainability—that is to say, embracing the economic, social, and environmental impacts of producing and consuming a product, and designing it in such a way that its impacts are minimal, or reduced in comparison to mainstream alternatives.

Green design can be applied to the built environment, the design of consumer goods and products and other physical objects, or can transcend the tangible and extend to the design of systems and methods of doing things in a way that is superior in terms of sustainability, compared to conventional methods.

Design is not just about looking at the product being designed, it is about considering the wider impact that the manufacture, use, and eventual disposal of that product has on the wider society. Designing different "modes and methods of consumption" is as much the green designers role as specifying environmentally sound materials and processes.

Working With Nature

In the past, the view was held that the role of science and technology was to subjugate nature and control it. In recent years, it has become more widely acknowledged that the best approach is not to try and "conquer" nature, but to work in harmony with it, learning from the way things in the natural world are designed, borrowing from it where necessary, and trying to design things that are more "in harmony" with the world.

Biomimicry

The term *biomimicry* describes an approach to design, looking at the way that nature ingeniously solves design problems, and attempting to mimic those approaches in

One goal of design for environment is that the product does not pollute at the point of use. Here, at a U.S. Department of Agriculture research center in Maryland, a tour bus fueled by soy-based biodiesel passes a soybean field ready for harvesting.

Source: Keith Weller/U.S. Department of Agriculture

designed products. The word comes from the two roots: *bio*s, meaning life; and *mimesis*, meaning to imitate.

An example would be the design of photo-chemical solar cells. Unlike photovoltaic cells that rely on expensive crystalline silicon, or a combination of exotic metal oxides, photo-chemical solar cells rely on chemical processes more akin to those that take place within plants to design a technology that is more cost-effective.

Chemists have "designed" pharmaceuticals by looking at natural remedies, identifying the active component, and making "nature identical" copies of the chemical.

Technical Design Improvements

Quality, Durability, and Designing Out Obsolescence

Wherever possible, products should be designed to have a long service life, fulfilling their function for as long as possible before replacement is required. This ensures that the maximum possible "value" is extracted from the input of materials and energy that were initially required to manufacture the product. This is a hard objective to reconcile with commercial goals. In a society where consumers are often led by price, there is a temptation to deliver products at a lower price point, that possibly require periodic replacement; either as a result of wearing out or designed-in obsolescence (i.e., by changing formats periodically or restyling a fundamentally similar product in order to take advantage of changing fashions).

Designing for Whole Life Cycle

Life Cycle Analysis

Life cycle analysis, also called life cycle assessment, LCA, ecobalance, or cradle-to-grave analysis, means to look at the environmental impact of a designed product throughout the stages of its life cycle, from when a product is created to when the product is disposed of. Life cycle analysis is integrated into international standards, such as the ISO 14000 environmental management standards.

Life Cycle Cost Analysis

Sustainable technologies can sometimes appear disproportionately expensive compared to cheaper, less ecofriendly technologies when capital costs are compared. However, to fully appreciate any cost advantage where present, it is important to appraise cost over the life cycle of the installation (and any subsequent decommissioning). In a narrow sense, cost can be viewed as purely economic; however, if you wish to take a broader view, cost can be viewed in terms of triple bottom line.

Substituting Finite Resources With Renewable Ones

There is a wide range of different materials and resources available to designers. For example, fast-growing softwoods regenerate relatively quickly, whereas others, for example, plastics derived from oil, come from a resource that takes millions of years to form. We can classify materials as being renewable (i.e., they regenerate quickly, or at least on a human time scale), or finite, meaning that the earth possesses a fixed amount of these materials, or they regenerate on time scales that are many (thousands) of multiples of the human life span.

Ecomaterials

Materials that have minimal impact on the natural environment while still offering material performance equivalent to the traditional material it is replacing can be termed an *ecomaterial*. Natural materials can often be absorbed back into the biosphere and "recycled" by the degradation processes, whereas synthetic materials and metals can only be recycled by man-made processes.

Design for Whole Life Cycle

A key concept in designing for the environment is looking at the product impact over the course of its life cycle: what environmental impact does producing the product have; how can we work toward minimizing requirements for packaging and transportation; what will be the effect of using the product in terms of resource and energy consumption and at the end of its useful operating life; and can the product profitably be repurposed, reused, or recycled.

Waste Management Hierarchy

The waste management hierarchy should be considered when designing products to be green. It arranges different categories for the handling and disposal of waste in order from the most preferred environmental option to the least favored option. The waste management hierarchy is:

1. Prevention of waste

2. Minimization of waste

3. Reuse of waste

4. Recycling of waste

5. Energy recovery from waste

6. Disposal of waste

The waste management hierarchy greatly impacts the green design process. First, products should be designed in order to prevent waste. Designing a product that is "durable" will ensure it has a long useful service life. It may also be a case that "obsolescence" is designed out of the product, by making it in such a way that it is either "timeless," or so that it can be upgraded easily as new innovations permit.

Waste can then be minimized, through minimizing packaging or making the product physically smaller. It may be that extraneous material not fulfilling a useful function can be removed from the design, however, this must be weighed carefully against the impacts that this will have on product durability.

Design for Reuse and Recycling

Increasingly, products are being "designed for disassembly," which means that the product is designed so that it can be taken apart easily, and broken down into its constituent materials, permitting easy recycling.

Products should be designed in such a way that after they have finished their useful service life they can either be easily repurposed for another application or easily recycled. An example would be a vehicle. At the end of its useful life, it can taken apart to be reconditioned, its components can be sold, or parts of components can be separated into different material streams that can then be recycled.

By implementing "closed loop" recycling schemes in products, the environmental impact of using certain materials can be drastically reduced, as in many cases the embodied energy content of a recycled material can be as little as 10 percent of the equivalent value for virgin material.

The sustainable design community has developed approaches to deal with products that have not been designed for reuse or recycle, by creatively repurposing them. An example would be in the design of Earthship homes, which are primarily constructed using old car tires.

Cradle-to-Cradle Design

Cradle-to-cradle design is sometimes also called "regenerative design" or is abbreviated as C2C. The process, similar to biomimicry, attempts to emulate the way in which waste products from one natural process are often used to nourish another natural process. Cradle-to-cradle design challenges the conventional wisdom of green design, which says that items are "recycled" at the end of their life, by arguing that many materials in the process of being recycled are "degraded." A cradle-to-cradle design ethos aims to constantly improve materials by "upcycling" them into better and improved products.

Upcycling

Upcycling is a term that was introduced by William McDonough and Michael Braungart in their book *Cradle to Cradle: Remaking the Way We Make Things*. They define upcycling as "the practice of taking something that is disposable and transforming it into something of greater use and value."

Energy Recovery

As a last resort before the landfill, when materials cannot be recovered from a product, a product can be incinerated or gasified, recovering energy from the waste material. This results in some recovered energy, and a greatly reduced volume of material to landfill. However, there are concerns about emissions from energy produced in mixed-waste plants. By designing products with fewer toxins, it is possible to reduce the impact of recovering energy from any remaining fraction of a product that cannot be recycled.

Embodied Energy and Embodied Carbon

To create products, materials must be extracted and processed. These raw materials must then be machined and formed, processed, and assembled into the final usable item, and in the process of doing so, energy is consumed and carbon emissions are generated. Embodied Energy and Embodied Carbon are concepts that can be used to appraise a product so that it can be examined in terms of the amount of energy and carbon that are needed to turn raw materials into the product. In designing for the environment, designers aim to reduce the amount of energy and carbon incorporated in their products through the process of manufacture. Unfortunately, there is no universally accepted standard for accounting for embodied energy and carbon, however, awareness of it as a concept is important in designing effective green products.

Green Design Fields

Green architecture seeks to construct buildings with a reduced environmental impact, through careful selection of materials and a switch to natural and recycled building materials. The energy demands of the building may be reduced by adopting passive processes for heating and cooling, using solar-driven processes to heat the interior space with effective shading to prevent the building from overheating, while devices such as Trombe walls may be used to create air flow when a building is too warm.

Green automotive design seeks to design vehicles that have a reduced impact on the environment. This may be achieved through vehicle weight reduction (resulting in improved fuel economy), improved aerodynamics to reduce vehicle drag, and by using new drive train technologies such as electric, hybrid, plug-in hybrid, or hydrogen drive trains. Mainstream manufacturers are focusing on improving internal combustion engine vehicles for the mainstream market; however, niche vehicle manufacturers are beginning to produce vehicles of a design that challenges the automotive orthodoxy.

Green electronics design seeks to reduce the power consumption of portable devices (a side effect of which is also that battery life, and hence the usability of a device improves). There is a move in many countries to ban the use of solder that contains lead, necessitating a move to alternatives. In addition, electronics designers may look to renewably powering devices either using renewables or "human power."

Barriers to Green Design

While green design makes environmental and often economic sense, there are barriers to green design that prevent more people and organizations from adopting the practices. The first is knowledge—the human race has only recently become collectively cognizant of the impact human activities have on the environment. While localized pollution has been relatively easy to identify, some of the Earth-scale environmental problems have only been detected with the advent of sophisticated monitoring and data processing techniques. Another barrier to green design is a lack of effective incentives. There is no fee levied to use the world's resources that have been long held as "commons" (the land, the air, and the water). Thus, environmental degradation is accepted as an "externalized cost" of doing business. Regulations are being put into place to make businesses pay for these externalized costs, which are passed on to consumers.

Products that are designed poorly and use resources inefficiently will cost the customer more to operate and make the product less attractive. The introduction of labeling regimes, which state, for example, product energy ratings or fuel consumption, can encourage consumers to make informed choices about the products they purchase. Certification schemes are being set up that mandate that a product has reached a given environmental criteria. Designing for the environment may take into account the requirements of a certification scheme, to ensure that the product meets the criteria set out by that scheme.

See Also: Biomimicry; Cradle-to-Cradle; Recycling, Business of; Social Return on Investment; Sustainable Design.

Further Readings

Benyus, Janine M. *Biomimicry: Innovation Inspired by Nature*. New York: HarperCollins, 2002.
Hawken, Paul, et al. *Natural Capitalism: Creating the Next Industrial Revolution*. New York: Back Bay Books, 2008.
McDonough, William and Michael Braungart. *Cradle to Cradle: Remaking the Way We Make Things*. New York: North Point Press, 2002.
Walker, Stuart. *Sustainable by Design: Explorations in Theory and Practice*. London: Earthscan, 2006.
Williams, Daniel E., et al. *Sustainable Design: Ecology, Architecture and Planning*. Hoboken, NJ: John Wiley & Sons, 2007.

Gavin Harper
Cardiff University

DISCOUNTING

Discounting is a method used to adjust the monetary values of economic goods from different time periods to a standard basis for comparison. It involves converting the value of cash flows occurring at different time periods to a base year value, generally the year in which the analysis is being conducted (i.e., present value). The calculations use a discount rate, which reflects the time value of economic resources by discounting future monetary flows to the base year or present value.

The concept underlying discounting is known as time value or time preference, which refers to the tendency for people to prefer economic benefits to accrue sooner rather than later. This preference is well supported by empirical evidence, having been a topic of discussion in economic literature for nearly two centuries. However, it wasn't until the 1960s that debates on how to determine the appropriate discount rate started to dominate economic discussion. Fifty years later, the question has still not been conclusively settled.

When costs and benefits associated with a particular economic activity occur at different points in time, using a simple summation of cash flows is not appropriate because this ignores the effects of monetary inflation, changes in specific markets or the overall structure of the economy, or changes in pertinent technologies—in short, all of the risks and uncertainties of the future. Discount rates attempt to quantify the combined value of these effects. Because they deal with unknown future events, discount rates necessarily involve estimation, and the object in constructing a discount rate is to predict as accurately as possible the trends for all important factors. To the extent the effort is successful, discounting is an invaluable tool with which to compare different courses of action, such as weighing the relative costs (also known as opportunity costs) associated with making any one investment rather than any other. Discounting is an integral part of cost benefit analysis, which is used, among other things, to decide on investments in environmental technology, conservation, and pollution control.

Controversies among experts on how best to construct the discount rate have included the following: (1) whether the discount rate should reflect the private investment rate, social opportunity cost rate, the social time preference rate, or some other rate; (2) whether public and private sector projects should be discounted using the same discount rate; (3) whether discount rates should vary by type of investment, as well as by sector; (4) whether differential discounting should be used for costs and benefits; (5) whether different discount rates should be used for projects with unequal life expectancies; (6) how risk and uncertainty should be included in the discount rate; and (7) whether the nominal discount rate or the real discount rate should be employed.

There are numerous analytical methodologies that use the principle of discounting including net present value, internal rate of return, discounted payback period, cost-benefit analysis, and life cycle analysis. The choice of discount rate can profoundly affect the outcome of an analysis since even small variations in the discount rate can significantly bias a cost study by tilting the scales away from one project to another or even changing a positive net present value to a negative net present value. For this reason, it is imperative for business managers and investors to understand the discount rate being used in the analyses that form the basis for their operating and investment decisions. Life cycle analysis was developed explicitly to incorporate both future direct and indirect costs into business models by using a discount rate.

In the absence of standardized rules on the discount rate, careful consideration should precede any decision on constructing a rate. A useful starting point is to review the literature on similar projects, as well as bank and treasury rates in the project's home country. For example, the U.S. Department of Energy annually publishes energy price indices and details on discount rates utilized to evaluate federal energy projects.

Once an approach is selected, sensitivity tests should be performed with a range of discount rates. This process should not be used to deliberately manipulate the discount rate for political or other motives in order to make proposed investments appear feasible or unfeasible. Rather, it should be used to verify the robustness of results under different assumptions, for example, to estimate the impact of various rates of inflation on cash flows.

Furthermore, analysts need to pay careful attention to the issue of real and nominal rates and values. Real rates are rates that have been adjusted to remove the effect of inflation, whereas nominal rates have not. Real and nominal values must never be combined in the same analysis. The recommended approach is to use real values, and to reflect the anticipated future price effects. Real prices can be estimated by deflating nominal values using a general price index such as the Consumer Price Index or any other appropriate price index. Nominal prices can be estimated by inflating all cash flows. The choice of which to use should rest on which provides the clearest understanding for the intended audience.

The choice of a discount rate is at the center of the debate about how aggressively to respond to the threat of climate change. The Stern Review on the Economics of Climate Change, released in October 2006, became front-page news because it called for world governments to immediately begin investing an amount equal to 1 percent of global gross domestic product (GDP) in order to reduce the anticipated impact of global climate change. The British report predicted that procrastination would lead to future economic damages equivalent to a reduction of up to 20 percent in global GDP, a figure substantially higher than earlier estimates of costs associated with climate change.

Environmental economists have subsequently studied the 700-page report to try to determine why the numbers are so large. Yale University professor William Nordhaus took issue, among other things, with the Stern Review's choice of the social rate of time preference, the rate used to compare the well-being of future generations to the well-being of those alive today. Nordhaus ran a model similar to the one used in the Stern Review, but used a 3 percent social discount rate that slowly declined to 1 percent over 300 years (rather than the stable 0.1 percent discount rate used in the Stern Review) and found that preventive measures, like a tax on carbon emissions, though still required, were of a much lower magnitude than those recommended the Stern Review.

There is no definitive answer to the question of whether the discount rate used in the Stern Review or in Professor Nordhaus's analysis is the "correct" one because the choice requires a value judgment weighing and quantifying the well-being of people living today against future generations. But for business owners, investors, or consumers it is critical to understand the assumptions implicit in whatever discount rate is being used, and to be sure that they reflect as objectively and accurately as possible the long-term implications investment decisions have for sustainable development.

See Also: Cost-Benefit Analysis; Life Cycle Analysis; Social Return on Investment.

Further Readings

HM Treasury. "The Green Book: Appraisal and Evaluation in Central Government." http://www.hm-treasury.gov.uk/d/green_book_complete.pdf (Accessed January 2009).

New Zealand Treasury. "Cost Benefit Analysis Premier." http://www.treasury.govt.nz/publications/guidance/costbenefitanalysis/primer (Accessed January 2009).

Nordhaus, William. "Critical Assumptions in the Stern Review on Climate Change." *Science*, 317 (2007). http://nordhaus.econ.yale.edu/nordhaus_stern_science.pdf (Accessed January 2010).

Rushing, Amy and Barbara Lippiatt. "Energy Price Indices and Discount Factors for Life-Cycle Cost Analysis." Washington, D.C.: U.S. Department of Commerce, 2009. http://www1.eere.energy.gov/femp/pdfs/ashb09.pdf (Accessed January 2009).
Stern, Nicholas. "Stern Review: The Economics of Climate Change." London: UK Treasury, 2006. http://www.hm-treasury.gov.uk/stern_review_report.htm (Accessed January 2010).

Nadini Persaud
University of the West Indies, Cave Hill Campus

DISTRIBUTED ENERGY

Distributed energy is both complementary and alternative to "centralized power generation," addressing many of the shortcomings of centralized systems. Key qualities of distributed energy are that it is decentralized, diverse, and close to the end user. As such, distributed energy resources (DERs) are increasingly used to address common problems of centralized power systems such as peak power, backup power during power outages, increased power quality, as well as lower cost and lower energy consumption for electrical, heating, and cooling needs. Distributed energy systems commonly used today include diesel and fuel generators, solar power, fuel cells, wind turbines, biomass, micro-turbines, load reduction technologies, and battery storage systems. Though very basic distributed energy systems, such as diesel-operated backup generators, represent a significant exception, more and more distributed energy technologies provide cleaner, faster, and more efficient energy than centralized power systems. All distributed energy systems have the built-in advantage of generating power nearer to the point of consumption, thus reducing losses inherent in the transmission and distribution system of centralized power stations.

In one type of distributed energy, the concentrating solar plant, the sun's energy is focused onto a point using mirrors or lenses, generating intense heat. This solar-energy dish-collector field is in Shenandoah, Georgia.

Source: Sandia National Laboratory

The nation's overall energy portfolio is increasingly reliant on distributed energy technologies, for they have, as many reports clearly indicate, the important potential "to mitigate congestion in transmission lines, reduce the impact of electricity price fluctuations, strengthen energy security, and provide greater stability to the electricity grid." Along the same lines, industry observers have likened distributed energy systems to the historical evolution of computer systems: once highly centralized and expensive, they have become, in short order, more efficient, more decentralized, cheaper, and more user friendly. And, of course, primarily as a result of these developments, both distributed energy systems and computer systems find an ever-widening range of uses.

The basic idea behind distributed energy systems, of course, is not at all new. Indeed, it was not until the advent of the industrial revolution, the creation of urban centers, and the building of the first large-scale power plants that energy production became centralized. During the early years of the American republic, people had their own furnaces and their own wind or water-generated mills—very much a decentralized or distributed energy system. Despite the fact that there is still no clearly agreed-upon definition for distributed energy, however, it is different from such historical examples of decentralized systems in that it is part of a larger system, or network. Indeed, many European countries now pay costumers for feeding energy back into the grid from their own energy generators, assuming they are zero carbon emissions. In some cases, more per kilowatt is fed into the grid from these homes than what is taken out of the grid.

DERs have become particularly significant in the information technology industry, which is dependent on a steady and reliable supply of electricity, without any tolerance to variations in energy quality, much less complete power outages. And since a large and growing part of U.S. business transactions are based on information technology, DERs are increasingly viewed as vital to the overall well-being of the economy.

As socially responsible investing, carbon footprints, and general environmental stewardship models geared toward sustainability are rapidly gaining significance within American society, and among U.S. businesses in particular, DERs provide a number of opportunities for companies and institutions, large and small, to make positive contributions to both their carbon footprint and their bottom line. The top 20 Green Energy Partners recognized by the U.S. Environmental Protection Agency (EPA) in 2009 are reported to have a combined usage (and generating power) per year of more than 736 million kWh of on-site green power, equivalent to the amount of electricity needed to annually power more than 61,000 American homes.

All top-20 energy producers are DERs, and generate green power on-site from renewable resources, such as solar, wind, geothermal, biomass, biogas, and low-impact hydropower. The highest ranked on the EPA list of Green Energy Partners is the Texas-based corporation Kimberly-Clark, as they generate nearly 193 million kWh of biomass power annually, which, according to the journal *Distributed Energy,* equates to 7 percent of Kimberly-Clark's electricity use in the United States. Not only did the company receive a top ranking on the EPA's Green Partners list, it was also commended for its ongoing operational efforts to increase energy efficiency and reduce greenhouse gas emissions. This feat was accomplished by using a DER facility that utilizes biomass fuel from lumber mills and clearing operations to generate electricity.

There are several synonyms for distributed generation, including decentralized energy, decentralized generation, dispersed generation, distributed energy, and embedded generation or on-site generation. In all cases, however, they refer to independent, decentralized power units that typically produce somewhere between 3 to 10,000 kWh of energy.

Technologies for Distributed Energy

Increased political and consumer demand for greener energy production has led to a growing array of distributed energy systems and sources. Though by no means a comprehensive list of available, smaller-scale decentralized systems that fit the definition of DERs, the following list encompass the primary commercially available and currently economically relevant system types (separated into sustainable and fossil-fuel dependent).

Sustainable	Fossil Fuel Dependent
Photovoltaic and Other Solar: photovoltaic (PV) technology involves converting solar energy directly into electrical energy by means of a solar cell.	*Internal Combustion:* engines burn liquid or gaseous fuels to convert chemical energy into mechanical energy in the form of moving pistons, which in turn spin a shaft and convert the mechanical energy into electrical energy via an electric generator.
Wind: wind turbines convert kinetic energy in the wind into mechanical power that can be converted into electrical energy with a generator.	*Microturbines:* customer-site electric user applications with an output power of between 30 kW and 400 kW, variously fueled by gasoline, diesel, kerosene, or natural gas.
Biomass: a renewable energy source derived from live or dead biological matter that can be turned into heat or electric energy.	*Fuel Cells:* an electrochemical device that produces electricity without any intermediate power conversion stage.
Hydro: the use of the gravitational force of water for generating electricity.	*Flywheel:* stores energy in the form of the kinetic energy of a spinning mass and then converts into electric energy.

History of Decentralized Power

In the early days of electric power, isolated power stations were built which in turn connected to small groups of consumers or a large industrial consumer of power. This model seemed logical in the early days of the electricity industry. As the demand for electricity grew, the economies of scale of connecting ever-increasing numbers of consumers to large, centralized generating plants made increasing economic sense. This required standardization—competing interests, each with their own proprietary voltages (some supplying power as direct current)—with those who produced alternating current supplying at a range of frequencies. Eventually, nations settled on standardized voltages, alternating current became the standard (because of the ease of producing transforming equipment), and frequencies were standardized. This standardization permitted the interconnection of all generators and consumers of electricity together in a large synchronous grid.

The advantages of centralized power generation were that generators could be concentrated in locations where there was a ready supply or good connection to distribution infrastructure for fossil fuel or large hydropower resource, while consumers could be in any location within reach of "the grid." The term *coal by wire* became popular in the United Kingdom to signify that electricity is just a "vector" for power produced by other means. This paradigm remained unchallenged throughout the latter part of the 20th century. Electricity was produced centrally in large power stations that enabled inputs of primary energy to be concentrated in a small geographical area, often near the source of supply, while consumers were dispersed throughout a wide geographical area. By concentrating power generation outside major towns and cities, the pollution produced from combustion of coal could be kept away from highly populated urban areas.

There have been a number of drivers to reevaluate the dominance of centralized power generation:

- Centralized power stations only transform a relatively small amount of power into electricity—the rest is wasted as heat (the efficiency of traditional steam-generating plants is between 30 and 42 percent, with modern combined cycle gas turbine power plants achieving up to 60 percent).
- Greater integration of generation from renewable energy sources into the grid will change the shape and form of the power distribution infrastructure.
- Centralized power stations rely on a steady supply of fuels in finite supply for their operation. With international competition for scarce resources, the cost of many of these commodities has risen sharply. Decentralized energy, while requiring a greater capital investment in equipment, does not have the ongoing running costs of providing fuel for primary energy.
- With the growing awareness of climate change, and the potential that carbon could come under taxation in future green laws, centralized generation based on burning fossil fuels begins to look expensive when the externalized cost of carbon is included as the product of taxation.

The Structure of Our Power Utilities

Traditional centralized power utilities are structured and operate in a way that is hierarchical, essentially creating a one-way road of power supply that leaves consumers, whether private or business, dependent on energy sources, reliability, and quality provided by supplier. All four of the following essential stages of the journey of energy from source to end user are traditionally controlled by power utilities:

- Generation: large generating sets powered by fossil fuel, nuclear power, or large-scale hydropower "supply" network to the transmission system.
- Transmission: the transmission system is the network of high-voltage power lines that transmit electricity over long distances.
- Distribution: the distribution network is the "local" part of the power utility.
- Consumer: the consumer receives an incoming connection at a relatively low voltage (110v/60Hz AC in the United States, 230v/50Hz AC in the European Union).

Distributed generation challenges this paradigm by connecting generating capacity to the distribution network of the grid at much lower voltages than a centralized generating plant. This results in a different structure, with the unidirectional power flow of hierarchical networks from source to sink being replaced by multidimensional flows between consumers and producers.

From "Consumers" to "Consumer Producers"

In the traditional model, in short, energy only flows one way, and end users are passive consumers of energy. In a distributed energy network, the role of the consumer changes. This has a number of effects: as more power is generated near the point of consumption, a lighter, less infrastructure-intensive power distribution network may be possible to share power between producers and consumers. Furthermore, as consumers become producers of energy, producing power becomes a tangible act that happens locally rather than an intangible activity that occurs at a distance. This is said to have a psychological effect on the act of consuming energy, and may incentivize consumer producers to reduce their energy use or invest in efficiency measures.

Reimbursement for Decentralized Energy

When consumers also become producers, they effectively begin an energy trade within the grid, which requires clear contractual agreements with whoever owns the grid. As the economic viability of DERs often depends primarily on whether, and how much, the small decentralized consumer-producers get paid for the energy they feed back into the grid, the commercialization and frequency of use of DERs in turn depends on the contractual and pricing structure of each respective market. There are several types of metering contracts:

- "Avoided cost" metering is where the utility pays the cost it avoided by not having to generate the electricity contributed. In some cases, this can be as low as the price of the fuel that was not used (ignoring the inefficiency of plant, distribution losses, etc.).
- "Net metering" is where the supplier subtracts the amount of energy the consumer–producer contributes to the grid from any power they may have purchased to cover any shortfall in power generation.
- "Renewable energy feed-in tariffs," "feed-in tariffs," or "REFITs" refers to a system where a regulator establishes the value of a unit of electricity from a given source. That value takes into consideration the reduced environmental impact of generating from renewables, the savings in transmission losses, and the value of encouraging the spread of decentralized energy. It is generally the most generous form of recompense for renewable energy, and as such has rapidly spread in countries like Germany that have adopted this scheme.

In order to get around the sometimes low prices paid for renewable energy, some enterprises have established private wire networks.

Feed-in tariffs attempt to provide a level playing field for competing distributed energy sources by acknowledging that generation from fossil fuels and nuclear plants carries with it an implied environmental burden that is externalized without a financial reflection of this externalized cost in the cost paid for a unit of cheap, polluting electricity. By acknowledging that there are costs to society that are not reflected in the price of a unit of electricity from conventional sources, it becomes possible to begin to construct the value that society should be willing to pay for incentivizing "clean" electricity. Feed-in tariffs stipulate the price that an electricity provider should be willing to pay for a unit of clean electricity generated by a given technology, based on a combination of the environmental desirability of that technology and the need to encourage it. This price is always above the market rate paid for electricity from conventional sources, and the burden of paying this additional cost for electricity is spread to all of the consumers that buy from the utility, therefore the premium paid per customer is small.

For example, energy generated from photovoltaic cells is inherently costly, so the REFIT price paid for a unit of electricity from this technology is higher than, for example, a unit of electricity from large wind power. In that way, REFIT tariffs have been shown to be effective in promoting the growth of distributed energy sources in those countries where they have been adopted, as a small penalty for every customer in the price paid per unit results in a large incentive to developers of renewable energy schemes.

An alternative to feed-in tariffs for encouraging distributed energy development is a quota-based system. In a quota-based system, a government or regulating body allocates a set percentage of the total electricity supply that must come from renewables. Often quotas are set on a sliding scale, which mandates an increasing year-to-year proportion of energy from renewables. Quota-based systems, however, have been found to be less effective than REFIT tariffs in encouraging the uptake of renewable energy and distributed generation.

A private wire network is a local electricity network that is separate from the main utility. Electricity is "traded" internally between a group of producer–consumers. This block of consumer–producers may then in turn trade with a larger (national) grid. Usually, there is a diverse range of consumers, requiring power services over different times of the day, and the supply and demand profiles are optimized to ensure the minimum of external trading with the grid. The advantage of a private-wire network is that it can often help to overcome the poor market prices paid for power from clean sources, especially where there are no REFIT tariffs in place. By trading excess energy produced internally, and also by trading heat from com-bined heat and power (CHP), the full value of the energy is realized with value being retained.

Challenges and Opportunities

The energy needs of the world economy will continue to rise, and the price for energy is very likely to rise with it. Fossil fuels as an energy source will become increasingly expen-sive, as energy-hungry nations are inevitably depleting the world's fossil resources, and environmental concerns are very likely to grow in significance. In this scenario, DERs will undoubtedly gain in importance, particularly for businesses (medium to large), but also for hospitals, educational institutions, research facilities, or even the military. Not only do DERs increasingly provide a host of cost-effective options, but they also can substantially help businesses improve their green profile.

See Also: Clean Fuels; Cogeneration; Ecoefficiency; Ecological Footprint; Green Building.

Further Readings

Patterson, Walt. *Keeping the Lights On: Towards Sustainable Electricity*. London: Earthscan, 2007.
Patterson, Walt. *Transforming Electricity: The Coming Generation of Change*. London: Earthscan, 1999.
Wood, Janet. *Local Energy: Distributed Generation of Heat and Power*. London: Institution of Engineering and Technology, 2008.
World Alliance for Decentralized Energy. http://www.localpower.org (Accessed May 2009).

Gavin D. J. Harper
Cardiff University

Dow Jones Sustainability Index

Launched in 1999, the Dow Jones Sustainability Index (DJSI) monitors the environmental and financial performances of leading corporations that promote and employ sustainable practices. In comparison to broad-based stock market indices, such as the Dow Jones

Industrial Average, the DJSI is a socially responsible index (SRI) that enables investors to match their social or ecological criteria with companies supporting those beliefs. The DJSI resulted from an increasing interest by the financial community in the broad area of sustainability and ethical investments. In addition to the world DJSI, several specialized DSJI have been created, based on a world region (Asia/Pacific, Europe, North America, and the United States), the dominant currency exchange, sustainability subtype, or other criteria.

In the last few years, much research has been done to establish a relationship between environmental performance and financial performance, notably the stock returns of publicly traded companies. For investors to understand and compare this relationship across multiple companies, standard indices like the DJSI have been created; furthermore, some mutual funds are now labeled as "green" or "socially responsible" for the environmentally conscious firms they represent. There have also been a few for-profit efforts that evaluate and rate corporate environmental performance and link it to stock returns, including rankings provided by the DJSI and other sources, such as the Innovest Strategic Value Advisors. The concept of sustainability and its many variations (i.e., sustainable development, sustainable use, and sustainable growth) have been addressed by all the disciplines from science, economics, and social sciences, but the concept of sustainability still remains conceptually weak and no universally accepted corporate definition exists. Consequently, the comparison of different sustainability indices must be made with caution.

Whereas the Dow Jones Index looks at the economic performance of the companies listed on the stock index in general, the similar DJSI focuses only on companies committed to sustainability. Investors need a value (preferably in financial terms) and a global, rational, consistent benchmark of their sustainable investments. Dow Jones index publishers, STOXX Limited (a leading European index provider), and SAM (an international pioneer group in sustainability investment) came up with the sustainability index DJSI to track the financial performances of the leading companies believing in the principle of sustainability. DJSI is intended to create a "hard" or measureable benchmark for corporations interested in sustainability issues, rather than just compete with "soft" issues of sustainable development.

DJSI focuses on corporate sustainability and sustainability investment. As defined on the DJSI website, "corporate sustainability is a business approach that creates long-term shareholder value by embracing opportunities and managing risks deriving from economic, environmental and social developments." The concept of providing measurable indicators of corporate sustainability performance is motivating investor interest into companies promoting sustainable practices. It is believed that most of the corporate sustainable leaders gear their strategies toward sustainable products and services while at the same time reducing and avoiding sustainability costs and risks. Awareness among investors is growing and the trend is to invest in companies that set industry-wide best practices in sustainability. Most investors view the concept of corporate sustainability as a means for successful, long-term management, superior performance (by the companies participating in the index), and favorable return profiles. The index thus provides investors and companies an insight into the sustainable practices of the companies against a standard benchmark. In the long term, it also indicates that sustainable companies deliver more predictable results (i.e., less negative returns). In a way, when investors look for sustainable companies, they are not necessarily looking for outsized performance, but for a more constant above-average growth they can rely on.

DJSI mainly consists of leading companies that value sustainability around the world. The index captures the top 10 percent of the 2,500 companies worldwide mainly based on three main criteria: economic, environmental, and social performance. In addition to these three core principles, it is also believed that DJSI superior performance of proactive, cost-effective, and responsible corporations is directly related to their commitment to five corporate sustainability principles: the use of innovative technology of products and services; corporate governance including management, organizational capability, corporate culture, and stakeholder relations; shareholder relations based on sound financial returns and long-term economic growth; industrial leadership by demonstrating commitment; and social well-being. These components are selected after a systematic corporate sustainability assessment that includes looking at the general as well as industry-specific sustainability trends, and more specifically, examining the climate change strategies, energy consumption patterns, supply chain standards, human resources development, labor standards, knowledge management, relationships with stakeholders, and corporate governance; however, the criteria assessed are based on the publicly available information only. DJSI lists all the industries and users of DJSI, and the investors can then apply filters against certain sectors to get the results they are looking for. The assessment criteria are reviewed frequently, and all the different components have a different weight attached to them to come to a final comparative index. Furthermore, the DJSI methodology has the ability to be customized for different geographic regions and specific sustainability criteria.

Since the launch of DJSI, the number of companies participating in the index has steadily increased. The participating companies have also improved their sustainability results on a wide variety of cross-industry criteria, such as risk and crisis management and human capital development. Progress has been made in several industry-specific areas, including climate change mitigation strategies in the utility industry, raw material sourcing in the food sector, and supply chain standards in textiles, technology, and telecommunication. At the same time, substantial room for improvement in sustainability (in the areas of environmental and social reporting, as well as labor practice indicators and innovation management) remains on the corporate agenda.

Parallel to DJSI, there are other initiatives as well, such as the Global Reporting Initiative (GRI, backed by the United Nations Environment Programme), and its Sustainability Reporting Guidelines. The idea for GRI originated when the Coalition for Environmentally Responsible Economies noted that a lot of stakeholders/shareholders were demanding information about the environmental and social performance of companies. Individually reporting back to these queries is very time consuming, and in the absence of any template and guidelines, reports are very inconsistent and incomparable. These factors have led to more systematic guidelines for reporting that include environmental performance and social and economic dimensions of sustainability. Just as financial statements reflect the economic activity report of the company for the shareholders, the GRI addresses the other aspects of economic performance and impact.

Many indices exist, but all of them suffer from at least two common (and probably unavoidable) problems: insufficient data and uncertainty. Resources can be dedicated to gather more data, which can reduce (but not eliminate) the first issue. It is the second problem of uncertainty that defines, and at the same time limits, understanding of sustainability. Although there are a few indices reporting on sustainability performance of companies, investors and shareholders are wary of them, and the message is that the results, in most cases, are seen as greenwashing.

It can be concluded that DJSI will improve transparency and benchmarking (since it includes beneficial social, environmental, and economic performances of the company) and will contribute toward development of better methodologies regarding existing screening processes toward sustainability. The intention of the DJSI is well surmised by Peter Dobers and Rolf Wolff as "The point is not, that the Dow Jones methodology is perfect or correct. The point is, that one of the global players in the financial market gives legitimacy to issues that were previously treated as 'soft.'" The new index will influence companies to make their practices transparent, report and evaluate continuously, and communicate their measures in sustainable framework.

See Also: Ceres Principles; Corporate Social Responsibility; Socially Responsible Investing; Social Return on Investment.

Further Readings

Aignier, Dennis J., Jeff Hopkins and Robert Johansson. "Beyond Compliance: Sustainable Business Practices and the Bottom Line." *American Journal of Agricultural Economics*, 85/5, Proceedings Issue (2003).

Cerin, Pontus and Peter Dobers. "What Does the Performance of Dow Jones Sustainability Group Index Tell Us?" Paper presented at Business Strategy and Environment Conference, University of Leeds, UK, September 18–19, 2000.

Dobers, P. and R. Wolff. "Competing with Soft Issues From Managing the Environment to Sustainable Business Strategies." *Business Strategy and the Environment*, 9/3 (2000).

"Dow Jones Sustainability Indexes." http://www.sustainability-indexes.com (Accessed November 2009).

Knoepfel, I. "Dow Jones Sustainability Group Index: A Global Benchmark for Corporate Sustainability." *Corporate Environmental Strategy*, 8/1 (2001).

Laufer, William S. "Social Accountability and Corporate Greenwashing." *Journal of Business Ethics*, 43/3 (2003).

Mayer, Audrey L., Hale W. Thurston and Christopher W. Pawlowski. "The Multidisciplinary Influence of Common Sustainability Indices." *Frontiers in Ecology and the Environment*, 2/8 (2004).

Willis, Alan. "The Role of the Global Reporting Initiative's Sustainability Reporting Guidelines in the Social Screening of Investments." *Journal of Business Ethics*, 43/3 (2003).

Velma I. Grover
United Nations University–Institute for
Water, Environment and Health (UNU–IWEH)

ECOEFFECTIVENESS

Ecoeffectiveness is a strategy that seeks to minimize waste, pollution, and resource use when designing and manufacturing products. The goal is to ensure that they use as few inputs—materials and energy—as possible, and that ideally their outputs become inputs for other products, making production far more sustainable.

In 1992, the World Business Council for Sustainable Development (WBCSD) used the term *ecoefficiency* to widen the term *efficiency*—which, in manufacturing, referred to the production of a set amount of goods for the least economic cost—to include considerations of resource usage, waste production and pollution discharge. Ecoefficiency, according to the WBCSD, seeks to produce "competitively priced goods and services that satisfy human needs and bring quality of life while progressively reducing environmental impacts of goods and resource intensity throughout the entire life cycle to a level at least in line with the Earth's estimated carrying capacity."

While ecoefficiency can be applied to existing products in a gradual or incremental manner, ecoeffectiveness seeks to redesign products and manufacturing processes from the bottom up with total sustainability as the primary consideration. It involves new ways of thinking, a change of mindset, and new design paradigms, not simply an adjustment to business as usual. Ecoeffectiveness, if widely adopted, would represent a new industrial revolution.

Traditional industrial production views manufacturing as a one-way flow from source to disposal, from cradle-to-grave, while ecoeffective production aims for a circular flow from cradle-back-to-cradle. Examples of ideal ecoeffectiveness include waste discharged from factories that is cleaner than the water that entered the factory, buildings that produce more energy than they use, and products that are not discarded when they are no longer useful but are fed back into production cycles.

The concept of ecoeffectiveness was developed by William McDonough and Michael Braungart in their book, *Cradle to Cradle: Remaking the Way We Make Things*. McDonough and Braungart use natural systems as their model for industrial ecoeffectiveness. In nature, the waste of one species becomes the food of another, and nutrients flow indefinitely in cycles of birth, decay, and rebirth. In nature there is no waste. This provides the basis for one of their key principles: "waste equals food."

In industry, products can be made of natural materials that biodegrade when they reach the end of their useful life and can be composted so they return their nutrients to the soil. McDonough and Braungart refer to such materials as "biological nutrients." Alternatively, if they are made from synthetic materials, safe and nontoxic materials should be used so that they can be reused to make other useful products. These are called "technical nutrients," and feed into industrial closed-loop cycles, just as "biological nutrients" feed into nature's cycles.

Recycling tends to be economically inefficient because the materials degrade each time they are used and can require large amounts of energy to be reused. For example, steel has to be melted down and becomes weaker in the process. However the aim of ecoeffectiveness is to use materials that can be reused forever without these drawbacks.

Ecoeffective production would utilize renewable energy sources, such as solar and wind power, rather than those that consume coal, gas, or oil, or have dangerous by-products, such as nuclear power.

Achieving ecoeffectivenss involves five steps:

1. Replace materials in the product and production process that are toxic or harmful to the environment: for example, mercury or chlorine.

2. Examine each material, and where economically feasible, find more sustainable replacements.

3. Examine each material and assess for their ability to be incorporated into natural or industrial cycles and replace those that will not require redesign of the production process.

4. Ensure that all materials can be incorporated into natural or industrial cycles without regard to disruption of production methods, that is, they are all technical or biological nutrients.

5. Reinvent the relationship between the product and the consumer: for example, by considering alternative products or services that could meet the same needs, or by considering communal ownership of a product that can be shared.

An example of ecoeffectiveness as a design principle is used in Climatex Lifecycle™ fabric. The fabric is designed so that it will biodegrade at the end of its useful life and can even be used as mulch. The fabric combines wool from free-range sheep and ramie—a fiber made from the nettle plant—and is both comfortable and moisture wicking. Dyes, fixatives and other additives were chosen that would not only perform well but would cause no harm to people, animals, or the environment.

The $2 billion redesign of the Ford Motor Company production facilities in Dearborn, Michigan, were inspired by the concept of ecoeffectiveness. It includes a living roof covered with 10 acres of sedum plants that will keep the factory cooler, skylights and giant windows to utilize natural light, and wetlands surrounding the factory to filter rainwater that flows off of the factory roof.

Solar-powered carpet manufacturer Interface claims that it has increased its revenue by 20 percent and its earnings by 30 percent by adopting the goal of ecoeffectiveness. It "dematerialized" the product line by only using material input that could be recycled into new carpet fiber. Their carpets became "technical nutrients" that could be fed into cradle-to-cradle cycles.

Proponents of ecoeffectiveness argue that it is not only good for the environment, but also reduces operating costs, lowers risks, provides a competitive advantage, differentiates a company's products for environmentally-minded consumers, and enables companies to go beyond regulatory compliance. Costs are reduced because there are fewer wasted resources and less inputs used. There is less waste to be disposed of, and less need for pollution control equipment or waste treatment. Because the materials are not toxic, there is less risk of harm to the workforce, community, or the environment.

A quote from Albert Einstein is often used to summarize the principle of ecoeffectiveness: "The world will not evolve past its current state of crisis by using the same thinking that created the situation."

See Also: Closed-Loop Supply Chain; Compliance; Cradle-to-Cradle; Industrial Ecology.

Further Readings

McDonough, William and Michael Braungart. *Cradle to Cradle: Remaking the Way We Make Things*. New York: North Point Press, 2002.
McDonough Braungart Design Chemistry (MBDC). "Transforming Industry: Cradle to Cradle Design." http://www.mbdc.com/c2c_home.htm (Accessed February 2010).
Schmidheiny, Stephan and the World Business Council on Sustainable Development. *Changing Course: A Global Business Perspective on Development and the Environment*. Boston, MA: MIT Press. 1992.

Sharon Beder
University of Wollonong

ECOEFFICIENCY

The World Business Council for Sustainable Development (WBCSD) claims to have coined the term *ecoefficiency* in 1991. They defined ecoefficiency as

> The delivery of competitively priced goods and services that satisfy human needs and bring quality of life, while progressively reducing ecological impacts and resource intensity throughout the life cycle, to a level at least in line with the Earth's estimated carrying capacity.

The principal objectives are as follows:

- Reducing the use of resources (energy, water, land, and materials), enhancing the recycling and reuse of lost resources, and closing material use/loss loops
- Reducing the impact on the environment by minimizing air emissions, water discharges, waste disposal, dispersion of toxic materials and heat, and fostering the sustainable use of natural resources
- Increasing product or service value by providing more benefits to customers through product functionality, flexibility, and modularity, providing additional services and focusing on selling the functional needs that customers actually want—raising the possibility of the customer receiving the same functional need with fewer resources used and lost in producing the product

Many companies use an environmental management system as a means to ensure that all the risks and opportunities relating to ecoefficiency are properly identified, efficiently managed, and made part an integral part of the way the business is operated. According to the WBCSD, the key elements that organizations can use to improve ecoefficiency are to (1) reduce energy, water, land, and material intensity; (2) reduce the dispersion of toxic substances and heat; (3) enhance the recyclability and reuse of the lost resources; (4) maximize the sustainable use of renewable resources; (5) extend product durability standards; and (6) increase service intensity of products and services.

Companies can achieve ecoefficiency by using various methods, including improving their processes to reduce the consumption of resources, reduce resource loss, and minimize operational risk, which should also help reduce costs. Process changes may also be made in other parts of the product life cycle, including supplier operations, distribution, customer use, and end-of-life considerations for the product. Additionally, companies can try to find by-product value in their wastes by seeking out companies that can use their lost resources as feedstock. Many companies have set zero-waste targets to create increased resource productivity and create additional cash benefits. It could be argued that this is ecoefficient because it allows for the creation of more value with fewer resources and less waste. Companies can also redesign their products in an attempt to dematerialize the product. Some have called this process "design for the environment."

Although these ideals of business ecoefficiency deserve considerable merit, they have not yet become fully employed. However, mainstream attitudes regarding ecoefficient business practices began to alter in 1987 when the U.S. Congress voted to amend the Stevenson-Wydler Technology Innovation Act of 1980. The amendment called for the establishment of the Malcolm Baldrige National Quality Award, a presidential recognition for business performance excellence that includes criteria on demonstrating sustainable and efficient practices. The award was established at a time when the leadership of American manufacturing quality and cost was being challenged by foreign competition, and U.S. manufacturing productivity growth had improved less than that of other countries for two decades. Congress determined that poor quality was costing companies as much as 20 percent of their sales revenues, and that improved quality of manufactured goods and services would go hand in hand with improved productivity, lower costs, and increased profitability. Since this amendment, approximately 70 other countries have instituted performance excellence frameworks.

Key in most of these performance excellence frameworks are criteria for process management. These criteria examine how an organization designs, manages, and improves its central processes for implementing its work systems to deliver stakeholder value and achieve organizational success and sustainability. From the design-side point of view, ecoefficiency requires addressing issues such as adjusting and establishing economical work processes to meet all the major requirements from customers, suppliers, partners, and other stakeholders; incorporating new technology, organizational knowledge, and flexibility into the design of more efficient business operations; and integrating cycle time, productivity, cost control, and other efficiency and effectiveness factors into the work process.

The frameworks direct that businesses should demonstrate methods that are able to meet the design requirements and day-to-day operations of the work process; provide indicators and in-process measures for the control and improvement of work processes; minimize overall costs; prevent defects, service errors, and rework and minimize warranty costs or customer productivity losses; and improve work processes to achieve better performance, reduce variability, and improve products and services while maintaining current business needs and directions. In many cases, the language of the performance excellence criteria was much more adaptable to business practices than concepts such as biomimicry, "cradle to cradle," design with nature, and ecoaudits and has led to overall increased business efficiency, including many ecological sustainability initiatives.

In order to implement these processes and improve business performance, several approaches can be taken, such as the Lean Enterprise System, the Six Sigma methodology, the International Organization for Standardization (ISO) 9001:2008 standards, the Plan-Do-Check-Act methodology, or other process improvement tools. These approaches involve examining resource productivity and/or intensity. Resource productivity is the quantity of good or service (outcome) that is obtained through the expenditure of unit resource. This can be expressed in monetary terms as the monetary yield per unit resource. Resource intensity is a measure of the resources (e.g., water, energy, land, and materials) needed for the production, processing, and disposal of a unit of production or service or for the completion of a process or activity. It is a measure of the efficiency of resource use and is often expressed as the quantity of resource embodied in the unit cost (e.g., liters of water per $1 spent on product). Resource productivity and resource intensity involve very different approaches and can be viewed as reflecting, on the one hand, the efficiency of resource production as outcome per unit of resource use (resource productivity) and, on the other hand, the efficiency of resource consumption as resource use per unit outcome (resource intensity). The sustainability objective is to maximize resource productivity while minimizing resource intensity.

Resource management guidelines can be found in ISO 14001:2004, requirements that help an organization examine the potential for ecoefficiency and resource productivity by having it consider the resource use and loss associated with all of its activities, products, and services. The organization must examine its core processes (i.e., those directly responsible for the products and services), its supporting processes, and other nonproduct processes (e.g., business processes, infrastructure processes, grounds maintenance, and so on). Once all of the environmental aspects are identified (i.e., the interaction of resource use and/or loss with the environment), an organization uses operational risk management to prioritize the resources used and lost based on their potential environmental impact. Ecoefficiency can be then be improved using existing business methodologies, such as performance excellence frameworks, operational risk management standards, ISO 14001:2004 guidelines, and process improvement.

See Also: Ecoeffectiveness; Environmental Assessment; Environmental Risk Assessment; Sustainable Development.

Further Readings

Esty, Daniel and Andrew Winston. *Green to Gold: How Smart Companies Use Environmental Strategy to Innovate, Create Value, and Build Competitive Advantage.* Hoboken, NJ: John Wiley & Sons, 2009.

National Institute of Standards and Technology (NIST). "Malcolm Baldrige Performance Excellence Framework." http://www.quality.nist.gov/History.htm (Accessed August 2009).

Schmidheiny, Stephan. *Financing Change: The Financial Community, Eco-Efficiency, and Sustainable Development.* Cambridge, MA: MIT Press, 1998.

Suh, Sangwon. *Handbook of Input-Output Economics in Industrial Ecology (Eco-Efficiency in Industry and Science).* New York: Springer, 2010.

U.S. Environmental Protection Agency. "Design for Environment." http://www.epa.gov/dfe (Accessed August 2009).
World Bank Council for Sustainable Development (WBCSD). *Eco-Efficiency: Creating More Value With Less Impact*. Geneva: WBCSD, 2000.

Robert B. Pojasek
Harvard University

Ecoindustrial Park

An ecoindustrial park (EIP) is a cluster of businesses networking and cooperating with each other as well as with other stakeholders, such as the local community, in an attempt to efficiently share resources to reduce overall waste and pollution. The goal is to achieve sustainable development and improve environmental quality, while also increasing economic gains. Members work together in what is often a contiguous location where coinvestors seek superior environmental, economic, and social performance through symbiotic collaboration in resource management and the supply of ecoefficient services, such electricity, transportation, and waste removal. EIP designs usually include individually managed buildings as well as shared support facilities.

In Kalundborg, Denmark, a symbiosis network allows DONG Energy's Asnaes Power Plant to heat 3,500 local homes. Steam from the power plant is also sold to a pharmaceutical and enzyme manufacturer, and fly ash and clinker from the plant are used for road building and cement production.

Source: lcl/Wikipedia

The ecoindustrial park concept is a culmination of several fields of research and practice including industrial ecology, clean production, urban planning, and architecture. At a basic level, the model builds from a business concept known as economies of scale, or reductions in unit cost and increases in efficiency gained from large-scale or aggregate production. For example, it can be costly for one firm to build transportation infrastructure, but spread across a dozen firms the expense per firm is reduced. Furthermore, options such as public transportation routes or carpooling are increasingly feasible when business partners cooperate. By augmenting savings through cost sharing, the park may be able to host support services such as a training center, a cafeteria, a daycare center, or research and development (R&D) facilities that may otherwise be unaffordable.

The EIP concept has evolved along with innovation in patterns of intercompany relationships and organization design, but the unique contribution of the EIP is the goal to advance ecologically friendly development. This objective is particularly notable because industrial clusters or corridors have historically been linked to pollution and degradation of the local environment. In contrast to many industrial parks, an EIP should fit into its natural setting and take a proactive role in rehabilitating or at least maintaining the surrounding ecological systems. A number of EIPs have adjacent forest, wetland, or coastal areas under sustainable management plans. In some instances, attunement to the local environment can contribute to reductions in operating costs, such as when firms collect and utilize rainwater in drought-prone environments. Other advantages come from efficiencies in lighting and heat. Some parks develop on-site renewable energy sources, such as wind and solar. An EIP goal may be to become totally energy independent of fossil fuels or outside electricity, and the EIP may be created adjacent to existing or expanding power plants for this reason. The park can also include an efficient material flow system by anchoring development around resource recovery companies that are set up for this purpose. For example, a water flow system can be designed such that processed water from one plant may be reused by another.

The process by which companies collaborate to utilize each other's by-products and share resources has come to be known as "industrial symbiosis." For example, in Kalundborg, Denmark, a symbiosis network links a coal-fired power plant that heats 3,500 local homes to a nearby fish farm, whose sludge is then sold as a fertilizer. Steam from the power plant is sold to a pharmaceutical and enzyme manufacturer; this reuse of heat reduces the degree of thermal pollution discharged to a nearby fjord. Moreover, fly ash and clinker from the power plant are used for road building and cement production. This industrial symbiosis, like most others, evolved gradually over several decades, and was not created as a top-down initiative. As environmental regulations became more stringent, companies were motivated to cooperate and create synergies that would reduce the costs of compliance and make productive use of by-products. Industrial parks built around the notion of symbiosis can aim to have zero emissions and essentially become closed-loop manufacturing centers.

The EIP concept gained popularity in the United States and Canada in the 1990s, although a significant number of proposed park projects stalled prior to construction. In the United States, many EIPs were promoted by the U.S. Environmental Protection Agency (EPA). A popular model was linked to brownfield redevelopment, where state agencies sought to rehabilitate lightly contaminated zones in areas of economic distress. A number of EIPs were funded in 1996 by the President's Council on Sustainable Development and found supplemental support through tax incentives or grants. However, when state funds dried up, some EIPs did not continue. Those that did continue were often linked to local government agencies, community development associations, or enterprise commissions.

State and local agencies often support EIP projects to create new high-technology hubs, generate employment, and develop tax revenue. Some EIPs were largely economic partnerships, although members could capture some ecological benefits through green building technologies and other on-site environmental initiatives. Most EIPs host one or more anchor businesses with goods or services marketed as sustainable, although in some instances it was initially planned that all park residents would be green businesses. Some clusters planned as EIPs reverted over time to become conventional industrial parks with narrow ecological gains, rather than zero emissions sites hosting frontrunners in new green technologies, as was initially proposed.

Some EIPs are organized around a top-down model, with strict ecological guidelines imposed on tenants. In these instances, the involvement of tenants and local communities may remain limited. However, successful EIPs have generally made concerted efforts to constantly educate member populations and maintain ongoing outreach programs in surrounding communities. Some EIPs are member-run. Cabazon Resource Recovery Park is an EIP set up by a Californian Native American tribe currently running a biomass electricity generation plant and a recycling-manufacturing firm making products of used tires. Devens EIP, located in Massachusetts, was created out of a 4,400-acre military base and Superfund site closed under the Base Realignment and Closure (BRAC) program. Today, Devens operates an award-winning Eco-Star program, which is a set of internal standards to minimize and reduce environmental impact through interfirm cooperation. A more residential model can be found at the Coffee Creek Center in Indiana. This EIP is a multiuse neighborhood made up of residential, office, and retail buildings with shared public spaces that integrate the natural and built environment.

There has been a recent growth of EIP model in developing countries, particularly in Asia, with mixed results. Focus on cost savings often dominates decision making and inadequate attention to other benefits, including the ability to boost the firm competitiveness of small and medium enterprises, are overlooked. The EIP construct creates many opportunities to contribute to the triple bottom line, but in general many industrial parks are not taking full advantage of a wide range of possible synergies that would support broader social programs and promote green technologies.

See Also: Best Management Practices; Biological Resource Management; Closed-Loop Supply Chain; Corporate Social Responsibility; Design for Environment; Ecoefficiency; Ecological Footprint; Environmental Management System; Industrial Ecology; Recycling, Business of; Remanufacturing; Smart Energy; Sustainable Design; Waste Reduction.

Further Readings

Belt, Mike. "ECO2 Commission's Report Outlines Industrial Park Plan." *Knight Ridder/Tribune Business News* (April 25, 2008).
Chertow, Marian R. "'Uncovering' Industrial Symbiosis." *Journal of Industrial Ecology*, 11/1:11–30 (2007).
Chung, Yi-Chan, Yau-Wen Hsu and Chih-Hung Tsai. "The Effect of Ecodesigning on the Development of New Products in High-Tech Firms: An Empirical Study." *International Journal of Management*, 25/3:403–17 (2008).
Ehrenfeld, J. and N. Gertler. "Industrial Ecology in Practice. The Evolution of Interdependence at Kalundborg." *Journal of Industrial Ecology*, 1/1:67–79 (1997).
Geng, Yong, Pan Zhang, Raymond P. Côté and Tsuyoshi Fujita. "Assessment of the National Eco-Industrial Park Standard for Promoting Industrial Symbiosis in China." *Journal of Industrial Ecology*, 13/1:15 (2009).
Hewes, Anne K. and Donald I. Lyons. "The Humanistic Side of Eco-Industrial Parks: Champions and the Role of Trust." *Regional Studies*, 42/10:1329 (2008).
Kim, Heungsoon. "Building an Eco-Industrial Park as a Public Project in South Korea: The Stakeholders' Understanding of and Involvement in the Project." *Sustainable Development*, 15/6:357–79 (2007).

Ecolabels   159

Lowe, Ernest. "Eco-Industrial Parks: A Foundation for Sustainable Communities?" http://www.globallearningnj.org/global_ata/Eco_Industrial_Parks.htm (Accessed April 2009).

Lowitt, Peter C. "Devens Redevelopment: The Emergence of a Successful Eco-Industrial Park in the United States." *Journal of Industrial Ecology*, 12/4:497–510 (2008).

Park, Hung-Suck and Jae-Yeon Won. "Ulsan Eco-Industrial Park: Challenges and Opportunities." *Journal of Industrial Ecology*, 11/3:11–13 (2007).

Zhang, Xiangping, Anders H Strømman, Christian Solli and Edgar G. Hertwich. "Model-Centered Approach to Early Planning and Design of an Eco-Industrial Park Around an Oil Refinery." *Environmental Science & Technology*, 42/13:4958–87 (2008).

Abhijit Roy
University of Scranton

ECOLABELS

Ecolabeling systems are usually voluntary label systems for consumer products that designate the environmentally friendliness of a product based on an evaluation by a third party using a set of environmental standards. The designation of "green products" developed in response to growing consumer interest in purchasing products that meet an acceptable standard for environmental protection or social preference. Common ecolabels will include terms and icons identifying products as recyclable, natural, energy efficient, green, environmentally safe, and ecofriendly, among others. Ecolabeling systems are opt-in, requiring an application for certification and meeting various requirements to demonstrate that the product or policy has a reduced negative environmental impact. The systems are regulated by various governmental and consumer advocacy organizations (e.g., Consumers Union, publisher of *Consumer Reports*) to prevent greenwashing, the deceptive business practice of advocating products and policies as environmentally beneficial without verifiable certification. Ecolabels serve as a market-based approach to encourage businesses to improve their environmental performance by requiring information disclosure, and are found in many business sectors, including appliances, services, community infrastructure, and some aspects of the food industry.

This Energy Star label, a program backed by the Environmental Protection Agency, can be found on energy-efficient appliances.

Source: U.S. Environmental Protection Agency

Several categories of ecolabels exist depending on the regulating body, the number and type of product or policy attributes covered, and the level of enforcement. Different types of organizations including governmental and nongovernmental organizations (NGOs) have established ecolabel programs (e.g., Energy Star, LEED). Ecolabels may address a single product characteristic (e.g., chlorine-free) or be more comprehensive, often using broader guidelines and terminology to characterize the product as green (e.g., natural). The regulation of ecolabel standards is dependent on the enforcing body and the whether or not compliance with a particular ecolabeling program is mandatory or voluntary; mandatory ecolabeling systems are often government regulated.

Organic certification and some other food labels are generally not considered subsets of ecolabeling; the requirements for organic certification overlap with, but are not exactly the same as, environmental concerns. On the other hand, the "Dolphin Safe" label systems used for canned tuna are environmentally motivated. There are several, formulated in response to consumer demand when public attention was drawn to the harm routinely caused to dolphins by what were then common tuna fishing practices. Different label systems have different requirements, and not all of them require independent verification (through unannounced inspections of canneries and tuna boats). For this reason, Greenpeace and other groups have cautioned that Dolphin Safe labeling is unreliable without the consumer researching the requirements of the specific label in use.

A more reliable program is the Marine Stewardship Council's (MSC) "Certified Sustainable Seafood" label, a blue label that can be displayed on any seafood that comes from a verified sustainable source. It is used in about 60 different countries. Fisheries that want MSC certification are certified by an independent accredited agency that investigates the fishery and compares its practices to those prescribed in the MSC standards, along with chain-of-custody certification from the boat to the point of sale.

The Global Ecolabelling Network is an international nonprofit group for networking among 25 ecolabeling programs throughout the world. Members include ecolabel groups in Australia, Brazil, Croatia, the Czech Republic, the European Union, Germany, Hong Kong, India, Indonesia, Japan, Korea, Canada, the United States, the Philippines, Russia, Taiwan, Singapore, Sweden, Thailand, Ukraine, and the United Kingdom. The U.S. GEN member is Green Seal, a nonprofit ecolabel standards developer and certifier founded in 1989. Green Seal–certified products are required to demonstrate that they have less impact on the environment and human health than other products in their class; this is determined through life cycle assessment. Life cycle assessments examine the cumulative environmental impacts of a product from the raw materials phase through to eventual disposal or recycling. Furthermore, a Green Seal–certified product must be competitive and at least as good as other products on the market.

A program similar to Green Seal in Canada is the Environmental Choice Program, established in 1988 for 300 product categories. The symbol of Environmental Choice certification is the Eco-Logo, a green maple leaf design formed by three interlinked doves. With the increase of green brands in the 21st century, both Green Seal and Environmental Choice are becoming more visible and relevant programs.

In the case of many products, the principal issue at hand when it comes to green certification is energy efficiency. In the United States, there are both state and federal "green stickers." California issues such stickers (affixed to the license plate) to automobiles and all-terrain vehicles that meet certain emissions and fuel-efficiency standards, which are greater than those mandated by the U.S. Environmental Protection Agency (EPA). At the federal level, the EPA introduced the Energy Star program in 1992, which proved to be one of the agency's biggest successes—the Energy Star standards have since been adopted by Canada, Australia, New Zealand, Japan, and the European Union.

Energy Star

Designed to promote energy efficiency in order to reduce carbon emissions associated with power plants, the Energy Star program is a voluntary labeling program that requires a product-type-specific energy-efficiency gain of about 20–30 percent, depending on type. When it was introduced in 1992, it was used for computer products, which with the growing ubiquity of home and office computers were becoming major contributors to energy use, and which by their design could stand a great deal of streamlining. The program was first expanded in 1995, with labels for new homes as well as for residential heating and cooling systems. In the subsequent years, more and more product types were added, with specifications being updated periodically with new standards.

On July 1, 2009, Energy Star 5.0 specifications went into effect for computers, the oldest Energy Star product type. The new specifications require the use of 80 PLUS Bronze or better power supplies. "80 PLUS" is a rating of energy efficiency in computer power supply units, calling for at least 80 percent energy efficiency (meaning that less than 20 percent of the energy is wasted in the form of heat) at 20 percent, 50 percent, and 100 percent load levels; "80 PLUS Bronze" is a step up from 80 PLUS, requiring 82 percent, 85 percent, and 82 percent efficiency at those load levels.

Energy Star–certified refrigerators are required to be 20 percent more efficient than the minimum standard. Refrigerators have been a source of criticism for the program, because when manufacturers test their refrigerators, they're allowed to disable the built-in icemaker, which constitutes an efficiency gain that won't be reflected in home use. Furthermore, the inherently inefficient side-by-side category of refrigerator, which consumes substantially more power than other refrigerators of the same volume, is given an Energy Star category, which some environmentalists find counterproductive. This is all part of a broader overall complaint about Energy Star: a lack of sufficient oversight. Manufacturers are allowed to conduct the tests themselves, and spot-checking by the program is sporadic, leading to the possibility that error or fraud could go unnoticed.

Energy Star–certified fluorescent lightbulbs use 75 percent less energy than incandescent lightbulbs and last up to 10 times longer. They have been one of the most successful green consumer products in recent years. Similar requirements are set for other types of energy-efficient lighting, such as LEDs.

The Energy Star certification for several home appliance categories is currently being reevaluated in response to a 2006 federal court ruling in a complaint brought against the EPA by 14 states. The ratings were ruled misleading, and products are expected to be recertified (or have their certification revoked) sometime in 2011. A more recent complaint, at the end of 2008, alleged that the program has based some of its certification ratings on inaccurate data pertaining to greenhouse gas emissions (erring in the favor of the products certified).

One of the interesting aspects of the Energy Star program has been the certification of buildings and the rise of "green building" (much of which was done during the housing boom of the first decade of the 21st century). Energy Star standards exist for residential homes, banks, courthouses, hospitals (acute care and children's), hotels/motels, schools, offices, college dormitories, retail stores, supermarkets, refrigerated warehouses, and unrefrigerated warehouses, as well as automobile assembly plants, cement plants, corn refineries, and municipal wastewater treatment plants. The Energy Star certification for residential homes requires a 15 percent gain in energy efficiency over standard homes, and Energy Star appliances, lighting, and water heaters. The gains are usually achieved with efficiency-minded tight construction, high-performance windows, energy-efficient heating/cooling systems, and carefully installed insulation. It's generally more feasible to build a new Energy Star–compliant home than to convert an existing home to meet the standards, but it depends on the age and construction of the home. At the height of the housing boom, about one-eighth of new construction was Energy Star certified—a fairly high number for a program that was then only 10 years old.

Ecolabels have made an important impact on business marketing, consumer spending, and environmental protection. Green is now considered a mainstream, marketable attribute and more business practices and products are conforming to ecofriendly standards set forth by governmental and nongovernmental organizations. Ecolabel programs have enabled consumers to comparison shop. However, not all ecolabel programs are comparable and some have yielded criticisms regarding changes in product performance, labeling consistency, and/or misrepresentation during different stages of the product life cycle.

See Also: Certification; Clean Technology; Corporate Social Responsibility; Green Retailing; Greenwashing; Voluntary Standards.

Further Readings

Aldy, Joseph E. and Robert N. Stavins, eds. *Architectures for Agreement: Addressing Global Climate Change in a Post-Kyoto World.* New York: Cambridge University Press, 2007.
Hinrichs, C. Clare and Thomas A. Lyson, eds. *Remaking the North American Food System: Strategies for Sustainability.* Lincoln: University of Nebraska Press, 2009.
Ikerd, John E. *Crisis and Opportunity: Sustainability in American Agriculture.* Lincoln, NE: Bison Books, 2008.
Stewart, Richard B. *Reconstructing Climate Policy: Beyond Kyoto.* Washington, D.C.: AEI Press, 2003.

Bill Kte'pi
Independent Scholar

ECOLOGICAL ECONOMICS

Ecological economics is a relatively new field of study that cuts across all of the major scientific disciplines required for the knowledge, vision, and tools to help restructure society toward sustainability that relies upon the vital but finite services provided by the natural environment. Ecological economics is defined as the integrated or transdisciplinary field of study of the sources and nature of interactions between ecological and economic systems to promote their joint sustainability and socially accepted welfare and distributional outcomes. In this entry, the key features of this new approach to environmental problem solving and its relevance to sustainable business operations will be examined.

Historical Context

Ecological economics (EE) emerged in the mid to late 1980s partially as a response to growing dissatisfaction with the efforts of mainstream or neoclassical economics (NCE) in accounting for environmental welfare effects. Economics did have well-developed links to environmental or natural resources by the mid-20th century, with the establishment of the field of natural resource economics. Here, natural resources were the center of interest, but the primary original focus was upon how to optimize (in present value monetary terms) the intertemporal allocation of natural resources. There was limited concern for external environmental costs and benefits, and broader issues of system-wide sustainability.

The 1960s and early 1970s marked the birth of significant global environmental awareness with notions such as environmental toxicity, limits to growth, and steady-state economics. This provided an intellectual hotbed for both neoclassical economics-based environmental economics and the more radical perspective encapsulated by EE after the mid-1980s. Environmental economics originally focused on waste or pollution market failures with a traditional incrementalist view aligned with neo-Keynesian market economics that merged free market and government intervention (for market failure) ideas. The theoretical and applied scope of environmental economics was gradually extended, and by the late 1980s the field had opened up substantially and become popular in its supporting role to the rapid uptake of sustainable development notions (for example, the Brundtland Report and the work of David Pearce). The status of environmental aspects was raised significantly in economic reasoning, with much wider recognition of the potential for integration and the application of market-based policy for environmental management and sustainability. Arguably, this work was the main predecessor to the influential 2006 Stern Report highlighting the immense magnitude of the economic consequences of climate change.

However, many scientists in the environmental field believed that the close association of NCE with mainstream market economics placed too many paradigmatic restrictions for an effective sustainability science. They resisted the assumption and optimism of NCE that partial fixes via market correction and subsequent pricing and substitution would be sufficient to sustain nature's support of humankind. A consequential meeting of sympathetic economists, ecologists, and other key interested scientists at the Wallenberg Symposium in Stockholm in 1982 led to the establishment of the new transdiscipline of ecological economics. By the late 1980s, EE was well established, with the creation of the International Society of Ecological Economics (ISEE), the influential journal *Ecological Economics*, and many university courses and programs throughout the world.

Like NCE, this new approach also has strong roots in zero-growth and steady-state economy ideas, but EE adopts an overriding focus upon biophysical limits to the economy that are explained by thermodynamic principles. Social limits are also accepted in regard to the full welfare consequences of human-induced material and energy flows. In comparison to NCE, EE is more revolutionary and much broader in its coverage of relevant natural and social sciences, and leads to the recommendation of more profound and radical sociocultural and institutional change. Ecological economics also reflects classical economic ideas in areas such as the importance attached to natural resources for welfare, and the absolute scarcity and material growth restrictions from limited "supply" from nature. EE is united by its critical view on the neoclassical economics view of environmental scarcity.

The Scope of Ecological Economics

The concept of "ecological" economics can be seen as somewhat of an oxymoron, given the historical view that the perspectives and goals of economics and ecology are fundamentally opposed. However, the term makes eminent sense if one accepts the idea that the human economy is fundamentally dependent upon ecosystems. Support comes not only from the notion that economic activity needs nature's physical inputs and waste assimilation functions, but that the environment also provides a whole range of welfare-enhancing services in its natural state. This is a new way of thinking about economic–ecology relations in which interactions between the two fields are upheld to be much more profound than in environmental economics with its strong neoclassical economic base.

Ecological economics tends to embrace a bigger picture view of the economy and society than that of most business decision makers. Despite this extensive scope, its philosophy and approach provide a common scientific basis for sustainability across all major relevant actors, including business, government, consumers or citizens, and the community in general. The ecological economic framework and perspective forms a shared scientific and policy platform for communication, analysis, planning, and action for socioeconomic and related ecological change toward sustainable development. This guidance about the path to sustainable societies appears most relevant at broader collective levels. However, its insights and logic can inform individual people and smaller groups and communities as producers and "consumers" and help in better understanding and shaping their behavior, planning and options to effectively respond to likely future changes in lifestyle and demands, resource costs, efficiency and competitiveness, and community goals and relations.

The sustainability themes and logic of EE can also be applied directly within the environment and resource planning activities of private firms, public enterprise, and community organizations. Its assumptions and emphases about the relationship between the economy, society, and nature explain and link together the range of emerging green business imperatives encapsulated in moves toward ecoefficiency, triple bottom line reporting, ISO 14000 standards and environmental management systems, sustainable consumption, life cycle assessment, cradle-to-cradle, and related approaches. Despite a somewhat bleak premise about the limits of nature, EE also offers hope by outlining the potential for simultaneously attaining a better quality of life and reducing pressure on the natural resource flows and processes upon which human well-being depends.

Although EE emphasizes the relationships between economics and ecology, it necessarily extends beyond the economic and biological sciences in an effort to provide more accurate, holistic analyses for the understanding and identification of potential changes required to meet high-level societal goals—notably sustainability. Unlike the value-free or positive stance taken in mainstream economics, which focuses on the description and explanation of economic behavior and phenomena and avoids value judgments, EE adopts distinct value judgments about the way society should be. A detailed focus upon desired end states and outcomes for society also introduces the need to better understand well-being as a policy target and to investigate how changes in values, motives, and behavior can help bring about desired end states with appropriate and acceptable means. Hence, EE must also encompass the suite of disciplines pertinent to (1) understanding the linkages between underlying sociocultural patterns and economic phenomena (and environmental pressures); (2) the accurate and comprehensive assessment of well-being; and (3) discourse about related consequences and acceptability of distributional aspects across local, regional, and national communities, international and global classes, current and future generations, and human and other life forms (and arguably inanimate phenomena).

The cross-disciplinary analytical perspective of EE can be related to the core of its essential assumptions. These assumptions also help define its unique character as perhaps the penultimate environmental (trans)discipline. First, in EE, the stock of natural assets and related services is considered as finite—there are biophysical and thermodynamic limits to material and energy flows available. Second, the human economy is depicted as highly dependent on nature as a source of its welfare and reproduction. Apart from social and affective relationships, nature forms the ultimate source of meeting human needs and wants, and there are marked constraints on the ability to substitute nature's essential services with human-made capital (which itself depends on natural resources), and between many of the key services provided within nature.

The third essential assumption of EE is that the scale of human activity on the Earth's surface is now so large that the economy and ecology are actually jointly determined systems. Humans are capable of perturbating carbon, water, and other natural cycles to the extent that critical thresholds may be exceeded, thus jeopardizing not just conventional economic production but also essential ecosystems and life support services.

Given these axioms in EE, the connection between the economy and ecology is considered fundamental to assuring the future flow of essential environmental services required for human well-being. To achieve sustainability, it becomes necessary to understand how the relevant economy–ecology connections work and relate in physical, sociocultural, and economic terms. Just as the human economy is seen to be part of, and embedded in, the ecosphere, ecological economic models of the economy must be grounded within the parameters of biophysical reality.

Businesses interested in becoming more green can use the tools, as well as the broad point of view provided by EE, to analyze their current resource use and environmental impact and develop means to move closer to sustainability. For instance, researchers at the Gund Institute of Ecological Economics at the University of Vermont have consulted with agriculture specialists to evaluate and plan sustainable agriculture programs in Iowa, and to design watershed management strategies in Mexico, while researchers affiliated with the Beijer Institute of Ecological Economics in Sweden have collaborated on studies to help optimize economic and ecological functioning of aquaculture systems.

Material and energy flows from nature, into and through society, and back to nature, provide an appropriate focus for studying many economy–ecology connections. They form the pathways and flows of society's "throughput" or "metabolism," and help connect the economy to the biophysical universe. The same connections enhance the strategic capability of ecology by revealing how it can incorporate and jointly analyze the economic consequences and trade-offs so important as pressures and facilitators of change in ecosystems. Thus, a primary method of connecting the various disciplines traversed in EE requires having an explicit focus upon identifying, understanding, and tracking key processes, material and energy flows, and other system changes (often within a spatial framework). This provides a basis for mapping out the primary cause–effect linkages between the economy and ecosphere. Ecological economics focuses upon major cases of unsustainability, such as climate change and biodiversity loss, and uses the biophysical connections between the range of relevant disciplines specializing in socioeconomic activity and ecology to help identify pathways toward more sustainable societies.

Two Unique Features

The background on the definition, scope, and goals of EE introduces two of its major unique features: its transdisciplinary and problem-solving approaches.

With regard to the first feature, the need for an intensive cross-disciplinary, pluralistic dialogue between an extensive range of scientific disciplines is a central and defining feature of EE. This approach is often labeled as "transdisciplinary" in order to distinguish it from the simple appending of disciplinary knowledge blocks in multidisciplinary perspectives, or the partial linking of ideas on overlapping areas of interest in interdisciplinary approaches.

For an ecological economist, adopting transdisciplinarity means doing research within a context (often team based) where the analysis and modeling of environmental problems is premised upon and incorporates the full context of all relevant disciplines and their specific theories, methods, and knowledge. An important step in this process is to develop a common set of concepts and analytical tools by the adoption of knowledge and skill areas from the social and natural sciences. Ideally, transdisciplinarity represents a closer, shared paradigmatic approach rather than an ad hoc collaboration of scientists from different fields. The idea is to begin with the focal problem and integrate and synthesize the different relevant disciplines to provide a coordinated and comprehensive analysis of major environmental issues. The analysis and design of effective strategies to address such problems might resolve back to micro-level causes and potential solutions at the level of localized catchments, ecosystem communities or bioregions; individual people, institutions, and human settlements; and specific economic activities and technologies.

Of course, tackling complex environmental problems with an integrated transdisciplinary scientific framework is ambitious and ecological economic analyses face a difficult task in deriving cohesive methodologies that can effectively meet this goal. Despite such challenges, EE has proven to be a useful platform and bridge for interaction and complementarity of the array of disciplines required to address sustainability issues. Individual disciplines, with their paradigmatic blinkers, simply cannot deliver. The complex and interconnected nature of environmental problems means that analysts must draw upon skill sets, and analytical or problem-solving methods from across the gamut of natural and social sciences. Not the least of these is Joseph Schumpeter's idea of a disciplinary approach's "preanalytic vision," or its perspective on the way the world functions, what content and problems are important, the main concepts and influences emphasized, theories linking these aspects, and its vision for desired solutions and change.

Ecological economics aims to join these filtered versions of reality via a common scientific interface and guiding ideology, as no single discipline is considered adequate to the task. Ecological economists must have a reasonable understanding of the complex interrelationships between economic and ecological systems, and an appreciation of related ethical notions, which requires familiarity with each discipline's general approach and view as well as how they overlap and interconnect, providing an operational medium for effectively analyzing problems that cover much of biophysical and socioeconomic reality. For example, even an introductory delving into the climate change issue suggests we would need to use engineering, physics, and chemistry to understand the processes and technology of vehicular, industrial, and agricultural greenhouse gas emissions; climatology, biology, ecology, and numerous other natural sciences to assess absorption capabilities and net atmospheric concentration changes; and economics and geography to assess the total (including embodied) emissions associated with transport, industry, housing, and other contributing economic activities.

A second unique feature of EE is the means used to solve problems stemming from the economy–ecology link. At a superficial level, it can be difficult to distinguish EE from neoclassical environmental economics (NCE) in terms of policy strategies, instruments, and informational needs. However, in EE, tools based on internalizing external costs and benefits into market transactions are only one part of comprehensive strategies that begin with, and are framed within, the limits of the ecosphere together. This is supported by stronger collective recognition and action concerning the full consequences of economic activity on people's well-being and the need to inform and sway people's values, choices, and behavior associated with consumption, lifestyles, production patterns, and technologies. Thus, EE is concerned not only with changing external incentives and quantitative and system rules, but also with the mind-sets of the main human actors.

In NCE fixing, apparent market failures linked to the environment are viewed as an adequate objective and approach. However, in EE, open access and common pool environmental resources (including the oceans, atmosphere, life support, and most critical ecosystem services) are considered of great importance to sustained human well-being, but are poorly addressed by partial market internalization or pricing fixes. Their public good services can only manage with dynamic, society-wide strategies. Similarly, the broad, holistic, and interrelated nature of ecosystem limits requires integrated strategic guidance. Ecological economics perceives the fundamental requirement for sustainability as the institutional and policy capability to guide economic activity and associated technology impacts in a way that ensures that the overall throughput or metabolism of society can be maintained within the biophysical limits (or carrying capacity) of nature.

The market can certainly help achieve the goal of staying within ecological limits in an efficient manner, but EE takes a much broader stance based on the "ends–means spectrum." In this conceptual framework, economics and technological aspects are only intermediate dimensions between the ultimate ends and means of achieving these goals (the latter based on materials and energy from nature), and all ecological economic problems can be analyzed in terms of this spectrum.

Hence, in ecological economics:

- The extended interest in "ends" sets the context for greater interest in well-being and subjective envisioning of desired outcomes for the community or society.
- There is less emphasis upon piecemeal microeconomic fixes than on comprehensively addressing system-wide problems with full cognizance of the need for integrating values and ethics. Of course, micro-level firm and consumer changes would be a key part of adaptive, system-wide evolution.
- The limits of a positive, value-free science are acknowledged, societal outcomes are evaluated, and a marked role is attached to conducive and fundamental changes in human values and mores.
- A major requirement for decision making and policy is to understand and measure the full welfare effects of specific economic activity and technologies across their life cycle and input–output relations with other economic activities (ideally with comprehensive spatial and temporal boundaries).
- There is a call for ecosystem, carrying-capacity consistent, demand-side, and consequent production changes spanning private and public institutions.

Major Component Disciplines and Approach

As noted earlier, the two obvious disciplines under the rubric of EE—ecology and economics—need to draw upon the approaches of many other fields. Of the extensive range of contributing sciences, one major focus is ethics and other tools that help define desirable outcomes for people within society. The roles played by ecology, economics, and ethics are often denoted by the concepts of scale, allocation, and distribution (SAD), respectively.

Ecology (Scale)

Ecology is perhaps best described as the science of ecosystems or the relationships between organisms, their communities, and the rest of the living and nonliving environment. Typically, humans are not a focus of the organismic–environment relationship study (if they were, ecology would tend to encompass environmental economics). Instead, humans are viewed as part of the broader ecology. The economy is depicted as an open system embedded within the natural environment that is essentially a closed system with the exception of limited energy and material inputs from the sun, celestial, and deep Earth crustal sources.

A primary role for ecology in EE is to identify the sustainable scale for society. This involves analyzing and assessing the ecosystem parameters and capabilities that limit the economy's long-term biophysical demands upon nature. As a scientific field, ecology has the relevant foci and perspective to provide knowledge about how much nature can tolerate in terms of human intervention, and to specify the ecological and sustainability impacts of different economic activities, technologies, concentrations and levels. While ecology is a holistic approach that covers much of the relevant content required for this task, it focuses upon biological interactions, and must draw upon many other scientific disciplines, such as physics, chemistry, climatology, geography, geology, and soil science.

Nature provides a diverse range of services that contribute to human well-being. Many of these do not involve significant intervention in material and energy cycles and flows. However, material and energy exchanges between the economy and nature are a primary aspect circumscribing the limits to the biophysical growth of the open economic subsystem within an ecosphere, which has a finite capacity to provide and assimilate human-induced flows.

In EE, a significant part of these system limits are derived by reference to the first and second laws of thermodynamics and the associated notion of entropy. The first physical law describes how in a closed system, energy and matter must be conserved—they cannot be created or destroyed. The second law says that closed systems move toward higher entropy or the dissipation and reduction in the potential work and useful energy that can be extracted from overall system sources. The economy is thought to "feed" upon low entropy sources that provide high heat or energy return, such as fossil fuels. Hence, the finite energy being consumed by humans can only be augmented and made sustainable by accessing sources outside of the ecosphere. Ultimately, this is limited to the sun, gravitational, and crustal sources. In EE, sustainability can only be achieved by creating a balance between the entropy requirements of humans and low entropy sources from external, long-term (typically renewable) sources.

Thus, ecology can play a key role in assessing ecological and entropy limits to the services that can be sustainably utilized by society. In comparison to neoclassical economics and environmental economics, this represents a far more pronounced influence attached to the existence of absolute scarcity in the fundamental natural sources of welfare for humans. Nature is the ultimate source of material, energy, and, arguably, many other forms of human welfare, and its "natural capital" stock and derived services are limited in a complex interdependence of geographic, biological and thermodynamic constraints. Unlike neoclassical economic views, EE holds little faith in the ability and possibility of markets and technology change forcing substitution of scarce natural capital by human-made capital, or between many of the principal environmental services. Belief in the need for strong sustainability stands in contrast to the general environmental economic premise that market and other policy can effectively induce new substitutes for critical natural capital (weak sustainability).

Economics (Allocation)

Ecology provides society with an understanding of the likely and often precautionary levels of sustainable scale for human pressures upon the environment. For EE, the next logical step is to engage economics as the relevant discipline for getting the best overall human welfare levels within these constraints.

In its essence, economics is simply the study of the way societies allocate their scarce resources. In a more normative and prescriptive sense, economics studies how scarce resources should be allocated to achieve optimal or desired levels of community economic welfare. Mainstream economics has centered upon describing how properly functioning markets (assuming coverage of all costs and benefits) will generate pricing signals that allocate scarce societal resources in a way that maximizes total welfare. This is considered an optimal outcome, as maximum output (or utility) can always be redistributed ex post by other political criteria.

A strength of mainstream economics is its specialized focus upon identifying the knowledge, mechanisms, and tools that are useful for the allocation and efficient use of scarce resources to bring about desired welfare outcomes. Having overarching constraints on inputs (and emissions) in order to stay within biophysical limits does not impede this capability. Ecological economics draws quite heavily upon economics' suite of techniques for helping quantify social costs and benefits and correct markets, as well as its policy tool strategies based on market behavior and incentives. These tools do have some distinct advantages for helping bring about sustainability.

The economic dimensions of EE differ from mainstream economics in at least two significant ways. First, market allocation and economic activity and growth are physically restrained so that associated environmental resource pressures stay within sustainable scale limits and with acceptable and just distribution outcomes. Price signals reflecting full welfare impacts can then lead to efficiency or best welfare per unit throughput or resource consumption, but EE believes that there are thermodynamic limits on efficiency and material production efficiency so that related output and material-based welfare will reach some limit given that environmental services are finite.

A second unique economic dimension of EE extends beyond its use of conventional market tools. Economics can also provide the appropriate framework for examining the material, energy and other environmental and well-being consequences of specific activity sectors, technologies, lifestyles, and consumption patterns. This mesoeconomic level (versus micro or macroeconomic foci) reveals the environmental intensity of well-being or link between economic structure and the resource flows that confront sustainability limits. This analytical level provides many strategic insights into successfully achieving the transformation to sustainable societies. An important aspect of this view, and one with profound policy implications, is the highly conditional nature of individual consumer sovereignty as the optimal impetus behind the economic system. Ecological economics is open to the use of collective economic, technological, infrastructural, and sociocultural strategies that override consumer sovereignty where necessary, given the extensive market failures linked to environmental resources as a major source of well-being. This is reinforced once long-term concerns and the welfare of future generations are brought into account.

Ethics (Distribution)

There are many other fields of enquiry that feed valuable information into ecological economic analyses. For example, while economics and ecology form the most visible core of EE, one should not underestimate the critical role of ethics. Ethics considers questions such as "what is just and fair," "who should get what," and "how should people live," with particular regard to their impact on others. It requires expounding moral assumptions and reasoning to identify criteria for what is deemed as "right" or "wrong" with economic and environmental management outcomes. Distributional justice forms a central theme in ethical analyses, and for EE, this involves assessment of both means and end in terms of their full welfare consequences across (1) individuals within a community or nation; (2) nations in the present; (3) current and future generations; and, possibly, (4) humans and other life. Poverty is a factor in the ability to achieve sustainability, and the distribution of economic and environmental costs and benefits influences efficiency and overall well-being conditions across societies.

This perspective aligns with the underlying ecological economic view that the economy is only one, if very significant, part of society. Although economic structural forces are powerful influences on motives, decision making, and actions at individual to collective levels, there are many other important sociocultural, political, and ethical and informal institutional factors at work.

Ecological economics also has a growing interest in the range of scientific disciplines capable of examining the nature of human well-being, including psychology, welfare economics, neuroscience, economics and ethical/religious studies, which are related to ethics as distributional and justice issues often pivot around differential welfare outcomes. It is difficult to refute that a laudable goal of societal policy is sustained high levels of (subjective) well-being, rather than just sustained life, with basic physiological needs fulfilled from environmental functions. At deeper levels, the internal drivers for human motives and actions, and the nature of their links to well-being lies at the heart of the ultimate means–ends spectrum that forms the breadth (and complexity) of the ecological economic approach. To complement the strong biophysical roots of ecological economics, growth of interest in the sociological, ethical, and political influences on economic–environmental interactions has been encompassed in the newer movement of "socioecological" economics.

Conclusion

Although the field is relatively new and still subject to quite rapid development, EE has been defined here as the integrated or transdisciplinary study of the sources and nature of interactions between ecological and economic systems. This requires a common framework that facilitates extensive and coordinated research across many natural and social sciences. The three closest fields are probably (1) ecology, to help set sustainable scale or carrying capacity limits; (2) economics, to efficiently enhance net economic–environmental welfare outcomes within these limits; and (3) ethics, to assess the justice of economic activity means, ends, and outcomes. Better understanding of well-being and the need to complement biophysical limits emphases with social dimensions is also becoming an important focus.

Unlike mainstream economics (and many aspects of its derivative field of environmental economics), central issues in EE include the pervasive theme of long-term, sustainability analysis and goals; the need to extensively inform, guide, and sometimes override consumer sovereignty; and a holistic analytical framework that considers the key interconnections between an economy embedded within broader society (and beliefs, norms, values, and motives), and a natural environment with thermodynamic and ecological limits.

Ecological economics uses many research approaches from environmental and market economics in general, such as valuation techniques and the general cost-benefit analysis approach. However, it draws upon a much wider set of analytical, decision making, and policy tools that reflect its integrated, transdisciplinary approach to environmental problem solving which covers the full social, economic, and environmental dimensions required for sustainability. Many of these tools reflect the need to specify and assess the biophysical connections between economic activity and ecological processes and services. A limited set of examples includes multiple criteria analysis, material and energy flow analysis, life cycle analysis, integrated environmental and economic modeling/methods, and policy evaluation.

In terms of policy tools, emissions trading or "cap-and-trade" approaches reflect the logic of EE by setting ecological service scale limits and using economic tools to help stay within these limits while maintaining or improving society's overall welfare. Nonetheless, EE typically proposes the need for much more extensive and holistic socioeconomic change directed toward a guiding vision for society, and full recognition of the highly interdependent nature of effects across all economic activity and ecosystems. The aim is to create societies that are fundamentally sustainable in terms of people's goals, preferences, lifestyles, urban form, production, and technologies. It is an ambitious goal, requiring a complex and pluralistic new integrated methodological approach.

See Also: Environmental Economics; Externalities; Gross National Happiness; Industrial Metabolism; ISO 14000; Steady State Economy.

Further Readings

Daly, Herman E. *Beyond Growth: The Economics of Sustainable Development*. Boston, MA: Beacon Press, 1997.

Edwards-Jones, G., et al. *Ecological Economics: An Introduction*. Oxford, UK: Basil Blackwell, 2000.

Farley, Joshua, John Erickson and Herman Daly. *Ecological Economics: A Workbook for Problem-Based Learning*. Washington, D.C.: Island Press, 2005.

Sahu, N. and A. Choudhury, eds. *Dimensions of Environmental and Ecological Economics*. Hyderabad, India: Universities Press, 2005.

Sahu, N. and B. Nayak. "Niche Diversification in Environmental/Ecological Economics." *Ecological Economics*, 1/11:9–19 (1994).

Schumpeter, Joseph A. *The Theory of Economic Development: An Inquiry Into Profits, Capital, Credit, Interest, and the Business Cycle*. Piscataway, NJ: Transaction Publishers, 1982.

Peter Daniels
Griffith University, Australia

ECOLOGICAL FOOTPRINT

The term *ecological footprint* refers to the overall human impact on the ecosystem and measures the amount of land and ocean area required to sustain the consumption patterns and absorb the wastes on an annual basis of individuals, nations, or industries. The term was coined in academic publications by William Rees of the University of British Columbia and his doctoral candidate Mathis Wackernagel in the early 1990s and later expanded in their 1996 book *Our Ecological Footprint: Reducing Human Impact on the Earth*. The figurative use of "footprint" was inspired by a computer technician's reference to a new computer's small footprint and calls to mind the prescription, attributed to indigenous peoples, to "walk softly on the Earth." Other related "footprint" terms have followed.

Typically, the ecological footprint is used in reference to the entire population of Earth, and is measured in "Earths." For instance, the Global Footprint Network, established in 2003 to actively promote awareness of the ecological footprint concept, measures resources and consumption every year, based on data collected by the United Nations. There is a three-year lag in the collection of data, so that at the end of 2008, the available measurement pertained to consumption as of 2005. According to that data, the ecological footprint in 2005 was 1.3 Earths—meaning that the human population was consuming resources 130 percent faster than the Earth can replenish them. Clearly, that's an unsustainable rate of consumption, which is exactly the purpose of raising awareness: by consuming resources at this rate, we literally risk destroying the very foundation upon which life depends.

The ecological footprint can also be calculated on a per capita basis. Websites offer tools to estimate one's personal ecological footprint—within an understood margin of error—and footprints can be calculated for particular businesses, regions, or industries, which is a useful tool for identifying the relative consumption of different aspects of modern life. The Global Footprint Network makes available a set of ecological footprint standards for calculation at www.footprintstandards.com.

Typically, per capita ecological footprint measures are taken of national populations, in order to compare their ecological lifestyles. That 2005 data reveal, for instance, that the footprint per capita in the United States was 9.4 global hectares, compared to 2.1 global hectares in China; 2.1 global hectares per capita is also the Earth's "biocapacity"— the "1 Earth" measurement. In other words, although the global ecological footprint was 1.3 Earths, the U.S. lifestyle applied to the entire population would result in a footprint of 4.48 Earths, 448 percent more than the Earth can replenish. This tells us a number of things, including the fact that there must be a significant population consuming much less than the biocapacity threshold in order to balance out industrialized high-footprint countries like the United States—and indeed, many African countries have such low consumption levels. However, while rates of growth are being curbed in Europe, many of the lowest-footprint areas are increasing their consumption at very fast rates, including the African continent, but particularly countries with rapidly growing production and consumption levels such as China and India.

On the Global Footprint Network's list, only the United Arab Emirates (9.5 global hectares) has a larger per capita footprint than the United States; Kuwait (8.9 global hectares), Denmark (8 global hectares), Australia (7.8 global hectares), and New Zealand (7.7 global hectares) follow close behind. Every European country except Armenia and Georgia is above the global per capita average. The lowest-consumption countries are those that are developing and in several cases have faced significant social and political turmoil in the recent past: Afghanistan, Congo, Haiti, and Malawi, all at 0.5 global hectare per capita.

Ecological footprint accounting is new, and no one as yet claims it to be a perfect measure. Some new technologies are hard to account for, as are less utilized technologies like nuclear power, the environmental impact of which is difficult to measure. Above all, for example, how does one account for radioactive nuclear waste? Though the radioactivity of all nuclear waste diminishes over time, most elements used in the production of nuclear energy remain hazardous to human health for thousands of years. Proponents of nuclear power have called for a more in-depth look at the effect its widespread adoption would have on the ecological footprint, believing the evidence will show that nuclear energy is far superior to traditional sources of energy that, according to the World Health Organization (WHO), kill an estimated

3 million people worldwide from vehicles and industrial emissions, and another 1.6 million people through the indoor use of solid fuel.

Another continuous problem with ecological footprint accounts is that disasters, both natural and manmade, such as oil spills or undisclosed toxic waste dumping, are not accounted for (and cannot be accounted for) and so accelerate the depletion of resources faster than the numbers indicate. To make matters worse, there is hardly a day that goes by that scientists do not discover new adverse environmental effects of existing technologies—and, above all, the toxic effects of combinations of existing substances.

The expansion and refinement of ecological footprint measurements is thus an ongoing process. Unlike other environmental measures (such as the carbon footprint), however, the ecological footprint attempts to capture the totality of complex interrelationships between nature and the actions of human beings. Critics have pointed out, however, that more narrowly focused measures such as the carbon footprint have the advantage of highlighting the more immediate and pressing crisis of anthropogenic climate change, potentially leading to permanent damage to the ecosystem and human civilization, long before we will have had time to use up our natural resources.

Businesses of all sizes can generate their own ecological footprint assessment account. The Global Footprint Network has developed a set of internationally recognized Ecological Footprint Standards, providing standards and guidelines for product and organizational Footprint assessments. Through continuously refined methodologies, businesses can not only assess their overall environmental impact, but also use such analyses for investment decisions, productivity enhancement measures, marketing campaigns, and in efforts geared toward increasing the sustainability of operations and products.

See Also: Carbon Footprint; Environmental Accounting; Externalities.

Further Readings

Cui, Yujing. *Ecological Footprint and Environmental Sustainability Index: Instruments of Measuring Sustainable Development Case Study: Sustainability of Shandong (China).* Saarbrücken, Germany: VDM Verlag, 2009.

Dodds, Walter Kennedy. *Humanity's Footprint: Momentum, Impact, and Our Global Environment.* New York: Columbia University Press, 2008.

Ohl, Brian, Steven Wolf and William Anderson. "A Modest Proposal: Global Rationalization of Ecological Footprint to Eliminate Ecological Debt." *Sustainability: Science, Practice, and Policy,* 4/1 (2008).

Wilson, Edward O. *The Future of Life.* New York: Vintage, 2003.

Bill Kte'pi
Independent Scholar

Economic Value Added

Economic Value Added (EVA) is a technique of accounting that is aimed at providing the market, stakeholders, and shareholders with a less opaque and more trustworthy view of a company's prudent use of resources and overall economic health. EVA can also be called economic rent or economic profit. Although the application of EVA has been adopted by companies from all business sectors, businesses with an interest in sustainability and green enterprise have adapted it to include considerations that may have both direct and indirect environmental effects. A prominent example is the natural grocery store Whole Foods Market. Given the vast, but not uniform, adoption of EVA across the spectrum of the green sector, debates have raged concerning how value creation or destruction is defined, especially in light of the challenges posed by new paradigms such as ecoeffectiveness and extended product responsibility.

EVA was created as a proprietary system and trademarked by Stern Stewart & Co., a New York management consulting firm. EVA has been promoted as a system that helps to avoid accounting problems that in the past have allowed companies to produce statements and U.S. Securities and Exchange Commission (SEC) filings of profits while actually teetering on the edge of bankruptcy by hiding debt. Of course, the real estate bubble of the late 2000s and the use of mortgage-backed securities, as well as credit default swaps, demonstrate that assessing value gained is always difficult.

It is commonly held that, to determine the EVA for an enterprise, one must tally the weighted average cost of capital, multiplied by the total capital invested. This is then subtracted from the net operating profit after tax (NOPAT) to obtain the EVA, or the enterprise's profits adjusted for the expense of financing capital. Because the calculation of the NOPAT is essential to determining EVA, it is important to clarify what constitutes NOPAT.

NOPAT is a common calculation of generally accepted accounting principles, as well as a term common to green business. It is considered to be a company's operating profit after taxes have been deducted. Operating profit is the company's earnings in a given period before interest and taxes have been deducted. Calculating NOPAT is done by starting with earnings before interest and taxes (EBIT), converting accrual (accumulated revenues, which can include money owed to the company but not yet paid to the company) into cash, and capitalizing some expenses on the balance sheet to reflect the fact that they are investments. At that point, taxes are deducted. The difference between NOPAT and EBIT is essential to understanding the applications of EVA to multiple business settings.

Proponents of EVA argue that it is an accurate performance metric for measuring the success of senior leadership to increase value in an enterprise. For green businesses, or businesses that are including green elements in their management plans, EVA can be an additional method for showing vision and efficient use of resources. Other wealth metrics, like equity market capitalization, depend on the performance of the company's stock price and, therefore, are affected by market swings and macroeconomic trends. These methods essentially measure the success not of the company but of the stock price. Conversely, a performance metric like EVA reflects the company's actual performance and is the result of management decisions in a way that stock performance is not.

EVA views the intrinsic value of a company as consisting of its invested capital plus the current value of its future economic profits. EVA prizes "cash," which was often lacking by many of the allegedly profitable firms of the early 2000s that in fact were hiding debts and would later enter bankruptcy. Since the 1990s, EVA has been used internally by an increasing number of businesses and global corporations, and it has grown in its usage by green industries. Further, it has become a tool of prospective investors that can be used to determine a company's "true" value and to determine whether or not the company efficiently uses its resources. These items are of particular concern to green investors, and the information may not be readily apparent from the company's annual reports and SEC filings.

See Also: Ecoefficiency; Ecological Economics.

Further Readings

Stern, Joel M., John S. Shiely and Irwin Ross. *The EVA Challenge: Implementing Value Added Change in an Organization.* Hoboken, NJ: John Wiley & Sons, 2001.

Whole Foods Market. "Economic Value Added®."http://www.wholefoodsmarket.com/company/eva.php (Accessed January 2010).

Young, S. David and Stephen F. O'Byrne. *EVA and Value-Based Management: A Practical Guide to Implementation.* New York: McGraw-Hill, 2000.

Bill Kte'pi
Independent Scholar

ECOSYSTEM SERVICES

Ecosystem services is a collective term to describe the range of valuable benefits that the natural environment provides for people, either directly or indirectly. These services fall into four basic categories: (1) the provision of food, water, fuel, and other materials; (2) the regulation of natural processes such as climate, water supply, and water or air quality; (3) the cultural services that natural systems contribute to our quality of life or spiritual well-being by providing beautiful landscapes and opportunities for recreation or education; and finally, (4) the supporting services that are essential to all of the other ecosystem services, such as soil formation, photosynthesis or nutrient cycling, the maintenance of biodiversity, and the space to build and maintain our supporting infrastructure. Recognition of the economic value of ecosystem services or the cost of replacing them provides an opportunity to assess the wider impacts of a particular development.

Rudolf de Groot initially defined ecosystem functions as "the capacity of natural processes and components to provide goods and services that satisfy human needs, directly and indirectly," and identified 23 functions, grouping them into the four primary categories of regulation, habitat, production, and information. Each of these represent a subset of ecological processes and ecosystem functions, the result of natural processes and the complex interactions between living organisms and the chemical or physical components of ecosystems through the universal driving forces of matter and energy. A range of ecosystem goods and services were identified to result from each of these functions and the ecosystem processes and components involved. For example, the regulation function of soil formation involved the processes of weathering of rock and accumulation of organic matter, resulting in ecosystem goods and services such as the maintenance of productivity on arable land or the maintenance of natural productive soils.

Regulation functions relate to the capacity of natural and semi-natural ecosystems to regulate essential ecological processes and life support systems through bio-geochemical cycles and other biospheric processes, namely:

- Gas regulation
- Climate regulation

- Disturbance prevention (i.e., structures to buffer or limit environmental disturbances)
- Water regulation
- Water supply
- Soil retention
- Soil formation
- Nutrient regulation
- Waste treatment
- Pollination
- Biological controls

In addition to maintaining a healthy biosphere and ecosystems, these regulation functions provide many direct and indirect benefits to humans, such as clean air, water, and soil.

Habitat functions relate to the role of natural ecosystems to provide a refuge and reproduction habitat for wild plants and animals, namely:

- Refugium function (a suitable living space for wild plants and animals)
- Nursery function (a suitable reproduction habitat)

In this way, they conserve biological and genetic diversity and accommodate evolutionary processes.

Production functions relate to the process of photosynthesis and nutrient uptake, producing living organisms and a variety of materials for human consumption, namely:

- Food
- Raw materials
- Genetic resources
- Medicinal resources
- Ornamental resources

Information functions relate to the role of the natural environment as a reference point to the habitat where human evolution took place, providing an opportunity for cognitive development, primary health, and well-being, namely:

- Aesthetic information
- Recreation
- Cultural and artistic information
- Spiritual and historic information
- Science and education

Regulation and habitat functions are fundamental to natural processes, and they are a prerequisite for the production and information functions. From a human perspective, these functions thus represent not only valuable but essential goods and services. These values can be recognized as falling into ecological, sociocultural, and economic categories, mirroring the environmental, social, and economic aspects of sustainable development. Clearly, the capacity of many ecosystem services is limited by the necessity to maintain sustainable use levels.

In 2000, then United Nations Secretary-General Kofi Annan called for the first comprehensive evaluation of the state of the global environment. From 2001–05, more than 1,300 experts from around the world generated a state-of-the-art evaluation of both the consequences of human activities on the environment, and the consequences of ecosystem changes on human welfare. The report, titled "The Millennium Ecosystem Assessment," also provided recommendations for the conservation of the ecosystem, and the data on what sustainability would require. Perhaps their most significant overall finding was that "over the past 50 years, humans have changed ecosystems more rapidly and extensively than in any comparable period of time in human history, largely to meet rapidly growing demands for food, fresh water, timber, fiber, and fuel. This has resulted in a substantial and largely irreversible loss in the diversity of life on Earth."

While it is relatively easy, and logically indisputable, to argue that ecosystem services are critical to the functioning of all human endeavors, including all economic activities, it is much more difficult to translate such insights into the day-to-day operations of a business. For one, many ecosystem services are not easily valued, for they don't usually come with a price tag. While direct market valuation is possible where ecosystem services (or the goods they provide) are traded (water, land, oil, and so on), more often economists have to rely on so-called indirect valuations. Such attempts include valuing "avoided costs" (e.g., without the flood control of wetlands, what would the costs have been), "replacement costs" (e.g., water purification plants instead of forest or wetland services), "factor income" (direct economic benefit to businesses due to ecosystem improvements such as cleaner river water for fisheries or tourism), or "hedonic pricing" (what people are willing to pay according to quality associated with the good, as in identical houses fetching a higher price in a clean and safe environment than in a dangerous and toxic environment).

The services of ecological systems and the natural systems that produce them are critical to the functioning of Earth's life support system, contributing to human welfare and the total economic value of the planet. The total annual economic value of ecosystem services of the entire biosphere are conservatively estimated to be well above the total annual global gross domestic product.

While the direct implications for green business are not always clear or easily priced, it is clear that they will play an increasingly significant role. To the extent that government regulations and pollution taxes will grow in variety and volume, businesses of all sizes will have to find ways to lessen their harmful impact on the ecosystem, develop sustainable business models, including investment in technologies that increase the efficiency of use of ecosystem services, and above all develop business models that do not deplete and destroy, but rather protect, renew, and develop ecosystem services.

See Also: Corporate Social Responsibility; Environmental Assessment; Industrial Ecology; Sustainability; Sustainable Development.

Further Readings

Costanza, Robert, et al. "The Value of the World's Ecosystem Services and Natural Capital." *Nature*, 387/6630 (1997).

De Groot, Rudolf, et al. "A Typology for the Classification, Description and Valuation of Ecosystem Functions, Goods and Services." *Ecological Economics*, 41 (2002).

Millennium Ecosystem Assessment. *Ecosystems and Human Well-Being: Synthesis*. Washington, D.C.: Island Press, 2005.

Ninan, K. N. and Achim Steiner. *Conserving and Valuing Ecosystem Services and Biodiversity: Economic, Institutional and Social Challenges.* London: Earthscan, 2009.
Ruhl, J. B. *The Law and Policy of Ecosystem Services.* Washington, D.C.: Island Press, 2007.

Richard Alastair Lord
Teesside University

Ecotourism

Ecotourism (also known as ecological tourism) is a widely used, yet ill-defined term in today's tourism industry. In very broad terms, most observers would probably agree that ecotourism describes "the practice of low-impact, educational, ecologically and culturally sensitive travel that benefits local communities and host countries." In contrast to traditional tourism, ecotourism often involves travel to smaller and less developed destinations, fosters respect and educates participants, claims to economically benefit communities visited, and purports to protect the environment, natural habitats, and precious cultures. Other terms widely used are sustainable tourism, green tourism, responsible tourism, and nature-based tourism.

While there is obviously no central authority providing a set definition for any of these terms, all forms of environmental tourism appeal to a growing environmentally aware segment of the population that wants to learn and experience different cultures, peoples, and places, without, at a minimum, causing harm to such cultures, peoples, and places. With reported growth rates of up to 15 percent annually, ecotourism by now represents a major source of income not only for the tourism industry, but also for the national economies of favorite destinations such as Costa Rica, Antarctica, or New Zealand. Unlike traditional tourism that takes a top-down, consumptive approach to visiting people, places, and things, ecotourism appeals to travelers who are interested in the beauty of nature, including pristine and rarely visited locations, and who generally try to stay away from the hustle and bustle of traditional tourist sites. Ecotourist destinations tend to be those regarded as more authentic and natural, such as trips to the Amazon River, mountain climbing in the Himalayas, or rafting in Costa Rica. Ideally, the image is of traveler and natives engaging in mutually beneficial exchanges of time, resources, and experiences in ways that contribute to maintenance of both ecological and cultural diversity. Ecotourist destinations are generally areas rich in natural attractions (flora, fauna,

By emphasizing the value of preserving nature, environmentalists and tour organizers are able to bring money to places that need it, such as this slice of the Amazonian rainforest in Ecuador's Cuyabeno National Park.

Source: iStockphoto

and cultural heritage). Key to alternative forms of tourism are mutually beneficial exchanges, opportunities for local peoples to participate in and benefit directly from tourist visits, and not overwhelming communities by the presence of visitors.

Because of the large potential economic benefits of ecotourism for both providers and destinations, definitions of what ecotourism does, or should, encompass vary widely between environmentalists, industry standards, and governments. A progressive New York couple taking a one-week rafting trip to Costa Rica, paying what they consider fair wages for local rafting guides, and reading up on the history and culture of Costa Rica before departing, for instance, would undoubtedly be considered ecotourism by the tourism industry, and probably also by the Costa Rican government. From the perspective of an environmentalist, however, the roughly 4,400 mile roundtrip flight alone, generating approximately 40–50 pounds of carbon dioxide emissions per passenger each traveled mile, fails to meet the standard of "low impact" travel, as do hotel stays, cab rides, and various other amenities likely fail to meet the standard of ecologically and culturally sensitive travel that benefits local communities and host countries.

Ecotourism Criteria

Several attempts to define ecotourism have been made. According to the International Ecotourism Society (TIES), for instance, ecotourism is "responsible travel to natural areas that conserves the environment and improves the well-being of local people." It is intended to be responsible, sustainable travel that includes partnerships between governments, communities, and travelers and adheres to the following characteristics:

- Minimizes impact
- Builds environmental awareness
- Provides direct financial benefits for conservation
- Provides financial benefits and empowerment for local people
- Respects local culture
- Supports human rights and demographic movements

Lacking further specificity, such definitions can obviously encompass a wide range of different types of travel and destinations. Even before the term was initially coined, simple forms of *ecotourism* became popular in the United States, such as visiting natural parks in which visitors would experience, and not disturb, a preserved natural environment. Trips that are intended to support the preservation of threatened or endangered land and life through photography excursions, botanical study tours, wildlife treks, or bird-watching expeditions are other classical forms of ecotourism. Generally, ecotourism involves integrated planning and execution of tourism between travel organizations, local communities, governments, and travelers. This also requires ongoing monitoring of ecotourist operations once they are in place for assurance of maintenance of standards and minimization of cultural and environmental impact. Classical ecodestinations include volcanoes, rainforests, mountains, deserts, and rivers. Some of the top geographical locations include the Galapagos Islands, Antarctica, Madagascar, areas of South and Central America, India, Australia, Malaysia, and Thailand. Some forms of ecotourism are species-specific, and involve activities such as whale, dolphin, and big cat watching. In virtually all cases, the proclaimed goals include cooperation, sustainable use of resources, habitat preservation, and education.

Benefits of Ecotourism

The idea of ecotourism has been in place since the late 1980s. As a subsector of the tourism industry, it is quickly gaining popularity. The shift of interest from traditional tourism reflects perceptions of the value of preserving the natural environment, increased awareness of the environmental and cultural impact of typical tourism, valuing smaller group experiences, and an increasing desire of baby boomers and younger people to combine physically active travel with educational experiences and personal growth.

Ideally, ecotourism's goals are accomplished by travel programs that advocate ecological conservation, promotion of biodiversity to sustain all living beings of an area, and travel to an area negotiated with the involvement of the people living there, as well as of those who organize the travel experience. Many ecotourism destinations include countries where tourism, eco- or otherwise, is a major industry. For example, in countries such as Nepal, Kenya, Costa Rica, Ecuador, and Antarctica, tourism is a significant sector of local economies. By emphasizing the value of preserving and sustaining nature, environmentalists and tour organizers are able to bring money to places that need it, and rich natural experiences to travelers who desire it. According to estimates, in Europe and America alone, the ecotourism sector signs up over 15 million people each year for trips covering the entire globe.

Drawbacks of Ecotourism

While ecotourism encompasses a range of organized activities that are serious, environmentally and culturally sensitive, and in some cases even truly sustainable, the growing prospect of profits also brings an increased amount of greenwashing and false advertising. In order to appeal to ecologically conscious travelers, a growing number of vacation packages present themselves as ecosensitive even when they do not fulfill even the most basic criteria of what constitutes ecotourism. An example of such misleading advertising is Nature's Sacred Paradise, a self-described ecotheme park in the Mayan Riviera of Mexico. The park operators promise not only natural landscapes, rich biodiversity, and a wealth of outdoor nature activities, but many educational opportunities to learn about the culture and traditions of the Mayan peoples who have lived in the region for thousands of years. Not mentioned in the brochures are reports about the displacement of Mayan populations in order to set up the park, or the keeping and displaying of endangered species, or the impact of a large tourist park on local cultures. Reports of fraud, deception, and false advertising are spreading, and include eco-seeming adventures that in fact are a cover for organized hunting in remote areas such as in African game reserves, worker rights violations at ecotourism parks, and the introduction of environmental hazards due to massive building campaigns to accommodate tourists.

Even if the tourism is ecologically motivated and adheres to many of the criteria of ecotourism, the sheer numbers of people traveling in this way can negatively impact communities. For example, money that might be used for social welfare projects is sometimes directed toward public relations campaigns to attract tourists. While ecotourism can help preserve threatened land and species, dollars generated do not always reach the people who might be displaced by these protections. Although ecotourist groups are usually small, the demand for these experiences is high, which results in significant wear and tear on local resources, trails, and habitat. Regardless of the "take only photos, leave only footprints" philosophy of ecotourism, those footprints, over time, have significant environmental impact. The increase in the number of motor vehicles and foot traffic into previously less-visited areas has a significant environmental impact. Visitors also require restrooms, places to eat and stay, and generate a tremendous amount of trash. Animals that might have frequented an area either stay away because of loss of privacy or begin losing their fear of humans and thereby become viewed as threats to human safety.

In addition, as anthropologists have documented for decades, from the moment of contact, the visited culture is forever changed. People and places become objects to be viewed and examined under the guise of tourism, local culture is often treated as unusual or strange, and local customs and habits are disrupted. In some areas, the intrusiveness of ecotourism has in fact resulted in local people killing wildlife or destroying habitat to put an end to the tourism. Indeed, data suggests that disruption and displacement of native communities are frequent effects of ecotourism. In the case of East Africa's Masai people, for instance, local governments took over wide swaths of land for ecotourism purposes that had previously been used for Masai livestock grazing, thereby harming the local livelihoods and economy.

The multiple varieties of ecotourism, in short, represent a mixed blessing. While ecotourism can undoubtedly help provide jobs, inject money into needy communities, provide travel opportunities to, as well as educational experiences in, remote and ecologically threatened areas, and build forms of intercultural interaction that can improve sensitivity and understanding between peoples, the potential drawbacks are equally numerous. All too often, what began as concern and care for the diversity of environments and peoples devolved into yet another money-making scheme, exploiting the growing public interest in natural biodiversity, environmental protection, and cultural sensitivity. On the other hand, there are also growing entrepreneurial opportunities to satisfy intensifying consumer demand for ecological tourism by redefining tourism, away from carbon intensive travel and mere consumption of goods, services, peoples, and cultures. The models may differ a great deal, but will all have in common that the economic bottom line is not primary, that they are not resource intensive, and that they are organized in a way that gives all stakeholders a voice.

See Also: Corporate Social Responsibility; Ecological Footprint; Greenwashing; Natural Capital; Social Entrepreneurship; Sustainability.

Further Readings

Fennell, David A. *Ecotourism,* 3rd Ed. London: Routledge, 2007.

Higham, James E. S. *Critical Issues in Ecotourism: Understanding a Complex Tourist Phenomenon.* Woburn, MA: Butterworth-Heinemann, 2007.

Honey, Martha. *Ecotourism and Sustainable Development: Who Owns Paradise?* 2nd Ed. Washington, D.C.: Island Press, 2008.

McKercher, Bob. *Cultural Tourism: The Partnership Between Tourism and Cultural Heritage Management.* London: Routledge, 2002.

Patterson, Carol. *The Business of Ecotourism,* 3rd Ed. Bloomington, IN: Trafford Publishing, 2007.

Weaver, David B. *Sustainable Tourism.* Woburn, MA: Butterworth-Heinemann, 2005.

Debra Merskin
University of Oregon

EMISSIONS TRADING

Emissions trading is a policy measure aimed at reducing pollution. Originally referred to as tradable pollution rights, it enables firms to trade the right to emit or discharge specified amounts of particular pollutants. It has been used to control sulfur dioxide (SO_2) that contributes to acid rain, nitrous oxides and volatile organic compounds that contribute to smog, greenhouse gases, and nutrient discharges into waterways.

In cap-and-trade emissions trading, a limit is set for total amount of a specific pollutant, or set of pollutants, that are allowed to be emitted over a particular period—usually a year—by specific industries in a particular region. The limit or cap chosen is supposed to be within the estimated capacity of the environment to assimilate the pollutant (or at least a step toward achieving that goal), but in reality is often more an indication of what is considered economically feasible. This cap is then divided into allowances (or permits) that are allocated to firms.

Emissions trading schemes are normally limited to large firms in a particular industry sector with significant emissions. A participating firm can sell any allowances that are surplus to its requirements to another firm that needs extra allowances, or it can save them up for the future when they might be needed.

The two main ways of initially allocating allowances are usually referred to as "grandfathering" and "auctioning." Grandfathering involves allocating allowances to firms on the basis of their past emissions. Alternatively, allowances can be auctioned off to polluters.

Open market emissions trading allows companies to earn emission reduction credits for voluntary reductions in a particular time period of specified pollutants discharged from their plants. These can be either reductions from the usual emission rates for a particular facility or reductions below the regulated standards that the facility is required to meet, whichever is less.

In the United States, the Acid Rain Cap and Trade scheme is cited as a success. Part of the process involves collecting water samples for acid rain analysis, such as here in a Chesapeake wetland tributary.

Source: Mary Hollinger/National Oceanic and Atmospheric Administration

Firms that reduce their rate of emissions from a particular facility can then sell the credits they earn to other firms for whom buying credits is cheaper than reducing their emissions to comply with those regulations. Trading is usually open to all firms. The money that can be earned from selling credits is supposed to provide an incentive for firms to come up with innovative ways to reduce their emission rates.

Some open market emission trading schemes allow firms to gain credits from reducing pollution from a variety of small mobile sources such as old cars, leaf blowers, and lawn mowers. Credits can be exchanged between different types of sources and industries. Sometimes different types of pollutants are covered in one trading scheme so that reductions in emissions of one pollutant can be used as credits for increased emissions of a different pollutant.

Emissions trading is based on the idea that it is cheaper for some firms to reduce their emissions than others and therefore more cost effective to allow the market to decide where emission reductions will be made than for governments to require uniform reductions across an industry. Firms that find it expensive to reduce emissions are able to buy up emission permits instead. Those that can reduce emissions cheaply can in turn profitably sell their unneeded permits.

This might be acceptable if only limited pollution reductions were required—that is, if reductions can be limited to what can be done cheaply. However, emissions trading makes little sense if substantial reductions are required. If more expensive reductions have to be made, then there is little point in setting up markets that enable some firms to avoid making those expensive reductions so as to minimize overall costs.

This became evident in Germany when it considered implementing an acid rain emissions program. The aim of the German program was a 90 percent reduction in SO_2 between 1983 and 1998. In comparison, the aim of the U.S. emissions trading program for SO_2 permits was only a 50 percent reduction between 1990 and 2010. This meant that in the United States there was much more scope for power stations to find cheaper ways to reduce their emissions, whereas in Germany, every power station had little choice but to retrofit their plants with flue gas desulfurization and selective catalytic reduction for nitrogen oxides. This meant that there was no scope for trading in Germany.

The U.S. acid rain cap-and-trade scheme is consistently cited as a success because it achieved emissions reductions at minimal cost to the firms involved, but those reductions do not compare well with what was achieved elsewhere with traditional regulation. The United Kingdom Environmental Agency noted in 2003 that sulfur emissions in the United States exceeded those from the European Union (EU) member states by 150 percent.

Emissions trading can also result in a great deal of disparate environmental stress in some neighborhoods, as they may end up with a lot more pollution than others because the companies in their area are buying up allowances rather than reducing their pollution. In the case of mercury trading in the United States, power plants that buy up mercury emission credits put their neighbors at risk of brain damage. Even the trade of nonhazardous gases like carbon dioxide can cause hot spots of pollution because such gases have toxic copollutants that increase with the increase in carbon dioxide emissions.

When the EU emissions trading scheme for greenhouse gases was introduced, many governments were overly generous in allocating permits to local firms because they feared their local industries would be at a competitive disadvantage if they had to buy extra permits. A study by Ilex Energy Consulting for World Wildlife Fund that examined six EU countries found none of them had set caps that went beyond business as usual. Because allowances were not in great demand, the market opened at 8 euros per ton and settled around 23 euros a few months later, far less than necessary to provide an incentive to reduce emissions. When the recession hit at the end of 2008, the price went down again to around 10 euros.

Emissions trading schemes usually allow companies to offset any emissions they do not have permits for by paying for carbon reductions elsewhere. However, there are many questions about how effectively carbon offset schemes reduce greenhouse gases in the long term. It is up to those claiming carbon credits to explain how they are reducing greenhouse gas emissions and why these reductions would not have occurred without their investment. This means the carbon offsets can be rather debatable and often would have occurred anyway.

Offset projects favor cheap methods of reducing carbon emissions rather than renewable energy projects in developing countries. The use of tree plantations as carbon offsets is particularly problematic. First, there is no accepted method for calculating how much carbon is temporarily taken up by growing trees. Such trees may release their carbon early as a result of fires, disease, or illegal logging, but the necessary long-term monitoring is often not carried out. In many situations plantations are not sustainable because they use so much water and agrochemicals. Such plantations reduce soil fertility, increase erosion and compaction of the soil, and increase the risk of fire.

Emissions trading tends to protect very polluting or dirty industries by allowing them to buy emission allowances or offsets rather than requiring them to meet higher environmental standards. In this way, trading can reduce the pressure on companies to change production processes and introduce other measures to reduce their emissions.

See Also: Carbon Trading; Environmental Economics; Pollution Offsets; Waste Reduction.

Further Readings

Beder, Sharon. *Environmental Principles and Policies*. London: Earthscan, 2006.
"Carbon Trading: A Critical Conversation on Climate Change, Privatisation and Power." *Development Dialogue*, 48 (September 2006).
Sorrell, S. and J. Skea, eds. *Pollution for Sale: Emissions Trading and Joint Implementation*. Cheltenham, UK: Edward Elgar, 1999.
Tietenberg, Thomas H. *Emissions Trading: Principles and Practice*. London: RFF Press, 2006.

Sharon Beder
University of Wollongong

ENERGY PERFORMANCE CONTRACTING

Energy performance contracting, also known as energy savings performance contracting, is a contractual and financing arrangement, typically negotiated between a building owner and an energy service company, whereby future energy cost savings compensate the building owner for current energy conservation and efficiency improvements and additions. Energy performance contracting is an important innovation for green business as it provides an alternative method of financing crucial efforts toward sustainable development. Under a typical energy performance contracting agreement, the energy service company completes a comprehensive review and energy audit of the current status of the real property. Thereafter, the energy service company provides an integrated recommendation for implementation of energy-efficient measures for the property. These additions and modifications might include sustainable construction; energy effectiveness and efficiency; electric, water, and thermal energy conservation; and emission reduction. Typically, after this review and recommendation period, the energy service company arranges third-party financing, through structured agreements like operating or municipal leases, for the sustainable design and construction so that there is little to no cost up front for the building owner. Although most energy performance contracts are premised upon these types of financing arrangements, a small number of these projects do use an initial outlay of capital budget allocations.

Throughout the negotiated time period of the energy performance contract, the energy service company typically provides continued monitoring of the implementation of the recommended measures and verifies the economic savings of these conservation measures. This service of monitoring and verification provides for long-term controls and management of the energy performance contract itself. Additionally, pursuant to the energy performance contract, the energy service company also generally guarantees that the future energy cost savings will cover the costs of the energy conservation improvements. In most energy performance contracting arrangements, if these future costs savings do not meet this guarantee, the energy service company is contractually obligated to cover the monetary difference. If, however, the cost savings that occur as a result of the implemented conservation changes do meet the guarantees of the energy performance contracting agreement, the building owner is contractually obligated to compensate the energy service company with the negotiated proportional interest of the cost savings as stipulated by the energy performance contract. By combining cost-saving measures with sustainable development and environmentally conscious design, energy performance contracting can be a "win-win" situation for all of the parties involved in its negotiation and implementation.

Energy performance contracting began to emerge in the mid-1980s; from its emergence to the early 1990s, the idea of energy performance contracting made slow inroads primarily in private industry. Subsequently, energy performance contracting began to become an important strategy for energy conservation for state and federal governmental agencies. This governmental development of energy performance contracting was spurred by legislation such as the Energy Policy Act of 1992, which mandated energy reduction for federal facilities and which authorized federal agencies to enter into long-term energy performance contracts. For example, from 1996 to 1998, the U.S. Environmental Protection Agency (EPA) began its first energy performance contract, based in part on the mandate of the Energy Policy Act. From the late 1990s to the 2000s, energy performance contracting witnessed exponential growth in both the public and private sectors, fueled in part by the volatility of the energy market. By May 2009, 19 federal agencies had entered into over 450 energy performance contracts in almost every state in the United States. The projects covered by these contracts ranged from the Statue of Liberty to the U.S. Naval Station at Guantánamo Bay, Cuba. These projects were based on the authority of both the Energy Policy Act and the Energy Independence and Security Act of 2007 enacted in December 2007. The Energy Independence and Security Act reaffirmed federal agency authority to enter into energy performance contracts. Additionally, under this act, the secretary of energy was charged with the creation and administration of a Federal Energy Management Program, which would provide training to federal agency managers and other contracting personnel in the negotiation and management of energy performance contracts. In addition to the federal government, state governments throughout the 1990s and 2000s began to implement pilot programs and to debate legislation concerning the procurement and management of energy performance contracts. States like Florida, New York, Washington, and Texas passed legislation that authorized state agencies, educational institutions, or municipalities to procure energy performance contracts.

In both public and private energy performance contracting, there are several shared motivations to enter into this type of agreement. These motivations include a push for economic savings, a response to global climate change, a desire to move toward green construction and building, and a goal to update and modernize existing facilities. Still, energy performance contracting retains some limitations. These limitations include a lack of training for contracting personnel about the intricacies of energy performance contracting; a shortage of engineers and technicians required to accomplish the large-scale implementation of the energy-efficient measures at the core of energy performance contracting; and certain markets' hesitance to enter into the indefinite delivery, long-term arrangements of energy performance contracting. To overcome these constraints, energy service companies have begun to provide extensive education to the private industrial sector as well as to state and federal governments. Additionally, both the public and private sector have begun to provide increased training of their own contracting personnel, like the Federal Energy Management Program. Furthermore, in an attempt to expedite the negotiation and implementation of energy performance contracts, the Department of Energy has established the Super Energy Savings Performance Contracting Program, which is a regional program that allows an approved number of energy service companies the right to bid on federal energy performance contracts within the specific regions.

Based on these initiatives, it appears that energy performance contracting will be a substantial component of the future of green business. Energy performance contracting provides a cost-efficient way to achieve and implement sustainable design, modernization, and development; helps to achieve energy conservation measures, both self- and governmentally mandated; stimulates the green energy market to develop more innovative techniques to achieve environmentally friendly property; and promotes responsible environmental stewardship as well as long-term economic growth. Although energy performance contracting does have certain constraints, and aspects of this process may be the subject of litigation in the future, its meteoric growth in the years since 1985 and its prominence within federal and state legislation appear to signify that energy performance contracting will be a strong foundation for energy-efficient property development nationally and internationally, privately and publicly, throughout the 21st century.

See Also: Conservation; Corporate Social Responsibility; Energy Service Company; Environmentally Preferable Purchasing; Sustainable Design.

Further Readings

Energy Independence and Security Act, Public Law No: 110-140 (42 U.S.C § 17001–42 U.S.C. § 17386). http://www.govtrack.us/congress/bill.xpd?bill=h110-6 (Accessed January 2010).

Hansen, Shirley J. and Jeannie C. Weisman. *Performance Contracting: Expanding Horizons.* Lilburn, GA: Fairmont Press, 1998.

ICF International and National Association of Energy Services Companies. "Introduction to Energy Performance Contracting, Prepared for U.S. Environmental Protection Agency." http://www.energystar.gov/ia/partners/spp_res/Introduction_to_Performance_Contracting .pdf (Accessed May 2009).

U.S. Department of Energy. "Energy Savings Performance Contracts." http://www1.eere .energy.gov/femp/financing/espcs.html (Accessed May 2009).

U.S. Environmental Protection Agency. "Energy Savings Performance Contracts." http://www
.epa.gov/epp/pubs/case/espc.htm (Accessed May 2009).

Amanda Harmon Cooley
North Carolina A&T State University

ENERGY SERVICE COMPANY

An energy service company (ESCO) is an organization that delivers "energy services" like electricity or heat, whereby the mode of service provision is determined by the ESCO rather than the user. This is different from the traditional model of energy provision, where the supplier provides the energy and the user decides how to use it. With the rise of green enterprises, ESCOs have sought to enter the energy market by offering services that meet interests in sustainability and alternative energy sources.

The main principles and concepts underlying the business model and management of ESCOs include the following:

- Providing energy "services" rather than just providing energy
- Providing those services in the most efficient way and, often, on private networks
- Increasing the penetration of renewable energy sources and low-carbon technologies
- Distributing electricity with an accompanying heat main, which disperses excess heat produced in energy generation

Although the intent of this ESCO business model is to address green needs in obtaining energy, the implementation of these models has not gone without controversy. For example, ESCOs (such as that of the now defunct Enron) have been criticized for anticompetitive practices, as well as for using the complexity of their business models to hide debts that ultimately harm shareholders and possibly the environment.

From Energy Production to Energy Services

Most energy consumers are interested neither in the unit of power itself nor the medium through which it is transmitted. Instead, consumers typically are more concerned with the service that they derive from taking the raw energy carrier and transforming it into useful work. Essentially, users need certain amenities (like light and heat) and, as long as these services are delivered in a reliable way, the intermediate technologies through which they are delivered are of minimal concern. However, the rise of environmental interests may alter this attitude.

Given this status of energy consumer demand, supporters of the distributed generation model of power provision argue that, alongside the topographical changes that will occur in energy distribution networks, there should be an associated change from business models of "energy provision" to "service provision." ESCOs have capitalized on this argument. By moving to a more service-oriented architecture, ESCOs have become empowered to make decisions as to what is the most efficient way of providing the service as well as to consider other green factors demanded by the market or governmental regulations. Further, ESCOS have marketed themselves as service, rather than traditional utility, providers.

In a traditional "utility" arrangement, the providers of utilities make a profit on every unit of energy sold. The more gas and electricity they sell, the greater their profits. This provides little incentive for implementing energy-efficiency or sustainable measures. For example, in the construction of a new power station, a traditional utility has lots of spare energy (e.g., low-grade heat) that cannot be sold as there is no mechanism to sell this heat to the consumer. As a result, this useful energy is often wasted by being vented into the atmosphere.

In an energy service model of provision, the user purchases an "energy service" rather than the energy itself. They may purchase a "warm home at a given temperature," and it then becomes the ESCO's job to provide this amenity. This gives the ESCO a great opportunity to develop creative processes and to have the flexibility to invest in energy-efficient measures.

Technologies for Energy Service Provision

The aim of many green ESCOs is to produce energy in a cleaner, more efficient way than traditional centralized networks. This is often done by reducing the amount of primary fuel input. This can be accomplished by utilizing technologies that use fuel more efficiently (for example, combined heat and power/cogeneration where waste heat from electricity production is reused in other processes) or by replacing fossil-fuel installations with renewable energy sources implemented through local communities. These ideas are further enhanced by the deployment of energy-efficiency technologies, such as energy-saving compact fluorescent lightbulbs, green appliances, and cavity wall insulations to retain heat.

Another one of the distinguishing technological features of ESCOs' energy provision is the installation of a private wire network. It is possible for other generators to connect to public electricity networks in many countries with liberalized electricity markets. However, there is an associated cost penalty in connecting to the public network. When distributing electricity on a public network, there are often network charges for the use of the transmission network and distribution network, as well as other charges that only make sense when applied to a large centralized power distribution system.

Generally, when ESCOs only need to distribute electricity over a local section of the network, it is more cost effective to install private wires with their consumers and producers. This allows them to trade power between themselves internally—while the private wire is then able to trade with the power grid for any surplus or shortfall that may exist. These private wires help avoid network tariffs and make ESCOs more profitable.

Scales of Implementation for Energy Service Companies

Traditional utilities tend to be national or regional concerns, generating power centrally in large power stations which is then distributed over public wires. By contrast, ESCOs are often implemented on a community scale with power being distributed locally to an estate, community, or small conurbation. When sustainable energy is provided on a community level, there are often economies of scale that cannot be realized with single "consumer/producer" connections to the grid. For example, in the Woking area of London, Thameswey, the local ESCO has fitted a large number of buildings on several housing estates with photovoltaic panels. Rather than providing each property with an inverter, a D.C. network feeds all the power from individual arrays to a large, central inverter. This is more cost effective than having individual inverters for each house. Similarly, combined heat and power (CHP) installations have a minimum scale at which they are reliable and practicable.

While domestic CHP units are in development, it makes sense to implement this technology on a community scale, where only one unit requires servicing/maintenance. Both of these examples offer opportunities for green businesses to work with local governments and ESCOs to effectively meet the demands of the market.

ESCO Financing

The investments required for energy efficiency often take a long period of time to reap dividends. For an ESCO, this requires a commitment from the consumer to buy services for a greater period of time than the investment payback period in order for the operation to generate a profit. There is some tension between this model of energy provision and the wider movement toward "liberalized electricity marketplaces," where consumers are free to change providers in order to secure the cheapest rate of supply. Therefore, the long-term commitment by the consumer is offset by the cost-effective nature of the service provision by the ESCO, which is enabled by the installation of more efficient and likely more sustainable technologies.

The basic model for ESCO financing is that the ESCO will provide capital upfront for the purchase of equipment and material needed to deliver energy services in an efficient manner. The user will then consume these services over a number of years, paying for the service that has been used. The user pays a cost for the service, which is usually marginally less than the cost they would pay for receiving that service if it were delivered using conventional energy technologies. Consequently, the cost is a "perceived value" of the service, as opposed to a charge for the energy delivered. However, it is important to note that while less energy is used to deliver these services, an initial capital investment must be made to facilitate efficient energy use. The difference in price between the "perceived value" of the service (whose cost is based on provision using inefficient technologies) and the "actual cost" of delivering the service (which is less because it has been delivered using the most efficient cost-effective method) provides a profit for the ESCO, which then can be used to pay back the capital investment in equipment.

In sum, ESCOs likely have an important future in the development of green and alternative energy provision. Their decentralized approach will certainly continue to be popular in all communities that come together around green issues.

See Also: Energy Performance Contracting; Smart Energy; Sustainable Development.

Further Readings

Bullock, Gary. *A Guide to Energy Service Companies.* Upper Saddle River, NJ: Prentice Hall, 2001.

Jones, Allan. "Moving Toward a Sustainable Low Carbon London, Reducing the Carbon Footprint in the Built Environment, 2007." Institution of Engineering and Technology Seminar, IEEE, 2007. http://ieeexplore.ieee.org/xpls/abs_all.jsp?arnumber=4451671 (Accessed April 2009).

Patterson, Walt. "Transforming Our Energy Within a Generation." Chatham House Briefing Paper, 2007. http://www.chathamhouse.org.uk/files/9253_bp0607climatewp.pdf (Accessed April 2009).

Studebaker, J. M. *ESCO: The Energy Services Company Handbook.* Tulsa, OK: PennWell
    Books, 2001.
Vine, E. "An International Survey of the Energy Service Company (ESCO) Industry." *Energy
    Policy* (2005).

Gavin D. J. Harper
Cardiff University

ENVIRONMENTAL ACCOUNTING

Environmental accounting (also known as green accounting) is a style of accounting that includes the indirect costs and benefits of economic activity—such as environmental effects and health consequences of business decisions and plans. It is increasingly used by green companies interested in sustainability, ecoeffectiveness, and ecoefficiency, along with direct and more obvious market tools in strategic management. Environmental accounting collects, analyzes, assesses, and prepares reports of both environmental and financial data with a view toward reducing environmental effects and costs. This form of accounting is central to many aspects of governmental policy as well. Consequently, environmental accounting has become a key aspect of green business and responsible economic development.

The link between the environment and the economy is an inarguable one. In recent decades, environmental accounting has been taken seriously in the incorporation of environmental impacts into management decisions and strategic planning as well as in the reporting of national accounts. It is important to understand national accounts to give environmental accounting the proper context in a governmental sense. National accounts present a complete conceptual framework for the economic activity of a country in a given time period, constructed with double-entry accounting. National accounts are concerned with economic factors such as production, expenditure, and household incomes, which are the frameworks from which measures like gross domestic product (GDP) and purchasing power parity are figured. National accounts are generally divided into stocks and flows (sometimes called levels and rates, especially in older texts), assets that are accumulated over time (inventory, stock shares, and money in a bank account), and the changes that occur to those assets (inflows like deposits, interest, and new purchases and outflows like withdrawals, fees, and sales). National income accounts are at the center of macroeconomic environmental accounting.

In the United States, national accounts have been maintained since 1947. Most of Europe adopted the practices around the same time, and, today, the United Nations System of National Accounts (UNSNA) provides a set of standards that many countries use for construction of their accounts. The United States is one of the countries that utilizes independent standards rather than accepting the UNSNA standards. In addition to the UNSNA, the United Nations' Environmental-Economic Accounting Section works to standardize environmental accounting methodologies and practices. However, progress has been incremental and tentative. In 1993, it issued the *Handbook of National Accounting: Integrated Environmental and Economic Accounting* (SEEA) to stimulate discussion about environmental accounting and to focus the conversation at a time when some countries were starting to formulate their own policies on the matter. In 2003, the first revision, SEEA 2003, was released in the hope of harmonizing the environmental accounting methods then in use around the world. This provided additional guidance to companies in developing best practices. Three years later, the 37th session of the Statistical Commission endorsed the position to adopt SEEA 2003 as an international standard by 2010. The implications of these shifting international standards and comparisons for environmental accounting can be seen in the difficulty of changing a traditional paradigm to include new factors or considerations such as environmental effects.

Regardless of the governing standards for environmental accounting, all stocks, inflows, and outflows are applicable to environmental concepts such as atmospheric gases, emissions, and sequestrations; species populations, births, and deaths; and landfill volume, growth, and reduction. A truly green, environmentally informed system of accounting would include aggregates that incorporate such environmental data when applicable to management decisions. The GDP measure, for instance, is commonly used as an indicator in policy making, but it does not adequately account for positive or negative environmental consequences. This complaint is similar to others made about such aggregates in that they do not provide a full picture of the possible hidden harms that are being done now and that could be extremely costly in the future.

One environmental accounting suggestion in this area is that environmental costs—which are presumed to be paid by future generations either through the cost of repairing environmental damage or the cost of suffering its effects—should be deducted from present financial gains. Similarly, capital spent on preventing or repairing environmental damage yields gains that should be reflected in companies' annual reports as well as, at the governmental level, in national accounts.

One of the underlying principles of environmental accounting is that of natural capital, which extends the concept of capital to environmental stocks and flows. Stocks of plant and animal resources can replenish themselves naturally, but are sustainable only so long as outflows of use or loss through destruction do not exceed the inflows. Often, these considerations are quite complicated for companies, as the effects of environmental pollution only come to light many years into the future. However, accounting for natural capital requires keeping track of material recycling, water reclamation, waste treatment, erosion, deforestation, and resource conservation, just as much as production, employment, and income. As such, natural capital is purchased at a cost that is misleading if it is limited to the monetary dimension without planning for environmental effects. The monetary cost of drinking water, for instance, is not just the cost of the water itself; rather, it also includes the cost of distribution, treatment, and infrastructure.

Without the use of environmental accounting, the underreporting of cost is commonplace and usually makes natural resources seem artificially inexpensive. This creates a false price impression that can harm the environment and green enterprises as the use of any resource exacts a cost, particularly if it is nonrenewable (like fossil fuels) or consumed at a rate greater than the rate of its replenishment (like water, forests, and many other renewable natural resources). If prices do not reflect the true cost of natural resources, then green products will continue to struggle to compete against environmentally subsidized ones. Though environmental accounting advocates agree on the need to incorporate this concept of depreciation with respect to natural capital, there is no agreement yet on how to do so.

Additionally, natural and other resources can have a value that is not necessarily well represented by the way they are consumed. Collectability aside, an automobile's value is directly related to its use—it is manufactured for a purpose, bought for that purpose, and used for that purpose. Trees, however, have a value and make a contribution to the world before they are extracted, processed, and sold as a product. This is an intrinsic value that is unrelated to the value put on that good after manufacturing and on the use of that product later in its life cycle. Trees help to clean the air, contribute food and shelter to organisms in ecosystems, provide watershed protection, and help prevent erosion (particularly in old growth forests that take generations to replace). An authentic environmentally adjusted system would account for these losses and would appropriately include them in management decisions, even though this may be difficult to do.

Furthermore, there are environmental externalities that do not directly involve the raw materials used in the economic transaction. For example, while trees are obviously affected by the lumber industry, a good deal of economic activity— both industrial and agricultural— has an impact on the insect population, which has a value because of its contributions to crop health. (Not all insects are pests to be eliminated; many are helpful or necessary to pollination in productive farming.) These effects are even more difficult to reflect in traditional methods of accounting, because too often there is not sufficient data or even instruments to gather data to facilitate the appraisal of the environmental effects of industrial production, particularly in limited time spans.

Another question that rises here is whether environmentally adjusted aggregates found in environmental accounting (for example, a figure that adjusts the GDP or corporate profits in light of environmental gains and losses) should be reported alongside conventional aggregates or in lieu of these orthodox measures. Reporting the environmentally adjusted figures, pursuant to one of these options, has the advantage of being a more holistic perspective. Such an accounting also prevents stakeholders and shareholders from being able to ignore the environmentally adjusted data and environmental costs associated with public governmental or private managerial decisions. Consequently, the reporting of both traditional and green measures is a demonstration of the total cost of environmental impacts.

These questions carry over to debates between traditional accountants, managers, and economists and environmentally conscious accountants, managers, and economists who concentrate less on traditional measures and more on the development of new green measures of environmental accounting that can be of use to the public and private sectors. These interests are not limited to anthropogenic environmental impacts, as the real cost of natural disasters, like Gulf Coast hurricanes and California wildfires, can be better estimated with environmental accounting methods. Of course, it is difficult to complete these tasks, in that constructing a presentation of data that is transparent and meets the needs of various constituencies is complicated. These complications are compounded by the novelty of environmental accounting in all of its manifestations, from financial accounting to managerial accounting to macroeconomic accounting.

An international example adds further context to the evolving standards of environmental accounting. Norway's economy has long relied on natural resources like fish and forests, and it has been using various environmental accounting methods since the 1970s. Originally, this was geared toward monitoring and resource conservation. Over time, the Norwegian environmental accounting standard has expanded to include additional types of pollutants, has become an important tool in private sector management and finance, and is now included in the national account used by the government to plan economic growth strategies that are environmentally conscious and energy-efficiency cognizant. The country's comparatively long history of monitoring these data helps it anticipate the effect of different growth patterns and the impact of compliance with various international emission standards, both real and proposed. Further, increasing numbers of companies in Norway have accepted environmental accounting methods to meet their needs and to fulfill governmental requirements.

Since the late 1980s, more countries, like Norway, and more corporations have been noting the depreciation of natural resources when compiling national or private accounts in response to case studies that demonstrated how misleading economic growth figures were in countries that were consuming, and failing to replenish, their natural resources in order to achieve that growth. Many developing countries, conscious of the resource loss that has occurred in industrialized countries, have begun maintaining environmentally adjusted accounts, including at least some minimal natural resource tracking. For example, in Namibia, these methods have helped to inform policy governing fisheries, taxation, and the allocation of water resources, and they have provided information on the effects of livestock on the environment. In the Philippines, the government has gone a step further, adding a monetary value to its environmentally adjusted account to represent the recreational and aesthetic value of its unspoiled public and private lands. In each case, companies involved in these countries must make decisions based on the considerations and data provided by environmental accounting methods. For green businesses looking to expand into developing nations, the use of environmental accounting can be an important part of winning contracts in the countries, in making key managerial and financial decisions, and in meeting the needs of environmentally conscious stakeholders. Next, it is useful to discuss some of the types of environmental accounts that exist.

Just as national accounts can include production, income, expenditure, capital, and financial accounts, so, too, are there different types of environmental accounts. This calculation can be particularly important to companies in developing their environmental management systems. One type of environmental account is an emissions account, which tracks pollutant emissions categorized by the economic sector of their source. This is a common type of environmental account in Europe, where the European Union's statistics organization, Eurostat, helps members implement the National Accounting Matrix including Environmental Accounts (NAMEA) methodology. NAMEA data are used to quantify the environmental impact of different proposed growth strategies. It can be organized by type of pollutant and used to demonstrate transborder environmental effects, which is particularly important in Europe, given the large number of countries occupying a small continent. On this point, the implications for green practices among companies are substantial, as pollution from one country can easily migrate into another country.

As previously mentioned, new metrics are being developed to account for environmental concerns. Green GDP is an environmentally adjusted GDP, in which environmental costs are subtracted from the aggregate economic value of production while environmental gains are added. The greatest liability to the green GDP account is that, unlike unadjusted GDP, the method for deriving it has not been internationally standardized, which means green GDPs may not be consistent among countries using different methodologies. Another type of environmental account is a natural resources account, which tracks natural resource stocks and flows, whether impacted by human causes or natural processes. This includes fisheries, forests, water resources, agricultural acreage, and mineral resources. The data may or may not be monetized, depending on the practice of the country or group preparing the account. Valuing natural resource stock when the change is caused by market fluctuations or natural processes can be a tricky venture, though.

In sum, environmental accounting has seen an increase in its use by companies, governments, and nongovernmental organizations, which have become more aware of the need to develop sustainable development policies and practices. It is likely that the expansion of this use of environmental accounting, as well as in the sophistication of its techniques, will continue as the needs of the marketplace and demands of consumers expect green considerations to be taken seriously.

See Also: Balanced Scorecard; Certification; Corporate Social Responsibility; Cost-Benefit Analysis; Ecoefficiency; Ecological Economics; Ecological Footprint; Environmental Assessment; Environmental Audit; Environmental Economics; Environmental Impact Statement; Environmental Risk Assessment; Natural Capital; Triple Bottom Line.

Further Readings

Aldy, Joseph E. and Robert N. Stavins, eds. *Architectures for Agreement: Addressing Global Climate Change in a Post-Kyoto World*. New York: Cambridge University Press, 2007.

Gray, Robert H. and Jan Bebbington. *Accounting for the Environment*, 2nd Ed. Thousand Oaks, CA: Sage Publications, 2002.

Hecht, Joy E. *National Environmental Accounting: Bridging the Gap Between Ecology and Economy*. London: RFF Press, 2005.

Odum, Howard T. *Environmental Accounting: Energy and Environmental Decision Making*. Hoboken, NJ: John Wiley & Sons, 2005.

Bill Kte'pi
Independent Scholar

ENVIRONMENTAL ASSESSMENT

Environmental Assessment (EA), which is a term frequently used interchangeably with the terms *Environmental Impact Assessment* (EIA) and *Environmental Impact Statement* (EIS), refers to the process of assessing the environmental effects of proposed initiatives prior to implementation. The purpose of an EA is to prevent or minimize harm to the environment by evaluating the impact of new initiatives on the environment and ensuring that environmental implications are explicitly expressed and incorporated into decision making prior to the approval process. In principle, environmental assessment can be undertaken (1) for individual projects such as power plants, highways, or dams (EIA), which may require national, state, and sometimes even local assessment reports and often has consequences on regulations and requirements for the businesses involved in building the project; or (2) environmental assessment can take place for more comprehensive plans, programs, or policies (Strategic Environmental Assessment), such as fuel-efficiency standards for motor vehicles or environmental building codes.

The origins of formal EIA legislation in the United States are quite recent and can be traced to the National Environmental Policy Act (NEPA), which was enacted by President Richard Nixon on December 31, 1969. This legislation (42 U.S.C. 4321–4347) established the Council on Environmental Quality (CEQ) and set forth the requirements that advocates of proposed initiatives would need to adhere to when implementing programs, projects, or policies that involved federal tax dollars, federal land, and federal jurisdictions. The enactment of the NEPA legislation considerably influenced the global community's views on developmental initiatives and environmental issues. The 1970s witnessed several countries' (Japan, Singapore, Canada, Australia, France, the Philippines, and Taiwan) following the U.S. lead and enacting official environmental assessment legislation. During that same period, Britain, the Federal Republic of Germany, and most of the Nordic countries also embraced environmental assessment by adapting their well-developed land use policy systems; however, with the passage of time, these countries eventually enacted specific environmental legislation. Several other countries followed in the 1980s—South Korea, Indonesia, Thailand, Malaysia, and the rest of Europe. Today, EIA is applied in almost 100 countries, many of which have formal environmental assessment legislation and/or guidelines.

In addition to country-specific environmental assessment legislation and guidelines, several international institutions have also enacted their own specific environmental assessment guidelines. The European Community (EC) adopted its first environmental guidelines in 1973. Over the next 12 years, numerous environmental directives followed. However, these were superseded by 1985 European Council EIA Directive (85/337/EEC), which sought greater uniformity and standardization among EC member countries. This directive has since been superseded by the Strategic Environmental Assessment Directive (2001/42/EC). Equally important events that triggered further global interest in EA included the World Bank's public announcement in 1986 that EIAs were to become a formal requirement for the bank's project appraisal process, and the United Nations' "Brundtland Commission Report" in 1987, which further stimulated and accelerated interest in EIA outside the United States. Other prominent international bodies that have also embraced environmental assessment include the Organisation for Economic Co-operation and Development, the Asian Development Bank, and the World Health Organization.

The formation of the International Association of Impact Assessment (IAIA) in 1981 has also helped to improve and promote the use of EIA. As the premier organization in the field, this body plays a leading role in promoting EIA best practices by providing guidance to its members on EIA issues. The IAIA provides a forum for the exchange of ideas and experiences not only via membership networking but also via its journal *Impact Assessment and Project Appraisal* as well as its quarterly newsletters.

Over the last three decades, major strides have been made by governments and local and international organizations in promoting EIA. The literature abounds with information on environmental assessment. However, different terminology is often used by different countries and institutions when discussing how proposed initiatives may affect the environment. For example, Britain uses the terms *Environmental Assessment* and *Environmental Statement* with the word *Impact* removed. In addition, terminology among countries tends to vary both in terms of meaning as well as in application scope. In some countries, an EIS and an EIA are considered to be one and the same. In other countries, an EIS becomes an EIA only after official approval. There are other countries that view an EIS as only a part of an EIA framework. Moreover, different terminology may even apply within a particular country. For example, in the United States, California uses the term *Environmental Impact Report* rather than EIS. Regardless of terminology and/or scope, the ultimate goal of all of these studies is the same, namely, to ensure that proposed initiatives are environmentally sound.

In the United States, the predicted environmental effects of proposed initiatives are categorized into three thresholds—each has a different scope and results in a written report. The first threshold is Categorical Exclusion (CE). CE refers to minor agency projects that normally have no environmental impact. Projects classified as CE require no public involvement or formal circulation to agencies.

Environmental Assessment/Finding of No Significant Impact (EA/FONSI) is the second threshold. Projects in this category are generally those whose potential environmental impact(s) have not been clearly established, but whose impacts are believed to be minor and can be successfully mitigated. The purpose of the EA is therefore to gather sufficient evidence to ascertain the extent of the environmental impact(s). At a minimum, the EA report should contain (1) a need for the proposed action; (2) a description of proposed alternatives; (3) an analysis of environmental impacts; and (4) a list of public consultation/participation. Following the completion of the study, a public announcement is made to advertise the availability of the EA report and interested parties are invited to comment in writing within 30 days. If no request is received for a formal hearing and no adverse opposition is received or raised about any potential "significant" environmental impacts within the stipulated time frame, no further action is required (i.e., no EIS is required) and a FONSI is subsequently prepared by the sponsoring agency. The FONSI can either be prepared as a separate brief document or the EA can be appended as a part of the FONSI. The FONSI should indicate why an EIS was not necessary and should clearly outline any mitigation plan(s) if this was part of the decision. Following completion, an announcement is once again made to alert the public that the FONSI is available.

The final threshold is Draft and Final Environmental Impact Statements (DEIS and FEIS). Projects in this category are those that are expected to have "significant" impacts on the human environment. Additionally, projects that have not received a FONSI automatically qualify for a DEIS/FEIS. All issues addressed in an EA are also covered in this threshold; however, the DEIS/FEIS process is much more rigorous, detailed, and comprehensive compared to the EA process, and the final report is much more scientific and technical. Classifying a proposed initiative as having a "significant" impact is, however, quite controversial. Although the word *significant* appears in most national EIA legislation, this concept is highly subjective. To resolve this problem, some governments have attempted to provide guidelines so that proponents can be guided on when to prepare an EIS. For example, in the United States, the CEQ provides guidance on "significance" by defining it not only in terms of "context" (geographical setting), but also by "intensity" (severity of impact). Case law interpretation has also directly influenced the criteria for this concept.

In countries where the terms *EA*, *EIA*, and *EIS* are interpreted as meaning the same thing, only one procedure is conducted. The scope of this procedure generally involves (1) a systematic appraisal of the probable effects (long term and short term, direct and indirect, individual and cumulative, beneficial and adverse) that a proposed initiative such as a project, program, or policy may have on the environment (physical, chemical, biological, ecological, historical, cultural, social, economic, health, aesthetic); (2) an evaluation of the impact and consequences of those effects on the environment; (3) an analysis of the project alternatives and locations; (4) proposed measures that should be adopted to mitigate and/or eliminate any adverse effects on the environment; and (5) predicting if the proposed mitigation measure(s) will be sufficiently adequate to conserve the environment. Similar to the United States, public consultation is also a key element in the process.

The EA process is undoubtedly a technical task. However, various guidelines are available to provide guidance. For example, the IAIA has developed Principles of EIA Best Practice to provide guidance on EAs. The IAIA principles are broad, generic, and nonprescriptive and can be applied to all types of initiatives. The principles are divided into two tiers—basic and operating. The 14 basic principles are purposive, rigorous, practical, relevant, cost effective, efficient, focused, adaptive, participative, interdisciplinary, credible, integrated, transparent, and systematic. The operating principles describe how the basic principles should be applied to the main components of the EIA process, namely, screening, scoping, examination of alternatives, impact analysis, mitigation and impact management, evaluation of significance, preparation of EIS or report, review of the EIS, decision making, and follow-up. The Canadian Environmental Assessment Agency has also provided a useful seven-step model for conducting an EA: (1) determine if an EA is required; (2) identify who is involved; (3) plan the EA; (4) conduct analysis and prepare EA report; (5) review EA report; (6) make EA decision; and (7) implement mitigation and follow-up program as appropriate.

In general, major events such as the emergence of the green revolution in the 1970s, the Earth Summit Conference in 1992, and the *Exxon Valdez* oil spill on March 24, 1989, have undoubtedly spurred public interest in EA. In addition, legislation such as the Oil Pollution Act of 1990 and other EIA legislation have certainly played an important role in helping to protect the environment. Today, EAs are considered vital decision-making tools in many organizations concerned with sustainable development. Additionally, many governments now require EAs as routine legal requirements for certain types of proposed initiatives before they are granted development consent. Notwithstanding, environmental degradation is still very much a reality. Although legislation such as NEPA provides the impetus for federal, state, and municipal agencies and private institutions in the United States to carefully examine the consequences of proposed initiates on the environment, NEPA does not prevent institutions from implementing initiatives that may be harmful, nor does it impose any penalties for inaccurate reports. Consequently, many agencies/institutions have made environmentally disastrous decisions that have resulted in lawsuits initiated by public concern. Globally, many EAs continue to be criticized and questioned. Major criticisms include lack of transparency, failure to adequately advance environmental objectives, failure to address process and outcome fairness, insufficient attention to social impacts, and inadequate mitigation/monitoring measures.

The development of EIAs, however, creates both challenges and opportunities for businesses. Increasingly, and particularly when formal approval is required, business projects have to assess their potential environmental impact. Businesses interested in improving their environmental performance, furthermore, can find guidance in EIA standards, and are provided with tools such as sustainability surveys, life cycle assessment strategies, or environment risk assessment categories that can help them identify areas of poor performance and help them set goals for improvement.

See Also: Environmental Impact Statement; Environmental Indicators; Environmental Risk Assessment.

Further Readings

Burdge, Rabel. J. *A Community Guide to Social Impact Assessment*, Rev. Ed. Middleton, WI: Social Ecology Press, 1985.

Gilpin, Alan. *Environmental Impact Assessment: Cutting Edge for the Twenty-First Century*. Cambridge, UK: Cambridge University Press, 1995.

International Association for Impact Assessment. "Principles of Environmental Impact Assessment Best Practice." http://www.tucs.org.au/~cnevill/eia_principles.pdf (Accessed March 2009).

Lawrence, David P. *Environmental Impact Assessment: Practical Solutions to Recurrent Problems*. Hoboken, NJ: John Wiley & Sons, 2003.

Marriott, Betty B. *Environmental Impact Assessment: A Practical Guide*. New York: McGraw-Hill, 1997.

Nadini Persaud
University of the West Indies, Cave Hill Campus

ENVIRONMENTAL AUDIT

Environmental audit (EA) is an environmental compliance and management tool that aims to systematically, periodically, and objectively examine and document an organization's environmental management activities and operations. It is conducted to verify the compliance with environmental regulations, standards, and specific management practices; evaluate the effectiveness of the existing environmental management/control systems; and propose measures to improve environmental compliance and performance. Undertaking periodic EA can benefit organizations in numerous ways. It can help identify environmental problems and develop solutions, reduce operating contingencies, avoid excessive costs and fines or lawsuits, increase environmental awareness, meet the stakeholders' environmental expectations, and improve environmental performance as well as corporate image and public relations. Therefore, awareness about the nature, types, process of and constraints to a successful EA program is indispensable for "greening" organizations.

Michael McElhiney, from the U.S. Natural Resources Conservation Service, discusses waste management with dairy farmers in California in a typical portion of an environmental audit.

Source: Lynn Betts/U.S. Department of Agriculture Natural Resources Conservation Service

EAs emerged in the 1970s in response to growing environmental and health concerns and calls for stronger environmental control. The increasing quantity and complexity of both international and national environmental legislation and standards stimulated the proliferation of EAs. Many industrial companies started voluntarily conducting EAs to avoid noncompliance costs and civil/criminal liability, and to obtain better control over their environmental performance. EA was also encouraged by the regulators as a powerful tool to safeguard the natural and human environment. In the 1990s, as the sustainability concept called for increased social and corporate environmental responsibility, organizations recognized that EAs could promote their sustainability image and environmental credentials to the public and regulators. Initially emerging from pragmatic business rationales, EA has evolved to integrate sustainability values in the quotidian activities of both business and public organizations.

Currently, EA is widely institutionalized as part of the control and review function of environmental management systems—methodological and procedural frameworks that guide organizations' environmental performance (e.g., ISO 14000 series). During the last decade, a nonstandardized and ad hoc practice of Environmental Impact Assessment–related EAs has emerged. Overall, EA remains largely voluntary and does not replace regulatory inspections. Rather, it helps identify deficiencies, develop solutions, and improve environmental performance and management prior to official inspections.

Depending on its objectives, EA may include various types of activities, such as

- a compliance audit (checks against environmental legislation and organization policies);
- a performance audit (an evaluation of conformance of the actual environmental management to stated objectives);
- an issues audit (an evaluation of organizations' activities in relation to a specific environmental issue, e.g., waste or water pollution);
- a life cycle audit; or
- a health and safety audit.

The type and objectives of audits coupled with the organization's needs and capacities determine whether an EA is conducted in-house, or by external auditors, or both.

Frequency of audits varies among organizations. Based on the prescriptions of the environmental management systems in place and on the level of environmental risks and impacts, EAs are conducted from every six months to every three years.

The EA process typically includes three stages—pre-audit, on-site audit, and post-audit—each consisting of several steps. The pre-audit activities start with the determination of the goals and objectives of the EA, drawing upon a detailed consideration of the organization's needs and activities, management philosophy, corporate culture, organizational structure, and size. Communication of management commitments and audit objectives to internal personnel should occur to secure on-site cooperation and support.

The next step is to select an audit team that possesses complementary and/or overlapping skills as needed and will ensure objectivity and professional competence. When forming a team, thought should be given to language, ethnicity, and cultural background.

One important pre-audit activity is to determine the scope and focus of an EA. Given the limited resources availability and usually wide areas to audit, the audit program should clearly specify the organizational (e.g., unit/corporate-wide), functional (e.g., waste management, occupational health), compliance (e.g., national regulations, industry standards), and locational (e.g., adjacent sensitive areas) boundaries. After this, a specific facility to audit is selected and its management is contacted in advance to schedule a facility visit.

Following the site selection, the audit team collects and reviews the relevant regulations, policies, industrial standards, and background information such as facility layout, permits, operating manuals, records, and emergency plans. Reviewing this information prior to an on-site audit is essential to obtain the necessary insights into the facility operations, prepare a list of issues and questions to be audited further, ask relevant questions, and more easily identify noncompliance items. If the audit resources and the facility activities permit, a good way to collect primary information about the facility is an advance visit. During this step, it is also advisable to develop and send to the facility a pre-audit questionnaire/checklist with potential review areas.

The preceding steps serve as the basis for an audit plan, which is an outline of what activities are to be done, where and when, who is responsible for each activity, and how they are to be implemented. Simultaneously, on-site audit protocols and/or checklists with evaluation criteria (e.g., permissible emissions levels) are prepared to concentrate the audit efforts on potential areas of concern.

The final pre-audit step entails finalizing administrative and logistics arrangements, such as coordinating the availability of audit team members, setting a time/place for the audit team to meet before the audit, developing a reporting mechanism, and confirming arrangements with the facility managers. The pre-audit activities should not be underestimated as they largely determine the success of a usually short site visit.

The on-site EA commonly begins with an opening conference, which involves the audit team and site operational/environmental management. Usually run by the facility manager and audit team leader, it serves to introduce the team, audit objectives, scope and agenda, present the audit methods, and so on.

The audit team then needs to develop the understanding of the facility's internal control systems instituted for environmental matters. This is accomplished through information/document review, questionnaires, facility orientation tour, and provides a benchmark against which the compliance status can be determined. Alongside documenting the understanding of the internal management systems, the audit team needs to evaluate them in terms of adequate authorization system, proper record keeping, clearly assigned responsibilities, internal verification, staff capacity, and so on. In the absence of significant flaws, the audit typically focuses on the effectiveness of the internal environmental management; otherwise, it should examine the environmental problems and the associated management deficiencies.

The next on-site audit step, gathering of audit evidence, is conducted to collect materials on the identified weaknesses in the environmental management/control systems, to verify their consistent and adequate application, and to form a basis for audit analysis. Audit evidence can be documentary (e.g., manifests, orders), physical (e.g., sighting of oily ground or emissions), and/or circumstantial (indirect evidence like attitude of staff). The typical methods/tools include interviews, formal questionnaires, physical facility inspection, including remote areas, protocols/checklists, informal discussions, and verification tests.

The evaluation of audit findings is accomplished to compare and integrate the individual auditor's observations and findings, to determine whether they are properly substantiated with audit evidence or not, to agree upon which findings are to be urgently brought to the site management's attention, which evidence is to be included in the final report, and so on. To facilitate this step, the auditors should daily discuss interim findings with the site staff.

A closing conference completes the on-site audit and provides an opportunity to inter alia summarize and discuss the preliminary audit findings and recommendations, clarify misunderstandings, decide if additional information is needed, and communicate problems that require immediate remedial measures.

The first post-audit step is preparing a draft audit report with the audit team members usually drafting their parts and the team leader collating a full report. The report should contain a summary, overview of the facility, background to the audit, audit objectives, scope and methodology, findings, and recommendations. It is typically circulated and reviewed by site management, facility staff responsible for corrective measures, corporate management, and other stakeholders as agreed.

Simultaneously with producing a final audit report, an audit action plan is set out with targets and objectives for environmental improvement, distribution of responsibilities, implementation timetables, and financial allocation.

The EA is incomplete until a formal follow-up is initiated to ensure that the audit findings and recommendations are addressed. Responsibility for monitoring of audit follow-up may rest with either the audit team leader or facility management.

The major constraints to quality EAs are the increased resources requirements, including direct (e.g., salary, travel/living costs) and indirect costs (e.g., time spent by facility staff), disruption of facility operation, late or low-quality reports, lack of commitment or resources to implement audit recommendations, and insufficient follow-up. Despite these barriers, conducting EAs has become crucial for many businesses and public organizations that strive to improve their environmental performance and demonstrate their commitment to corporate social responsibility and sustainability.

See Also: Corporate Social Responsibility; Environmental Accounting; Environmental Assessment; Environmental Management System; ISO 14000; ISO 19011.

Further Readings

Cahill, Lawrence B., et al. *Environmental Audits*, 7th Ed. Rockville, MD: Government Institutes Publishing, 1996.

Greeno, J. Ladd, Gilbert S. Hedstrom and Maryanne Diberto. *Environmental Auditing: Fundamentals and Techniques*, 2nd Ed. Cambridge, MA: Center for Environmental Assurance, Arthur D. Little, Inc., 1987.

Humphrey, Neil and Mark Hadley. *Environmental Auditing*. Bembridge, UK: Palladian Law Publishing, 2000.

McGaw, David. *Environmental Auditing and Compliance Manual.* New York: Van Nostrand Reinhold, 1993.

Maia Gachechiladze
Central European University

Environmental Economics

Environmental economics loosely defines a body of academic research that applies the values and tools of mainstream economics to the task of integrating the environment in economic decision making. Originating mostly in the 1960s and 1970s, environmental economics developed simultaneous to an emerging public awareness of environmental crisis. Unlike much of the environmental movements around the globe, however, environmental economics has deliberately stayed away from more fundamental philosophical questions about the interaction between humans and nature, and instead concentrated on the practical requirements of policy analysis and recommendations within the given economic and political context. Environmental economics, in this sense, also needs to be contrasted to competing schools of economics such as ecological economics and green economics, both of which deviate substantially from the premises of mainstream economic thinking.

When it comes to the environment and public welfare in general, the market reveals major flaws: it ordinarily does not account for "external" costs, or costs that are not directly part of a market transaction (e.g., pollution caused by coal-fired power plants is not part of the cost of producing and selling energy). A critical component of environmental economics has been to develop systems and models that allow economists to assign prices ("valuation") to previously unappraised but scarce (external) environmental resources such as clean water or clean air.

Origins

Sometime between the early 1960s and the late 1980s, people in large numbers began to realize that the world faced an environmental problem. Though there continues to be much debate about the nature and scale of the problem, many citizens of industrialized nations remember specific incidents that forever buried the comfortable notion that the world's resources were unlimited, and humankind's actions too insignificant to spoil our ecosystem—a nuclear test extinguishing all life of a small island; a shoreline fouled up by an oil spill; thick smog sickening thousands of urban residents. In American history, Cleveland's Cuyahoga River became the poster child for the birth of environmental awareness—a river so contaminated with industrial waste that it self-ignited on June 22, 1969, burning for days. But whatever the problem—pollution, deforestation, global warming, depletion of the ozone layer—a growing realization began to emerge: the environment was central to society's well-being. As described in more detail below, this was true whether one chose to examine the environmental crisis using logic, hard science, subjective feelings of wellness, or economic cost-benefit analyses.

Economics and the Environment

A key concept in all debates about the environment is "growth"—growth of populations, economic output, and consumption. Due to the explosion of productive capacities in the wake of the industrial revolution, humans have managed to use up more resources since 1945 than in all of prior human history combined. Solid economic growth rates (in both capitalist and communist nations) were generated by increases in production and consumption of goods and services—much of which was resource intensive. Improvements in standards of living became virtually synonymous with increased use of resources—"better" became virtually synonymous with "bigger" and "more."

Indeed, the idea that economic growth is good, and that more growth is better, has been, and still is, widely celebrated—by citizens and economists alike. Since the very beginnings of the discipline, "growth" has made up the basic rationale for modern economics; it is what politicians and business leaders the world over strive for; it is the one and only parameter that informs our key economic indicators (the national income and product accounts, like gross domestic product [GDP]); it is widely perceived as the key indicator for measuring standards of living; and economists see it as an essential ingredient not only of success, but of the very survival of modern economic systems.

To be sure, the track record of economic growth over the last 200 years or so is impressive. Though greatly favoring richer over poorer countries, and, even within highly developed rich countries, greatly favoring the rich over the poor, the world has witnessed a historically unprecedented increase in human productivity, levels of education, life expectancies, and general standards of living. Historically, economic growth also seems to help, or perhaps even facilitate, things like democratic governance, peace, and women's rights. Though there are notable exceptions such as Nazi Germany or Chile under Pinochet, richer nations (i.e., nations in which there is less scarcity of essential goods and services) seem to be able to afford their citizens with a higher degree of security and civil rights.

There is one fundamental problem with this record, though: after a certain point of growth and development, additional growth no longer necessarily translates into progress. A growing body of research even suggests that additional indiscriminate economic growth reaches a critical point after which it begins to generate more bad than good outcomes (again, this is obviously not true for underdeveloped poor countries). Basic logic makes this readily understandable: the first glass of wine may have been greatly beneficial to you; that does not mean the tenth glass will be as well. Some logging is not only good but vital to the health of the forest; clear-cutting destroys the forest. The same logic applies to national economies: for Americans, the per capita amount of energy used, as well as the amount of trash produced, roughly doubled every 20 to 30 years in the 20th century. Several scholars have pointed out that, if we assume a relatively conservative national economic growth rate of 2 percent, an economy overall (and this of course includes massive economies such as the United States and China) would double in output every 35 years (or every 23 years at 3 percent growth rate). Given basic laws of physics and biology, or just given the resource intensity of modern economies, to articulate such a goal would seem absurd. And yet, economists and politicians the world over continue to do just that.

However hopeful one is about the prospect of new technologies and advanced science to bail us out, this logic also necessitates the conclusion that, at current rates of resource depletion and pollution, the question is no longer if humanity will run into existential problems, but rather when.

Highly developed nations, and above all the United States, until very recently contributed the lion's share of adverse environmental impact: burning up of nonrenewable fossil fuels, emission of greenhouse gases, and introduction of toxic chemicals. The rapidly growing economies of highly populous countries like China and India are now beginning to rival the environmental impact of the old economic power houses. Today, despite advanced technologies and increasing governmental regulations, growing economic output and increasing consumption levels of an expanding global population continue to translate into exponentially increasing levels of environmental depletion and destruction. Among natural scientists, a near total consensus is emerging: the only thing we now know to be "unrealistic" is to continue on the present path of economic growth, for the planet cannot sustain it. Even if we merely wanted to sustain current global levels of production and consumption, conservative estimates suggest that we would need about 1.3 planets. If everyone in the world lived liked Americans, we would need about five planet Earths.

A steadily growing body of work by psychologists, sociologists, and environmental economists has formulated another troubling insight: people's sense of well-being, happiness, and life satisfaction has long stopped following the upward curve of economic growth rates. In America over the last 40 years, there no longer appears to be a discernible correlation between wellness and economic growth. On the contrary, an increasing amount of growth is spent at the expense of wellness: less free time and more hours worked to keep up the same standard of living; increasing expenses in order to be able to enjoy basic things like safe neighborhoods, good public schools, clean air and water; and, of course, more and more "defensive" expenses, such as costs to prevent or ameliorate the effects of stress, crime, sickness, pollution, degradation, and depletion.

As the first full-view Earth pictures of Apollo 17 (1972) made so stunningly apparent, our planet is a rather small and finite organism with a finite supply of resources. This basic insight has dramatic economic consequences. A frequently cited article by economist Kenneth Boulding captured the problem well by likening economic activities on Earth to what we would have to do if we were confined to a spaceship: with a finite supply of resources, we suddenly face the existential need—to expand on a common environmental phrase of the early 21st century—to reduce, reuse, recycle, and to repair. When people run out of essential goods—clean air, water, land—there no longer is a possibility to go elsewhere and find more. Growth of population combined with exponential growth in production has pushed the Earth to its limits—or, to be more precise, since Earth will survive with or without human beings—it has reached (or perhaps already surpassed) the limits of the kind of human existence the planet can support.

Today, there seem to be only two alternative conclusions left. One, we can ignore environmental impact, for much of the more dramatic consequences still lie in the future, a future possibly past the life expectancy of current generations of earthly inhabitants. Two, and this is a view still propagated by a number of economists, we bank on innovations in science and technology to bail us out. After all, the argument goes, many scary predictions of the past have not panned out either (cities drowning in horse manure before the widespread use of cars, global cooling, the imminent depletion of oil reserves during the oil crisis of the early 1970s, to name a few).

Environmental Costs and Economic Accounts

Environmental distress not only poses grave problems to life on planet Earth; it is also, by any account, increasingly expensive. Whether on the level of small firms or national governments, however, how exactly to measure environmental impact—how to attach a price to depletion, erosion, or pollution—remains far from clear, despite significant advances. Most environmental impact has no price for the simple reason that it is not part of the initial market transaction. When you buy gasoline for your car, you do not pay for your contribution to greenhouse gas emissions and global warming. When a pharmaceutical manufacturer contaminates the river, it does not pay for the loss of recreational or fishery value of the water downstream, for it is not an internal part of the transaction between pharmaceutical producer and consumer. But in either example, who, if anyone, should pay? And how much?

For decades, such problems of "external" costs were considered to be marginal, of no great significance, given the seemingly endless supply of natural resources. If a frontiersman in the early 1800s killed a bison in the wide-open spaces of Oklahoma merely for target practice, or an early car belched out big clouds of unfiltered exhaust fumes in cities where horses were still the primary mode of transportation, the "external" costs seemed minimal. This drastically changed with the explosion of productive capacity among industrializing nations. When the "externalized" collateral damage of industrial production finally began to spoil air, water, and land, and in the process sickened and killed people in significant numbers, ignoring became more difficult.

Environmental economists consider environmental impact that is adverse to the common good a "market failure," for it is a clear indication that resources are not distributed in the most efficient manner. The underlying problem, according to environmental economists, is not the market itself, however, but rather the lack of sufficient knowledge and understanding among market players. Elaborated as early as 1968 by economist Garrett

Hardin, there is potential "tragedy" in the making when a few people, acting out of pure self-interest, spoil or destroy essential common goods (e.g., air, water, soil) for everyone, including those who reaped no benefits from the destructive behavior. What is required, therefore, are better instruments that help inform markets of the "true" costs, usually by including the so-called externalities describe above into a more complete set of accounts.

Ironically, whether the inclusion of externalities is brought about through regulations, taxes, or market-based instruments such as cap-and-trades, all such models fundamentally depend on government to articulate and enforce such public policies. In an unregulated market, without government intervention, the "commons" may fall victim to individual greed.

While many of the environmental movements emerging in industrialized nations began to ask fundamental questions about human interaction with planet Earth (in the process questioning many of the pillars of modern economies), and while traditional economists by and large continued to ignore the environment in their pursuit of economic growth, environmental economists began to articulate models that would allow environmental impact assessments to be included in traditional cost–benefit accounts. In part, this is an attempt to bridge the growing chasm between two groups of people—environmentalists and economists—who do not share much of a common understanding of, or even a common language describing, the world they inhabit. Above all, however, it is an attempt to find practical means to affect public policy making in ways that effectively integrate the environment.

Concepts and Tools of Environmental Economics

That the environment is not a separate entity from the economy is, by now, almost universally accepted. Every economic act has environmental consequences; every change in the environment bears economic impacts. The two are interrelated and interdependent—one affects the other. What environmental economists have fully realized, and attempt to translate into models for analysis and action, is that both economy and the environment must be fully integrated in decision making.

When describing in more detail the challenges faced, and models provided, by environmental economics, it is worth keeping in mind that any evaluation of the contributions of this field of study fundamentally depends on one's understanding of (1) the nature and scope of the environmental crisis; and (2) the range and extent of what is politically and economically possible. Someone who is confident that modern technology will provide solutions to all environmental problems will arrive at different conclusions than someone who believes the natural environment is at the breaking point; someone who thinks that "more" will always be a necessary precondition for "better" will naturally evaluate the contributions of environmental economics very differently than someone who believes that human communities need to abstain from traditional economic growth models to reach social wellness and environmental health.

A central premise behind environmental economics is that one does not have to give up on growth in order to protect the environment and reach sustainability. Another path, one already recognized by the father of America's national accounting system (Simon Kuznets), but then largely ignored by successive generations of national income accountants, is to ask not "whether," but "what" it is we are growing. Though it may appear obvious on second thought, all growth is not the same. Growing the amount of medication people have to take in order to get through the day is not the same as growing the wellness industry; growing the number of coal-fired plants is not the same as growing a renewable energy industry. Perhaps one could even imagine a growing cap-and-trade market of time, a still highly undervalued commodity. The critical distinction attempted here, of course, is between unsustainable and sustainable growth.

How to direct or regulate economic growth into a more sustainable direction, particularly when working only with the limited neoclassical tool set of market and price incentives, is one of the central questions of environmental economics.

Externalities, whether environmental or otherwise, are not a new phenomenon in economics. They describe the basic idea that transactions between two parties have effects on third parties—beneficial or detrimental—for which no price is extracted. How to conceptualize these effects, and particularly how to valuate them, however, continues to be a difficult and contentious issue. In standard economics, externalities are recognized as existent, but are largely neglected, or downplayed as insignificant. As mentioned above, when a motorist buys a gallon of gasoline for a certain amount of money, mainstream economics considers this a proper market transaction in which the buyer presumably pays for the full value of expenses plus profits of the seller (extraction, refinement, transportation, storage, and so on). That the burning of gasoline in turn leads to smog, which in turn leads to sickness, which in turn leads to various remedial expenses, is not an intrinsic part of the original transaction. Whether it is of concern to society or not, it is not valued by the original transaction for it has not been given a price. Rather, it is seen as "external" to the transaction, and, as such, either not important because it is not valued by the market, or, at best, a responsibility of governmental regulation.

The primary contribution of environmental economics is to find ways to internalize externalities by assigning a price to third-party effects of a market transaction—whether the effect is sickness, depletion, or pollution. This process of assigning artificial prices to external effects is called "valuation." Over the years, environmental economists have developed increasingly diverse and complex models for valuing externalities. Initially, they included efforts that ranged from finding a value for national parks, for instance, by pricing the costs of travel and accommodation people pay in order to enjoy the parks, to attempts to estimate implicit prices of noise pollution through questionnaires of Realtors of properties adjacent to airports. Today, such efforts include highly sophisticated mathematical input–output analyses and, a favorite among environmental economists, game theory. In all cases, the basic goal is to price the benefit of a certain good—clean water, a quiet environment, the beauty of a national park—by figuring out the cost people are (or might be) willing to bear in order to obtain (or maintain) the good.

As such, the field of environmental economics always represents an admission that market transactions are imperfect in that they either do not at all, or do not adequately, account for social costs (in this case primarily the social costs of environmental pollution, degradation, and depletion). It also always represents the argument—explicitly or implicitly—that the only effective way to account for the interdependence between human life, economic production and growth, and the environment, is to attach a price to each item (or have the market determine a price for each item) large enough to protect the item.

A simple theoretical example that bears some of the classic dilemmas of the valuation method might look like this: assume the production process of a commodity (say bed pillows) can be shown to have caused environmental contamination of the adjacent neighborhood which in turn demonstrably led to the death of a resident in that neighborhood. From the perspective of mainstream economics, that contamination is not part of the transactions between the corporation and the buyers of their pillows, nor is it part of the transactions between the company and its employees (irrespective of whether the fatally sickened resident was an employee or not).

How, then, do environmental economists figure out a way to attach a monetary value to something that is external to all the market transactions that happened in this example, namely the contamination? In this case, environmental economists would likely look at what the market would have yielded in value for the particular human life extinguished by the contamination. What earning potential did the person killed have? That is, what education, what professional experience, what age did this person have, and, based on all that, how much more was that person likely to earn for the rest of her or his working life? Whatever the amount such a calculation would generate would then be used by turning an "externality" of indeterminate monetary value—that is, the contamination—into a precise price that can then be made "internal" to the transaction. In the process, of course, a particular human life would end up being valued at, say, something like $1.3 million.

Clearly, such valuations are not for the sentimental, and plenty of criticism has been leveled against attempts to attach cold price tags to what many consider simply "priceless"— things of intrinsic and overarching value such as a clean environment, diversity of species, or human life itself. A standard response of environmental economists is that while such criticism may perhaps be understandable from a moral or spiritual point of view, it is also dangerously naive in political and economic terms, for it is likely to consign care for the environment to a marginal or even irrelevant status in terms of the articulation of public policy. According to environmental economists, in other words, to be effective one needs to understand, and then play by *existing* rules of the game, rather than by ideal-case scenarios. What is needed, according to this logic, are practical solutions, not lofty visions that lack all specificity.

Over the years, environmental economists have developed a series of public policy instruments geared toward achieving a more optimal environmental outcome. Broadly defined, there are so-called command-and-control options, as well as market-based public policy instruments. The former include technological requirements, legally set emissions controls, or legally binding fuel-efficiency standards. The latter include pollution taxes, deposit-refund schemes, and tradable pollution and resource permits. While environmental economists are not adverse to recommending the command-and-control options, they generally prefer all market-based instruments, for they are assumed to have two basic advantages: (1) they minimize the costs of compliance enforcement, and (2) they stimulate scientific and technological advances by helping producers reduce pollution-related expenses.

Since its early days in the 1970s, environmental economics has grown significantly. Almost every major university in the United States by now has a program in environmental or resource economics. Hundreds of academic practitioners across the nation are organized through various regional and two national professional organizations, and publish in six English-language professional journals specializing in environmental economics. Most importantly, though, environmental economists have become increasingly influential in shaping public policy, from implementing higher fuel-efficiency standards to articulating rules for carbon cap-and-trade to negotiating international emission standards.

Major Problems—Possible Future Paths

One prominent economist concerned about the environment, Richard Norgaard, wrote during the early days of the discipline that "environmental economics is a contradiction in terms." He was referring to a basic conceptual problem that has dogged the discipline from its inception, namely, how to capture a highly complex system of evolving interconnectedness (the environment) with the tools of a discipline mostly based on atomistic and mechanistic assumptions (economics). Things like consequences in the distant future, or consequences that qualitatively change in combination (such as global warming intensifying drought) are inherently difficult to measure and incorporate. Things like equity, fairness, or human happiness are persistently resistant to economic valuation attempts.

Some practical examples of resulting problems are as follows:

- Environmental problems that, most or all, lie in the distant future create a number of current conceptual and practical problems. For one, the magnitude of the future problem is entirely unclear. And, even if it were clear, it is difficult to figure out how much current generations should pay for the benefit of future generations, or, alternately, how much one can or should expect future generations to pay for the misdeeds of current generations. Also, within cost-benefit analyses of mainstream economics, future problems tend to be of marginal significance due to a technique called "discounting," in which far-distant costs, even though they might be substantial, become insignificant if expressed in present prices.

- Environmental problems that are "public" on a scale larger than the nation, such as global warming and climate change, but also things like acid rain, fresh water depletion, or border-crossing toxic fumes. When there is no clearly identifiable market, and no possibility of governmental regulation (and least not on a national level), valuation tends to become useless.

- So-called non-use value, that is, the potential cost of the depletion, destruction, or pollution of goods that were previously not part of the market economy, but have value independent of any use, present or future, that people might make of those goods. A good example for that would be the Alaska coastline that was despoiled by the *Exxon Valdez* oil spill. Only after the widespread destruction occurred, and even then only in small parts, was it possible to ascribe a value to a vital ecosystem.

Addressing many of the above problems (at least theoretically), one promising new path of research and economic modeling, geared toward the creation of a true sustainability standard, has come out of the collaborative and interdisciplinary work of a group of economists and ecologists, working under the leadership of Nobel Prize–winning economist Kenneth Arrow. What they coined *inclusive wealth* significantly deviates from all other economic models. It is an attempt to measure the change in value over time of all the critical capital stocks in an economic system—not just those that get traded in the market, but all of them: manufactured capital, natural resources, all ecosystems, and also essential things like human knowledge and human welfare. Everything, in short, that matters to economic development, now and in the future, is measured and counted as a "real price," or reflecting the actual cost it would take to replace the asset. It thus represents the first model that provides a comprehensive, or "inclusive," theory of sustainability: the value of a national economy, in *all* its forms, should not be allowed to decline over time.

The model does not address the more fundamental objections to attaching a price to environmental goods or aspects of human life. It also will require a lot of effort working out genuine substitution prices for everything. As a model for environmental economists, however, it has significant potential, and could finally provide a reliable indicator for whether or not we are moving in the direction of true sustainability.

Overall, much conceptual work has gone into the articulation and practical estimation of expanded cost-benefits analyses that include calculations of the price of pollution, depletion, and degradation, and thus help make the market more adequately reflect the short- and long-term societal costs of environmental damage. In doing so, the focus of environmental economics has largely remained on goals identical to the rest of the economics profession: utility maximization and economic growth, though in this case with accounting models that reflect costs arising from environmental damage, and incorporating them into the prices of goods and services bought by consumers, businesses, and governments.

As such, within the larger environmental debate, environmental economics is both celebrated and derided: celebrated as the only realistic tool to have resolutely impacted public policy making on behalf of environmental protection; derided as a narrow set of tools that fails to address the deeper problems behind the global environmental crisis.

See Also: Ecological Economics; Environmental Accounting; Externalities; Pollution Prevention; Sustainability.

Further Readings

Arrow, Kenneth, et al. "Are We Consuming Too Much?" *Journal of Economic Perspectives,* 18/3 (Summer 2004).
Boulding, Kenneth. "The Economics of the Coming Spaceship Earth." In *Radical Political Economy: Explorations in Alternative Economic Analysis,* Victor D. Lippit, ed. Armonk, NY: M. E. Sharpe, 1996.
Hanley, Nick, Jason F. Shogren and Ben White. *Environmental Economics in Theory and Practice*, 2nd Ed. New York: Palgrave Macmillan, 2007.
Hardin, Garrett. "The Tragedy of the Commons." *Science,* 162/3859 (1968).
Kuznets, Simon. "How to Judge Quality." *The New Republic* (October 20, 1962).
Norgaard, Richard B. "Environmental Economics: An Evolutionary Critique and a Plea for Pluralism." *Journal of Environmental Economics and Management,* 12 (1985).
Stavins, Robert, ed. *Economics of the Environment: Selected Readings,* 5th Ed. New York: Norton, 2005.
World Bank. *Expanding the Measure of Wealth: Indicators of Environmentally Sustainable Development.* Washington, D.C.: World Bank, 1997.

Dirk Peter Philipsen
Virginia State University

ENVIRONMENTAL IMPACT STATEMENT

The Environmental Impact Statement (EIS) process is the product of the 1970 National Environmental Policy Act (NEPA). This act established the EIS process as a way to hold federal agencies and federally funded projects accountable for potential environmental degradation. NEPA requires that U.S. federal agencies or anyone receiving federal funding prepare detailed analyses of any of their actions that significantly impact the quality of the environment, which includes almost all developments, such as roads, industrial parks, malls, or airports. It is, therefore, an important piece of U.S. legislation that obliges federal agencies to incorporate environmental values into their decision making. The significance of the EIS process does not simply lie in its ability to make federal actions more transparent but also in its capacity to instill an environmental value system into the federal decision-making process and, by extension, into the guidelines of national and international organizations and corporations that are either based in the United States, or routinely do business with or in the United States (such as the World Bank, the International Monetary Fund [IMF], or a transnational corporation like General Electric [GE]). Indeed, while NEPA and the EIS process based on its guidelines are regulated by the Council on Environmental Quality (CEQ), several states have adopted their own mini versions of EIS procedures, and most transnational corporations as well as international organizations now include versions of EISs as a routine part of the planning process.

The EIS Process

There are three potential documents that may be filed as a result of the environmental consideration on the part of federal agencies. Actions may require an Environmental Assessment (EA), an EIS, or a finding of no significant impact (FONSI) document. When it is unclear whether an action might cause environmental harm, an EA is conducted first. The EA process is a much shorter, less-intensive process than the EIS that helps determine whether an EIS should be conducted. The EA simply requires that sufficient evidence be presented to determine the need for a full EIS study. If the EA determines that the federal action will result in little or no effect on the human environment, a FONSI is submitted to the CEQ and the project is permitted to move forward without further environmental analysis. If, however, significant environmental impact will occur as a result of the federal action, a notice of intent (NOI) is issued to indicate that the EIS process will commence.

The NOI is published in the Federal Register and includes the following elements. First, the NOI must describe the proposed action and alternative actions that might be conducted instead. It then must describe the scoping process and include the contact information for the responsible party within the agency who can respond to queries related to the proposed action and the EIS. Any other background information related to the agency's project or to the EA should be included in this document as well. The NOI is an important part of the EIS process because it sets the stage for the scoping process.

Scoping is an exercise that must be conducted as a collaborative process between the federal agency and interested parties such as organizations with a stake in the project, tribes, and other interested parties. During these scoping meetings, the scope of the EIS is determined. This means that issues in need of in-depth investigation are identified. Similarly, issues that are well vetted in other EAs or EISs are given lower investigative priority. Isolated actions, connected actions, and cumulative actions are all considered during scoping. Other considerations that may be included in the scope are geographic considerations and the timescale of potential impacts. In other words, scoping allows interested parties to determine the depth and breadth of the EIS. During scoping, the public plays an important role in helping to determine the important issues that must be examined in the EIS. Involving the public and other interested parties is critical to the transparency of the process as well as to the quality of the process.

The responsible federal agency also has specific duties within the scoping process. The agency must consider alternatives to the proposed action including the consideration of the effect of taking no action. In addition, the agency must consider related impacts that are direct, indirect, and cumulative. The scoping process is a way to ensure that all actions, impacts, and alternatives are fully considered within the EIS.

Following the scoping exercise, preparation of the preliminary draft environmental impact statement (PDEIS) begins. This stage of the EIS process is when most of the reporting occurs. First a cover sheet, summary, and table of contents are prepared for ease of use. Following the organizational elements are the descriptive parts of the EIS. The purpose and need are the first elements to be included as they are able to serve as an introduction as to why the EIS is being conducted in the first place. A statement of the proposed action, the purpose of the action, and the need for the action is presented first. Within the introductory part of the section are items such as a summary of the scoping process and explanation of the dual purpose of the EIS if there is one. "Dual purpose" refers to the use of the EIS as documentation for some other regulatory process such as the Comprehensive Environmental Response, Compensation, and Liability Act (CERCLA) feasibility study.

Next, the alternatives to the proposed action are presented. This section is meant to be comparative so that the reader can readily assess the differences between the impacts of the proposed action and those of the alternatives. Evaluation of the different options should be rigorous so that there is a clear basis for choice among the alternatives. Within the alternatives section, the "no action" option must be presented in the same rigorous manner as the other alternatives so as to ensure that all possible impacts are accounted for within the EIS. Furthermore, it is not sufficient to simply present the alternatives. The PDEIS must also include a discussion of alternatives that were eliminated from the EIS. Typically, the preferred alternative is highlighted in this section but may not be confirmed until later in the process.

Environmental consequences are vetted next in the EIS. The different environmental impacts are assessed here including any effects that are unavoidable if the action is approved. Consequences might include resources that will be used, ecology that will be disturbed, and energy requirements for the action. If they exist, mitigation strategies for all of the impacts and alternatives are important to include as well.

Finally, any applicable laws, regulations, or permits that must be obtained in order to embark upon the project are included in a final section of the EIS. Discussion of other state and federal requirements is important because it helps to explain boundaries within the context of the project. These boundaries may be pollution thresholds or control prerequisites that require interagency coordination.

When the PDEIS is complete, it is reviewed, changed, and reissued to the public as a draft EIS. The draft EIS is specifically submitted to interested parties. For those interested in the EIS who did not receive a copy, instructions are published in the Federal Register as to how to obtain one. Comments are typically accepted from the public for up to 45 days following the release of the draft EIS. These comments are transcribed and answered within the final EIS by the preparer. When the final document has been edited and approved by the federal agency, it is considered complete.

The EIS is considered in conjunction with other factors such as economics, technical considerations, or an agency's statutory mission to make a final decision about a project. What this means is that the most environmentally preferable alternative presented in the EIS may not be the alternative that is ultimately chosen in the end. The final decision is presented in a document called the record of decision (ROD).

As of today, EISs have become a pervasive global phenomenon. As a premier international organization for a range of impact statements, the International Association for Impact Assessment (IAIA) defines environmental impact statements as "the process of identifying, predicting, evaluating and mitigating the biophysical, social, and other relevant effects of development proposals prior to major decisions being taken and commitments made." Literally hundreds of national and international, governmental, and nongovernmental bodies and organizations now provide services and guidelines for EISs, or, in the case of governmental bodies, require their constituents to generate EISs. From the World Bank to the Sierra Club, to national governments across the globe, to joint governmental bodies, such as the European Union, all provide a variety of guidelines and requirements for EISs.

Critiques

It is rarely contested that the EIS process in general (both nationally and internationally), and NEPA in particular, provide valuable legislative contributions. There are, however, critics who feel that the EIS process is too procedural: long on paperwork and good intentions, short on transparency and enforcement. The process generally requires that environmental impacts and mitigation strategies be reported; for instance, there are no mandates for avoidance of environmental degradation. The procedural nature of the EIS process also may deter interested citizens from participating in the process for fear that their comments are irrelevant. The purpose of not requiring the most environmentally preferable alternative to be used was to ensure that other considerations could be included in the decision-making process. The result, however, is that the EIS process has become a bureaucratic ritual rather than an insertion of national environmental values into the decision making of national agencies. Critics around the world have thus argued that the best way to reinsert the value dimension into the EIS process is to include regulations that require the avoidance of environmental damage or mandate that the most environmentally preferable options be chosen.

See Also: Environmental Assessment; Environmental Risk Assessment; Stewardship; Transparency.

Further Readings

Bregman, Jacob I. and Kenneth M. Mackenthun. *Environmental Impact Statements*. Boca Raton, FL: Lewis Publishers, 1992.

Kreske, Diori L. *Environmental Impact Statements: A Practical Guide for Agencies, Citizens, and Consultants*. Hoboken, NJ: John Wiley & Sons, 1996.

Lindstrom, Matthew and Zachary Smith. *The National Environmental Policy Act: Judicial Misconstruction, Legislative Indifference, & Executive Neglect*. College Station, TX: Texas A&M University Press, 2001.

Moore, Emmett B. *The Environmental Impact Statement Process and Environmental Law*. Columbus, OH: Battelle Press, 2000.

Heather M. Farley
Northern Arizona University

ENVIRONMENTAL INDICATORS

Environment indicators are a set of measurements or statistics that indicate the overall condition of the environment and provide evidence of the effectiveness of environmental management programs and business sustainability practices. Often environmental indicators are aggregated or grouped into a single weighted index. Indicators may show the improvement or worsening conditions of the environment as a trend line, starting from a current baseline to a projected state; thus, a judgment could be made regarding the relative impact and magnitude of a business practice to the environment. The number and type of indicators utilized are dependent on the company size, consumer base, company policy, governmental enforcement, and potential financial and reputation impact. Environmental indicators may incorporate information on resource management, energy usage, pollution emissions (e.g., air quality, transportation, soil, water), and waste production. The information provided from various environmental parameters raises awareness among the public, policy makers, and corporations regarding the health status of the planet, which then guides initiatives for environmental protection and improving sustainable business operations.

Level I indicators are the activities initiated by governmental agencies and/or environmental groups. Here, the U.S. Environmental Protection Agency's Millionth Mercury Switch program has dismantlers remove mercury switches from junked automobiles— a process that takes only minutes in most cases.

Source: U.S. Environmental Protection Agency

Xander Oltshoorn and others note that environmental data used for standardized indicators in the business sector generally correspond to one of five types: economic based, physical impact, linear programming methods, economic valuation methods, or business management review processes. Economic-based indicators scale environmental data according to the financial cost or market impact, such as the cost involved in switching from fossil fuel to renewable energy sources. Other indicators demonstrate the direct effects to the physical system, such as the global warming impact potential, pollution toxicity, or habitat loss. Linear programming methods determine the most efficient methods for conducting business with minimal adverse environmental impacts, with efficiency incorporating information on production, financial costs, and pollution, among others. Economic valuation centers around indicators that provide information on environmental quality in relation to societal benefits, often using a "net value added" approach. Management-related indicators demonstrate the ability (or inability) of companies to supervise matters that can have an effect on environmental performance, such as resource distribution, environmental policies, and regulation compliance.

Regardless of the standardization procedure used, environmental indicators are often reported as a value, showing the percent change; the total or mean amount used per unit time, volume, or person; or other quantitative data summary. Businesses focusing on minimizing environmental impacts from energy consumption may use environmental indicators that report information on ratio of the amount of renewable energy resources (e.g., solar, wind) used for production to fossil fuels (e.g., oil, coal). The effectiveness of waste management practices could be indicated by the amount of materials recycled or donated, landfill production waste, and money spent on waste removal. A business may also want to examine reducing the environmental impact of their employees with reference to pollution from waste production, commuting, and resource usage. For instance, companies may focus on increases in the percentage of employees walking or bicycling to work as opposed to driving as an indicator of reducing the environmental impacts of transportation pollution.

Environmental Indicator Providers

Numerous state, federal, international, professional, university-based, corporate, and nonprofit agencies and organizations monitor and provide information about the state of the environment.

One such organization, the U.S. Environment Protection Agency (EPA), sets national standards by writing and enforcing federal regulations for environmental protection. In 1982, the EPA developed six levels (Level 1 to Level 6) of environmental indicators, divided into two categories (administrative and environmental) according to their hierarchy. Among the administrative indicators, Level 1 indicators are the activities initiated by the governmental agencies and environmental groups; for example, permits written, or grants issued to reduce or prevent pollution or exposure to pollution. Level 2 indicators measure the responses of the community to the Level 1 actions in reducing the pollution entering air, water, or land. The Level 3 through Level 6 indicators were categorized as environmental indicators. Level 3 and Level 4 indicators measure the actual amount of pollutants emitted by the source and levels of pollutants detected in the local community, respectively. Level 5 indicators follow the chemical and biological changes that occur to some pollutants, such as concentration of a toxin as it moves up the food chain. Level 6 indicators monitor the changes in human health or ecological condition that we truly seek to attain. Based on the above indicators, the EPA has adapted a model called "Logic Model" for learning and managing an organization's environmental effect.

The United Nations Environment Programme (UNEP) initiated the Global Environment Outlook (GEO) project in 1995 to review the global environment by analyzing environmental change, causes, impacts, and policy responses. They have selected six priority areas of the environment as challenges of the 21st century. They are climate change, disaster and conflicts, ecosystem management, environmental governance, harmful substances, and resource efficiency. For example, in their 25th session in 2009, the governing council of UNEP laid out the strategy for the international chemical management of toxic heavy metals such as lead, cadmium, and mercury. The latest report (GEO-4) published a comprehensive assessment on the state of the world environment, development, and human well-being, for information and decision-making purposes. This report described the changes in the environment indicators from the local level through regional and global effects. The future well-being of our planet up to the year 2050 is projected in this report, based on the current social, economic, and environmental trends and each of the four possible scenarios of policy making: (1) markets first, (2) policy first, (3) security first, and (4) sustainability first.

The researchers at Regional Resources Center for Asia and the Pacific (UNEP RRC AP) have also developed a list of key environmental indicators to best reflect the environmental concerns in and across the subregions of Asia and the Pacific. The indicators are divided into three main categories (social, economy, and environment) with subdivisions for an environment category as shown in the table below.

		Environment			
Social	Economy	Land	Water	Air	Biodiversity
Population	Gross domestic product	Arable land per capita	Biological oxygen demand (BOD) level in major rivers	CO_2 emissions	Protected area
Human Development Index		Forest area		NO_2 concentration	Threatened plants
Population with income less than $1/day	Gross domestic product comparison	Forest cover change	Population with access to safe drinking water	SO_2 concentration	Threatened mammals
Infant mortality rate	Gross national income			Suspended particulate matter (SPM) concentration	Threatened birds
Life expectancy at birth	Gross national income per capita		Total water withdrawal	NO_x emissions	Wetlands of international importance
	Energy consumption per capita		Total water availability and access to safe sanitation	SO_2 emissions	

As shown in the table, indicators are measured on the basis of the items listed under them.

The United Nations Statistics Division (UNSD) also developed a similar list of environment indicators in 1995. These indicators are economic issues, social/demographic issues, air/climate, land/soil, water (fresh and marine), other natural resources (biological and mineral, including energy), waste, human settlement, and natural disasters. This organization provides data on the socioeconomic activities and events for those indicators and monitors their effects or impact on the Earth. They also record the human responses to improve the environment and the following outcome.

Another international organization, the Organisation for Economic Co-operation and Development (OECD), at their last report in 2008, selected two sets of key indicators based on pollution issues and existing natural resources and assets. They are listed as follows:

Pollution Issues

	Available Indicators*	Medium-Term Indicators**
Climate Change	CO_2 emission intensities Index of greenhouse gas emissions	Index of greenhouse gas emissions
Ozone Layer	Indices of apparent consumption of ozone-depleting substances (ODS)	Same, plus aggregation into one index of apparent consumption of ODS
Air Quality	SO_x and NO_x emission intensities	Population exposure to air pollution
Waste Generation	Municipal waste generation intensities	Total waste generation intensities, Indicators derived from material flow accounting
Freshwater Quality	Wastewater treatment connection rates	Pollution loads to water bodies

Natural Resources and Assets

	Available Indicators*	Medium-Term Indicators**
Freshwater Resources	Intensity of use of water resources	Same plus subnational breakdown
Forest Resources	Intensity of use of forest resources	Same
Fish Resources	Intensity of use of fish resources	Same plus closer link to available resources
Energy Resources	Intensity of energy use	Energy efficiency index
Biodiversity	Threatened species	Species and habitat or ecosystem diversity Area of key ecosystems

*Indicators for which data are available for the majority of OECD countries.

**Indicators that require further specification and development such as availability of basic data sets, underlying concepts, and definitions.

The above indicators are chosen because of their analytical soundness, measurability of environmental progress, policy relevance, and public awareness. This international organization is currently comprised of 30 member countries worldwide, seeking answers to common environmental problems, and deciding on policies relevant to domestic and international scenarios. The OECD, however, recognizes that there is a lag between the demands for environmental indicators, related conceptual work, and the actual capacity to mobilize the underlying data set.

The World Bank also publishes data on environment indicators in seven major categories: agriculture, forest and biodiversity, energy, emissions and pollution, water and sanitation, environment, and health. The data set can be chosen for a specific country, a region, or an income group. A joint collaboration of the Development Data Group and the Environment Department of the World Bank published a collection of international statistics of the world development indicators in 2009. The indicators were selected on the basis of the economic growth, urbanization, and greenhouse gas emissions for different regions and countries throughout the world. The World Bank's current initiative includes monitoring the following environment indicators: adjusted net saving, wealth estimates of a nation, country assistance strategies and the environment, poverty and the environment, indicators of rural sustainability, monitoring natural resources, and forest management.

The sustainability of human and other living organisms is dependent upon the quality of our environment. The rapid changes in environment in recent years have caused great concern to scientists, policy makers, and the public as well. The environment indicators are helpful to understanding the causes and effects of human activities on nature and thus guide us toward an appropriate sustainable future for our planet.

See Also: Carbon Footprint; Carbon Neutral; Clean Technology; Corporate Social Responsibility; Dematerialization; Ecoeffectiveness; Ecological Footprint; Environmental Management System; Green Design; Green Technology; Industrial Ecology; Natural Capital; Pollution Prevention; Smart Energy; Sustainability; Sustainable Design; Toxics Release Inventory; Waste Reduction.

Further Readings

Ash, Michael and T. Robert Fetter. "Who Lives on the Wrong Side of the Environmental Tracks? Evidence From the EPA's Risk-Screening Environmental Indicators Model." *Social Science Quarterly*, 85/2:441–62 (2004).

Brunklaus, B., T. Malmqvist and H. Baumann. "Managing Stakeholders or the Environment? The Challenge of Relating Indicators in Practice." *Corporate Social Responsibility and Environmental Management*, 16/1:27 (2009).

Clarke, David P. "Indicators of Progress: The State of the Nation's Ecosystems: Measuring the Lands, Waters, and Living Resources of the United States." *Environment*, 45/5:40 (2003).

Gallego, Isabel. "The Use of Economic, Social and Environmental Indicators as a Measure of Sustainable Development in Spain."*Corporate Social Responsibility and Environmental Management*, 13/2:78–97 (2006).

Länsiluoto, Aapo and Marko Järvenpää. "Environmental and Performance Management Forces: Integrating 'Greenness' Into Balanced Scorecard." *Qualitative Research in Accounting and Management*, 5/3:184–206 (2008).

Niemeijer, D. and R. de Groot. "Framing Environmental Indicators: Moving From Causal Chains to Causal Networks Environment." *Development and Sustainability*, 10/1:89–106 (2008).

Oltshoorn, Xander, Daniel Tyteca, Walter Wehrmeyer and Marcus Wagner. "Environmental Indicators for Business: A Review of the Literature and Standardisation Methods." *Journal of Cleaner Production*, 9/5 (2001).

Rao, P., A. Singh, O. la O'Castillo, P. Intal and A. Sajid. "A Metric for Corporate Environmental Indicators . . . for Small and Medium Enterprises in the Philippines." *Business Strategy and the Environment,* 18/1:14 (2009).

Roper, William E. "Environmental Indicator Assessment for Smart Growth." *International Journal of Environmental Technology and Management,* 5/2–3 (2005).

Rosenstrom, Ulla and Jari Lyytimaki." The Role of Indicators in Improving Timeliness of International Environmental Reports." *European Environment,* 16/1:32–44 (2006).

Seppala, Jyri, et al. "How Can the Ecoefficiency of a Region Be Measured and Monitored?" *Journal of Industrial Ecology,* 9/4:117–30 (2005).

Toffel, Michael W. and Julian D Marshall. "Improving Environmental Performance Assessment: A Comparative Analysis of Weighting Methods Used to Evaluate Chemical Release Inventories." *Journal of Industrial Ecology,* 8/1–2 (2004).

Mousumi Roy
Penn State University, Worthington Scranton

Jill Coleman
Ball State University

ENVIRONMENTAL JUSTICE

Beginning in the 1970s, there has been a multilayered and ongoing process between citizens and governmental agencies concerning basic questions as to how exactly to define "clean environment," how to regulate and enforce a reasonably equitable distribution of environmental benefits and costs, and how to rectify gross environmental injustices. Historically, environmental protections and civil rights were instituted separately until social movements and individual advocates began to connect these concerns and demand what has become known as environmental justice. The goal of an environmental justice framework is twofold (1) to confront the inequitable distribution of environmental risk within society and (2) to identify and address the structural causes of discriminatory practices and environmental hazards.

History

Concerns about environmental justice, and the cost-benefit analyses of environmental risk that accompany them, have become a central component of decisions about how businesses and industries are regulated. All industrial and agricultural production impacts communities and citizens up and down the commodity chain. To grow tomatoes, for instance, raises environmental justice concerns for communities that drink water from sources tainted with agricultural runoff from fertilizers, herbicides, and pesticides, for farm workers who are exposed to such toxic chemicals, and for consumers whose health may be adversely affected by eating tomatoes with high toxic chemical content. Yet these same populations also potentially benefit from the jobs, from economic growth, and from the food produced by the tomato industry. These relationships only increase in complexity in a world in which commodity chains flow across national boundaries and have a global environmental impact. Public policy, articulated in response to organizing or mobilization efforts on the part of any of the involved constituencies (farmers, agribusiness, farm workers, and consumers) can in turn serve to incentivize or discourage certain types of investment and production. Through fines and taxes, or pollution control caps and remediation efforts, public policy can, and has, significantly impacted environmental justice concerns of various communities and constituencies. Not surprisingly, however, environmental justice has not exactly been blind in the United States: according to studies, poor and minority communities are far more likely to reside in contaminated communities, and are far less likely (up to 10 times less) to be recipients of public and private remediation efforts.

The United States first began instituting rights to a healthy environment in 1970 with the creation of the U.S. Environmental Protection Agency (EPA), followed by a series of legislations such as the Clean Water Act, the Clean Air Act, and the National Environmental Policy Act. As people across the country developed a growing awareness of the dangers of pollution, contamination, and degradation, they began to demand effective environmental assessment and regulation. On a local level, the social justice and environmental concerns were often seen as connected. In Michigan in 1976, for example, members of environmental groups and workers unions gathered with local residents, farmers, and academics for a conference titled "Working for Economic and Environmental Justice and Jobs." At the national scale, however, environmental legislation did not engage questions of justice. Amid the wealth of research on the environment and standards of conduct for individuals and businesses being generated during the 1970s, one crucial aspect remained overlooked: the inequitable distribution of the burdens of pollution and of access to environmental goods along with the political process that governs them.

These environmental injustices slowly gained prominence, beginning with the first use of civil rights law challenging a particular site for a polluting facility. In 1979, residents of a middle-income suburb of Houston, Texas, filed a class action lawsuit—*Bean v. Southwestern Waste Management, Inc.*, represented by attorney Linda McKeever Bullard—against the construction of a sanitary landfill in their neighborhood. This lawsuit prompted local environmental sociologist (and husband of the aforementioned attorney) Robert D. Bullard to conduct one of the earliest studies of the impact and siting of waste facilities. His research identified systematic patterns of injustice: 90 percent of Houston's landfills and 75 percent of its garbage incinerators were located in predominantly black neighborhoods, though African Americans made up a mere 28 percent of the city's population. Bullard documented that such differential siting affected communities with a wide range of household income.

Additional investigations provided further evidence of discrimination along racial lines. A 1983 General Accounting Office report stated that three out of four hazardous waste sites in EPA Region 4 (the southern states) were located in predominantly black communities. A 1987 study by the United Church of Christ Commission of Racial Justice titled "Toxic Wastes and Race in the United States" found that African American and Latino communities were most likely to live near hazardous waste facilities. In the first recorded use of the term, this report described such patterns of exposure to toxic waste as "environmental racism."

In the early 1980s, a series of small, seemingly isolated environmental struggles began to coalesce into a larger movement for justice, one that would connect communities across geographic and cultural distances. Over 500 protestors, black and white, were arrested in Warren County, North Carolina, in 1982 while demonstrating against the construction of a landfill in a predominantly black residential community. The event has ever since been viewed as the harbinger, if not the official beginning of, the modern environmental justice movement. During the same decade, the Concerned Citizens of South Central Los Angeles defeated a city proposal to site a municipal solid waste incinerator in their neighborhood. The Good Road Coalition on the Sioux Rosebud reservation in South Dakota was successful in blocking a plan for a 6,000-acre landfill on the Good Road burial grounds for which the coalition was named. The Concerned Citizens of Choctaw in Philadelphia, Mississippi, likewise prevented the siting of a toxic landfill in their neighborhood. In 1989, the Navajo tribe of Dilkon, Arizona, formed a group called C.A.R.E. (Citizens Against Ruining our Environment) and succeeded in stopping a proposed hazardous waste incinerator in their community. Yet, despite similar disputes across the country, national environmental policy did not recognize the underlying patterns of injustice.

In 1990, Robert Bullard published a seminal text on environmental justice called *Dumping in Dixie*. He looked at patterns of operation for polluting industries in five predominantly African American communities in the south: Dallas, Texas; Alsen, Louisiana; Emelle, Alabama; Institute, West Virginia; and Houston, Texas. In each place, from Appalachia coal country to the so-called Cancer Alley, Bullard found that factories, incinerators, and landfills, and all the toxins that come with them, were imposed with no input from local residents. He argued that environmental racism is an extension of a broader institutional racism in society evident in segregated housing, unequal health and education services, and discriminatory law enforcement. Certain instances of injustice are targeted while others may be unintentional, but all take place in a context—our society's laws and customs—that enables discriminatory practices. Industries follow the path of least resistance when locating polluting facilities such as waste incinerators and landfills, which leads them to economically poor and politically powerless neighborhoods. Given the broader context of institutionalized racism, the most vulnerable of these neighborhoods are communities of color. Yet Bullard had a more profound goal than identifying the culprits behind individual cases of discrimination: he established a pattern of injustice under existing environmental protections and argued for a strategy that would effectively prevent environmental threats before they occur.

Environmental Protection Standards

The EPA's explicit commitment to environmental justice began in the early 1990s in response to pressure from environmental justice advocates. The first national gathering on environmental justice took place at the University of Michigan School of Natural Resources and Environment in 1990. Participants at this "Conference on Race and the Incidence of Environmental Hazards" sent a letter to the EPA, prompting the agency to establish an Environmental Equity Workgroup. It formally instituted the workshop in 1992 and renamed it the Office of Environmental Justice.

The creation of this office followed the "First National People of Color Environmental Leadership Summit" held in Washington, D.C., the previous year. This summit adopted a 17-point Principles of Environmental Justice that continues to guide many advocates for environmental justice to this day. The Office of Environmental Justice was charged with integrating environmental justice considerations into the agency's policies and programs. In 1993, the EPA also established the National Environmental Justice Advisory Council (NEJAC), consisting of members from nongovernmental and environmental groups, business and industry, academic institutions, and state, local, and tribal governments. The NEJAC meets annually to discuss issues of human and environmental health, with a particular focus on the concerns of minority and low-income populations, and to offer advice and recommendations to the EPA. In 1994, President Bill Clinton signed Executive Order 12898, which ordered federal agencies to abolish and prevent policies that lead to unequal distribution of environmental risk in communities of color and low-income communities.

The EPA has adopted the language of environmental justice and states that one of its top priorities is ensuring equitable protection from environmental and health hazards and equal access to the decision-making process governing a healthy environment for all citizens. However, critics argue the agency has not made adequate progress toward equity or justice. Internal and external studies have concluded that the EPA has repeatedly failed to develop a clear strategy for evaluating the success of its programs and policies geared toward environmental justice. A 2007 environmental health study reported that low-income and minority populations have been underrepresented, and increasingly so, in the EPA's Superfund program that cleans and reclaims hazardous waste sites. Republican EPA administrator Christine Todd Whitman has been heavily criticized for the changes she made in 2001 to President Clinton's Executive Order 12898: she removed the requirements for environmental legislation to take low-income and minority populations under special consideration. While many critics agree that the EPA has not been successful in designing or enforcing standards of environmental equity, some go further and contend that equity is not enough; environmental justice requires addressing the root causes of risk, not merely distributing it equally.

Green Business

The turn of the 21st century has seen a surge in interest, within both the private and public sectors, in developing environmentally sustainable ways of conducting business that also generate economic growth. This new political climate is reflected in the NEJAC's 2008 annual report, titled "Strengthening the Participation of Business and Industry in Environmental Justice, Green Business, and Sustainability." This report makes nine recommendations as to how the EPA might partner with the private sector to link efforts toward environmental justice, green business, and sustainability:

- Facilitate citizen participation in developing green business and develop substantive and technical solutions to the potential disproportionate impacts of global climate change
- Improve existing environmental justice programs and policies along these lines
- Educate other federal, state, and local agencies in these areas and increase the coordination of multiagency efforts
- Facilitate ongoing conversations in these areas among and between business and community organizations
- Maximize the co-benefits of environmental justice, green business, and sustainability efforts (an example of this could include building or upgrading affordable housing, which would in turn lower emissions and also reduce energy costs for residents)
- Support research by internal and external partners to identify and maximize the potential environmental justice benefits of green business efforts
- Form a separate NEJAC work group to address the concerns of indigenous communities related to green business, sustainability, and climate change
- Create an educational presentation to upper management in large businesses and corporations featuring the economic and environmental benefits to environmental justice policy
- Consider the potential for pollution and other harmful impacts to communities from green businesses and technologies

This report highlights a key feature of business in the 21st century: green jobs, green business, and green investments have the potential to generate both burdens and benefits for communities with environmental concerns. It will take rigorous, coordinated efforts from the private and public sectors on the local, state, and federal level to identify and mitigate the potential unjust consequences of so-called green technologies.

See Also: Corporate Social Responsibility; Green-Collar Jobs; Green Technology; Socially Responsible Investing; Sustainable Development.

Further Readings

Bullard, Robert. *Dumping in Dixie: Race, Class, and Environmental Quality.* Boulder, CO: Westview Press, 2000.
Bullard, Robert and Maxine Waters. *The Quest for Environmental Justice: Human Rights and the Politics of Pollution.* San Francisco, CA: Sierra Club Books, 2005.
First National People of Color Environmental Leadership Summit, Washington, D.C. "Principles of Environmental Justice." http://www.ejrc.cau.edu/princej.html (Accessed October 2009).
Moore, Richard. "Strengthening the Participation of Business and Industry in Environmental Justice, Green Business, and Sustainability." http://www.epa.gov/compliance/resources/publications/ej/nejac/green-business-letter-100908.pdf (Accessed October 2009).
O'Neil, Sandra. "Superfund: Evaluating the Impact of Executive Order 12898." *Environmental Health Perspectives,* 115/7 (2007).
Schlosberg, David. *Defining Environmental Justice: Theories, Movements, and Nature.* Oxford, UK: Oxford University Press, 2007.

Emma Gaalaas Mullaney
Pennsylvania State University

ENVIRONMENTALLY PREFERABLE PURCHASING

Environmentally preferable purchasing (EPP) refers to procurement policies that seek to minimize the environmental impacts of purchasing decisions. EPP, alternatively referred to as "green purchasing," has been implemented in various forms by governments at all levels with the intent of using their purchasing power to strengthen markets for environmentally preferable products, and to lead the way to environmentally responsible purchasing decisions by example. While private firms and individuals also engage in EPP, except when taking collective action, their ability to sway markets is comparatively limited. Governments expend vast amounts of money annually to very effectively augment private sector demand both by government purchases and by providing incentives to manufacturers for developing greener products.

In the United States, EPP became federal policy with the adoption of Executive Order 12873 in 1993. Two years later, the U.S. Environmental Protection Agency (EPA) released the Proposed Guidance on Acquisition for Environmentally Preferable Products and Services. In 1996 the EPA teamed with the Organisation for Economic Co-operation and

NASA was among first federal agencies that committed to environmentally preferable purchasing after President Clinton signed it into law in 1993. Here, a great blue heron flies near Kennedy Space Center in Florida, which shares a boundary with a wildlife refuge.

Source: Dimitri Gerondidakis/NASA

Development (OECD) to study the programs operating in OECD member countries. Based on that research, comments received on the 1995 document, and pilot projects conducted by the agency, the EPA released its Final Guidance on Environmentally Preferable Purchasing in 1999.

The EPA's EPP program rests on four primary guiding principles. The first is that environmental costs must be as a matter of course factored into any governmental purchasing decision. In decades past, price and performance were the only metrics considered. Ignoring externalized costs benefited certain vendors, giving them a competitive edge in the procurement process, but those environmental costs avoided by manufacturers were ultimately exacted on citizens and the government in the form of public health, remediation, or liability costs. The EPP approach requires all foreseeable costs to be documented using life cycle cost analysis techniques and transfers the competitive advantage to firms that operate in an environmentally responsible manner. The second principle of the EPP program is that pollution prevention should be integral to the procurement process, starting at the earliest decision point. Consideration must be given to whether acquisition of the product is necessary at all, whether an environmentally superior substitute is available. or, where no substitute exists, how the product can be altered to lessen any detrimental environmental impacts. The third principle requires comparative life cycle analysis. This entails examining the product from multiple perspectives, taking such attributes as toxicity, energy requirements, and impacts to sensitive habitats into account. The fourth principle calls for comparison and rank ordering of products relative to the type, scope, scale, and geographic extent of any impacts. Of primary concern is the impact to human health, including the health of the producer's employees, the end user, and the general public. The impacts must also be assessed on the basis of how broad an area is affected, how long the impacts can be expected to persist, and whether impacts are reversible.

Many national governments adopted EPP policies as part of their Agenda 21 sustainable development plans following the 1992 United Nations conference in Rio de Janeiro. The OECD and European Commission (EC) encourage governments at all levels to engage in green purchasing protocols. The EC adopted an Integrated Product Policy in 2003 that called for member states to add a product component to their sustainability plans. The aims of the document, while extending beyond public procurement, noted the relatively heavy weight that government purchases exert on the marketplace. It also called on governments to incorporate life cycle thinking into their purchasing policies, not only to ensure EPP but to set an example for the private sector. To advance adoption of EPP, the EC has produced a suite of tools that includes a model action plan, legal guidance on setting up a protocol, and practical examples for purchasing officers based on case studies. The EC's policy rests on many of the same principles as the EPA's, but inserts the additional dimension of social equity, calling for products to meet the fair trade standards established by the European Fair Trade Association. The EC promotes the use of the European Union Eco-Label, a program instituted in 1992, to identify products that are subject to life cycle analysis and rated on environmental performance metrics by an independent rating agency. Seventeen European nations had implemented programs and set targets for EPP as of June 2009.

The association ICLEI - Local Governments for Sustainability involves over 1,100 local governments from 68 countries, and has several programs that address EPP. The Procura+ Campaign was initiated in 2004 to facilitate and expand the adoption of EPP by governments at the local level. Twenty-seven cities in 10 European countries and 9 national governments currently participate. Through Procura+, ICLEI offers training and guidelines for EPP programs. Procura+ focuses on six product groups: building construction and renovation, bus purchases, cleaning products and services, electricity, food and catering services, and information technology products. For each area, standards are suggested; for example, that vendors of electricity should be certified as green generators.

ICLEI also participates in Smart SPP, a European pilot program intended to spur innovation in products generally procured by governments, such as lighting, fleet vehicles, and heating and cooling units. Participants in the project work with technology developers and suppliers in advance of tendering requests for services to foster development of new technologies and products. Smart SPP provides a guide to working with innovators and tools to facilitate incorporation of life cycle cost analysis and CO_2 reduction estimates into decision protocols.

A study released by the EC in 2009 indicates that EPP not only accomplishes environmental protection objectives, but on average translates to cost savings. The estimated reduction in CO_2 emissions attributable to EPP in the seven countries included in the study averaged 25 percent. Though capital costs were frequently higher for environmentally preferable products, these were generally offset by lower operating costs. This was found to be especially true in transportation and construction categories. The average cost savings across all expenditure categories for all seven countries included in the analysis was 1.2 percent. The United Kingdom realized the greatest savings—5.7 percent.

The effectiveness of governmental EPP in increasing the market share of environmentally preferable products or spurring innovation is less clear. Such programs are most effective, particularly in the latter regard, when government is the primary purchaser of the product. In sectors where governmental purchases constitute a small portion of demand and absent any other incentives, governmental purchases may crowd out other buyers, though that risk diminishes as the general demand for green products grows.

Private sector firms are increasingly incorporating EPP into their operations. Their reasons for doing so generally fall within four categories. For some it is first and foremost a matter of corporate philosophy to conduct business in an environmentally responsible manner. Other companies adopt EPP in response to a perceived or articulated demand from consumers. Some undertake EPP in order to differentiate themselves from their competitors, and others to remain competitive with those in their sector that have already done so. In some instances, companies adopt EPP as a cost saving measure, in other words, in order to reduce their energy costs.

Patagonia, a manufacturer of active lifestyle clothing, and the cosmetics firm The Body Shop are examples of companies that adopted EPP primarily for philosophical reasons. As early adopters of the practice, they also clearly differentiated themselves from their competitors and built a loyal following of clients who share their commitment to environmentally responsible actions.

Private sector EPP programs generally resemble governmental ones. Common features include avoiding any product containing certain chemical compounds, giving preference to products made from recycled materials, and purchasing energy-efficient office equipment. Reliance on life cycle analysis is becoming more prevalent, particularly among multinational corporations.

Consumer polls have shown steadily increasing support for environmental protection. Those firms that can document their efforts to use green components and limit their carbon footprints are generally rewarded with increased sales; those that lag behind their competitors can anticipate losing market share. After releasing environmental performance profiles for its automobiles in 1996, Volvo realized a 17 percent increase in sales in Japan over the next three years. Sony, on the other hand, saw sales in the Netherlands slump precipitously when a local magazine review published a less favorable environmental performance ranking than two of its competitors. Sony quickly took corrective actions. Consumers are ever more savvy about environmental matters and companies know that they will be held accountable not only for their own environmental policies but for those of the contractors who supply their components.

See Also: Body Shop, The; Carbon Footprint; Cost-Benefit Analysis; Life Cycle Analysis; Patagonia; Resource Management; Responsible Sourcing; Voluntary Standards.

Further Readings

European Commission (EC). "Buying Green: A Handbook on Environmental Public Procurement." Brussels, Belgium: EC, 2004. http://ec.europa.eu/environment/gpp/pdf/buying_green_handbook_en.pdf (Accessed February 2010).

Gloria, T. P., B. C. Lippiatt and J. Cooper. "Life Cycle Impact Assessment Weights to Support Environmentally Preferable Purchasing in the United States." *Environmental Science & Technology*, 41/2:7551–57 (2007).

Organisation for Economic Co-operation and Development (OECD). "OECD Act: Recommendation of the Council on Improving the Environmental Performance of Public Procurement." Paris: OECD, 2002. http://www.iclei-europe.org/fileadmin/user_upload/Procurement/Online_Procurement_information/OECD_Council2002.pdf (Accessed February 2010).

Shannon, Julie. *State and Local Government Pioneers: How State and Local Governments Are Implementing Environmentally Preferable Purchasing Practices*. Darby, PA, Diane Publishing, 2000.

U.S. Environmental Protection Agency (EPA). "The City of Santa Monica's Environmental Purchasing: A Case Study." Washington, D.C.: EPA, 1998. http://www.epa.gov/epp/pubs/case/santa.pdf (Accessed February 2010).

U.S. Environmental Protection Agency (EPA). "EPA'S Final Guidance on Environmentally Preferable Purchasing." Washington, D.C.: EPA, 1999. http://www.epa.gov/epp/pubs/guidance/finalguidance.htm (Accessed February 2010).

Susan H. Weaver
Independent Scholar

ENVIRONMENTAL MANAGEMENT SYSTEM

An environmental management system (EMS) provides an orderly method for making concern for the environment a part of an organization's plans and decisions. The growth and development of EMS as a best practice for businesses, nonprofit organizations, and governmental entities has substantially altered organizational behavior to include an emphasis on green factors and other sustainable principles, like extended product responsibility and ecoeffectiveness. ISO 14001 is an international standard for designing an EMS. However, an organization does not need to certify or register to ISO 14001 to use it for the purpose of establishing an EMS. Some organizations have their own EMS formats, while others use a compliance-focused EMS that has been developed by the U.S. Environmental Protection Agency (EPA). Still, most organizations typically benchmark their EMS against the format of ISO 14001.

Planning an EMS starts with a determination of its scope. While many organizations use a property boundary to define the scope, there are cases where only parts of an operation are included in the system. For example, a company may choose to implement an EMS only for its logistics system. It would make sense to define a scope that matches what is managed so that there will be a clear integration between the focus on the environment and the overall management of the operations that are responsible for the products and services. Next, the organization must plan to have a policy that clarifies how it will make the environment part of the way the organization behaves, makes decisions, and manages. EMS policies typically include commitments to compliance with legal requirements, to the prevention of environmental impacts, and to the continuous improvement of the overall operation of the organization.

The foundation of any EMS is the concept of an environmental footprint. To assess this footprint, an organization considers the effects of all of its products, services, and distribution. The organization then analyzes its other activities and associated entities. Many organizations use process mapping tools to identify the resources (like energy, water, land, and materials) that are used and lost as a result of these activities. Through this process, workers can be made aware of how their job directly affects the environment. This is one example of how EMS affects organizational behavior; here, management can express the importance of resource productivity as a means of enhancing the sustainability of the operation. Workers can then become the "environmental managers" of their jobs. The environment is managed at the source under normal operating conditions and under a number of nonroutine conditions. If the equipment is being maintained, this worker takes over as the environmental manager. When a new employee is hired, they are trained in the environmental and resource productivity measures that are in place. A similar event occurs when an employee is assigned new tasks. Many organizations see this source control of environmental factors as a form of operational risk management.

In order to drive the continual reduction of the environmental footprint, the organization will use risk management to prioritize the use of resources, to evaluate resource losses, and to identify ways to lower its environmental impact. The organization will create objectives (often which take the form of environmental goals) to address the need for improvement. A number of projects will be initiated within an improvement program to achieve the objectives of decreased environmental impact and greater sustainability. These projects will become part of the management structure used within the organization to alter behaviors from previous paradigms to new green practices. For example, Seventh Generation, in its 2008 Corporate Consciousness Report, highlighted its EMS objectives of decentralizing logistics to reduce greenhouse gas emissions in their product distribution, with a goal of reducing its delivery miles by almost half in 2009, thereby significantly decreasing its environmental footprint.

Operations with a significant risk are subject to the use of operational controls to handle situations where the lack of control could lead to deviation from the goals of the environmental policy and the EMS objectives. Operational controls are a key part of risk management. They help maintain risk at a level that matches the "risk appetite" of the organization. Emergency preparedness and response plans will address risks or potential accidents that could have significant consequences on the environment. An effective EMS creates a way of responding to these situations and events by using a multitude of resources. For example, organizations can turn to the United Nations' International Environmental Technology Centre (IETC), which has outlined a variety of proper EMS approaches to emergency preparedness. Private organizations also often have business continuity plans that use risk management techniques. It is important that the EMS prepare for all of these situations and determine which it will include in its plan.

Implementation of the EMS is often supported through programs already in place in organizations. These include the following:

- Budgeting of resources
- Assignment of roles, responsibilities, and authority
- Competence, training, and awareness
- Communication—internal and external to the organization and other stakeholders
- Documentation and the control of documents and records
- Purchasing and contractor management
- Scheduling and project management
- Continuity planning
- Site security
- Maintenance, infrastructure improvements, and grounds management

Each of these areas is examined for ways that they possibly affect the EMS, as they will all be required to support the environmental policy. When taking into account these considerations, the environment should become a part of every management decision, and should change organizational behavior to minimize negative effects.

With the EMS in place and linked to the essential elements of the organization's general management system, it is important that regular checks are made to determine how efficiently these efforts are integrated. This starts with the monitoring and measurement of key environmental indicators and metrics, which include the results of the EMS's objectives and implemented projects. Other key performance indicators (KPIs) are defined by environmental permits and other considerations related to the organization's "license to operate" (e.g., noise, odors, and visual impacts). To drive continual improvement, some companies measure "leading indicators" as well as lagging indicators or results. Organizations take affirmative steps to ensure that they comply with all legal and self-imposed environmental requirements. An internal audit program is created and operated to make sure the organization is conforming to its EMS. Any nonconformance or environmental incidents are addressed using a formal corrective action program with root cause analysis. A preventive action program is initiated to meet the commitment in the environmental policy to be proactive and not simply reactive in dealing with environmental impacts at the source. Many organizations combine their environmental management projects, corrective actions, and preventive actions into a single project management initiative that is closely monitored by their selected measurement program.

The entire EMS should be reviewed by senior management at regular intervals. At these meetings, the managers look at the specific approach that was used and examine the means by which the approach was deployed within the overarching management system. They should analyze the results from a constructive approach toward deployment activities. Managers can see how much improvement took place, and determine whether it is worthwhile to seek further improvement in each case. While the focus in these meetings is on the environmental impact, often other related management issues are also discussed. When the EMS becomes fully integrated and not just an added-on consideration, this suggests that the EMS is truly a part of the way the operation is managed—just as stated in the definition of the EMS in ISO 14001.

Even when operating in conformance with ISO 14001, EMS programs still come in many different operating modes. This is to be expected, given the diversity of organizations that make use of EMS. Regardless, the key to success is to keep these efforts from becoming a separate silo limited to the environmental managers and excluded from other considerations. When employees take over the environmental management of their work, environmental managers take on the role of supporting their activities and providing communication with a host of environmental regulatory agencies outside the operation.

Further, the EMS can bring considerable improvement to the organization in the form of risk management and resource productivity enhancement. It is possible to "infuse" the EMS with management practices from other programs to make it function more efficiently while further reducing operational risks. The first area for infusing the EMS involves developing process focus. Here, the organization should be asking:

- How has the process been planned?
- How is the process being executed according to that plan?
- How does the organization present objective evidence that the planned results are being achieved?
- How does the organization embrace innovation to drive continual improvement?

A second area of infusion initiative can be derived from business excellence programs like the Baldrige Performance Excellence Model and EFQM-Europe. These programs emphasize social responsibility, strategic thinking, knowledge management, and process improvement. They also offer ways to change behaviors at all levels in the organization to transform company culture and to move in positive and profitable directions. More than 70 countries now have business excellence programs, which consider results as the outcomes of performance. It is important to remember that results do not happen just because objectives are established by top management. Instead, excellence criteria drive the performance that is needed to deliver the results. These performance models work well alongside EMS programs.

Once the programs have been infused, the organizations can spend more time fully integrating the management systems to include risk management, business excellence, and process improvement. All of these elements contribute to the management of day-to-day operations. With such an approach, environmental management is extended to all functions, and everyone has a stake in improving the environmental performance of the organization. Environmental objectives will be set not only by environmental managers working in isolation, but through initiatives, such as lean enterprise events and business excellence action plans. The job of the environmental management staff will be to support these efforts as well as working directly with environmental management on key EMS programs.

Although executive management may view this type of integration as having the potential to build value, functional managers may be nervous about blurring functional responsibilities within the organization. In many cases, this worry is created by the reward structure of the organization. In order to overcome this barrier, sustainability and the environment must be seen as sources of competitive advantage in a green marketplace. Moreover, the organization must ensure that everyone will be rewarded for working cooperatively to create value in all programs aimed at improving the environmental effects of the organization and at moving toward sustainability.

To build a value case for the environment, cross-functional teams need to work with employees to create action plans that address operational risks and other aspects associated with the environmental footprint. Using the risk management standard (ISO 31000) can help put financial value on the risk reduction efforts. These are some additional important steps that will help improve the success of the integration effort:

- Describe financial value over a reasonable time horizon (i.e., do not restrict your discussion to the current annual business cycle).
- Make sure that value creation through integration is attractive to key people (this can be addressed using the EMS review process).
- Set realistic expectations in the EMS objectives.
- Create action plans to address each objective and provide links to KPIs and leading indicators.
- Market the integrated management system program to key stakeholders inside and outside the organization.

Given all of these considerations, an EMS can provide an excellent foundation for sustainability. It provides the capability needed to put in place a program aimed at meeting its goals and challenges and engaging with stakeholders. However, to be successful, the EMS must be a central part of the way the operation is managed. There will be a need to look at the social footprint and to make sure that the organization is contributing to the financial well-being of the community as well as to its own financial well-being. This amounts to a significant increase in the scope of the program. With this firm foundation, it will be much easier to accomplish. Many companies feel that a good strategy is all that is needed. In order to find an approach that helps operationalize the concept of sustainability, it is prudent to have a solid management system that handles all three responsibilities.

See Also: Ecological Footprint; Environmental Audit; Environmental Risk Assessment; ISO 14000; Seventh Generation; Sustainability.

Further Readings

Pojasek, R. B. "Introducing ISO 14001 III." *Environmental Quality Management,* 17/1 (2007).
Seventh Generation. "2008 Corporate Consciousness Report." http://www .seventhgenerationcsr.com/sgreport (Accessed January 2010).
United Nations Environment Programme Division of Technology, Industry and Economics. "Disaster Management." http://www.unep.or.jp/ietc/dm/index.asp (Accessed January 2010).
U.S. Environmental Protection Agency, National Enforcement Investigation Center. http:// www.epa.gov/oecaerth/neic/index.html (Accessed November 2009).

Robert B. Pojasek
Harvard University

ENVIRONMENTAL MARKETING

With increased global warming, the swift growth of the world population, and the depletion of natural resources, corporations and governments have been forced to give some consideration to environmental issues that they may have chosen to ignore only a few decades ago. As it becomes increasingly clear to consumers that the policy of "business as usual" is not sustainable for our planet, businesses are responding to pressure from consumers to become more sustainable: advertising one's environmental concerns and sustainable practices has become one among many marketing tools available to businesses. This article begins by addressing the historical perspective on environmental management, followed by a discussion of greening the internal environment (e.g., by harnessing human resources and organizational culture of the firm) and the external environment (e.g., encouraging green consumer behavior as well as reaching out to consumers via green advertising). Operational activities, including greening the supply chain, are used to connect the two environments as well as to determine specific strategic orientations of the organization, such as the role of innovation management and sustainability as a source of competitive advantage. The variability of greening attitudes by industry and markets and other related issues are also discussed.

Several companies, including General Electric, have implemented environmental marketing programs to improve their business offerings. GE is researching greener technologies, such as this Integrated Gasification Combined Cycle, the end result of which is lower emission of greenhouse gases.

Source: GE

What Is Environmental Marketing?

Environmental marketing refers to a broad range of activities designed to emphasize to the consumer and to other businesses that a company's activities and products are sustainable or environmentally friendly. Several companies have proactively implemented environmental marketing programs to improve their business offerings in several different ways. Some recent notable examples are as follows:

- British Petroleum (BP) has committed to spending $350 million on energy efficient products over several years and is aggressively promoting its environmental awareness programs.
- General Electric (GE) is spending $1.5 billion on its Ecoimagination campaign, which aims to research and implement less-polluting technologies and to promote them as emblems of GE's environmental concern.

- Starbuck's announced the donation of $10 million over five years for clean drinking water around the world through the sale of its Ethos bottled water. The company already offers coffees that offer fair pay for growers and environmentally sound cultivation.
- 3M encourages employees to participate in its "Pollution Prevention Pays" program, which seeks to reduce pollution through means such as product redesign and modification, and recycling. Since 1974, the program has offset over 2 billion pounds of air, water, and solid waste pollutants from the environment.
- The largest home and garden center chain, Home Depot, discontinued the sale of wood products from endangered forests in 2002.

In addition to firms, major global bodies, such as the United Nations, have promoted green practices for municipal governments, and even buildings, for example, the William J. Clinton Presidential Library in Little Rock, Arkansas, advertise that they were built using environmentally friendly construction. Several industry-based associations actively encourage green marketing programs. For instance, the U.S. Green Building Council is responsible for certifying and promoting environmentally responsible and high-performance buildings, while the Green Seal organization awards green seals to products that meet rigorous environmental standards, which in turn helps consumers identify products that are environmentally safe.

History

The major focus on green marketing began in the last three to four decades after the publication of Rachel Carson's *Silent Spring,* but the history of green marketing in the modern era can be traced back to the 18th century when Benjamin Franklin urged France and Germany to follow England's practice of switching from wood to coal in order to conserve what remained of their forests. Franklin also attempted to regulate waste disposal and water pollution in Philadelphia, Pennsylvania, and wrote a codicil to his will to donate funds to build a freshwater pipeline to Philadelphia. Franklin was ahead of this time in observing the link between contaminated water and disease, and his efforts were partly responsible for Philadelphia being the first city in America (in 1801) to supply drinking water to its inhabitants through a municipal water department.

Concern for the environment became a popular issue in the United States in the 1960s and 1970s. In 1975, the American Marketing Association (AMA) held the first conference on ecological marketing: this conference brought together academics, practitioners, and public policy makers interested in examining marketing's impact on the natural environment. Concern for green marketing has escalated since the 1980s and 1990s, fueled in part by growing awareness of environmental ills, such as ozone-layer depletion, oil spills, and overflowing landfills. While some corporate efforts to address environmental concerns are sincere, increased interest in environmental issues among the general public has also been exploited by some companies who make false claims for their products and practices. In response, the U.S. Federal Trade Commission (FTC) and the National Association of Attorneys General actively monitor green marketing claims and have developed an extensive body of documentation regarding green marketing issues.

Greening the Supply Chain

"Greening the supply chain" refers to the process of integrating environmental issues within the new product development process that includes predevelopment activities, suppliers' business practices, and product design and development. Environmentally responsive companies take a proactive posture in requiring a significant level of environmental responsibility in core business practices of their suppliers and vendors, and many even require ensuring compliance from their suppliers at the second- and third-tier levels.

As Harold R. Kutner, General Motors (GM) vice president for purchasing, commented in 1999, concerning GM's decision to follow Ford's lead and require suppliers to ship directly to its facilities, "Working together with our suppliers, we can accomplish much more to improve the environment than GM can alone." Proactive supply chain management roles can play an important role in corporate greening and environmental strategies, while suppliers also benefit through greater operational efficiencies, cost savings, enhanced value to customers, increased sales, positive media attention, and positive ratings from socially responsible investment groups. Many different business practices and corporate culture adjustments are associated with greening the supply chain: some include environmental responsiveness of top management, environmental policy, early product development activities, design for environment and life cycle assessment, cross-functional environmental coordination, supplier involvement, environmental database, specialist environmental knowledge, and environmental benchmarking.

A recent report from the suppliers' perspective on effective supply chain management strategies, prepared by the Business for Social Responsibility Education Fund, talked with representatives from 25 suppliers in four industry sectors: automotive, business services, electronics, and forest products. The authors found that an increasing number of companies are seeking to influence their suppliers' environmental practices with the types of environmental issues involved varying by industry. Firms in the automotive and electronics sector received the most environmental requests from suppliers, while those in the forestry sector received requests mostly to explain their environmental practices. The service sector received the fewest environmental requests and most such requests came from the larger firms.

Motivations for Environmental Marketing

Michael Polonsky identifies five major reasons for firms' increased use of green marketing:

1. *Lower Costs or Higher Profitability:* many firms have achieved substantial cost savings in the process of adopting more environmentally conscientious practices. For instance, efforts to minimize waste may lead to the discovery of more efficient production processes that eliminate the need for some raw materials, thereby reducing costs. Similarly, the firm may discover that its waste products can serve as another firm's input to production processes. Finally, developing more efficient industrial processes can lower costs and increase profits.

2. *Competitive Parity:* in order to keep up with benchmarks set by competitors, firms may try to emulate their environmental marketing position. For instance, a manufacturer may stop deforestation in a sensitive area in response to a major competitor's similar action in order to avoid being singled out as the "environmentally unfriendly choice."

3. *Commitment to Social Responsibility:* many firms are beginning to realize that they are members of a broader global community and therefore must integrate environmental concern with their profit-based objectives. As such, they take a proactive approach to

embracing a philosophy of environmental social responsibility in their firm's overall culture. Some companies, like The Body Shop and Ben & Jerry's, heavily promote their involvement in green marketing initiatives, while others like Coca-Cola and Walt Disney World practice this philosophy but choose not to widely publicize it.

4. *Consumer Demand*: several surveys have shown that both individual and organizational customers in most countries are becoming more concerned about their natural environment and are demanding firms to be responsive to these concerns. Xerox, for example, introduced a "high quality" recycled photocopier paper in an attempt to fulfill consumer demands for less environmentally harmful products.

5. *Governmental Pressure*: finally, governments play a role in regulating the production and disposal of harmful products and by-products, regulating or prohibiting consumers' and industry's use or consumption of harmful goods, and by ensuring that consumers have the information required to evaluate the environmental composition of goods. Both laws and licenses are used to control the production of hazardous products and by-products and the disposal of hazardous wastes. Governments can modify consumer behavior by regulations or by promoting voluntary programs (such as recycling, which remains optional in many cities), or by imposing fees, such as the deposit collected on soda cans and bottles in many cities, which is refunded when the cans and bottles are returned for recycling. In the United States, agencies such as the FTC protect consumers by regulating what information must be supplied with products, and investigating and penalizing companies that make misleading claims, thereby enabling consumers to make better-informed decisions. Many other countries have agencies devoted to this purpose as well.

Green Consumer Behavior

The impact of green marketing on both marketing performance and consumer behavior has been thoroughly investigated. For example, research has shown that the market value of a firm declines slightly when green marketing initiatives are broadcast. Announcements related to green products, recycling efforts, and appointments of environmental policy managers have an insignificant effect on stock price reactions, while promoting green marketing produces significantly negative stock price reactions.

Consumers who have a positive attitude toward ecologically conscious living and a negative attitude toward pollution are most likely to buy green products, but there is a large gap between attitudes and purchasing behavior. This results in what is known as the "4/40" gap: roughly 40 percent of consumers say they are willing to buy green products, but only 4 percent actually do. Environmentally conscious beliefs are most likely to be translated into concrete behaviors when consumers perceive that their actions are likely to make a difference. Green marketing can be a very useful and successful strategy for firms as long as they comprehend the underlying motivations of customers in choosing environmentally friendly products.

The growing popularity of ecological claims does not mean that all are valid. In fact, the prevalence of false environmental claims has lead to the creation of a new word to describe it. That term is greenwashing, which is defined by the 10th edition of the *Concise Oxford English Dictionary* as "disinformation, disseminated by firms so as to present an environmentally responsible public image." Firms engaging in greenwashing practices often promote their products through the use of image advertising and misleading product labels, such as "all natural" or "organic." Various public relations strategies may also create the appearance of environmental concern that are not representative of the company's regular practices, including issuing but not enforcing mission statements and codes of conduct incorporating language that supports environmentally friendly practices, publishing sustainability reports that offer only partial disclosure, hiring scientists to vouch for biased, industry-funded research, and feigning public support for hidden antienvironmental agendas. Because of the prevalence of these greenwashing practices, industry ombudsmen, such as the Green Business Network and Green Life, attempt to monitor and expose such practices and keep the public informed about what is legitimate and what is not.

Profiling Environmental Consumers

There are many ways of segmenting green consumers. For example, J. Ottman Consulting identifies three types of environmental consumers: Planet Passionates, Health Fanatics, and Animal Lovers. Planet Passionates are committed to maintaining a healthy environment and avoid waste and products with poor environmental records. Health Fanatics are most concerned with maintaining a healthy diet and lifestyle by taking precautions against toxic waste, pesticides, sun exposure, and other environmental problems that might impact their health. Animal Lovers aim to protect the rights of animals through vegetarianism, buying products labeled as "cruelty free" and "not tested on animals," and boycotting products like those made with fur (because it is associated with animal cruelty) and tuna (in response to methods of tuna fishing that can injure or kill dolphins).

The marketing research firm Roper ASW, which publishes the *Green Gauge Report* about consumers' willingness to pay for green products to help companies target their marketing strategies, divides the total population into five segments, out of which two segments are likely to buy green. Those two segments are the True Blue Greens (9 percent of the population) who are wealthy and educated, and include environmental activists and leaders, and the Greenback Greeners (6 percent of the population) who share the same characteristics, yet are less likely to sacrifice comfort and convenience for the sake of the environment.

Finally, Ginsberg and Bloom have suggested that companies should follow one of four green marketing strategies depending on how large a segment of the company's market is composed of green consumers, and how effectively the brand or company can be differentiated from competitors based on their environmental friendliness. They divide companies into four types based on how they use green strategies within their business and how much they emphasize green marketing. Lean Greens, such as Coca-Cola, are interested in green strategies mainly as a way of reducing costs and increasing efficiency. Defensive Greens, such as The Gap, use green marketing as a precautionary measure and to avoid alienating important segments of their market. Shaded Greens, such as Toyota, make substantial investments in environmentally friendly processes, but choose not to emphasize it in their marketing. Extreme Greens, such as Patagonia and The Body Shop, fully integrate environmental issues into their production and emphasize this philosophy in their public image.

See Also: Body Shop, The; Carbon Footprint; Cause-Related Marketing; Clean Production; Clean Technology; Corporate Social Responsibility; Cradle-to-Cradle; Ecological Footprint; Environmental Assessment; Environmental Indicators; Recycling, Business of; Sustainability; Sustainable Development; Waste Reduction.

Further Readings

Barbaro, Michael. "The Energy Challenge: Wal-Mart Puts Some Muscle Behind Power-Sipping Bulbs." *New York Times* (January 2, 2007).

Bennett, S. J., R. Freierman and S. George. *Corporate Realities and Environmental Truths: Strategies for Leading Your Business in the Environmental Era.* Hoboken, NJ: John Wiley & Sons, 1993.

Chamorro, Antonio and Thomas M. Banegil, "Green Marketing Philosophy: A Study of Spanish Firms With Eco-Labels," *Corporate Social Responsibility and Environmental Management*, 1:11–20 (February 2006).

Crane, A. *Marketing, Morality and the Natural Environment.* London: Routledge, 2000.

Di Meglio, Francesca. "It's Getting Easier Being Green." *Business Week* (July 15, 2005).

Fuller, D. A. *Sustainable Marketing: Managerial-Ecological Issues.* Thousand Oaks, CA.: Sage Publications, 1999.

Ginsberg, Jill M. and Paul N. Bloom. "Choosing the Right Green-Marketing Strategy." *MIT Sloan Management Review* (October 15, 2004). http://sloanreview.mit.edu/the-magazine/articles/2004/fall/46112/choosing-the-right-greenmarketing-strategy (Accessed January 2010).

Green Life. http://www.thegreenlife.org (Accessed November 2009).

Lockwood, Charles. "Building the Green Way." *Harvard Business Review* (June 2006).

Mathur, L. K. and I. Mathur. "An Analysis of the Wealth Effects of Green Marketing Strategies." *Journal of Business Research*, 50/2:193–202 (2000).

Saha, Monica and Geoffrey Darnton. "Green Companies or Green Con-Panies: Are Companies Really Green, or Are They Pretending to Be?" *Business and Society Review*, 110/2:117–57 (2005).

Abhijit Roy
University of Scranton

ENVIRONMENTAL RISK ASSESSMENT

Environmental risk assessment (ERA) is the procedure in which the inherent hazards involved in natural events (flooding, extreme weather events, etc.), technology, practices, processes, products, agents (chemical, biological, radiological, etc.), and industrial activities and the risks posed by these are examined. ERA involves a critical review of available data for the purpose of identifying and possibly estimating the risks associated with a potential threat, either quantitatively or qualitatively. It includes human health risk assessments, ecological, or ecotoxicological risk assessments and industrial risk assessments that can identify end points in community, biota, and ecosystems. The risks examined in the assessment can be physical, biological, or chemical in nature.

Uses of ERA

ERA is needed in case of risk identification of an emerging issue or prioritization of further action. It has provided the basis for most legislative and regulatory programs as well as for international agreements and policy implications to address the associated threats. It is performed on a spectrum of environmental issues, such as hazardous waste clean-ups, permitting activities for water and air discharges, for land and water management, forests, watersheds, estuaries, and for establishing environmental quality standards and guidelines.

This technique is used in a wide range of professions such as industry sectors, finance sectors, private sectors, government bodies, nongovernmental organizations (NGOs), and academia to examine risks of very different natures. ERA can be used in a number of ways:

- *Prioritization of Risks:* when an organization is faced with a number of potential environmental risks, ERA can be used to establish their relative importance; it thus provides a basis for prioritizing which risks should be dealt with first.
- *Site-Specific Risk Evaluation:* ERA can be used to determine the risk associated with locating facilities in specific locations or to determine the risks that affect a particular site.
- *Comparative Risk Assessment:* ERA is used to compare the relative risks of more than one course of action, for example, the risks posed by untreated water versus the risks posed by chemicals used to treat water.
- *Quantification of Risks:* ERA may be taken to a level where the risks are quantified in order to establish controls on the risks, for example, maximum acceptable concentrations of chemicals in ambient or drinking waters.
- *Qualitative Risk Assessment:* ERA can be used to explain the risk qualitatively within any facility.
- *Transportation Risk Assessment:* ERA tools can be used to assess the risks associated with hazardous substances carried by various modes of transportation.

Overview of ERA Methods

ERA in general contexts consists of four integrated processes: (1) identifying underlying sources of risk; (2) determining the pathways by which such risks can materialize; (3) estimating the potential consequences of these risks under various scenarios; and (4) providing the means for mitigating and coping with these consequences.

Several procedures and a host of tools, software, and databases have been developed for ERA over the past half a century. The tools of mapping material and energy flows are essential diagnostic tools to indicate likely sources and magnitudes of emissions and wastes in key processes. The accepted conceptual structure for conducting ERA includes the following activities:

- *Problem Formulation:* initially, the problem has to be defined and certain issues such as nature of risk, risk source, acceptable level of risk, location, focus area, and system boundaries must be clear before the assessment starts.
- *Hazard Identification:* this is not a distinct and separate process, and is sometimes integrated with other processes or in subsequent steps. Hazards can be identified by examining all routine operations, materials, and products of potential concern, prioritizing them by their relative hazard, and identifying the consequences of deviations from normal operation. Hazards are identified using tools such as hazard indices, Hazop, event tree analysis, failure mode, effect analysis, toxicological testing, and epidemiological studies. Specialized software and databases are also used to identify the nature of potential hazards.
- *Release Assessment:* includes a description of the types, amounts, timings, and probabilities of the release of hazards into the environment, and how these attributes might change as a result of various actions or events.

- *Exposure Assessment:* consists of describing and quantifying the relevant conditions and characteristics of human and environmental exposures to hazards produced or released by a particular risk source. It includes a description of the intensity, frequency, and duration of exposure through the various exposure media, routes of exposures, nature of the population exposed, and other related objects in food chain models. It attempts to determine the magnitude of the physical effects of an undesirable event identified in previous stages, and the pathways and transport modes of the hazard to the receptor, using various tools such as predictive exposure modeling, tools for aerosols, thermal radiation, and vapor cloud explosions, among others.
- *Consequence Assessment:* examines the consequences of the release or production of hazards to the specified population, and the quantification of the relationship between specified exposures to the hazard and the health and environmental consequences of those exposures. The data for consequence assessment is based on toxicity and ecotoxicity testing, epidemiology, and modeling, such as dose-response models.
- *Risk Estimation:* consists of integrating the results from the release assessment, exposure assessment, and the consequence assessment to produce an estimate of the overall risk of an activity. All safety systems, redundancies, and mitigation possibilities are considered, and detailed probability distributions for the hazards are identified and damages are calculated.
- *Risk Evaluation:* includes acknowledgment of the public perception of the risk and the influence it will have on the acceptability of risks and risk decisions.
- *Risk Characterization:* the integration of risk evaluation and risk estimation.
- *Risk Management:* involves action that is going to be taken after critical analysis of socioeconomic conditions, the availability of alternative technology, products, practices, processes, international comparisons, impacts and communication, and consultation with the public and stakeholders who will be affected by the proposed changes. Risk acceptance and risk reduction guidelines, process hazards management procedures, including emergency response procedures, structure financial and insurance provisions should be specified, and communication procedures with affected employees and the public should be established.

Future Challenges

Some of the challenges of environmental risk assessment are the existence of different methodologies, mixtures or multiple stressors, data deficiencies and gaps, and a lack of active participation of stakeholders. A few approaches that have the potential to overcome this are harmonization of assessment methods, protocols, and mutual recognition of data used in assessments to reduce uncertainty and limitations.

For impact and risk assessment tools to be effective, they must be capable of providing data that relates to ecologically significant processes. This requires a better understanding of particular biomarkers related to health status in order to improve their interpretative value in monitoring. As with bioavailability and uptake, exposure to pollutant mixtures with the possibility of complex interactions resulting in emergent and novel toxicities and pathologies and intense visible and ultraviolet radiation and hypoxia need to be considered, since these are likely to be important in terms of potentially harmful interactions with contaminants. There should be a clear understanding of mixtures or multiple stressors affecting at the population, community, ecosystem, and human health levels.

Assessing the impact of environmental disturbance on organisms requires understanding stress effects throughout the hierarchy of biological organization, from the molecular and cellular to the organism and population levels, as well as the community and ecosystem level. An integrated environmental management strategy must be truly cross-disciplinary if an effective capability for risk assessment and prediction is to be developed in relation to resource sustainability. Areas of collaboration need to include, among others, remote/satellite surveillance, interpretation of complex information, and predictive modeling. There needs to be an increased focus on precautionary anticipation of novel environmental techniques from biotechnology, molecular biology, and nanotechnology. Essentially, all the tools of industrial ecology, including life cycle analysis, the Environmental Management System, ISO 14000, and the Eco-Management Audit Scheme have important implications in promoting the reduction of environmental risks. These require supporting research into understanding physical, chemical, biological, and ecological processes related to environmental issues. Decision support systems also need to be developed to link existing models with our experience and knowledge of the environment, as well as to develop and use indicators of sustainability to stimulate it toward sustainable development, where there is a need to link environmental, social, and economic measures.

Integrating the supply chain to identify, mitigate, and manage risks requires integration with key business processes, measurement of results, and commitment from top management through government regulation. Increased participation in risk management and communication is a critical factor for a successful ERA. The focus of risk regulation should move away from government bureaucracy to local assessment between affected communities, armed with legitimate information on the risks they face and the companies giving rise to such risks. It is crucial to educate politicians, industrialists, and environmental managers about the long-term consequences of pollution. Increased consumer awareness of environmental risks will in turn exert pressure on the industry to make its products "environmentally friendly" in order to maintain existing markets and to improve their penetration into new markets.

See Also: Corporate Social Responsibility; Cost-Benefit Analysis; Cradle-to-Cradle; Ecoefficiency; Ecological Footprint; Environmental Assessment; Environmental Audit; Environmental Economics; Environmental Management System; Environmental Marketing; Exposure Assessment; Extended Producer Responsibility; Extended Product Responsibility; Externalities; ISO 14000; ISO 19011; Life Cycle Analysis; Quantitative Risk Assessment; Resource Management; Responsible Sourcing; Sustainable Development; Transparency.

Further Readings

Asante-Duah, K. *Public Health Risk Assessment for Human Exposure to Chemicals (Environmental Pollution)*. New York: Springer, 2008.
Calow, Peter P. *Handbook of Environmental Risk Assessment and Management*. Hoboken, NJ: Wiley-Blackwell, 1998.
Kleindorfer, Paul R. "Industrial Ecology and Risk Analysis." http://www.opim.wharton.upenn.edu/risk/downloads/01-23-PK.pdf (Accessed April 2009).
Paustenbach, Dennis J. *Human and Ecological Risk Assessment: Theory and Practice*. Hoboken, NJ: Wiley-Interscience, 2002.
Suter, Glenn W., II. *Ecological Risk Assessment*, 2nd Ed. Boca Raton, FL: CRC Press, 2006.

Gopalsamy Poyyamoli
Rasmi Patnaik
Pondicherry University

ENVIRONMENTAL SERVICES

Environmental services is a concept from the ecological economics literature. A well-known definition of environmental services is "any functional attribute of natural ecosystems that are demonstrably beneficial to humankind." The concept basically tries to capture in broad terms the idea that the natural environment provides particular important uses or benefits that can be captured under the concept of "services." As such, it aims to frame the interdependence of humans and nature in ways attuned to mainstream economic discourse and as a logical "asset" to take into account in business and private sector decisions. This article gives a very general overview of the concept of environmental services, placing particular importance on its history, its current importance, and some of the arguments for and against the use of the concept.

Before moving on, however, some conceptual clarifications are in order. First, the concept of environmental services closely resembles the concept of ecosystem services, to the extent that they are often used interchangeably in the literature. The difference between the two is that the former is a more overarching, all-encompassing concept, while the latter specifically refers to a particular, geographically bounded ecosystem (even when talking about the global ecosystem).

Second, environmental services are often framed by conservation scientists as one of two ways in which biological diversity and "the environment" in a broad sense contribute to human welfare and development. Environmental services, then, represent the "indirect values" of nature to humans by supporting the ecological processes that help regulate the natural environment that humans are part of and depend on. So-called material goods represent the "direct values" of nature to humanity, for example, in the form of sources for medicines, new varieties of food, green energy, and so forth. Another main difference between environmental services and material goods is that the former often go directly to individuals or private organizations (i.e., producers or consumers) whereas environmental services accrue to societies at large, which makes it much more difficult to quantify them in economic terms—a point we shall return to below.

Third, it should be noted that the concept of environmental services is—often implicitly— connected to a particular worldview characterized by instrumentalism, (economic) rational choice, and a belief in progress and "designability." Hence and while often portrayed as a rather "neutral" concept, it is important to note that "environmental services" carries particular connotations to which we will return below. Next, however, we will outline some examples of environmental services.

Examples of Environmental Services

As stated above, the environmental services concept supposes that humans are dependent upon the goods and services provided by dynamic ecosystems for their existence and prosperity. Timber, medicines and drugs, new varieties of food, fodder crops, and fuel wood are just some examples of the material goods we extract from nature. Compared to material goods gained from these systems, the "indirect" environmental services encompass, for example, the cultural, recreational, and aesthetic values biodiversity provides. Additionally, the environment provides humans with critical life-support services such as the recycling of nutrients that are important to humans (oxygen, nitrogen, carbon, phosphorus, and sulfur); regulating or contributing to a stable climate; the provision of clean air and water; detoxification and decomposition of pollutants and waste; conservation of soil and soil fertility; pollination of agricultural crops and natural vegetation; and controlling pests, to mention but a few.

The provision of these services has never been as threatened by human activity as it is now. These services are threatened, by among others, habitat degradation or loss, the loss of species and ecosystem diversity, and increased changes in the climate. These are linked to the resilience of an ecosystem—the ability of an ecosystem to recover from disturbance, in other words, to resist stresses and disruptive changes. If the resilience of an ecosystem declines, so does its ability to provide environmental services.

To give another practical example, we will focus on the environmental services provided by forests, as these benefit both local and global communities, and can thus be classified as public goods. Biodiversity conservation, carbon sequestration, watershed protection, and landscape values are the major environmental services provided by forests. Watersheds conservation/protection is important on an ecological as well as a human scale. Degradation of watersheds has potentially devastating environmental and socioeconomic consequences, as it plays a vital role in the reduction of the risk of flooding, increase of groundwater tables, and reduction of soil erosion and maintenance of soil fertility. These have indirect benefits for improved agricultural productivity, on- and off-site fishery productivity, and improved quality and quantity of water supplies.

Biodiversity conservation in forests has as a result that ecosystems are more resilient to shocks, such as sudden changes in the weather, for example. Also associated with biodiversity conservation is the landscape value that has indirect benefits for recreation and generating income from ecotourism. The main benefit from forest ecosystem services is linked to the current debate around climate change and global warming. Forests contribute significantly to the sequestration of carbon and through this can reduce the threat of global warming. This has benefits not only for local communities but also for low-lying countries that would be affected by a rise in sea level if the current rate of global warming continues unabated.

Historical Notes

Humans have always to greater or lesser extent depended on nature and the environment for their development. Yet, with the advent of capitalism in the 16th to 18th centuries, this dependence took on distinctly new dynamics, which—as Karl Marx famously noted—had consequences for both the relation between humans and nature and for human nature itself. As such, nature started to be more literally seen and framed in terms of its use value for people and, in fact, as a "natural resource" to be tapped by humans and used in the ways they see fit. However, it soon appeared that nature was not boundless and that humans' capacity to destroy the natural environment was so great that it might actually endanger human societies on a grand scale.

Obviously, the idea of the breakdown of human societies as a result of their environmental impact is not solely attributable to capitalism; it occurred in many precapitalist societies also. However, the scale of environmental destruction proved to have become infinitely greater due to the new possibilities afforded by mechanized production and consumption, and the continuous drive to increase this production and consumption. In tandem, human societies started more consciously thinking about how to conserve natural resources and biological diversity, particularly nature's capacity to provide those services that humans ultimately depend on for their survival, well-being and—perhaps most importantly—economic opportunities.

During colonial times, for example, the idea of environmental services became popular among colonial powers, with consequences that are still seen today. The Germans were especially interested in the environmental services provided by forests and watersheds to support the colonial, state-building effort. They were already keenly aware of the support structures provided by forests—especially as watersheds—and that these were important both for general human development as well as commercial interests around the timber industry. As such, they set up forest reserves in Tanganyika (now Tanzania), Cameroon, and other areas, and so influenced later conservation dynamics in these countries.

The Germans were obviously not alone in these calculations. Debates within other colonial powers, but also in the United States, increasingly started recognizing the importance of environmental services, especially in relation to economic concerns. And while the context around this debate has changed dramatically, the link to economic concerns is arguably the most important reason why "environmental services" has become such an important concept today.

The Contemporary Debate Around the Concept of Environmental Services

The contemporary debate around environmental services has both ardent proponents and staunch critics of the concept. A first important reason why proponents value the concept of environmental services is because it makes clear what biological diversity and the environment mean to humans. It shows in simple terms that humans and their environment are interdependent, and that there is thus a strong argument to engage in conservation.

Second, proponents state that in a world characterized by strong linkages between economic globalization and environmental degradation, it makes sense to take environmental services into account in economic decision making and no longer treat it as classical economics would have it, namely as an "externality." As a result, they advocate calculating the economic value of environmental services in monetary terms. Closely related to this argument is the belief that such incentive-based governance mechanisms are more efficient and effective than "top-down, command-and-control" government regulations. They enable organizations and individuals to find the most cost-effective way to deal with environmental degradation by taking it up into the "normal" workings of the market.

One of the most popular recent ways of doing this is payments for environmental services (PES). According to S. Wunder, a PES scheme should comply with five criteria; PES is "(1) a voluntary transaction in which; (2) a well-defined environmental service (or a land use likely to secure that service); (3) is "bought" by a (minimum of one) buyer; (4) from a (minimum of one) provider; (5) if and only if the provider continuously secures the provision of the service (conditionality)." Arguably the most crucial element in making a PES scheme work is to be clear about what an environmental service exactly entails. Almost always, this involves quantification of some sort, which can be incredibly complex when dealing with the many interlocking and interdependent components of a particular environmental service, such as an estuary or a wetland. Yet, this quantification is necessary, say proponents, in order to make concrete the (economic) value of environmental services to decision and policy makers.

As for the critics of the concept, various points can be noted. First and most fundamental, critics point toward the worldviews underlying the concept of environmental services, as it defines nature in what it means for human (its use value) rather than its inherent or intrinsic value. As noted above, this worldview is often characterized by instrumentalism, (economic) rational choice, and a belief in progress and "designability." It believes that nature and its environmental services can be fully understood by humans and subsequently "managed" and controlled in such a way that it serves both human (economic) development and nature conservation.

Critics, however, point out that these assumptions are false. They argue that the perfect information needed to make this work is never possible, and that humans are not merely "utility-maximizing" *homos economicus* that respond to incentives in the same way. Furthermore, critics point out that this worldview is very partial and incomplete, especially in the values and norms it takes into account. They argue that there is a bias toward economic values, whereas other (cultural, social, political) values are left out of the model.

Related, critics argue that the science underlying the drive toward defining environmental services is very problematic. In the drive to better understand, quantify, and "marketize" environmental services, critics point out that scientists condone particular types of knowledge that work for further capital accumulation, while other types of knowledge are not taken as seriously, or are even brushed aside. Most notably, this relates to local indigenous knowledge systems.

A more practical point of criticism relates to the many difficulties in translating market-based theory into practice. In order for markets around environmental services to work, so it is argued, not only do proponents have to deal with the complexity of answering the question of what an environmental service exactly entails, but also what is called an "enabling environment": the "right" rules and regulations that guarantee private investments in environmental services, and that they can be bought and sold in standardized "packages."

An example here is "commercial wetland mitigation banking," which is a regulatory arrangement that enables private companies to restore "compensatory" wetlands, parts of which can be sold as "wetland credits" to other private firms that have to compensate for their share of the destruction of wetlands. Proponents of PES systems have enthusiastically embraced this idea, but critics argue that wetland banking is deeply problematic. They have shown that—besides the more fundamental problems—there are problems of measurement: how can functionally integrated wetlands be compensated through functionally disintegrated wetland credits? Concretely, they ask, for example, whether the destruction of a bird colony that depends on a particular wetland area can ever be compensated for by "buying" individual birds in multiple other wetlands. This is what is called the problem of "commensurability": it can never be completely ascertained that the destroyed wetland can be compensated, simply because the value of a wetland is place specific and can never be completely compensated by another wetland in another place.

The Future of Environmental Services

With the global climate and the environment under great stress, the urgency to maintain essential environmental services has never seemed so great. Yet the concept of environmental services cannot be seen outside the system that propelled it to fame, that of capitalism. The concept thus harbors a major contradiction: it tries to ameliorate the negative environmental effects of a particular system by using ideas that strengthen that same system. No doubt, this contradiction will be hotly debated for the foreseeable future.

See Also: Conservation; Deposit Systems; Ecological Economics; Ecosystem Services; Pollution Offsets.

Further Readings

Bellamy Foster, J. *The Vulnerable Planet: A Short History of the Environment*. New York: Monthly Review Press, 1999.

Costanza, R., et al. "The Value of the World's Ecosystem Services and Natural Capital." *Nature,* 387:253–60 (1997).

Daily, G. C. *Nature's Services: Societal Dependence on Natural Ecosystems*. Washington, D.C.: Island Press, 1997.

Heynen, N., S. Prudham, J. McCarthy and P. Robbins, eds. *Neoliberal Environments: False Promises and Unnatural Consequences*. London: Routledge, 2007.

Landell-Mills, N. and T. I. Porras. *Silver Bullet or Fools' Gold? A Global Review of Markets for Forest Environmental Services and Their Impact on the Poor*. London: International Institute for Environment and Development, 2002.

Marx, K. *Capital: A Critique of Political Economy*. New York: Penguin Books, 1976.

Myers, N. "Environmental Services of Biodiversity." *Proceedings of the National Academy of Sciences of the United States of America*, 93:2764–69 (1996).

Robertson, M. M. "The Neoliberalization of Ecosystem Services: Wetland Mitigation Banking and the Problem of Measurement." In *Neoliberal Environments: False Promises and Unnatural Consequences,* N. Heynen, S. Prudham, J. McCarthy and P. Robbins, eds. London: Routledge, 2007.

Wunder, S. "The Efficiency of Payments for Environmental Services in Tropical Conservation." *Conservation Biology* 21/1:48–58 (2007).

Bram Büscher
Institute of Social Studies, the Netherlands
University of Johannesburg

Stacey Büscher
HIVOS, the Netherlands

EQUATOR PRINCIPLES

The Equator Principles (EP) are a set of voluntary guidelines to promote corporate social responsibility (CSR) in project finance. According to the EP website, in project financing "a lender looks primarily to the revenues generated by a single project both as the source of repayment and as security for the exposure." Project finance is often defined as involving "the creation of a legally independent project company financed with equity from one or more sponsoring firms and nonrecourse debt for the purpose of investing in a capital asset." As these projects involve independent companies, they are not automatically covered under the involved financial institutions' own CSR policies. The EP were launched in June 2003 by a group of private sector banks to address this lacuna; in 2006, they were revised.

As issues of CSR have become more relevant for firms, a range of different industries have taken up this challenge, often stimulated by an increased social pressure. Codes of conduct and other forms of private regulation are being developed that do not rely directly on the law or on government. The financial services industry is one industry in which such voluntary codes have been developed. Firms within this industry comply with various codes of conduct and policies with respect to issues of CSR; examples include the United Nations Global Compact or the Global Reporting Initiative. The EP are another example of private regulation in this sector.

The EP focus on the assessment of social and environmental risks and aim to promote CSR in project finance. These principles are a specific case of private regulation as they are targeted at an entire industry at the global level. Although the financial services industry does not engage in production itself, it is an important industry for issues of CSR. As B. J. Richardson notes, "The financial services sector is increasingly recognized by scholars and policy-makers as an economic sector that has a significant bearing on sustainable development. Comprising primarily lenders, investors and insurers, the financial sector is environmentally significant . . . due to its indirect environmental effects through its loans to and investments in other businesses." The EP were devised as an answer from the industry to external calls for increased sustainability. According to the EP website, financial institutions "that adopt the principles ought to be able to better assess, mitigate, document and monitor the credit risk and reputation risk associated with financing development projects." They are closely linked to a set of guidelines these banks were facing, set by the International Finance Corporation (IFC) of the World Bank. In 2006, the EP were revised to meet changes in the IFC environmental and safeguard principles and to incorporate comments from a variety of stakeholders. By early 2009, over 65 financial institutions declared that they had adopted the EP.

The EP now consists of 10 guidelines to assess negative external effects of project finance for projects of $10 million or more, often located in developing countries. The principles are a set of short statements, arguing that financial institutions that adopt the EP (the so-called EPFIs) will only provide loans to projects that conform to the first nine principles, the 10th principle being a responsibility to the EP itself:

1. Review and Categorization

2. Social and Environmental Assessment

3. Applicable Social and Environmental Standards

4. Action Plan and Management System

5. Consultation and Disclosure

6. Grievance Mechanism

7. Independent Review

8. Covenants

9. Independent Monitoring and Reviewing

10. EPFI Monitoring

In short, the EP begin with a need to categorize the social or environmental risk of a project in one of three groups: A, B, or C, based on screening criteria of the IFC. For projects in groups A and B (with A being the most risky projects), an environmental and social impact assessment has to be made. The impact assessments should also include mitigation and management measures. Stakeholders have to be consulted and several requirements are imposed on borrowers, which often are large syndications. In category A projects, an environmental management plan also needs to be prepared and implemented, regulated through a loan covenant. In addition, periodical reporting obligations apply, of which expert reviews have to be part. The results of this process are outlined in the covenant and the process needs to be monitored and reviewed. Within the assessments, a distinction is made between standards to apply to projects situated in high-income Organisation for Economic Co-operation and Development (OECD) countries and elsewhere.

The EP thus can be seen as an instrument to guide CSR in project finance and to demonstrate the financial institutions' involvement with the issue. Although adopters commit to finance only projects that meet the EP, they are also accused by critics of window dressing or greenwashing. For instance, the EP then are claimed to be used as "signaling devices for demonstrating positive credentials, with the aim of strengthening corporate reputation and organizational legitimacy more generally." Critics pointed at the voluntary character of the EP, leaving room for "free riders" in the financial services sector to neglect the principles, and the lack of transparency in the accounting and reporting systems. Despite these serious concerns, the number of institutions adopting the EP has grown over time. This leads to questions whether the EP actually do result in performance differences between banks.

B. Scholtens and L. Dam compared banks that adopted the EP with nonadopters, and found that adopters' CSR policies "are rated significantly higher than those financial institutions that did not sign up." They also found signees to be bigger firms. According to these authors, adoption of the EP might be related to economies of scale and expected reputational effects may indeed be involved in stimulating adoption, but data are lacking to draw strong conclusions. However, like many examples of codes of conduct, the overall effects of the EP remain difficult to assess. All in all, Richardson suggested that the EP are a step in the right direction, but publicly accountable and transparent decision making would be required "to engage in meaningful change."

See Also: Corporate Social Responsibility; Environmental Economics; Greenwashing; Transparency.

Further Readings

BankTrack. "Equator Principles Re-Launched: Improvements Made, but Principles Fail to Live Up to Their Potential." (July 7, 2006). http://www.finanzaseticas.org/pdf/EquatorPrinciplesRelaunched_BankTrack_2006.pdf (Accessed April 2009).

Equator Principles. "Frequently Asked Questions About the Equator Principles." http://www.equator-principles.com/faq.shtml (Accessed March 2009).

Esty, B. C. "Why Study Large Projects? An Introduction to Research on Project Finance." *European Financial Management*, 10:213–24 (2004).

Richardson, B. J. "The Equator Principles: The Voluntary Approach to Environmentally Sustainable Finance." *European Environmental Law Review*, 14/1:280–90 (2005).

Scholtens, B. and L. Dam. "Banking on the Equator: Are Banks That Adopted the Equator Principles Different From Non-Adopters?" *World Development*, 35/8:1307–28 (2007).

Wright, C. and A. Rwabizambuga. "Institutional Pressures, Corporate Reputation, and Voluntary Codes of Conduct: An Examination of the Equator Principles." *Business and Society Review*, 111/1:89–117 (2006).

Frank G. A. de Bakker
Vrije Universiteit Amsterdam

E-Waste Management

Due to the increase of global production of and demand for consumer electronics, the question of what to do with e-waste has become an important and complex issue of international environmental policy. E-waste consists of used or broken electronic or electrical devices, like computers, televisions, cell phones, spare wires, and circuit boards. Careful consideration must be given to these devices, as they are a significant source of toxic pollution into national and international waste streams. In 2007, consumer electronics comprised approximately 2 percent of the United States' solid waste stream, totaling approximately 2.5 million tons, according to the U.S. Environmental Protection Agency (EPA).

Consumers upgrade electronic devices often, because companies release new models every few years. Electronic waste, or e-waste, contains many harmful elements, such as battery components.

Source: AvWijk/Wikipedia

E-waste management encompasses personal, business, and governmental strategies that can be used to responsibly handle decommissioned electronic components and devices. Depending on the goals of the user and the condition of the Waste Electrical and Electronic Equipment (WEEE), material can be donated for reuse, refurbished for donation or resale, sourced for parts, recycled into component materials, or deposited into a landfill. Successful e-waste management strives to avoid landfill disposal and to reduce environmental pollution at all stages of a product's life.

Effective e-waste management plans consider the functionality of the used electronic or electrical device and provide for several alternatives for these used products, along with e-cycling or landfill disposal. These options include donation plans, where still-functioning devices, like cell phones with expired service contracts, can be donated to other potential users. Donation plans can help to build an initial bridge across the digital divide, while simultaneously reducing e-waste. Another viable e-waste management option for working devices is a cash or incentive trade-in system, via either a physical site or an Internet forum.

For nonfunctional WEEE, there are a variety of options for recycling, or e-cycling. Municipalities and nonprofit organizations often host collection days for these materials. Typically, these events are intended for individual consumers, limiting the number of items that can be processed to discourage companies from using the opportunity to discard large amounts of e-waste at little or no cost. E-cycling programs are not limited to governmental and nonprofit efforts. Growing numbers of electronics producers and retailers are allowing consumers to return items at the end of their life cycles as a demonstration of corporate social responsibility. For example, Staples collects a wide variety of used WEEE at its stores; Dell recycles Dell products for free and will recycle other brands for free with the purchase of a new Dell, and Apple Inc. will accept any brand of computer, display, MP3 player, or mobile phone for recycling when the item is mailed to their recycling program address.

In developed countries, there are two general methods of e-cycling. First, devices can be dismantled by hand into their various component parts. This allows workers to sort the materials to find working parts that can be reused before the remaining parts are recycled in bulk. The second method requires the entire commingled WEEE collection to be fed into a large, industrial separator. The goods are shredded into small pieces, and separated by a complex system of magnets, screens, and currents. The entire process takes place within an enclosed system with dust collectors and scrubbers to minimize the exposure of workers and the environment to pollutants.

While recycling of e-waste is preferable to disposal, e-cycling still presents several complications. Any major recycling effort requires substantial investments in the collection and sorting of the variety of items that compose WEEE. This can be a costly and time-consuming process, given the variety of materials in electronic and electrical devices, and taking apart these materials into their distinct components, a necessary step in one method of recycling operations, is difficult. Once disassembled, some of these components, like cathode ray tubes from televisions and monitors, are inherently difficult to recycle. These nongreen design flaws have only recently started to be addressed by principles of ecoeffective design and extended producer responsibility.

Further, while recycling WEEE keeps toxic chemicals and heavy metals from leaching out of landfills, the recycling process also potentially exposes workers and communities along the recycling chain to these toxins. While the toxicity of any of these materials varies, there is always some level of risk present. For example, the lamps of flat screen televisions contain mercury, which is toxic even in extremely low amounts and the exposure to which can cause brain or kidney damage. Other heavy metals found in electronics include lead and cadmium. Nonmetal WEEE toxins include carcinogenic substances and hormone disruptors.

E-Waste Disposal Industry

The high cost of processing WEEE has led to a thriving industry of shipping mixed electronic waste to less developed countries, which results in a number of negative outcomes for communities that receive the materials. The cost of labor is one of the driving factors for the outsourcing of e-waste. Another contributing factor to the growth of this industry is that collected WEEE can contain a wide range of items, some of which are profitable to restore and some of which are useless and very expensive to recycle. Finally, the international waste destination sites often have very low environmental health and safety standards, which significantly reduces the cost of recycling or disposing of the goods in a developed country. This results in an accumulation of e-waste in areas with low labor costs, negligible environmental and safety regulations, high repair capabilities, and high demand for any external revenue.

Proponents of this international trade in WEEE believe that this system is the most economically efficient method. They argue that, without exporting, e-cycling would be economically unfeasible and disfavored for landfill disposal. Further, these proponents claim that this system provides jobs for people in lesser-developed countries and places raw materials closer to where they are needed. Also, this view provides that while refurbishment has become an industry in parts of Asia and South America, there is neither expertise in refurbishing used devices nor the demand to purchase refurbished goods in the United States.

Critics of international trade in WEEE believe that it is unethical to saddle other countries with the toxic waste from the United States and other developed countries, arguing that it removes the evidence of the need to reduce consumption and invest in ecoeffective or sustainable design. Moreover, the workers—including children—in overseas operations are often directly exposed to the toxic materials present in WEEE as they dismantle devices using rudimentary tools and little safety equipment. Further, the surrounding community is exposed indirectly, as waste products are freely dumped into local water supplies and often left in unlined garbage pits to seep into the soil. Additionally, the open burning of plastic wire casings to obtain copper releases highly toxic dioxins and furans that contribute to air pollution.

Regulation of E-Waste Disposal

Given the status of the e-waste disposal industry, the international community has provided guidelines for the reduction, recycling, and disposal of e-waste. The Basel Convention on the Control of Transboundary Movements of Hazardous Wastes and Their Disposal works to regulate international movements of hazardous waste, particularly from developed to less developed countries. The Basel Convention took effect in 1992 and had 172 signatory countries. Of these 172 parties, only Afghanistan, Haiti, and the United States have yet to ratify the convention.

Despite its intent, the convention's restrictions on the international shipping of waste caused traders to continue this process with simply changing the labeling of their cargo as that of a recycling program, thus making it a commodity and not waste. To close this loophole, 65 parties to the convention have ratified the Basel Ban Amendment, which bans the export of hazardous waste for any reason, including recycling. However, because the Ban Amendment is not yet in force as it requires additional signatory ratification, the European Union implemented the amendment's prohibitions in its Waste Shipment Regulation. Further, the European Union has developed the Waste Electrical and Electronic Equipment Directive (WEEE Directive), which became law in 2003 and which requires EU countries to comply with its sustainable e-waste management targets. This e-waste management legislation incorporates principles of extended producer responsibility, as it requires manufacturers to implement take-back systems for the return of their used products.

The United States lags behind other developed nations in regulating the exporting or domestic recycling of WEEE. There is scant federal guidance on the exporting of hazardous e-waste, and there is little federal attention paid to the practices of WEEE recycling companies. Further, there is no national policy to make manufacturers responsible for taking back their products at the end of their life, as is becoming common in other countries. The federal government has attempted to raise public awareness of the importance of electronics recycling, through programs like the EPA's "Plug-In to e-Cycling." Additionally, the EPA has created a voluntary set of guidelines, "Responsible Recycling," for electronics recyclers. However, the only two environmental groups represented in the process of the formation of these guidelines withdrew their participation over concerns that they were too business friendly and an example of greenwashing.

States are thus left to create a patchwork of e-waste management laws and regulations. California has led the way with the passage of e-cycling legislation in 2003; since then, 18 other states have passed electronics recycling laws. Each of these states requires that manufacturers take back their used products and pay to recycle them, except California, where consumers pay a fee when purchasing a covered item. Eleven states and New York City have WEEE item lists, such as computers, that cannot be disposed of in the normal waste stream.

Strategies for Sustainable E-Waste Management

The diligent work of nonprofit organizations and increasing corporate awareness are improving the development of sustainable e-waste management strategies. Organizations, such as the Electronics TakeBack Coalition, can identify which states mandate recycling and can help locate responsible recyclers. Additionally, the reemergence of environmental awareness in mainstream society has led to environmental features being touted in the marketing campaigns for consumer electronics.

For responsible e-waste management to become the norm, the financial balance will have to be shifted in the continued debate on recycling and disposal. Several possibilities exist. The first is that the demand for the recycled materials will increase, either due to a shortage of raw materials or consumer demand for green electronics. Another option is that the price of dismantling and recycling will be internalized into the initial purchase price of each device. This would cover the cost of management, but incentives might still be necessary to persuade consumers to return their equipment in an environmentally sustainable way. If consumers are willing to return used, still functioning electronic and electrical equipment, then reuse and refurbishment will become a more viable portion of any country's e-waste management. When possible, reuse and refurbishment are more sustainable than recycling as they do not require energy-intensive deconstruction and manufacturing processes. Of course, data security of these devices will have to be ensured in order for this sector to thrive.

In crafting e-waste management plans, it is important to remember that—despite all of the focus on donating, refurbishing, and recycling—the cycle begins long before these processes. The best e-waste management strategy is to identify opportunities to use fewer electronic devices, to extend their useful life through repair, and to make ecoeffectiveness a priority. Sustainable choices include those devices that are refurbished, are made from recycled content, incorporate fewer toxic components, are designed to be easy to repair, or are designed for easier recycling. The EPA's Electronic Product Environmental Assessment Tool, or EPEAT, provides a basis for identifying sustainable devices, and organizations, such as the Electronics TakeBack Coalition, list additional environmental considerations. Standardizing collection of electronics at the end of their life to divert them from the landfill and vetting potential recyclers are also important steps in any effective e-waste management plan.

While navigating the varied options for e-waste management can be daunting, it also presents market and consumer responsibility opportunities. Communicating a company's due diligence in enacting its environmental goals can distinguish that company from competitors. Additionally, questioning suppliers about their electronics take-back policies or recycling practices demonstrates demand for greater responsible social action. This type of consumer demand will be a driving force in making sustainable e-waste management a standardized practice.

See Also: Environmentally Preferable Purchasing; Extended Producer Responsibility; Recycling, Business of; Remanufacturing; Voluntary Standards; Waste Reduction.

Further Readings

Bily, Cynthia A. *What Is the Impact of E-Waste? (At Issue Series)*. Farmington Hills, MI: Greenhaven Press, 2008.

Carroll, Chris. "High-Tech Trash." *National Geographic Magazine* (January 2008).

Electronics TakeBack Coalition. http://www.electronicstakeback.com (Accessed October 2009).

Grossman, Elizabeth. *High Tech Trash: Digital Devices, Hidden Toxics, and Human Health*. Washington, D.C.: Island Press, 2007.

Slade, Giles. *Made to Break: Technology and Obsolescence in America*. Cambridge, MA: Harvard University Press, 2007.

U.S. Environmental Protection Agency. "Frequently Asked Questions." http://www.epa.gov/waste/conserve/materials/ecycling/faq.htm (Accessed October 2009).

U.S. Government Accountability Office. "Electronic Waste: EPA Needs to Better Control Harmful U.S. Exports Through Stronger Enforcement and More Comprehensive Regulation." http://www.gao.gov/new.items/d081044.pdf (Accessed October 2009).

Claire Roby
Independent Scholar

EXPOSURE ASSESSMENT

Because thousands of industrial and commercial chemicals are in use today and distributed globally, many of which pose a potential health risk to organisms and the environment, exposure assessment is a critically important process. Exposure assessment is a method of measuring or estimating an organism's level of exposure to a toxic compound or group of compounds. Exposure is usually defined as the amount of a chemical in the environment that may affect an organism. In the recent past, exposure assessment has become an integral part of green enterprises because of their interest in environmental sustainability. Additionally, exposure assessment is a major component of the risk assessment process to prevent liability for the use (and, perhaps, misuse) of toxic chemicals. Exposure assessment also integrates information about observed and reported adverse health effects in organisms and communities. As such, exposure assessment is key to the maintenance of a safe workplace.

The risk assessment process, originally formalized by the National Academy of Sciences' paradigm in 1983, has four components: hazard identification, dose response evaluation, exposure assessment, and risk characterization. Exposure assessment is central to this risk analysis process and asks questions like "What are the types and levels of exposure, as well as what are the pathways of exposure and bioavailability of the toxin?" In other words, exposure assessment is the process of measuring or estimating toxic chemical concentrations, duration, frequency of exposure, and pathways to an organism that arise from chemicals released into the environment. For example, a company that is introducing a new type of carpet glue would conduct such an analysis to make sure that the product does not contain levels of toxins that could later harm people who are exposed to them in homes or workplaces.

Clearly, in these assessments, first examined is the element of exposure to a toxin, which is the process by which a chemical becomes available for intake and/or absorption by individuals in a population. The concentration of the chemicals is a vital consideration in this assessment. It requires a calculation of how much toxin reaches an identified target (be it a specific population, animal, plant, organ, or cell). This calculation is expressed in quantitative terms for various scenarios that include contingencies for different pathways or durations of exposure.

Although exposure differs from the related concept of dose—in that exposure is the quantity of a chemical in the environment that is in contact with an organism while the dose refers to the level of the chemical actually absorbed by the individual—both exposure and dose are usually reported in exposure assessments. Exposure levels depend on a number of physical factors such as volatility, partition coefficient, molecular weight, polarity, and related concerns. Types of exposure range from deliberate (food additives) to accidental (industrial accidents that release toxic chemicals) to incidental (mercury via the environment). To perform an exposure assessment, the properties of the chemical, such as water solubility, the ability to migrate through the soil, or the ability to bind to airborne particulate matter, are characterized to identify potential routes to the target. Possible exposure pathways include the routes by which a contaminant reaches an individual via inhalation, ingestion, or dermal contact.

Exposures can be classified as acute or chronic. Acute exposure refers to different adverse effects that might occur following a limited time of exposure to relatively high concentrations of a chemical, as can happen in manufacturing accidents. The acute adverse effect exhibits itself in a relatively short time scale—either immediately or within a few days. Conversely, chronic exposure refers to a consistent exposure period of weeks, months, or years, and the adverse effects are seen only after a relatively long period of time. An example of chronic exposure is the existence of lead paint in a home and its subsequently harmful health effects on the home's occupants.

Physical settings and the time of exposure may impact the exposure's bioavailability, and, as such, also need to be assessed. Local climate, meteorology, vegetation, hydrology, and soil type are used to characterize the physical setting and can lead to variation in exposure concentration. Since exposure and dose are interrelated concepts, the concept of exposure as a potential dose that is in contact with outer membranes (dermal, respiratory, or oral) is a useful way to view the exposure levels. Building on this concept, multimedia and multipathway exposure assessments have begun to emphasize the total exposure from all pathways. These types of assessments will use a variety of evaluation methods—both direct (personal monitoring or biomarkers) and indirect (questionnaires, diaries, and environmental chemical monitoring).

A comprehensive exposure assessment, in addition to delineating the physical setting, spatial distribution, potential exposure routes, and temporal distribution, will also identify the target or sensitive subpopulation—the group of organisms—that will be exposed. Special attention is paid to vulnerable or hypersensitive groups, such as pregnant women, children, the elderly, and the sick (like people with cardiovascular or respiratory diseases). Plant, fish, and animal populations, including threatened or endangered species, also need to be considered in such an assessment. Activity patterns of organisms are important in understanding the duration and frequency of the exposure (whether it is acute, continuous, intermittent, or chronic). An example of a comprehensive exposure assessment is the Centers for Disease Control and Prevention's National Health and Nutrition Evaluation Survey (NHANES) study. This survey measures contaminant exposures and is linked to the U.S. Census. NHANES was one of the first surveys to include analysis of biological fluids for chemical substances and later to use biomarkers of exposure.

The exposure route is also important in answering the questions about the harmful effect of the toxin. The pathways of exposure include oral, respiratory, and dermal. They are considered in a transport analysis that views air, soil, groundwater, surface water, and food systems (global versus local or subsistence) as transport media. Contaminants released in the environmental matrix often arise from a point source. Some examples include oil wells, a line source like a road, or an area source like a harbor or farming region that contains pollution. Air, water, and soil are components of the environmental media and the contaminants can be transformed by processes along the way that are physical (volatility and solubility), chemical (oxidation), or metabolic (methylation). The assessment, therefore, needs to identify the following:

- Principle mechanisms for change in a media
- Natural removal from transport (sinks) or degradation
- Occurrence of biomagnification during transport
- Average lifetime or dwell time in the media

In exposure assessment, a model is commonly developed to estimate exposures to the target population. Two important factors affecting exposure models are the source term and the dispersion pattern. The source term is the composition of the mixture that is released from a defined area. Models often extrapolate from the chemical's toxicity data to acceptable exposure level. For nonstochastic effects, extrapolation is usually performed using the no observed adverse effect level (NOAEL). The NOAEL is often adjusted using uncertainty (safety) factors in order to determine an exposure limit for the model. It is more difficult to model health effects like cancer because of the complexity of the probabilistic mechanisms involved in the development of the health effect.

Fate and transport models for exposure assessment vary with the contaminant and geographic scale (global, regional, or local). Inhalation from the air leads to intake by the respiratory system. Dermal absorption comes from soil or liquid media. Background or steady-state levels need to be determined as well. Exposure assessment of individual contaminants, such as mercury or arsenic, in a single pathway is relatively straightforward, but this process becomes very complex in multimedia, multipathway exposure assessments. Exposure assessment in such a complex scenario usually includes the following components: scope and purpose, design detail, background review, forecast of study population, preliminary studies, protocol establishment, compliance reviews and permission, field study surveys, quality assurance, data analysis, and interpretation. Often, the final exposure assessment report is peer reviewed to ensure that the disciplinary best practices have been achieved.

Overall, exposure assessment has become an increasingly important part of green business and an especially vital aspect of companies' environmental management systems. These considerations of harmful effects of products and their manufacture are central parts of planning for sustainability and the minimization of negative environmental effects. Clearly, exposure assessment is a crucial part of these risk assessment and operations decisions and in the developing paradigm of green management.

See Also: Environmental Impact Statement; Environmental Risk Assessment; Persistent Pollutants; Toxics Release Inventory.

Further Readings

Byrd, D. M. and C. R. Cothern. *Introduction to Risk Analysis.* Lanham, MD: Government Institutes Press, 2000.

Committee on the Institutional Means for Assessment of Risks to Public Health/National Research Council. *Risk Assessment in the Federal Government: Managing the Process.* Washington, D.C.: National Academies Press, 1983.

Duffus, J. H. and H. G. Worth. *Fundamentals of Toxicology for Chemists.* Cambridge, UK: Royal Society of Chemistry, 1996.

Hemond, H. E. and E. J. Fechner. *Chemical Fate and Transport in the Environment.* New York: Academic Press, 1994.

U.S. Environmental Protection Agency (EPA). "Guidelines for Exposure Assessment." http://www.epa.gov/ncea/pdfs/guidline.pdf (Accessed January 2010).

Lawrence K. Duffy
University of Alaska Fairbanks

EXTENDED PRODUCER RESPONSIBILITY

The notion of extended producer responsibility makes producers retain some degree of responsibility for their products long after they have left the factory gate. It has also become the subject of regulation in a number of jurisdictions—for example, the European Union's (EU's) End-of-life Vehicle Directive—though the precise way the concept is implemented still varies, with the concept of product-service systems perhaps a logical outcome.

The notion of sustainable consumption and production (SCP) is attracting the interest of government and industry in the context of moves toward more sustainable societies. The SCP concept featured prominently at the United Nations' Johannesburg Summit and is part of the subsequent United Nations Environment Programme–driven Marrakech Process. Industry is consistently seen as a key player in this. Although it has traditionally blamed consumers for unsustainable product choices, arguing that there is no consumer demand for greener products, this view has now largely been discredited. American academic Stuart Hart wrote an influential article on this notion in *Harvard Business Review* in 1997 in which he put the responsibility firmly back in industry's court. Even prior to this, society increasingly considered industry responsible for its products long after they left the factory gate.

The notion of extended producer responsibility thus grew out of the concept of corporate social responsibility (CSR) and corporate environmental responsibility (CER), a concept also known as social "license to operate." It has become increasingly important and has begun to inform government regulation in a number of jurisdictions. One such example is the EU's End-of-life Vehicle Directive and also the similar Waste Electrical and Electronic Equipment (WEEE) Directive for electronic products. These EU directives required member states to introduce legislation requiring carefully controlled take-back and processing of products that have reached the end of their useful lives. The original manufacturer or its agents—such as importers in some cases—have to meet the cost of this process as it is deemed they were responsible for creating the product in the first place and therefore also for creating the potential waste problem resulting from this. A growing number of countries impose a take-back levy on new products at the point of sale, administered by the retailer on behalf of the manufacturer.

Extended producer responsibility appears to move away from the pure concept of "the business of business is business"; in reality, a number of successful companies have created profitable business models around this concept. Few would challenge the Rank Xerox product stewardship model as an example of a successful business model. This firm decided to move away from a model whereby it sold its photocopiers to the end user to a model whereby the copiers are instead leased under a full maintenance contract; the end user is simply billed for the number of copies made. The copier remains the property of Rank Xerox, which is therefore responsible not only for its production, but also for its maintenance, repair, and refurbishment, as well as end-of-life processing, recycling, and reuse. The firm determined that its customers do not necessarily want to own the machine; instead, they merely want the use of the machine for the service it provides. This business model has since been copied in a number of other areas and has become known as a product-service system, whereby the product itself becomes merely a means of delivering a service to the customer.

This product-service model is increasingly considered to be a more sustainable model, closer to the principles of SCP, than the still dominant model of "fire and forget" after a product has left the factory gate. This older, though still dominant, model encourages manufacturers to produce goods with rapid obsolescence—either in technical terms or fashion terms—in order to encourage the customer to buy a replacement from which the manufacturer then earns more income. In a product-service system, the product itself becomes a cost to the business, rather than a money earner. It is, therefore, in the interest of the product-service provider to optimize the life cycle of the product. Its premature replacement, for example, would increase the cost to the business. Within the context of SCP, therefore, such a product-service approach leads to more efficient, more durable products. In addition, it removes a burden from the end user who no longer needs to worry about purchase, maintenance, repair, or replacement of the product. In this respect, it provides a better service than the more established business model of selling product.

Some extend this idea even further, arguing that designers and manufacturers should take the entire life cycle into account in the concept and design process, and that they should also think beyond this about how the product might provide an input into future natural or manmade processes. A variation on this idea is the concept of leasing raw materials. Under this concept, raw materials are leased from a raw materials supplier and then made into a manufactured product. After the life of that product, it is processed so that the raw materials are extracted, returned to the supplier, and can then once again be leased to another manufacturer for another product.

Over time, a concept that was once regarded by most as a potential burden and cost on business is being turned around to inform new business models and new principles to change the way we do business and use and produce goods. Extended producer responsibility has thus become one of the key principles for sustainable consumption and production.

See Also: Corporate Social Responsibility; Cradle-to-Cradle; Green Design; Stewardship; Sustainability; Take Back.

Further Readings

Bohr, Philipp. *The Economics of Electronics Recycling: New Approaches to Extended Producer Responsibility.* Saarbrücken, Germany: VDM Verlag, 2008.
Hart, S. "Beyond Greening: Strategies for a Sustainable World." *Harvard Business Review* (January–February 1997).

Maslennikova, I. "Rank Xerox Product Stewardship." In *The Durable Use of Consumer Products: New Options for Business and Consumption*, M. Kostecki, ed. Dordrecht, Germany: Kluwer, 1998.

McDonough, W. and M. Braungart. *Cradle to Cradle: Remaking the Way We Make Things.* New York: North Point Press, 2002.

Tojo, Naoko. *Extended Producer Responsibility as a Driver for Design Change: Utopia or Reality?* Saarbrücken, Germany: VDM Verlag, 2008.

Paul Nieuwenhuis
Cardiff University

EXTENDED PRODUCT RESPONSIBILITY

The prevention and reduction of environmental pollution, as well as the promotion of sustainable consumption and production (SCP), are fundamental to safeguard the Earth's capacity to support life, to respect the limits of the planet's natural resources, and to ensure a high level of regenerative development. Consumption and production of goods contribute significantly to global warming, pollution, material use, the increase of hazardous waste, and the depletion of the planet's natural resources. Extended product responsibility establishes a framework that supports and encourages SCP. In contrast to Extended Producer Responsibility, which places the burden on the producer, Extended Product Responsibility stresses that all the actors in a product chain should work together to ensure more efficient use of natural resources, thereby reducing waste and energy during all stages of a product's life, from extraction of raw materials to manufacturing, transport, consumer use, disposal, and end-of-life treatment. Specifically, extended product responsibility places the initial impetus of sustainability on the designers of a product to select raw materials that come from regenerative sources; assigns responsibility to the product manufacturers to make choices that are in line with sustainable design; calls for environmentally friendly methods of marketing and distribution; and aims for consumers to choose products based on green factors, and then give the materials new life through proper recycling.

Traditional industrial systems and modern manufacturing are based upon a linear economy in which natural resources and valuable materials contained within a product are discarded rather than recovered once that product has been used. In the United States, an example of a linear economy, it is commonly held that cradle-to-grave (nonsustainable) designs, which are responsible for up to 90 percent of all extracted raw materials, end up as waste, according to W. McDonough and M. Braungart in their 2002 research. In response to this growing problem of linearity, the Brundtland Commission published the report "Our Common Future" in 1987, an important contribution to the international objective of sustainable development that aims to ensure that natural resources and finite materials be managed to sustain future generations. In this sustainable management of resources, it is not only the reduction of end-use waste that is a cause for concern but also the wasting of resources. In contrast to linear economies, closed-loop economies make an important contribution to sustainable resource management and positive patterns of production and consumption by (1) promoting the utilization of existing resources by reintroducing materials within society into the market; (2) encouraging increased efficiency through better and more ecologically sound design; and (3) adopting cleaner production technologies. The main objective of a circular economy is to achieve no waste at any point in the life cycle of a product through environmentally responsible product development and product recovery.

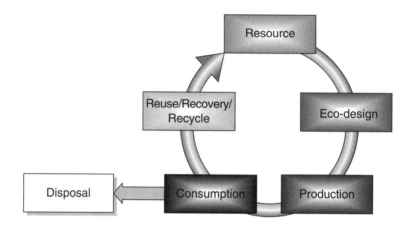

From this background, it is clear that extended product responsibility can be considered as a product systems approach to resource conservation, waste reduction, and sustainability. It is a term that is prevalent in U.S. discussions on the management of resources and waste in business and industry, but it has not been integrated into law and policy frameworks in Europe. In the European context, the related principle of extended producer responsibility has been considered an important mechanism in securing patterns of SCP. Extended producer responsibility focuses almost exclusively on producers. This concept aims to make producers responsible for their products throughout a product's life cycle. The doctrine encourages producers to facilitate the repair, reuse, disassembly, recycling, and environmentally sound disposal of their products. Most importantly, extended producer responsibility promotes better design and production of goods since the design and development of a product is evaluated by its environmental impact throughout the product's lifetime. The main objective of related ideas in ecodesign, also referred to as design for environment (DFE), is to design products that minimize product life cycle impacts on the environment. To this end, life cycle assessment is the standard procedure used by producers, engineers, and environmental scientists to investigate and assess the environmental impacts of a product and its component parts from raw material extraction to the recycling and disposal of the product. These concepts enable businesses involved at every stage in the product chain to select the most environmentally favorable options early in the design process and to avoid designs that have detrimental environmental consequences.

Fundamentally, SCP is dependent upon behavioral changes in producers and consumers to move from polluting products and services to sustainable and energy-efficient ones. Therefore, the challenge is to create a virtuous circle that incorporates life cycle thinking, that stimulates demand for better products, that supports ecoinnovation, and that encourages consumers to make sustainable choices. This goal is recognized by extended product responsibility along with the broader notions that encapsulate programs, such as extended producer responsibility and life cycle assessment. The common element to all these concepts is that they aim to alter the existing patterns of production and design priorities of producers in order to achieve developmental and technological advancement in a manner that takes into account the impacts and uncertainties associated with globalization, population growth, and climate change.

A key distinction that sets extended product responsibility apart from narrower, although similar, principles is the emphasis it places on collaboration with stakeholders, the creation of product chain partnerships, and shared responsibility among all actors in a product's life cycle. This shared responsibility is envisaged by encompassing entire product systems and asking how each player in the product chain—from raw material extraction through product design, manufacturing, distribution, consumption, recovery, recycling, and disposal—can work together to reduce environmental impacts throughout the product's life cycle. In this way, extended product responsibility focuses on product systems rather than on individual polluters by addressing the upstream pollution externalities associated with consumer products, which, according to some experts, are substantially more serious than disposal externalities.

Incentivizing changes to production and consumption can be difficult without the introduction of some form of regulatory structure, such as the Pigovian emissions tax that sets its rate equal to the marginal environmental costs of emissions at the level where cost to benefit is considered equal. Incentive-based policies consistent with extended product responsibility have the ability to shift the much used producer take-back requirements, most often used in extended producer responsibility measures, upstream in the product chain. For example, the upstream combination tax/subsidy (UCTS) combines a tax, levied per pound, on producers' intermediate goods, such as sheets of steel, with a subsidy granted to collectors of recyclables who sell the goods for reprocessing. UCTS is materials focused and acts to harness discarded valuable resources for recovery, recycling, and reuse.

The U.S. Environmental Protection Agency claims that businesses are beginning to embrace extended product responsibility in a variety of ways as a means to save money, drive product innovation, develop clean technologies, deliver products for increasingly green consumers, and secure their long-term future by gaining competitive advantage. It is likely that, as environmental laws and regulations become ever more stringent in response to adapting to and mitigating the effects of climate change, businesses will need to acquire new capabilities, minimize their use of resources, and reposition themselves as green enterprises to compete in a changing market.

See Also: Cradle-to-Cradle; Design for Environment; Life Cycle Analysis; Triple Bottom Line; Waste Reduction.

Further Readings

Brundtland Commission. *Our Common Future.* Oxford, UK: Oxford University Press, 1987.
Davis, Gary A., Catherine A. Wilt and Jack N. Barkenbus. "Extended Product Responsibility: A Tool for a Sustainable Economy." *Environment*, 39/7:10–15 (1997).
McDonough, W. and M. Braungart. *Cradle to Cradle: Remaking the Way We Make Things.* New York: North Point Press, 2002.

Palmer, Karen and Margaret Walls. "Extended Product Responsibility: An Economic Assessment of Alternative Policies: Discussion Paper 99-12." Washington, D.C.: Resources for the Future, 1999.

U.S. Environmental Protection Agency (EPA). "Extended Product Responsibility: A Strategic Framework for Sustainable Products." Washington, D.C.: EPA, 1998.

Hazel Nash
Cardiff University

EXTERNALITIES

Though largely unknown in the wider public discourse, externalities represent a key concept in economics, particularly as it concerns the environment. At its core, the term *externalities* captures a simple, basic idea: virtually every market transaction accrues either benefits or costs that are not directly involved in the initial transaction. Another way of stating this is that neither "positive externalities" nor "negative externalities" are "internal" to the transaction (they produce what economists call a "spillover effect"). A simple example for a beneficial externality: a consumer pays for, and receives, a flu shot. When they subsequently stay healthy, everyone around them benefits (even though they did not pay for the shot, nor in any other way were part of the original transaction). A simple example for a cost, or negative externality: a consumer buys a tomato from a farmer who uses toxic fertilizers that leak into the groundwater and end up killing fish and sickening water consumers at an adjacent lake; the expenses of water filtration, loss of fisheries, and healthcare costs are borne neither by you nor the farmer; the cost is effectively "externalized" to water consumers and the community. Economists frequently differentiate between "private costs" (those that are part of the transaction) and "social" or "socialized" costs (when society picks up the tab). A classic recent example for the latter distinction would be the government bailout of AIG or Goldman Sachs: as long as their creative financial dealings were profitable, the gains were private (kept by the company), but the subsequent huge self-inflicted losses were essentially externalized or socialized.

What we learn in grade school, namely that you fix what you break, in short, does not apply in our modern economy. Indeed, almost everything people generally associate with what is broken in the environment—resource depletion, pollution, waste, degradation—represents, in economic parlance, an externality, for it was, originally, a consequence external to the original economic transactions. The purchaser of the tomato, for instance, did not pay for (or even know about) the water contamination downstream. Similarly, a carpenter who buys lumber did not pay for the potential costs associated with clear-cutting, such as soil erosion, a decline in biodiversity, or the loss of biosequestration of carbon dioxide (CO_2).

An example widely debated in a series of recent documentaries, books, and articles on fast food can serve to clarify the significance of externalities, as well as illuminate some of the most daunting conceptual questions occupying scholars, economists, and environmental activists alike. Today, the price of a hamburger at any major fast food franchise is less than $4.00. According to a variety of studies, however, the real cost is significantly higher—if the cattle were raised in pasture cleared from rainforest, according to the Indian Centre for Science and Environment, the cost could be considered as high as $200, that is, if all externalities were to be included in the price. While this may sound absurd, let us examine what went into the calculations. For one, fast food franchises each sell hundreds of millions of hamburgers a year in the United States, and the carbon footprint of the energy used to produce and cook these hamburgers, according to conservative estimates, runs up to at least 2.7 billion pounds of carbon dioxide emissions that would cost hundreds of millions to clean up.

Estimates for additional externalities are somewhat more difficult to ascertain with precision, but it is easily demonstrable that they include massive water use, soil contamination and erosion, emission of pollutants during the cooking process, and, of course, a variety of health related issues, ranging from obesity to high cholesterol and diabetes, all of which have been connected to the consumption of fast food, and hamburgers in particular. Add it all up, and several science reports conclude that if the costumer were to pay the "real" cost of production, consumption, and environmental impact of the hamburger, the price could indeed run up to 20 times as high as it currently does. Again, it should be noted that as many of the associated costs are effectively "socialized," consumers already pay a much higher price than they think.

But before continuing with this example, several problems with externalities already become evident. The first and most obvious is so important that it has spawned an entire subdiscipline of economics: the problem of pricing, or valuation. What exactly is the value, expressed in money, of a contaminated river, a toxic landfill, a destroyed family farm, an obese diabetic child? Since most externalities are not goods commonly traded in the marketplace (such as good health, clean air and water, or vibrant communities), they do not have a price. Furthermore, since in the American economy, nothing that does not have a price is considered to have value (which leads to what economists call a market failure), environmental economists have labored to develop valuation models that allow them to attach prices to goods that are not exchanged in the market, and thus do not have a market price attached to them. The basic logic behind such efforts is that if we consider something valuable, we need to attach a price to it, or the market will ignore it. In the example above, most of the costs of producing and consuming a fast food hamburger are not priced—not the CO_2 emissions, not the soil erosion or water degradation, nor the exploitation of cheap labor.

The second significant problem is that those costs that do come with a price (obesity, heart failure, and diabetes) are extremely difficult to value in direct correlation with the consumption of, for instance, a hamburger. As decades of mostly unsuccessful antismoking litigation suggests, even when the connection meets a high burden of proof in a court of law, simply tying the basic health costs of a lung cancer patient who used to smoke to the price of cigarettes faces a mountain of legal, political, and cultural hurdles.

But the fast food hamburger example contains further externalities, or what might be called front-loaded hidden externalities. To wit, the other big reason why the hamburgers we purchase are so cheap is because their production and distribution were heavily subsidized. This starts with the cow, which is likely owned by a subsidized agricultural corporation that feeds its cattle with the most heavily subsidized crop in history—corn (subsidized at an estimated total of over $65 billion since 1995 in the United States alone). The roads and highways on which cattle is transported from farm to slaughterhouse to franchise, furthermore, are largely paid for by tax payers—not by the fast food franchise. And, of course, the general public further subsidizes the low-wage workforce at fast food establishments through Medicare, food stamps, free school lunch programs, and other social programs.

Taken altogether, the cost of $200 for a fast food hamburger may no longer appear as overestimated.

To summarize, externalities reveal that

- prices do not adequately reflect costs (what an individual consumer pays for a fast food hamburger is only a fraction of what society collectively ends up paying via subsidies, health costs, environmental degradation, and so on);
- profits are often possible only because they are both subsidized and because they successfully externalize costs; and
- markets are neither free nor always rational—increasingly, they fail to protect the environment and its inhabitants by inadequately reflecting both social and environmental costs of production and consumption.

This, in turn, is highly relevant for green business. To follow our example one step further, a healthy, organically grown alternative to a fast food burger today faces the major obstacle of being much more expensive, leading to the oft-asked question, "Why does the food item with the laundry list of ingredients cost so much less than the original, pesticide- and corn syrup–free original?" The answer follows from the fast food example: ordinarily, the organic alternative is not heavily subsidized, does not rely on low-wage labor, and is not part of the corporate food growing and processing industry. Even more importantly, however, as much as the negative externalities of the fast food industry are not added to their prices, the beneficial externalities of the community-based organic food are not subtracted from its prices. Here again, environmental economists can provide increasingly sophisticated models as to how to price the community benefits of sustainable farming, fewer transportation miles, and, significantly, much improved health among consumers. If such social benefits were added up and subtracted from the price of organically grown food, it would not only become much cheaper, but it would also suddenly be a far more affordable alternative to fast food.

What is true in the specific example of food is equally true for green businesses in general: to the extent to which environmental awareness translates into environmental regulations (including fees and taxes) that in turn will inevitably increase prices of practices detrimental to environmental health, community-grown and organic food will not only become much more economically viable, but sustainable energy (solar, wind, or biomass), green architecture and urban planning, alternative medicine, and sustainable green cosmetics and apparel industries will also benefit and become more affordable.

By internalizing externalities, some of the worst market failures can be rectified, and goods (or rather "bads") can be priced much more realistically. There is little doubt among environmentally concerned scientists and economists that pricing and internalizing externalities is a big, if not essential, step in the right direction. Advocates have argued that there is nothing that would shift economic choices as quickly as true pricing: a much more accurate price of $12 per gallon of gasoline, for instance, not only would automatically wean most people from SUVs, but it would also immediately provide large volumes of capital for sustainable energy resources and, in turn, make them significantly more affordable—all without having to rely on fines, fees, or regulations.

What remains questionable is whether the pressing crisis of global warming and resource depletion can be effectively addressed solely by attaching prices to values such as good health, clean air and water, or the life of communities and individuals themselves. And when it comes to the political challenge of how to develop and enforce a more realistic pricing system, another big question is whether corporate capitalism is able to operate when it actually has to shoulder the costs it has historically externalized to the larger society—particularly when it can no longer count on the many direct and indirect subsidies it has also historically counted on.

See Also: Ecological Economics; Environmental Economics; Genuine Progress Indicator; Sustainability.

Further Readings

Antoci, A. "Environmental Degradation as Engine of Undesirable Economic Growth via Self-Protection Consumption Choices." *Ecological Economics,* 68:1385–97 (2009).

Coase, R. "The Problem of Social Cost." *Journal of Law and Economics,* 3:1–44 (1960).

Fisher, W. H. and F. X. Hof. "Relative Consumption, Economic Growth, and Taxation." *Journal of Economics (Zeitschrift für Nationalökonomie),* 72:241–62 (2002).

Keohane, N. O. and S. M. Olmstead. *Markets and the Environment.* Washington, D.C.: Island Press, 2007.

Liu, W. F. and S. J. Turnovsky. "Consumption Externalities, Production Externalities, and Longrun Macroeconomic Efficiency." *Journal of Public Economics,* 89/5–6:1097–29 (2007).

Rosser, J. B., Jr. "Complex Ecologic-Economic Dynamics and Environmental Policy." *Ecological Economics,* 37:23–37 (2001).

Zak, Paul. *Moral Markets: The Critical Role of Values in the Economy.* Princeton, NJ: Princeton University Press, 2008.

Dirk Peter Philipsen
Virginia State University

Factor Four and Factor Ten

Factor Four and Factor 10 are two examples of targets proposed for concomitant increases in ecoefficiency and reductions in material usage. Factor Four sets forth a target of doubling welfare while resource consumption simultaneously halves. Factor 10 proposes that material flows should be reduced by 90 percent over a period of 30–50 years.

Both Factor Four and Factor 10 were first proposed in 1994. Factor Four was set forth by Ernst von Weizsäcker and Amory Lovins, while Factor 10 was introduced in the Carnoules Declaration published by the International Factor 10 Club. Each proposal rose from the concern that worldwide stores of natural resources were being drawn down at unsustainable rates and the conviction that changes in production methods coupled with technological advances could be instituted to not only stem the overdraft, but to reverse the damage already done. Both proposals contended that these changes could be adopted without reducing living standards, rather that they would be improved. Factor Four (republished in English in 1997), forms an extended counterargument to the claims that adopting environmental protection measures would be too expensive and would result in drastically reduced economic prospects. Other factors ranging from the original four to as high as 50 have subsequently been put forth by various groups. In January 2010, von Weizsäcker and several coauthors released an update to Factor Four titled "Factor Five: Transforming the Global Economy Through 80 Percent Improvements in Resource Productivity."

Factor Four and Factor 10 have been more readily accepted as government programs in Europe than anywhere else. Their value lies in encouraging resource conservation through technological innovation. The case studies highlighting the successes of early adopters of technologies to reduce the use of material inputs are valuable in recruiting other businesses to do the same or to undertake innovation in their own industry.

Factor Four

Factor Four and its successor Factor Five contain case studies of individual businesses and sectors to show technological innovations that have been implemented to improve resource productivity. Factor Four contends that society can live twice as well while using half as many natural resources. Among the suggested ways to reduce the drain on natural resources are incorporating solar features into homes and offices, car sharing, rethinking the use of wood in construction, increasing efficiency in farming, lengthening the active work life of individuals by 15 years, and shifting economic emphasis from production to service. Many of the innovations highlighted have already made their way into widespread use; some technologies are far more advanced today. Factor Four points out the need to reexamine government, which often subsidizes wasteful resource use. Though it makes a strong argument for redesigning our production methods and consumption patterns to be more frugal, it does not succeed in showing how Factor Four can be applied across the whole of the economy. Independent researchers applying total materials reductions or material intensity per unit service measures to the examples given in Factor 4 have been unable to find factor reductions approaching four (much less 10). Despite the lack of empirical evidence that Factor Four can be achieved, several countries have adopted Factor Four as a technology-forcing policy.

Factor Ten

Factor 10 sets a goal of reducing material flow by unit by 90 percent within 30–50 years. It was highlighted in the United Nations Environment Programme's (UNEP's) Global Environmental Outlook 2000, which called for its adoption in order to ensure long-term equity between developed and developing nations. For the planet as whole, UNEP estimates that resource use must be cut in half in order to reach a sustainable level. However, since developed countries represent 20 percent of the world population and currently consume 80 percent of the natural resources, it is argued that these nations should bear the brunt of reducing resource consumption. Analysts calculate that by 2050, per capita, per annum consumption of nonrenewable resources should not be in excess of 5–6 tons.

The degree to which a country would need to dematerialize its processes to reduce its resource consumption depends on the rate at which that country currently consumes materials and energy. Japan's necessary reduction factor is estimated as six, Germany is estimated at 10, the United States is estimated at 15, and Finland is at 19.

Factor 10 suffers the same criticisms as Factor Four: there are no documented cases of process innovations that have reduced resource use by 90 percent. There is no rigorous explanation of how the factor should be measured or against which benchmarks.

Resource Productivity

Both Factor Four and Factor 10 call for radical increases in the productivity of natural resources. Since the industrial revolution, enormous increases in industrial productivity have been realized, mainly by reducing the amount of human labor required to produce goods and services. Proponents of Factor Four and Factor 10 argue for extracting more and more value from less materials and energy input. Rather than making ever-fewer people more productive, more people and fewer resources should be employed to deliver the same quality of life. The approach calls for a combination of the best available technology, advanced production methods, and improved engineering in designing products that do more with less, which proponents contend will prove more profitable for business while still providing the general public with a high standard of living.

See Also: Dematerialization; Ecological Economics; Economic Value Added; Material Input per Service Unit (MIPS).

Further Readings

Factor 10 Club. *The International Factor 10 Club's Statement to Government and Business Leaders: A Tenfold Leap in Energy and Resource Efficiency.* Wuppertal, Germany: Wuppertal Institute, 1997.

Reijnders, Lucasx. "The Factor X Debate: Setting Targets for Eco Efficiency." *Journal of Industrial Ecology,* 2/1 (1998).

Schmidt-Bleek, F. "The Factor 10 Manifesto." http://www.factor10-institute.org/files/F10_Manifesto_e.pdf (Accessed October 2006).

Schmidt-Bleek, F. "The Factor 10/MIPS-Concept: Bridging Ecological, Economic, and Social Dimensions With Sustainability Indicators." Paper presented at the United Nations University Zero Emissions Forum, 1999. http://www.unu.edu/zef/publications_e/ZEF_EN_1999_03_D.pdf (Accessed October 2009).

von Weizsacker, Ernst, et al. *Factor Four: Doubling Wealth, Halving Resource Use—The New Report to the Club of Rome.* London, Earthscan, 1998.

Gavin D. J. Harper
Cardiff University

Fair Trade

Fair Trade is a system of private governance that aims to improve the conditions under which producers of certain goods interact with the global economy; the primary intention is to render terms of involvement more socially just and environmentally sustainable. Having said this, Fair Trade is subject to various discursive and practical interpretations and as such, the concept, its practice, and subsequent results are not homogenous. While there is considerable support for Fair Trade, there has also been much criticism of the approach. In particular, some have questioned the power and spread of Fair Trade discourse on the grounds that empirical understanding remains only modestly (if not increasingly) developed.

Although Fair Trade is usually considered to have emerged at the end of World War II, more recent scholarship has suggested that in some countries, such as the United Kingdom, the movement only really began in the 1970s. Irrespective of the time scale, the literature shows that Fair Trade developed as a multitude of institutions sought to channel greater benefit from the international economy to producer communities seen to be otherwise disadvantaged. These activities became known as the alternative trade (AT) movement, which had two primary aims (1) to link materially poor producers in the developing world with markets in richer countries as a means to facilitate resource transfers and (2) to organize this interaction around the aim of maximizing the benefit to these initial producers. Operationally, the concept of Alternative Trade was based on the idea of a partnership between socially orientated northern alternative trade organizations (ATOs), responsible for the purchase and import of goods, and southern producer organizations that provided services to their members in marketing, product development, financing, and distribution services. In order to facilitate equality, producer and trade organizations structured their operations and interaction on predefined socially orientated norms. These principles were then used as part of the

Fair Trade aims to improve conditions for producers of goods in today's global economy. Here, a worker labels Fair Trade bananas at a cooperative in Ecuador.

Source: Didier Gentilhomme/Fairtrade Labelling Organizations International

marketing of the final product, the "alternative" identity, which was legitimized by the social reputation of retailers. It was also considered that by embedding products with information about conditions of production, consumers and producers were brought closer together in links of solidarity. While alternative trade mainly dealt with handicraft items, ATOs also moved into a range of basic commodity crops including coffee.

However, the scope and impact of alternative trade proved to be limited by its nature as an "alternative" niche market, worsening financial conditions and increasing competition from mainstream and profit-orientated commerce. As a response, the Mexican cooperative Union de Comunidades Indigenas del Region del Istmo (UCIRI) and a Dutch nongovernmental organization (NGO), Solidaridad, developed a system of external third-party governance and certification. This gave rise to the Max Havelaar mark (launched in 1988) that guaranteed that certified coffee had been bought by importers direct from cooperatives for a price of up to 10 percent higher than world market prices; that importers had underwritten legitimate additional costs and provide prefinancing of up to 60 percent of the final price; and that producers were maintaining long-term relationships with producer communities. For the first time, alternatively traded coffee could be commercialized through mainstream channels and sold by conventional retailers.

This model proved tremendously popular and gradually other national certification systems began to emerge; for example, the Fairtrade Foundation in the UK (founded in 1992) and Transfair in the United States (founded in 1998). While the number of products eligible for certification widened, national organizations slowly assimilated under a unitary set of regulations, represented by a single certification "brand" and administered by the Fairtrade Labelling Organizations International (FLO). While certification requires producer organizations to produce outputs in a socially just and sustainable way, it also requires initial buyers to support the process by providing up to 60 percent up-front credit where requested; engage in long-term and predictable relationships; pay minimum guaranteed and above market prices; and pay an additional social premium. Certification is available for 16 different product categories, most of which are food commodities.

Another system of Fair Trade certification was developed by the International Federation for Alternative Trade that was founded in 1989, renamed the World Fair Trade Organization (WFTO) in 2009. In contrast to FLO governance, this system certifies whole organizations and guarantees that 10 Principles of Fair Trade (including generating economic advantage for the poor, capacity building, providing fair prices and wages, gender equality, and environmental sustainability) are all upheld. As such, the WFTO mark is closer to the original ATO principles in that certification refers to a type of relationship, rather than the fulfillment of minimum requirements. For example, buying organizations must provide fair prices through dialogue and participation (instead of fixed minimums) and must more actively contribute to building business and community capacity (in place of additional financial payments).

While Fair Trade practice and certification have been associated with many positive outcomes for producer communities (such as higher income, more stable livelihoods, better health, and infrastructure), the benefits of Fair Trade are far from agreed. Many question the scope and efficiency of this mechanism for reducing poverty and facilitating development. As part of this argument, it is often stated that instead of spending extra money on commodity foods, consumers should provide more efficient assistance by providing direct resource transfers to poor communities. As part of this critique it is argued that FLO governance forces producers to adopt cooperative business structures that retard the development of good quality and promote exploitation and corruption, worsening the inefficiency of consumer investment. However, this argument assumes consumption to be a direct substitute for charitable giving and fails to recognize the wider benefits over and above the financial return that engagement in trade brings to marginalized communities. While all Fair Trade–ensured producers have the opportunity to learn better business practice by undertaking the process in a supportive environment, many Fair Trade companies work even more closely with groups in order to develop their capacity to generate secure livelihoods. In the case of the company Divine Chocolate in the UK, this process has even included the transfer of shares in Divine to the farmers growing cocoa to increase financial returns for the producers via dividends, as well as provide them with greater control over decisions made higher in the value chain. Supporters of Fair Trade also see cooperative organization as more democratic than conventional business structures with no more or less inherent propensity to deliver opportunities for corruption or waste. Furthermore, it is argued that cooperatives also increase the power of small producers to bargain with buyers over prices as well as collectively own infrastructure that would otherwise be unavailable to them.

Another allegation made particularly against FLO Fair Trade is that it continues to isolate the neediest producers due to stringent entry requirements and an arguably substantial financial cost for certification (while these costs were initially borne by buyers, producer organizations are now required to pay their own auditing and certification charges). In response, supporters draw attention to the activities of organizations such as TWIN trading in the UK that ready producers for involvement in Fair Trade. This need to facilitate the entry of the most marginalized is also being considered to be recognized by the FLO in their introduction of new contract standards that transfer the burden of development to third-party sponsors. A final defense is that while critics take costs in isolation, a more appropriate way to evaluate these expenses is in the context of potential future returns (just as any investment decision should be evaluated against its future benefits). When costs are seen in this way, it is argued, such investments can be seen as more than justified.

Having said this, another of the primary criticisms of Fair Trade has been that while certification applies to all output of an individual product type, there is no guarantee that producer groups will find markets for goods produced in this way. In some cases producers only manage to sell a few percent of their output as Fair Trade, and, as a result, are forced to commercialize the rest through conventional markets for conventional prices. It is on this basis that supporters argue for the necessity to invest in marketing and education as a means to increase consumer demand for Fair Trade goods in northern markets (an investment that is often derided by critics as a use of higher consumer spending that does not directly benefit producers, or that consumers motivated by the desire to help producers are necessarily aware of).

However, perhaps the most complex debate around Fair Trade is grounded in the criticism that support, and in particular guaranteed minimum prices, will discourage the diversification necessary if producers are to escape reliance on low income and unstable commodity markets. Critics argue that Fair Trade primarily operates for products established under colonialism, and that such a system blunts the incentives for resources to be invested in more profitable areas (it is often cited that wealthy countries have only achieved success by moving away from low-value agriculture and into higher-value manufacturing). While Fair Trade certainly focuses on commodity agricultural and handicraft goods, other analysis has questioned the assumptions made by highly theoretical economic analysis and argues that while critics compare Fair Trade governance to perfect free markets, this is not the choice facing producers and consumers in the real world. Instead, it is argued that Fair Trade can be beneficial to diversification, because although shifting to more manufacturing is desirable in the long run, many countries have little short-term option but to invest in agriculture as a means to accumulate the resources needed for this transformation.

Another criticism has been the decreasing producer control of governance systems, and one particular problem has been relatively static minimum prices despite rapid inflation in producer living costs. While producer representation and some prices have been increased by the FLO, some say that the response has been far from timely or adequate in its adjustment. This is particularly problematic when significant price premiums at the retail level are still lost to intermediaries with price increases at the farm gate remaining modest. Indeed, as Mexican producers have developed their own domestic Fair Trade certification—Comercio Justo México—they have included maximum markups for intermediaries and an insistence that direct purchasing be used wherever possible. Other supporters of Fair Trade, however, argue that this distribution of benefits should be expected with the mainstreaming of Fair Trade and that as the market has become more competitive, experience has shown that retail margins have fallen to reduce the cost to customers. Other studies suggest the same at the farmer end, where prices above the agreed minimums are paid for higher quality output.

A final area of discussion has been the degree to which Fair Trade defetishizes commodity products and reconnects producers and consumers in solidarity. While some have seen Fair Trade as contributing to greater reciprocal understanding, others highlight the commodification of poverty, the construction of a stereotypically romanticized poor and honest producer, and that the views of certified farmers do not always reflect the ideas of Fair Trade projected in consumer countries. One factor that constantly underlies the discussion of Fair Trade is the necessity to better understand both the processes and the outcomes of the system, and for this reason, debate remains very much open on the merits and shortcomings of the Fair Trade system.

See Also: Certification; Supply Chain Management; Sustainable Development; Voluntary Standards.

Further Readings

Hayes, Mark. "Fighting the Tide: Alternative Trade Organizations in the Era of Global Free Trade—A Comment." *World Development*, 36/12 (2008).

Le Mare, A. "The Impact of Fair Trade on Social and Economic Development: A Review of the Literature." *Geography Compass*, 2/6 (2008).

Raynolds, Laura T., Douglas Murray and John Wilkinson. *Fair Trade: The Challenges of Transforming Globalization*. London: Routledge, 2007.

Smith, Alastair M. "The Fair Trade Cup Is 'Two-Thirds Full' Not 'Two-Thirds Empty.'" ESRC Centre for Business Relationships, Accountability, Sustainability and Society. Cardiff, UK: Cardiff University, 2008.

Alastair M. Smith
ESRC Centre for Business Relationships,
Accountability, Sustainability & Society

GENERAL ELECTRIC (GE)

General Electric (GE) is one of the largest global business conglomerates, with 2007 revenue of $183 billion. It has over 300,000 employees in 100 countries, with 50 percent of employees outside the United States. The American company, founded by Thomas A. Edison, has operated for 120 years. GE is the only company listed in the Dow Jones Index that was in the original index in 1896.

For the 21st century, GE has adopted a new green philosophy, which more or less coincided with the ascension of Jeffrey R. Immelt as chief executive officer in 2001. Its green initiatives have intensified since 2005, when GE coined the term *ecomagination* for its various efforts. GE is now a major force in renewable energy with green initiatives, including solar and wind power; diesel–electric locomotives; lower emission aircraft engines; sodium-based batteries; energy-efficient lighting; water purification; biomass, hydro, geothermal, and other renewable power projects; and financing for other companies' green energy efforts.

GE has been named on the Dow Jones Sustainability World Index as one of the world's leaders in environmental, social, and economic programs, and applauded by Greenpeace for its commitment to incorporate Greenpeace's climate-friendly technology into refrigerators to be

Perhaps better known for making Energy Star–rated home appliances and compact fluorescent lighting, General Electric has long produced energy-efficient, wide-body aircraft engines, such as this GE90-115B model.

Source: GE

sold in the U.S. market. But it has also been awarded its share of dubious distinctions, placing number six on a list of U.S. companies ranked by releases of toxic chemicals in 2005 as compiled by the University of Massachusetts' Political Economy Research Institute (PERI).

Environmental watchdog groups frequently accuse the company of greenwashing, that is, using misleading advertising to project an unwarranted image as an environmental leader.

There is no question that over the years GE has had its share of environmental problems. It continues to manufacture pollution-generating, global-warming products (appliances, plastics, fossil-fuel-fired turbines, diesel locomotives), and to invest in the petroleum industry, nuclear energy and coal-fired plants. Nor can its past history of dumping 1.3 million pounds of polychlorinated biphenyls into the Hudson River in the 1960s and 1970s be ignored. The company's response to the continuing problems associated with their dumping has been widely criticized as inadequate and defensive.

The company faces ongoing pollution issues both at home and abroad. It released more than 4 million tons of toxics into the air, and more than 7 million tons of toxics into rivers and lakes in 2004. GE's website does not list toxic release amounts, but does provide metrics for the number of releases related to global operations—98 in 2007 and 70 in 2008. For those same years, they report fines of $236,000 and $96,000, respectively. GE corporate policy is to comply with all applicable laws.

The 2005 ecomagination campaign aimed to double revenues from environmentally friendly products. After reporting $14 billion in revenues from ecomagination products and services in 2007, GE increased its annual sales revenue target for 2010 from $20 billion to $25 billion. To expand the product line, GE reinvests ecomagination revenues in "cleantech" development at triple the rate of its research and development in general. GE began with 17 ecomagination products in 2005. Today, the portfolio exceeds 60 products and pervades every one of GE's businesses.

Renewable energy constitutes the Energy Financial Services (EFS) Division's fastest-growing business. EFS hopes to further tap into the $60 billion market by doubling its investment in EFS to $6 billion by 2010. The GE businesses more deeply vested in the initiative have performed correspondingly better—GE Infrastructure generates the lion's share of ecomagination products and in 2007, its revenues grew by 23 percent, earnings by 22 percent, and product orders by 26 percent. In 2008, EFS led all other GE business units by increasing profits 26 percent.

Some of GE innovations include GEnx engines, which use less fuel, generate fewer emissions, and are quieter than typical aircraft engines. Its Evolution series locomotives, capable of hauling heavy loads of freight, produce more horsepower while using less fuel and fewer emissions than their predecessors, according to GE. A relatively new line of business for GE is water purification, with engineers working on new filters that can strip impurities and salt from water more efficiently. In the past three years, the wind business, purchased from Enron for $358 million, has experienced sales growth of 300 percent. It operates supersized turbines for offshore wind farms in Ireland and elsewhere, and smaller windmills in China where energy demand is booming.

A major 2009 initiative is a "smart grid" technology to save power with a digitally enhanced power delivery system that allows more precise monitoring and control of electricity distribution. GE could earn up to $4 billion a year in revenue from smart grid technologies. The United States is barely generating enough energy to meet demand, which is predicted to grow 26 percent over the next 20 years. Power shortages are expected, leading to the need for not only new energy sources, but for a new method of distribution. The national energy grid was not designed to move power from one region to another. With smart grid technology, utility companies can better manage demand and receive return energy from customers who use alternative methods, such as solar energy. In 2007 and 2009, the U.S. Congress authorized a total of $5.5 billion for smart grid technology.

Abroad, GE is using what the company calls "reverse innovation" to rapidly accelerate growth in global and emerging markets. Reversing its traditional business model, GE is developing local technologies in these regions and then distributing them globally, rather than following the traditional path of developing high-end products and adapting them for emerging markets.

GE does not use life cycle cost analysis methods, but is a member of the Climate Action Partnership, which advocates strong government-mandated greenhouse gas (GHG) emissions reduction requirements. The company monitors and reports on some environmental indicators for its internal operations. The company set a goal to reduce its total GHG emissions 1 percent and to improve energy efficiency 30 percent by 2012. GE claims that given its projected growth, those emissions would otherwise have risen by 40 percent. GE also indicates that it is ahead of schedule, having reduced 2007 GHG emissions by about 8 percent from the 2004 baseline, but critics point out that these figures do not entirely reflect GE's pro rata share of the GHG emissions from the coal-fired power plants that it is invested in.

While GE may not be the greenest of the green, it appears its commitment to addressing the environmental challenges of the 21st century is at once honest and pragmatic. The stance taken on GHG emissions by its CEO was not universally cheered by stockholders or customers, but Immelt has persevered. GE continues, as it has since 1890, to both respond to and lead in the marketplace.

See Also: Appropriate Technology; Corporate Social Responsibility; Dow Jones Sustainability Index; Leadership in Green Business; Smart Grid; Superfund.

Further Readings

Greenpeace. "Greenpeace Congratulates GE on Plans to Bring Climate-Friendly Refrigeration to United States." Washington, D.C.: Greenpeace, October 2008. http://www.greenpeace.org/usa/press-center/releases2/greenpeace-congratulates-ge-on (Accessed January 20210).
Halpern, Stephen. TheStockAdvisors.com. http://www.bloggingstocks.com/2008/05/20/ecomagination-going-green-with-general-electric-ge (Accessed April 2009).
Kranhold, Kathryn. "GE's Environmental Push Hits Business Realities." *Wall Street Journal* (September 14, 2007).
Political Economy Research Institute. "The Toxic 100: The Top Corporate Air Polluters in the U.S." Amherst: University of Massachusetts, 2008. http://www.peri.umass.edu/toxic100 (Accessed January 2010).

Francine Cullari
University of Michigan–Flint

GENUINE PROGRESS INDICATOR

The Genuine Progress Indicator (GPI) is a "system of national accounts" (SNA) suggested as an alternative to the nearly universally used SNA, gross domestic product (GDP). SNAs are used to measure economic activity in a political unit such as a nation. Because GDP, the market value of all final goods and services produced in a country, is viewed as an indicator of economic health and well-being, it is closely watched by business, government, and the media. Critics of GDP argue that it is a flawed measure that undercounts or ignores certain contributors to economic well-being, and that it includes many forms of economic activity that harm societal and individual welfare. As such, these critics have attempted to promote the replacement of GDP with GPI or other SNAs.

Increasing economic growth as measured by GDP is a goal of politicians in almost every nation. Various policies are adopted in order to increase GDP, and potentially negative impacts on GDP are frequently cited as a reason to oppose policies aimed at improving environmental quality or other social goals. This often results in adopting policies that may maximize GDP, but suboptimize overall societal well-being. Flaws in GDP serve to blind the business sector to the negative impacts of traditional economic activity on the environment and reinforce the impression that environmental quality and economic growth are incompatible.

The GPI was developed in 1995 by Redefining Progress, a California think tank. GPI uses as its basis financial transactions that contribute well-being. These are then adjusted for economic elements that are not addressed within GDP. GPI is, therefore, much more a measure of economically driven well-being than of economic activity per se.

First, GPI adjusts personal consumption expenditures using a measure of income inequality. Dramatically increased wealth among the rich masks, within GDP, the impacts of stagnant or declining incomes among the poor and middle class. Second, GPI includes the value of nonmonetary work, such as parenting and volunteerism, as well as of services provided by past expenditures, such as infrastructure (dams, highways, streets, etc.) and consumer durables (cars, appliances, etc.). Third, GPI subtracts three groups of expenditures that are included in GDP: defensive expenditures that attempt to maintain household quality of life in the face of declines in social or environmental conditions (e.g., home security systems or guards, household water filtration systems, repairs due to auto accidents); depreciation of natural resources and other environmental assets (e.g., habitat loss, soil erosion, depletion of minerals); and social costs (e.g., divorce, losses to crime, loss of leisure time, pollution-related health and property damage). Finally, while GPI views investments in capital stock as contributors to well-being, it subtracts money borrowed from foreign sources. If foreign borrowing adds to capital stock, the effect on GPI is neutral, but if used to finance consumption, the effect is negative.

Given these modifications, GPI paints a very different picture of U.S. economic activity than does GDP. Although GPI is lower in both per capita and total terms than GDP (in year 2000 constant dollars), growth rates of the two measures were similar from 1950 to the mid-1960s in total terms and until about 1970 in per capita terms. A significant difference between growth rates—both total and per capita—began to occur in about 1975. Per capita GPI has been essentially stagnant or even declining since about 1978, when per capita GDP was $22,987 and per capita GPI was $14,595. By 2004, per capita GDP was $36,596, while per capita GPI was $15,036. It is clear that growth in GDP has served to hide stagnation in economic well-being for roughly two decades, assuming that GPI measures what it purports to.

While GPI is perhaps the most well-known alternative SNA, others have been proposed. The most notable of these is the Index of Sustainable Economic Welfare (ISEW). The economies of several nations including Austria, Chile, Germany, Italy, and the United Kingdom have been reanalyzed using the ISEW. GPI has been applied to the United States and Australia. The Asian kingdom of Bhutan, meanwhile, has drawn media attention for its attempt to deemphasize GDP in favor of Gross National Happiness (GNH).

A 2006 Redefining Progress report notes and responds to a number of specific criticisms that have been raised about GPI. A 2003 examination of the theoretical foundations of alternative SNAs such as GPI and ISEW conducted found them to be fundamentally sound, but needing further work on accurate valuation techniques in order to be politically acceptable.

While several alternatives to GDP have been proposed, little headway has been made in replacing GDP. Even were this widely viewed as desirable, it would be very difficult to do so without a transition period utilizing dual SNAs. Over a period of years, this would enable the two accounts to be compared and improvements made in the alternate SNA. Tentative steps have been made in this direction in Canada, which has measured progress toward achieving federal policy goals under its Environmental and Sustainable Development Indicators Initiative. Minnesota examined the creation of a Minnesota Progress Indicator in the late 1990s but did not enact it. The U.S. Commerce Department in 1994 considered adjusting GDP to reflect depletion of nonrenewable minerals, but dropped the initiative when members of Congress objected.

Were GPI or a similar SNA to be adopted as the official national measure of economic activity, implications for businesses would be indirect but profound. As impacts of proposed policies on GPI were analyzed, support would be provided for those advancing environmental protection and withheld from those promoting socially and environmentally damaging economic activity. This would likely lead to enacting new laws and regulations that collectively would require business to act much more sustainably.

See Also: Environmental Accounting; Environmental Indicators; Gross National Happiness; Steady State Economy.

Further Readings

Lawn, Philip A. "A Theoretical Foundation to Support the Index of Sustainable Economic Welfare (ISEW), Genuine Progress Indicator (GPI), and Other Related Indexes." *Ecological Economics*, 44/1 (2003).
Lawn, Philip and Matthew Clarke. *Measuring Genuine Progress: An Application of the Genuine Progress Indicator*. Hauppauge, NY: Nova Science Publishers, 2006.
Talberth, John, Clifford Cobb and Noah Slattery. "The Genuine Progress Indicator 2006: A Tool for Sustainable Development." http://www.rprogress.org/publications/2007/GPI%20 2006.pdf (Accessed June 2009).

Gordon P. Rands
Pamela J. Rands
Western Illinois University

GLOBAL REPORTING INITIATIVE

The Global Reporting Initiative (GRI) is the most relevant framework for voluntary reporting on sustainable development performance by business worldwide, making it a

Shell is one of over 1,000 companies worldwide that participated in editing a sustainable development report. Shell produces and sells biofuels, as at this station in São Paulo, Brazil.

Source: Shell

key element for green business. First conceptualized as a sustainability reporting framework under the auspices of the Boston-based CERES group, GRI became its own organizational entity in 2001 and moved its headquarters to Amsterdam. Partnering with the United Nations Environment Programme, GRI has since established global sustainability guidelines that have synergies with the United Nations Global Compact and the Earth Charter Initiative. In 2008, 1,044 companies worldwide edited a sustainable development report according to the GRI's guidelines, including firms such as Deloitte Touche (accounting and consulting), Tata (car industry), Sony (equipments), Shell (petroleum), and Novo Nordisk (pharmaceuticals). At the beginning of its development, the GRI was supposed to offer a "win-win" solution for sustainable development, enabling firms to know more about their environmental impact while providing societal groups with information on corporations' behavior. However, nowadays, the GRI is more a tool for sustainability, trademark management, and reputation building by companies than an instrument of control in the hands of environmental nongovernmental organizations. Considering the short period of time since its launch, there is no doubt that the GRI will continue to be used and developed for better corporate transparency on sustainable development objectives.

Initially, the creation of the GRI was based on three principles. The first one is that corporate reporting, as an attempt to provide data on industrial externalities, encourages good corporate behavior and, therefore, ameliorates economic results. The information contained in corporate reports is able to empower societal actors that demand appropriate sustainable performance from companies. The power of information disclosure is a key principle behind the GRI, acting as a potential instrument of civil regulation.

Second, at the time of its creation, the GRI wanted not only to enforce the logic of reporting for corporations worldwide, but also to innovate in proposing evaluation standards to measure business environmental impact, in the same way as financial performance was already subject to corporate reports. The second principle of GRI was to extend corporate reporting beyond financial performance, to render environmental and social objectives as visible as economic ones.

The third objective of GRI was to harmonize the field of corporate reporting. Indeed, many codes of conduct have been produced in the last two decades by individual firms, business associations, or intergovernmental organizations—one of the most well known being the Organisation for Economic Co-operation and Development's guidelines for multinational companies. In parallel to these initiatives, GRI's major contribution to the field of reporting was to propose a participatory, multistakeholder process for international guidelines in order to harmonize the confusing domain of corporate reporting.

Based on these three principles, historically, the GRI was in 1997 the result of the mobilization of economic actors, in particular, the Coalition for Environmentally Responsible Economies (CERES) in the United States. This coalition based in Boston, Massachusetts, gathered firms, investors, labor, environmental, religious, and public interest groups. The model proposed by CERES was based on the well-established U.S. financial reporting system. However, CERES sought to expand its scope from only social concerns to economic and environmental performance indicators in accordance with the triple bottom line principle, to give it a global impact, and to go deeper into its multistakeholder basis.

In 2000, CERES joined the United Nations Environment Programme to create GRI and publish the first sustainability reporting guidelines in June the same year. The first guidelines were used as a pilot document for a handful of companies. Immediately afterward, the GRI created the Measurement and the Revision Working Groups to assist in revising these guidelines. After their analysis and a multistakeholder process, a second version was presented at the Johannesburg Summit in August 2002. Many elements changed between the first version of the guidelines and the second, such as the number of indicators, their conceptualization, and the consideration of integrative indicators. Moreover, three principles have been chosen as the basis of the new reporting framework: transparency, inclusiveness, and audit.

Achievements and Critics

With regard to its achievements, since its modest beginnings, GRI has been in several ways a successful institutionalization project. It has contributed to improved transparency, rigor, and quality of data related to business sustainable development reports. The guidelines do provide a comprehensive framework for business sustainable development reporting divided into seven sections including the definition of report content and quality, indicators, and disclosure on management approach. One of GRI's greatest contributions has been to raise the awareness of clients for disclosure of industrial data, while making companies keener on collaborating for such information disclosure. In many cases, GRI's reports provide a starting point for further inquiry and dialogue with companies.

Despite its progressive development, critics have emerged concerning the formulation of the guidelines as well as the general functioning of GRI.

Regarding the text of the guidelines, the first problem observed is the lack of an explicit definition or reference to a definition of sustainable development. Sustainable development is indeed a rather vague notion that can encompass a broad range of objectives. The absence of definition of other important notions, such as the one of stakeholder, a notion central to the GRI, does also pose difficulties for defining the reporting boundaries. For instance, as a consequence, company approaches to stakeholder dialogue vary importantly from one company to another.

Moreover, the guidelines do not propose any precise implementation procedure, which allows for their partial application, leading to a soft approach to sustainability. Partial implementation of the guidelines indeed means that a company can focus on just one of the dimensions of sustainability. It also means that organizations can focus on those activities that provide a better reputation for their organization, while neglecting other important aspects of their externalities. The unbalance between the different categories of performance indicators proposed as well as the lack of integrated indicators exacerbates the imprecision of the guidelines. As a consequence, the data presented in business reports that follow the GRI's guidelines are somehow incomplete and complementary sources of information are often needed in order to picture the global sustainable development practices of companies.

Small and medium-sized enterprises have never participated in the GRI. This point is problematic when we consider that the GRI's success is measured according to the number of reporting organizations following the guidelines, mainly corporations, which means that business pressures represent a key aspect for the definition of the guidelines' content. The influence of economic actors in comparison to civil society organizations is reinforced by the fact that the initial support from foundations received by GRI has been nowadays replaced by mostly private sources.

A surprising explanation for the poor involvement of civil society actors, in parallel to resources constraints, is societal actors' lack of interest in the GRI's guidelines. Whereas these have been conceived primarily as a tool for corporate accountability, the information happened to be hard to use by activists. Indeed, the rather vague character of the data provided impedes strong societal control. As a result, the GRI's guidelines tend to become an element legitimizing the status quo and acting as a barrier to change.

This lack of involvement is susceptible to contaminating the business community as well. A kind of report routine is indeed sensitive among the users of the GRI's guidelines, and companies do poorly communicate among themselves on their reports. Just as do civil society organizations, the primary target audiences of reporting companies—financial institutions, employees, and shareholders—seem to show little interest in the reports.

These critiques are somehow constructive and lead to the possible development of solutions to improve the GRI. A sectorial version of the guidelines could be developed in order to ameliorate the precision of sustainable development indicators for each industrial sector. Regarding enforcement, the use of external examinations could be extended to all the users of the guidelines. To adopt these changes, the GRI's guidelines will have to continue to develop in the future by involving a wide range of stakeholders and disciplines in many countries around the world.

See Also: Ceres Principles; Corporate Social Responsibility; Environmental Assessment; Industrial Ecology; Triple Bottom Line.

Further Readings

DiPiazza, Samuel A. and Robert G. Eccles. *Building Public Trust: The Future of Corporate Reporting.* Hoboken, NJ: John Wiley & Sons, 2002.

Moneva, Jose M., Pablo Archel and Carmen Correa. "GRI and the Camouflaging of Corporate Unsustainability." *Accounting Forum,* 30 (2006).

Szejnwald Brown, Halina, Martin de Jong and David L. Levy. "Building Institutions Based on Information Disclosure: Lessons From GRI's Sustainability Reporting." *Journal of Cleaner Production,* 17 (2009).

Willis, Alan. "The Role of the Global Reporting Initiative's Sustainability Reporting Guidelines in the Social Screening of Investments." *Journal of Business Ethics,* 43/3 (2003).

Amandine J. Bled
Université Libre de Bruxelles

GLOBAL SULLIVAN PRINCIPLES

The Global Sullivan Principles (GSP) arguably constitute the best-known template for addressing social responsibility within a business organization's governance documents, codes of conduct, and operating procedures. Based on the United Nations (UN) Universal Declaration of Human Rights, GSP promote the adoption of just and equitable policies for all affected by an organization's activities. They extend corporate responsibility beyond mere respect for the law to responsibility for community well-being, which is a key sustainability practice. GSP take their name from Leon H. Sullivan, a civil rights leader who rose to international prominence by waging a campaign encouraging U.S. corporations operating in South Africa to stand in opposition to that country's oppressive racial policies.

Born in 1922 and raised in Charleston, South Carolina, Rev. Sullivan became pastor of Zion Baptist Church in Philadelphia in 1950. In 1964, he founded the Opportunities Industrialization Centers, a self-help training program that eventually spread to 76 centers in the United States, and 33 centers in 18 other countries. In 1971, Rev. Sullivan became the first African American member of the Board of Directors at General Motors. Early in his tenure, Rev. Sullivan pressured the corporation, then the single largest employer of blacks in South Africa, to place ethical and moral imperatives above the corporation's pecuniary interests. He urged GM to take a principled stand against South Africa's segregation laws, a system known as apartheid, by refusing to treat its black employees any differently than white ones. He had hoped to change the South African structure from within, but faced with an obdurate South African government and mounting public demand for government-imposed economic sanctions, GM and more than 100 other multinational corporations (MNCs) ceased doing business in South Africa until after apartheid was dismantled in 1991.

Rev. Sullivan unveiled the GSP in November 1999 at a special meeting of the United Nations called by then Secretary-General Kofi Annan. The GSP were the culmination of a three-year collaboration between Rev. Sullivan and a group of concerned international business leaders. They saw the need for a set of transnationally valid ethical precepts to guide corporate behavior in a global economy that is not governed by the laws of any one nation, but is held accountable in the forum of world opinion by transnational nongovernmental organizations (NGOs) and an unrelenting international media that closely monitors MNC operations in less-developed nations (LDNs). The objectives of the Global Sullivan Principles are (1) to show respect and support for human rights by treating employees, the communities where operations are conducted, and clients fairly and equitably; (2) to offer employees equal opportunities for improvement and advancement and refrain from exploiting vulnerable groups; (3) to respect employees' rights to freely associate; (4) to provide fair compensation to employees; (5) to safeguard workers' health and that of the community by engaging in sustainable development; (6) to act conscionably with respect to laws and intellectual property rights; (7) to positively contribute to communities where operating; and (8) to advocate the propagation of these principles.

The eight overarching principles point the way for companies to purse their legitimate business objectives in the global marketplace while operating in a socially responsible manner. The Leon H. Sullivan Foundation (LHSF) maintains a list of organizations that have made a commitment to operate under the GSP. Endorsing companies commit to upholding these standards in both their domestic and foreign operations. There is no fee for registering as a GSP endorser; a company simply completes the Endorser Registration Form online at the LHSF website. Once registered, an endorser may display the Global Sullivan Principles Endorser Seal on its own website and include it on letterhead, business cards, and other materials. LHSF requires each endorser to file an annual restatement of endorsement and a report on progress made toward achieving GSP-based goals. LHSF may remove an endorser if commitment to the GSP is not apparent in words or deeds.

The November 1999 introduction of the GSP was well timed, as there was growing public demand to impose a code of ethical conduct on business. That same month, mass demonstrations disrupted the World Trade Organization (WTO) meeting in Seattle, Washington—reactions emblematic of public disenchantment that had been growing for a number of years because of the increased reliance of American MNCs on moving operations offshore or outsourcing problematic operations to contractors in LDNs. The North American Free Trade Agreement (NAFTA) had been in effect for a decade, and critics charged that NAFTA gave U.S. manufacturers license to flee the restrictive laws of the United States for LDNs, where cheaper labor and lax environmental laws made operations more profitable. The GSP filled a void, and companies responded to public pressure to become endorsers.

The GSP are still relevant in 2010. Chevron and Shell face corporate dilemmas in their operations in Nigeria, not dissimilar from those companies that acquiesced to apartheid in South Africa. Amnesty International and environmental groups claim that the companies have caused significant environmental degradation of the Nigeria Delta region, and that they have not lived up to their human rights responsibilities, having allowed conditions of high unemployment and a lack of basic resources, such as fresh water and electricity, to persist.

Environmental Sustainability and the GSP

Many of the LDNs that attract outsourcing MNCs have no laws in place to protect the environment, which has caused transnational NGOs—an emerging global civil society—to view MNCs operating offshore plants with the hermeneutics of suspicion. Hence, it is not just coincidental that the GSP commit those endorsing its precepts to protect the environment in Principle #5, which addresses both protection of the human environment and promotion of sustainable development. Today, being a steward of the environment is integral to the notion of good corporate citizenship. Moreover, the international community is reaching a consensus about the universal applicability of certain sustainability guidelines; for example, the Ceres Principles, ISO 14000, GRI Sustainability Guidelines, and the UN Environmental Principles (UNEP). Each of these sustainability guidelines reflects uniformity of purpose, in terms of the environmental stewardship goals being advanced.

The GSP encourage companies to manage their environmental performance as carefully as they manage quality, product/market development, and customer relationships. But unlike the environment-specific guidelines, GSP is not a science-based system, and provides no metrics against which to measure accomplishment. Some analysts feel that the lack of metrics in all areas poses a systemic problem of GSP. At the least, the applications of GSP, whether in South Africa, Nigeria, or elsewhere, show the need for independent scrutiny. Rev. Sullivan sought to establish external reporting requirements, as well as a system for monitoring an endorser's adherence to the GSP; however, he died in April 2001, prior to accomplishing this goal.

See Also: Ceres Principles; Corporate Social Responsibility; Global Reporting Initiative; Voluntary Standards.

Further Readings

Alexis, Gwendolyn Yvonne. "Coming Home to Roost: Offshore Operations From an In-House Perspective." In *Controversies in International Corporate Responsibility*, J. Hooker, ed. *International Corporate Responsibility Series*, Vol. 3. Pittsburgh, PA: Carnegie Mellon University, 2007.

Gentile, Carmen. "Analysis: Chevron Taking Hit in Nigeria." United Press International, 2007. http://www.upi.com/Science_News/Resource-Wars/2007/05/09/Analysis-Chevron-taking-hit-in-Nigeria/UPI-26241178757663 (Accessed January 2010).

Sethi, S. Prakash Sethi and Oliver Williams. "Creating and Implementing Global Codes of Conduct: An Assessment of the Sullivan Principles as a Role Model for Developing International Codes of Conduct—Lessons Learned and Unlearned." *Business and Society Review*, 105/2 (2000).

Sullivan Foundation. "Global Sullivan Principles." http://www.thesullivanfoundation.org/gsp/default.asp (Accessed January 2010).

Gwendolyn Yvonne Alexis
Monmouth University

GREEN BUILDING

A green building is a structure designed to reduce its environmental impact as compared to a standard building of its type, based on the construction and operations of the building. This impact includes the use of resources such as energy, water, and materials, as well as the reduction of outputs such as storm water runoff, construction waste, and indoor emissions. Because of the holistic nature of this approach, green building projects generally require closer collaboration between designers and builders than traditional projects. Green buildings have grown rapidly since 2000 as a result of voluntary certification systems such as Leadership in Energy and Environmental Design (LEED) and Green Star. Because of their technological rather than social focus, these structures are also sometimes called high-performance buildings. Cutting-edge architecture is now focusing on producing carbon-neutral buildings, where a structure produces zero net carbon emissions.

According to the nonprofit U.S. Green Building Council (USGBC), buildings within the United States produce nearly 40 percent of all carbon dioxide emissions and 30 percent of waste output while consuming 72 percent of electricity and 40 percent of raw materials and 14 percent of potable water. In most cases, there is relatively little that building inhabitants can do to reduce these figures once the structure is in place. The responsibility for producing more environmentally friendly buildings is therefore placed on the shoulders of those who design and construct them. At the same time, there is tremendous potential for reducing resource consumption and waste production by better designing buildings and construction practices.

Buildings in the United States are the source of nearly 40 percent of U.S. carbon dioxide emissions and use 72 percent of the country's electricity. Efficient underfloor heating, as shown here, is one aspect of green building.

Source: iStockphoto

There is no single definition of what makes a building green. The LEED standards of the USGBC use six categories in their definition: sustainable sites, water efficiency, energy and atmosphere, materials and resources, indoor environmental quality, and design innovation. Similarly, the U.S. Environmental Protection Agency (EPA) includes energy efficiency, water efficiency, environmentally preferable materials, waste reduction, toxics reduction, indoor air quality, and smart growth in its explanation of what makes a building green. The idea is to go beyond simply using less water or energy, but to consider how multiple components of a building's design work together in a mini-ecosystem: the site and its relationship to surrounding networks of transportation, wildlife habitat, and urban space; the indoor environment, including emissions from paint and carpet; and the sources and sinks for the materials that go into the building and the waste that comes out. The complexity of a green building means that architects, engineers, and construction managers have to work closely together to achieve the goals of the project.

Globally, programs such as the Building Research Establishment Environmental Assessment Method (BREEAM) in the United Kingdom, Comprehensive Assessment System for Building Environmental Efficiency (CASBEE) in Japan, Green Star in New Zealand and Australia, and the Green Globes in North America serve a similar function as the USGBC's LEED. The LEED standards themselves have been adapted by over 40 countries, including Brazil, China, and India, while BREEAM International has been employed across Europe and Asia. The major differences between the systems are in how certification is approved (on-site inspection for BREEAM, documentation for LEED) and in percentage reductions from a baseline (LEED) as compared to quantitative targets (BREEAM). The World Green Building Council acts as a clearinghouse for these different systems, sharing information about building techniques, policies, and certification schemes.

Green Building Methods

There are three general approaches to greening buildings, each with different implications in terms of capital requirements, marketability, and effectiveness. The first involves relying on existing vernacular building practices and materials, such as adobe structures in desert climates to take advantage of large temperature fluctuations or raised floors in hot and humid climates to allow for air flow. This is a particularly important trend to encourage in developing countries, where recent large-scale projects have focused on looking as "modern" as possible, resulting in buildings that require extra energy or water to maintain. However, vernacular structures are generally smaller in size, and this approach is therefore largely restricted to homes rather than other types of buildings.

The second approach is high tech, relying on new materials and processes to reduce resource use and minimize waste. This incremental, technological approach has been criticized for making too small of steps toward sustainability and for maintaining "business as usual" rather than encouraging new consumption patterns. For example, reducing the production of carbon emissions from heating and cooling at a new building is desirable, but locating it on the edge of a metropolitan area, thus requiring extensive driving on the part of its inhabitants, can be considered counterproductive. Nevertheless, the technological approach dominates the industry at the present time, as seen in the term "high-performance buildings."

The third approach is biomimicry, which means imitating natural processes and using natural materials as much as possible. While similar to the high-tech approach in its incremental change from current practices, it espouses a more holistic mind-set toward building with nature, including the use of bioswales, green roofs, and other technologies designed to integrate the human and built environments.

Within these approaches, green building techniques and methods vary widely from the high tech to the low tech. Energy efficiency can be achieved through passive techniques such as taking advantage of the exposure of a site to the sun, reducing the need for winter heating or summer cooling, or through more active means such as incorporating alternative energy infrastructure like photovoltaic panels. Reducing water usage can be done via fixtures that use less water, incorporating native plants into the landscaping, or recycling storm water or gray water on-site. Reducing storm water runoff is often done as part of water recycling by developing green roofs or using bioswales or rain gardens. Waste reduction is emphasized not only for building users but for contractors; the traditional construction process produces large amounts of waste, and green building certification often depends in part on recycling or reducing that waste. The indoor environment can be improved via the use of low-emission paints, carpets, and sealants, among other materials, by letting in as much daylight as possible, and by allowing building occupants to control the temperature of their immediate surroundings. Finally, using environmentally responsible materials, such as certified-renewable wood, draws on other programs and organizations such as Green Seal, which incorporate their own criteria with regard to harvesting materials.

Aside from definitions and methods, there are also variations in the means by which buildings are designated as green. Most green building certification systems are point based, designed to be flexible, and to allow project owners to decide the level of certification and which points or credits they want to earn to achieve that level. This flexibility is one of the reasons green building standards have been widely embraced—because they allow a project to be tailored to its physical and economic environment. Various categories of certification focus on different building types such as schools, hospitals, and homes in order to broaden the range of categories and include special uses such as laboratories that may have unusual energy or water requirements. While buildings that achieve green certification are largely new construction, there are also programs in place to encourage the renovation of existing structures to meet green standards.

Reasons for Building Green

Despite the voluntary nature of green building guidelines, many governmental jurisdictions from the local to national levels are moving to encourage or require that certain types of development (e.g., public, private, or receiving public funds) achieve a specified level of certification for certain building sizes or uses (e.g., residential, retail, or government). Some jurisdictions use incentives such as expedited permitting or tax breaks rather than requirements, creating a patchwork of regulation from one place to the next. Others require that all public buildings, all buildings subsidized with public funds, or all buildings regardless of ownership, meet minimum standards. For example, the city of Los Angeles requires that all private development projects larger than 50,000 square feet earn the basic level of LEED certification, with expedited permitting for those projects that aim for the Silver level. This allows municipalities and other units of government to set a good example through their own activities, as well as encouraging local builders and designers to acquire the skills necessary to produce green buildings if they want to be eligible for projects.

There are a number of different motivations for building owners to go green, including financial as well as environmental. Studies within the United States have found a dual-edged misperception: green buildings are perceived to be considerably more expensive than conventional structures, while cost savings from energy efficiency are usually underestimated. Data on finished buildings indicate an average increase in costs of around 5 percent, which is usually earned back within a couple of years through increased energy and water savings. Additionally, employees have been shown to have fewer sick days and schoolchildren to have higher test scores when they work and learn inside certified green buildings. Other studies have found that not only do green buildings in operation use less energy as predicted, but they also bring in higher rental value (whether this is because of the green image or lower operating costs has yet to be determined).

In some locations, green building is increasingly tied to green-collar jobs: manufacturing and installing materials such as wind turbines, low-emissions paint, and photovoltaic panels. Particularly within the United States, concerns over the outsourcing of manufacturing and the concomitant loss of jobs for skilled laborers is part of the green-collar job movement. Green building standards contribute to this movement by, for example, designating credits for using materials produced within 500 miles of the building site, meant to encourage economic development as much as environmental responsibility. In particular, installation of green materials and renovation of existing structures cannot be outsourced in the same way that manufacturing or service-sector jobs can.

Criticisms of Green Building

There are a number of criticisms of green buildings, including a failure to incorporate life cycle analysis, although that is changing with revisions of the certification systems. In particular, the "cradle-to-grave" approach to environmental responsibility is being replaced with a "cradle-to-cradle" approach, especially given the relatively short life span of most new commercial buildings. Designing structures that can be disassembled and their components reused in 20 or 30 years is one of the challenges currently facing architects. Another criticism of green buildings is their perceived higher cost, although that is under debate as described above. Especially within the residential market, green buildings are sometimes seen as elitist and not compatible with affordable housing. However, government policies that encourage or mandate certification for public buildings go some way toward spreading the benefits of green buildings throughout a community. Finally, a third major criticism relates to epitomizing ecological modernization by continuing business as usual with a green facade. Although the certification programs make it easier to back up claims to be "green," the fact remains that the requirements only have to be met when construction of the building is finished, not as it is used by its inhabitants.

These concerns are being addressed through revision of the LEED standards and by developing public policies to reward builders for producing truly sustainable projects. Initial studies about the economic and ecological effects of existing green buildings in order verify that they really do have a reduced environmental impact without necessarily costing more. In the meantime, green buildings are gaining a greater share of the market worldwide and are at least in a small way reducing the ecological impact of the built environment.

See Also: Biomimicry; Certification; Cradle-to-Cradle; Green-Collar Jobs; Green Design; Green Technology; Voluntary Standards.

Further Readings

Kibert, Charles. *Sustainable Construction: Green Building Design and Delivery*. Hoboken, NJ: John Wiley & Sons, 2008.

Langdon, Davis. "Cost of Green Revisited: Reexamining the Feasibility and Cost Impact of Sustainable Design in the Light of Increased Market Adoption." http://www.davislangdon .com/upload/images/publications/USA/The%20Cost%20of%20Green%20Revisited.pdf (Accessed April 2009).

U.S. Green Building Council. http://www.usgbc.org (Accessed April 2009).

Yudelson, Jerry. *The Green Building Revolution*. Washington, D.C.: Island Press, 2008.

Julie Cidell
University of Illinois

GREEN CHEMISTRY

Green chemistry (GC), also known as benign or clean chemistry, refers to the field of chemistry dealing with synthesis, processing, and design of products and processes by following the principles of biological compatibility, renewability, biodegradability, and the use of closed-loop systems that reduce or eliminate the use and generation of hazardous substances. The focus is on minimizing the hazard and maximizing the efficiency of any chemical choice, using the precautionary principle. Hence, GC is an ideal solution for chemical pollution.

As a chemical philosophy, GC derives from organic chemistry, inorganic chemistry, biochemistry, analytical chemistry, and physical chemistry. However, in practice, green chemistry tends to focus more on industrial applications than on theory.

A plethora of terms were developed during the early years of GC research: clean chemistry, design for environment, benign by design, inherently safe, environmentally benign, and others. The revolution in engineering design and operation paved the way for a new branch of engineering and a new approach to pollution prevention called green chemistry that was a fusion of applied chemistry and traditional engineering. The term *green chemistry* was coined by Paul Anastas in 1991.

Since the late 1990s, GC has rapidly evolved in the field of chemical sciences, with strong support from major chemical companies, trade and professional associations, and the U.S. Environmental Protection Agency. In recent years, it has become widely accepted as a concept meant to influence education, research, and industrial practice. There are GC institutions, centers, and networks in over 20 countries around the world.

Overview of the GC Approach

Twelve guiding principles of GC were proposed in 1998 by Paul Anastas and John C. Warner, by which to assess the greenness of a chemical, a reaction, or a process. The tenets are (1) it is better to prevent waste or pollution than to have to remediate; (2) design synthetic methods to maximize the use of inputs (atom economy); (3) minimize both the use of toxic materials and the production of toxic products and by-products; (4) minimize the use of solvents or other auxiliary agents, or when these are essential to process, the most innocuous substances should be used; (5) minimize energy use (i.e., operate at ambient temperature and air pressure); (6) use renewable inputs whenever practicable and economically feasible; (7) avoid the use of derivative methods; (8) use catalytic rather than stoichiometric reagents; (9) constantly monitor for and proactively control the production of hazardous substances; (10) design with accident and incident prevention in mind; (11) product design should provide both effectiveness and minimized toxicity; and (12) product design should ensure innocuous degradation so that constituents will not accumulate in the environment.

Together, the 12 principles constitute the goal of GC: the design, manufacture, and use of chemical products to intentionally reduce or eliminate hazards and waste. A key approach to this goal is to prioritize the use of renewable and reusable materials, including agricultural waste and biomass as primary starting blocks. Chemical reactions with these materials are typically less hazardous than when conducted with petroleum products. Other principles focus on prevention of wastes, use of less potentially toxic chemical syntheses, and designing safer end products, including safer solvents. Today, many plastic products are made from plant sugars and renewable crops like corn, potatoes, and sugar beets instead of nonrenewable petroleum products. Unlike petroleum-based plastics, they are also biodegradable.

Many other current industrial applications fulfill several, if not all, of the 12 principles of GC at the same time. Several industrial processes use significantly fewer chemical inputs due to the use of supercritical carbon dioxide (computer chip manufacture) or condensed phase carbon dioxide (dry cleaning). The application of GC to processed wood products has eliminated the need for highly toxic arsenic and chromium. Large reductions in lead pollution followed the replacement of lead additives in paint with safe alternatives, and by the development of cleaner batteries. Polychlorinated dibenzodioxins and polychlorinated dibenzofurans have been greatly reduced in the pulp and paper industry by replacing chlorine with chlorine dioxide as the principal bleaching agent.

Several tools, techniques, software, and databases are used in this field. Some of the assessment tools include system analysis, process flow sheet, mass balance, life cycle analysis, economic analysis, regulatory analysis, and uncertainty analysis. Bioengineering is seen as a promising technique for achieving GC goals.

Future Challenges

Great challenges face the field of GC. First, chemists need to comprehensively incorporate environmental considerations into their decisions concerning the reactions and technologies to be developed in the laboratory. These questions need to become as important as those associated with the selectivity of the technology. Data confidentiality issues with industry and difficulty of data acquisition should be addressed properly. It must be ensured that chemical producers generate, distribute, and communicate information on chemical toxicity, ecotoxicity, uses, and other key data. Disclosure of hazard information will enable businesses, consumers, and policy makers to choose the alternatives that provide maximum protection of human health and the environment. This information should improve the prospects for businesses seeking to market GC alternatives. Hazard and tracking data together will help identify and prioritize substances of greatest concern. Addressing this challenge will require grounding in the insights, desires, and uncertainties of an interdisciplinary group of stakeholders.

Green chemists should place their activities within an industrial ecology context. Adding a life cycle perspective to GC enlarges its scope and enhances its environmental benefits. Many of the guidelines and tools developed for the evaluation of assembled products and of industrial processes can be adopted for use in GC as supplements to the manufacturing-oriented guidelines that already exist. Developing assessment tools for identifying suitable alternatives to chemicals of concern and designing standards and technical specifications will help to overcome specific technical barriers. Approaches for achieving GC have to be interdisciplinary, focusing on chemical creativity and innovation.

It is important to teach the values of GC to tomorrow's chemists. They should learn to assess hazards and to adopt more sustainable chemical practices throughout their academic and industrial careers. Industrial chemists must not only be knowledgeable about the manufacture of chemicals (with special reference to ecological and economic issues), but also must be aware of types of hazardous wastes generated during product synthesis, essential toxic substances that need to be handled by the workers during production, issues of regulatory compliance, and liability concerns that exist with the manufacture of the product.

GC is being taught in universities around the world, but it is still not integrated into the majority of chemistry curricula. As a result, graduates have little awareness of how the chemicals they design will affect human health and the environment. GC courses are becoming more available and more popular, but more needs to be done to speed up this process. Education in GC and sustainability can ensure a skilled workforce, and should be integrated across academic disciplines and included in the curriculum from elementary through graduate-level education.

Legislation and public policies can be important drivers for wider adoption of GC. Clean environmental goals can be achieved more quickly with active political will. A comprehensive chemicals policy should include information-based strategies, direct regulation, extended producer responsibility, technical assistance, market-based incentives, and public support for research and education. There should be flexibility in regulation to lease, recover, purify, and reuse processing chemicals. Government tools should be strengthened for identifying, prioritizing, and mitigating chemical hazards. Research and development, technical assistance, entrepreneurial activity, and education in GC science and technology should be provided by the government.

Although some leading businesses adopted sustainable practices, the vast potential of GC remains untapped. Partnership of all stakeholders to educate, discover, develop, apply, and promote GC should be developed in all its forms. Campaigns that demand greater transparency on product ingredients from product manufacturers should be promoted. Besides promoting free online newsletters on GC and demonstration projects of best business practice, attempts should be made to quantify the greenness of a chemical process as well as chemical yield, the price of reaction components, hardware demands, energy profile, and assessment of safety in handling chemicals, and ease of product workup and purification.

GC issues are here to stay. The most successful chemical companies of the future will be those that exploit its opportunities to their competitive advantage, and the most successful chemists of the future will be those who use GC concepts in research and development, innovation, and education.

See Also: Appropriate Technology; Best Available Control Technology; Clean Technology; Green Technology; Life Cycle Analysis; Precautionary Principle; Sustainable Design; Sustainable Development; Toxics Release Inventory; Waste Reduction.

Further Readings

Anastas, Paul T. and John C. Warner. *Green Chemistry Theory and Practice*. New York: Oxford University Press, 1998.
Cann, Michael C. and Marc E. Connelly. *Real-World Cases in Green Chemistry*. Washington, D.C.: American Chemical Society, 2000.
Ryan, Mary Ann and Michael Tinnesand, eds. *Introduction to Green Chemistry*. Washington, D.C.: American Chemical Society, 2002.
U.S. Environmental Protection Agency. "Twelve Principles of Green Chemistry." http://www.epa.gov/gcc/pubs/principles.html (Accessed January 2010).

Rasmi Patnaik
Gopalsamy Poyyamoli
Pondicherry University

GREEN-COLLAR JOBS

As business, government, and the community have sought to address a growing list of environmental issues, their efforts have created a new kind of work. Employment has traditionally been divided between the categories of "blue collar" (i.e., laboring) and "white collar" (i.e., administration). "Green collar" is a new category that has been created to describe laboring, semi-skilled, and trade jobs that contribute to improvements in environmental quality. These can be found in many sectors of the economy, for example, laborers reforesting a degraded landscape, farmers growing organic food, waste sorters in a recycling facility, builders of ecoefficient homes, pollution-reduction technicians, and many others.

Green-collar jobs generate three kinds of benefits: environmental, economic, and social. On the environmental side, they reduce the harmful side effects of economic development, help to repair damaged ecosystems, and improve the quality of the environment. In terms of the economy, they offer a new growth area for employment, income, and government revenue. They also provide the labor and skills that enable employers to make more efficient

Green collar jobs include labor, semiskilled, and trade jobs that contribute to improvements in environmental quality, such as these two builders installing rooftop solar panels.

Source: iStockphoto

use of resources and develop environmentally friendly products or services. In social terms, they offer an opportunity to lift people out of poverty by providing employment opportunities, they offer good wages and working conditions, they provide a career path, they reduce the public health impacts of production, and they improve public amenity. Green-collar jobs may involve new kinds of work in emerging green industries or they may develop from older occupations as established employers make the transition to more environmentally sustainable practices.

Historically, there have been many jobs that have had a positive impact on the environment. In pre-industrial Europe, for example, gamekeepers maintained natural habitats to support healthy populations of animals and birds. Since the industrial revolution and the rapid growth of cities, more specialized jobs have been created to improve local environments. Workers who began the construction of urban sewerage systems during the late 19th century are a case in point. The rapid spread of industrialization around the world in the second half of the 20th century accelerated the expansion in this kind of work. Increasingly strict environmental laws in the 1960s and 1970s, for example, led to a growing number of technicians being employed by industry to reduce pollution. In the 1980s, links were drawn between environmental, social, and economic problems, and sustainable development was proposed as a systematic solution. The idea was that if industrial development were undertaken with more care, it could still generate economic growth and social benefits (such as employment) as well as conserve a healthy environment for present and future generations. By the 1990s, it was becoming evident that actions to protect the environment could create new jobs and shift employment from older damaging industries into cleaner production. In 1999, Alan Durning coined the term *green-collar jobs* in a study of employment changes in the northwest of the United States. The concept has since been expanded by researchers, such as Raquel Pinderhughes, and community organizations, such as the Apollo Alliance, to cover other types of work and develop the social dimension in more detail.

Social Benefits of Green-Collar Jobs

Several studies of green-collar jobs have identified some unique social benefits. First, they tend to employ local labor because they often address localized issues (such as cleaning up a river, revegetating a piece of land, decontaminating a landfill, installing pollution filters on a factory, or retrofitting a building to make it more energy and water efficient). This feature makes the work difficult to export, although the equipment and materials used could still be manufactured elsewhere. Second, green-collar jobs tend to have low barriers to entry because they often involve manual labor. These two features combined provide an opportunity for residents of poorer, more environmentally degraded regions to gain local employment even if they have little experience or no qualifications. Third, once in the job, significant training is offered that can lead to career advancement. A new recruit who starts by collecting water samples, for example, might later be trained to take acidity and dissolved oxygen measurements. This may lead to a laboratory technician's certificate which, in turn, could count toward qualification in environmental chemistry or a job as a lab manager. The fourth point is that green-collar jobs tend to be secure and long term, particularly as demand for this work has been growing steadily over several decades in the public, private, and community sectors. Finally, these jobs often have better wages and conditions than equivalent blue-collar jobs. Green-collar employers are usually committed to the principles of corporate social responsibility and so they are prepared to treat their workers well. In a nutshell, green-collar jobs have major social benefits because they give people living in poor and highly degraded regions the opportunity to gain meaningful jobs and improve their local environment.

In macroeconomic terms, expanding and diversifying productive activity via green-collar jobs contributes to economic growth, generates more income, and adds to government revenues. Further, these jobs conserve, protect, and repair the natural capital base of the economy (i.e., natural resources such as fresh water, fertile soils, timber, plants for food and medicine, etc.). At the microeconomic level, they correct market failures by reducing the negative economic costs of environmental damage (known as externalities) for present and future generations. They also provide the labor and expertise needed to increase the economic efficiency of employers. A green-collar worker trained to conduct an ecoaudit of a manufacturing plant, for example, can identify improvements that reduce raw material, energy, and water use as well as cut pollution and waste. The manufacturer saves money because they spend less on inputs and waste disposal. These improvements reduce the costs of compliance with environmental laws because firms that implement ecoefficiency measures can more easily meet tighter emission standards. Less pollution and waste also reduces the risk that local residents might sue the company for the damage to their health or property. This, in turn, reduces the cost of commercial insurance. Firms that voluntarily improve their environmental performance also benefit from good public relations, can charge slightly higher prices for their products, and develop stronger customer loyalty. The market for organic food is a case in point. They also tend to find it easier to raise finance as ethical investment funds provide an extra source of capital that is not available to environmentally damaging firms. In short, green-collar jobs have a net economic benefit for employers and for society.

Obviously, green-collar jobs provide the benefit of improved environmental quality— one of the key features that differentiate them from other types of work. As the example used in the previous paragraph indicated, reducing the use of raw materials, energy, and water helps to conserve natural resources and improve air and water quality. The elimination of hazardous substances and the cleanup of contaminated waste sites is another potential area of work. Green-collar jobs may also protect biodiversity and repair ecosystems. A laborer working to replant vegetation in a wetland area, for example, is helping to improve water quality, reduce soil erosion, and create habitats for migrating birds and other species. The protection of environmental services provides further green-collar job opportunities. Inspectors locating and destroying chlorofluoro-carbons (CFCs), for example, prevent damage to the ozone layer that shields the planet from ultraviolet (UV) radiation. Alternatively, laborers helping to install wind genera-tors reduce the greenhouse gas emissions that disrupt the planet's climate regulation systems. All of these kinds of green-collar jobs help to improve the environment and make society more sustainable.

Growth of Green-Collar Jobs

There are several conditions that encourage the spread of green-collar jobs. National environmental policies, laws, and regulations create opportunities in government agencies as well as in the businesses and community organizations that have to comply. Local planning initiatives directed at urban renewal can encourage improvements in environmental quality, the development of green small businesses, and community engagement. Public funding for training programs can assist employees to improve their skills and encourage employers to get involved. Local community support and the willingness of potential workers to participate are important factors, but businesses also need to be willing to change what they do and to take on green-collar workers. Finally, active community organizations can encourage both government and business to create green-collar jobs. All these factors are significant.

It should be noted, however, that the concept does have its critics. One issue is where to draw the line between a green-collar job and a blue-collar job. If a company is manufacturing solar cells, the production-line workers can be classed as green-collar, but what about the people who clean the factory offices at night? Some critics have even argued that there is no such thing as a green-collar job, only blue-collar or white-collar workers. Is a mechanic a green-collar worker when he services a hybrid car, for example, but not when he services a conventional vehicle? The third problem is whether the whole concept might be used as greenwashing, where firms make token gestures while continuing to damage the environment. A company might fund a tree-planting program, for example, while still polluting. Finally, the focus on generating local employment creates some concern that this concept will be used to encourage protectionism and to reduce trade. There is a problem in defining the limits of green-collar jobs, and there is always the risk that the idea will be misused, but these issues are not unique or fatal. The divide between blue-collar and white-collar work, for example, is often not well defined. Does a blue-collar worker temporarily change to a white-collar worker when filling in as a production supervisor? Further, greenwashing and protectionism were issues long before the concept of green-collar jobs was ever developed and, while potentially damaging, are no more fatal than for other changes to the economy. They have not, for example, been able to prevent the rise of genuine green businesses and ethical investment funds around the world.

Finally, the idea of green-collar jobs has some significant theoretical underpinnings. The concept of sustainable development (mentioned earlier) advocated cleaner production and meeting the needs of the poor first, as well as intragenerational and intergenerational equity. The rise of the environmental justice movement raised issues about the equitable distribution of environmental problems, the need to clean up poorer communities, and the need to generate more benefits for low-income earners. Finally, the idea that the forces of the market can be redirected to produce benefits for people, the environment, and the economy has been explored by natural capitalism, industrial ecology, biomimicry, and ecological modernization. Together, these theoretical underpinnings provide a foundation for green-collar jobs as well as for the policies, plans, and programs that encourage them.

See Also: Corporate Social Responsibility; Ecoefficiency; Environmental Justice; Sustainable Development.

Further Readings

Apollo Alliance and Green for All With the Center for American Progress and Center on Wisconsin Strategy. "Green-Collar Jobs in America's Cities." 2008. http://www .apolloalliance.org/downloads/greencollarjobs.pdf (Accessed October 2009).

Durning, A. *Green Collar Jobs: Working in the New Northwest.* Seattle, WA: Northwest Environmental Watch, 1999.

Goodstein, E. *The Trade-Off Myth: Fact and Fiction About Jobs and the Environment.* Washington, D.C.: Island Press, 1999.

Hargroves, K. and M. Smith, eds. *The Natural Advantage of Nations: Business Opportunities, Innovation and Governance in the 21st Century.* London: Earthscan, 2005.

Hatfield-Dodds, S., G. Turner and H. Schendi. "Growing the Green Collar Economy." Commonwealth Scientific and Industrial Research Organisation, 2008. http://www.csiro .au/resources/GreenCollarReport.html (Accessed October 2009).

Hintenberger, F., I. Omann and A. Stocker. "Employment and Environment in a Sustainable Europe." Sustainable Europe Research Institute, 2002. http://www.seri.at/index.php? option=com_frontpage&Itemid=56 (Accessed October 2009).

Pinderhughes, R. "Green Collar Jobs." Report funded by the City of Berkeley Office of Energy and Sustainable Development, 2007. http://bss.sfsu.edu/raquelrp/documents/ v13FullReport.pdf (Accessed October 2009).

United Nations. "Agenda 21." New York: Division for Sustainable Development, 1992. http://www.un.org/esa/dsd/agenda21/index.shtml (Accessed October 2009).

United Nations Environment Programme, International Labour Organisation, International Organisation of Employers, International Trade Union Confederation. "Green Jobs: Towards Decent Work in a Sustainable, Low-Carbon World." Washington, D.C.: Worldwatch Institute, 2008. http://www.unep.org/labour_environment/features/greenjobs -report.asp (Accessed October 2009).

Michael Howes
Griffith University, Australia

GREEN DESIGN

Green design is design that is environmentally conscious and has the goal of sustainability. It can refer to the design of consumer goods and products, technologies, the built environment, or even systems and methods of doing things. Other terms used include ecodesign and sustainable design. Green design embraces the economic, social, and environmental impacts of producing and consuming a product with a view to minimizing its environmental impacts and nonrenewable resource use, and helping people live in harmony with nature.

In architecture, the goal of green design is to construct buildings with a reduced environmental impact, through careful selection of materials and a switch to natural and recycled building materials. The energy demands of the building may be reduced by adopting passive processes for heating and cooling, using solar-driven processes to heat interior spaces, using effective shading to prevent the building from overheating, while devices such

The Toyota Prius is a full hybrid electric car, introduced in Japan in 1997 and worldwide in 2001. Pictured here is an early model of the car in the Toyota Museum in Japan.

Source: Gnsin/Wikipedia

as Trombe walls may be used to create cool drafts when the building is too hot.

The environmental benefits of green design should be self-evident. These impacts are visible on a local level (i.e., the product doesn't pollute at the point of use, or the factory that produces the product doesn't have a negative impact on its local environment) and on a larger global scale. Well-designed green products help the human race as a whole transition into a more sustainable future, and do not exacerbate global-scale problems, such as climate change and ozone depletion.

In addition to the environmental improvements gained as a result of employing green design, businesses can also benefit economically by reducing costs as a result of more efficient processes. They can improve their efficiency, and reach out to the rapidly growing segment of the market that purchases environmentally responsible products. According to the Hartman Group, 75 percent of U.S. consumers prefer greener alternative products. In this way, it is often possible to achieve many synergistic "win-wins" through green design.

Many traditional products and technologies achieved a much better integration with the natural world than modern technologies do. Tools, transportation, and buildings evolved to fit the local conditions. For example, mud huts in the French Cameroon had a hemispherical shape that was the most efficient shape for keeping the insides cool. The vertical ribs holding the huts up also guided rainwater away, and could be used as scaffolding during construction, as well as for later repairs.

Design methods changed with the changing requirements of industrializing nations. The traditional evolution of forms was no longer developing quickly enough to keep up with the constant demand for new products in rapidly expanding economies. The design process was removed from the site of manufacture to the drawing board, where scale drawings were made. The designer now had a few hours on a drawing board to achieve what had once taken centuries of adaptation.

At the same time, economic considerations became all important. In some cases, economic considerations also serve environmental goals. For example, the minimization of materials used means resources are saved. However, they may be saved at the expense of the length of the operating life of a product.

Today, the environmental impact of a product or project is often an afterthought, leaving the environmental consequences to be dealt with subsequently, often by others. Environmental considerations are marginalized in the design process. The environmental considerations that should be considered at the design stage of every product and project—such as the choice of materials, layout and processes, and the implications that follow from these—remain neglected.

This treatment of environmental impacts would not be tolerated if it were applied to economic impacts. It would be akin to designing each part of a project without concern for how much the project was going to cost, and then at the end doing an economic impact statement to see whether it was going to be too expensive. Not only would this be inefficient and result in a needlessly expensive project, but any attempts at that stage to cut down on costs would interfere with the integrity of the design.

However, as public awareness of environmental degradation and imminent crises grows, as the cost of energy increases and resources such as clean water become scarcer, the need for green design becomes more compelling, and there are more opportunities for green businesses to profitably utilize green designs.

Designing for Whole Life Cycle

A key concept in green design is looking at a product's impact over the whole of its life cycle; that is to say, what environmental impact does producing the product have, how can we work toward minimizing requirements for packaging and transportation, what will be the effect of using the product in terms of resource and energy consumption and at the end of its useful operating life, and can the product profitably be repurposed, reused, and recycled.

Life cycle analysis, also called life cycle assessment, ecobalance, or cradle-to-grave analysis, analyzes the environmental impact of a designed product throughout the stages of its "life cycle," from when the product is created in a factory to when the product is disposed of. Life cycle analysis is integrated into international standards such as the ISO 14000 environmental management standards. It is also possible for a designer to look at a "section" of the product's life cycle, for example, "cradle to [factory] gate," in more detail.

Sustainable technologies can sometimes appear disproportionately expensive, compared to cheaper, dirtier technologies when capital costs are compared. However, to fully appreciate any cost advantage where present, it is important to appraise cost over the life cycle of the installation (and any subsequent decommissioning).

Designing Out Waste

The waste management hierarchy should be considered when designing green products. It arranges different categories for the handling and disposal of waste, in order from the most preferred environmental option to the least favored option. Starting with most preferred, the waste management hierarchy is as follows:

- Prevention of waste
- Minimization of waste
- Reuse of waste
- Recycling of waste
- Energy recovery from waste
- Disposal of waste

The waste management hierarchy impacts greatly on green design thinking. First, products should be designed in order to prevent waste. Wherever possible, products should be designed to have a long service life, fulfilling their function for as long as possible before replacement is required. As part of this drive comes integral quality and durability—ensuring that the product lasts, does not wear out, or require periodic replacement. This ensures that the maximum possible "value" is extracted from the input of materials and energy that were initially required to manufacture the product.

This is a hard objective to reconcile with commercial goals. In a society where consumers are often led by price, there is a temptation to deliver products at a lower price and that possibly require periodic replacement, either as a result of wearing out or designed-in obsolescence. However obsolescence can be designed out of the product, by making it in such a way that it is either "timeless," or so that it can be upgraded easily as new innovations permit.

Waste can then be minimized: this could be through minimization of packaging, or making the product physically smaller. It may be that extraneous material that is not filling a useful function can be removed from the design—however, this must be weighed carefully against the impacts that this will have on product durability.

Increasingly, products are being "designed for disassembly," this means that the product is designed so that it can be taken apart easily, and broken down into its constituent materials, permitting easy recycling.

Products should be designed in such a way that after they have finished their useful service life, they can either be easily repurposed for another application or easily recycled. An example would be a vehicle that is designed in such a way that when it is no longer operable, it could be returned to the factory, easily taken apart to leave its constituent components that could then be reconditioned and sold as parts; or where it is impossible to reuse a component, the parts of the component could be easily separated into different material streams, which could then be recycled.

By implementing "closed-loop" recycling schemes in products, the environmental impact of using certain materials can be drastically reduced, as in many cases the embodied energy content of a recycled material can be as little as 10 percent of the equivalent value for virgin material.

Cradle-to-cradle design challenges the conventional wisdom of green design, which says things are "recycled" at the end of their life, by arguing that many materials, in the process of being recycled are "degraded." A cradle-to-cradle design ethos aims to "improve" materials constantly by "upcycling" them into better and improved products. Upcycling is a term that was introduced by William McDonough and Michael Braungart in their book *Cradle to Cradle: Remaking the Way We Make Things*. They define upcycling as "the practice of taking something that is disposable and transforming it into something of greater use and value." This is an opportunity in green design—identifying product waste streams that will have harmful effect in the environment.

Designing for Minimal Energy Use

In creating products, we use materials that must be extracted and processed. These raw materials must then be machined and formed, processed and assembled into the final usable item; and in the process of doing so, energy is consumed and carbon emissions are generated. "Embodied Energy" and "Embodied Carbon" are concepts that can be used to appraise a product, whereby it can be examined in terms of "How much energy and carbon were needed to turn raw materials into this product?" In designing for the environment, designers aim to reduce the amount of energy and carbon "incorporated" in their products through the process of manufacture. One thing that is problematic is that there is no universally accepted standard for accounting for embodied energy and carbon; however, awareness of it as a concept is important to designing effective green products.

Green automotive design seeks to design vehicles that have a reduced impact on the environment. This may be achieved through vehicle weight reduction (resulting in improved fuel economy), improved aerodynamics, reducing vehicle drag, and by using new technologies, such as electric, hybrid, plug-in hybrid, or hydrogen drive trains.

Green electronics design seeks to reduce the power consumption of portable devices (a side effect of which is also that battery life, and hence the usability of a device, improves). In addition, electronics designers may look to renewably powering devices either using renewables or "human power."

Design Context

In addition to changing the physical configuration and composition of products, it is also possible to view design problems as "social," taking a bigger picture systemic look at the problem. For example, while it is possible to design automobiles that are more environmentally friendly, if we question the root cause of the environmental problem caused by automotive emissions, we see that we design built environments that necessitate the need to travel long distances by car, compounded by poor public transportation systems, that results in the need to consume such a large quantity of cars, which in turn results in air pollution.

Service substitution is a different design of product provision. Where the building of a company or business is designed in such a way that that business is selling a "product" to the consumer, the relationship is geared in such a way that the onus is on the supplier to sell more product to generate more profit. Service substitution recognizes that this loop creates a culture of consumption that is unsustainable. An alternative is to offer a "service" that the customer consumes according to their needs and pays a regular fee for.

Examples could include models of service substitution, where a model of "buying" a product, and replacing it periodically—for example, carpet or floor covering—can be replaced with a greener model of buying the service that is floor covering, with the service provider.

We live in an increasingly globalized world, where products are manufactured cheaply in developing economies and shipped many miles to wealthy consumers. Many question the sustainability and wisdom of practices that have become increasingly commonplace and believe that the discourse of "globalization" is open to challenge on the grounds of environmental and social sustainability. There is an associated environmental impact with shipping products long distances, and the "process" through which products are manufactured and supplied to consumers are as much a part of the design as the product materials selection or configuration.

When designing products, we have to confine our frame of reference to a narrow view of the world in order to process information and make appropriate design decisions. However, decisions made to reduce impact in one area may have unintended consequences in another.

As an example, take biodegradable bags, which are seen as a solution to traditional plastic bags that take a long time to decompose in landfill sites. Biodegradable bags from biopolymers were developed as an alternative that would reduce the problem of how long it takes plastic bags to degrade in the natural environment. However, they have the unintended consequence of releasing methane into the atmosphere as they degrade—with methane being 21 times more active as an agent of global warming than carbon dioxide. To try and reduce the possibility of unintended consequences occurring as a result of green design, it is necessary to take a broad and systemic view of the impact of products and services.

Barriers to Green Design

In some cases, we need to evolve and develop "new" knowledge on how to deal with the various environmental effects that result from use of different materials and chemicals. In other cases, approaches to green design can be learned from technologies that were once seen as obsolete, which can gain a new meaning when viewed through the lens of sustainability. To take an example from UK energy generation, before the advent of the national grid, there were many small hydroelectric power stations in Scotland, which produced electricity for local people or businesses. With the advent of the national grid, which was perceived as a more modern and efficient way to procure electricity, many of these sites were abandoned and seen as defunct. Now that the inherent inefficiencies in centralized generation and the impact of burning fossil fuels in power stations are known, there is a movement under way to return to these abandoned technologies as a clean solution for the future.

Another barrier to green design is the lack of effective incentives. There is no charge to use the world's resources that have long been perceived as the "commons" (land, air, and water). As there is no charge for the use of these resources, environmental degradation is accepted as an "externalized cost" of doing business. The cost of environmental impact has long been left out of corporate balance sheets, and therefore uptake of green design has been low.

Increasingly, there are systems in place to make businesses pay for these externalized costs, which are then passed on to consumers. In addition, the impact of a product goes beyond design and manufacture—once consumers have purchased a product, they must purchase the materials and energy to maintain that product and make it work.

The Future of Green Design

In evaluating green design, it is apparent that the trend is growing rapidly, driven by two forces: the first is environmental safeguards, and the second is the constant pressure to cut costs. Together, these forces are causing companies to accelerate their plans for green designed products, with pressure often coming from large retailers like Wal-Mart. Retailers are increasingly pushing suppliers for green products with reduced packaging, that are more energy efficient, and that have reduced toxicity. All this must be accomplished with minimal or no increase in prices. In the future, products that are badly designed and use resources inefficiently will cost the customer more to run and make the product less attractive. Introduction of labeling practices, which state, for example, product energy ratings or fuel consumption, will increasingly encourage consumers to make informed choices about the products they purchase, which ultimately will provide incentive for companies to design even more green products.

As time goes on, more far-reaching certification schemes will be set up requiring that products have reached given environmental criteria. Successful green design practices will increasingly be necessary to meet the requirements of these certification schemes.

See Also: Biomimicry; Certification; Clean Technology; Closed-Loop Supply Chain; Cradle-to-Cradle; Ecoefficiency; Ecolabels; Green Technology; ISO 14000; Life Cycle Analysis; Pollution Prevention; Sustainable Design; Upcycle; Waste Reduction.

Further Readings

Benyus, Janine M. *Biomimicry: Innovation Inspired by Nature*. New York: HarperCollins, 2002.

Hawken, Paul, Amory Lovins and Hunter L. Lovins. *Natural Capitalism: Creating the Next Industrial Revolution*. Boston, MA: Back Bay Books, 2008.

McDonough, William and Michael Braungart. *Cradle to Cradle: Remaking the Way We Make Things*. New York: North Point Press, 2002.

Walker, Stuart. *Sustainable by Design: Explorations in Theory and Practice*. London: Earthscan, 2006.

Williams, Daniel E., David W. Orr and Donald Watson. *Sustainable Design: Ecology, Architecture and Planning*. Hoboken, NJ: John Wiley & Sons, 2007.

Gavin Harper
Cardiff University

GREEN RETAILING

Having been described as the "big middle" of the value chain between producers and final consumers, retailing enjoys a powerful position from which to influence sustainability in two directions: backward up the supply chain and forward to customers and postconsumer processes. Green retailing is driven by ambitious goals, such as the following:

- Achieving zero waste in all its functions and facilities
- Using only renewable energy
- Carrying only products that are socially and environmentally sustainable
- Educating and facilitating customers in green consumption and postconsumption behaviors

Few, if any, retailers are truly green. Currently, however, retailing is undergoing a significant and sustained process of greening, working steadily and systematically toward the above goals. The impetus for change often begins with public relations concerns. Large, successful, and highly visible retailers, such as Wal-Mart and Nike, have often drawn intense criticism for their practices with respect to workers, communities, and the environment. Continued organizational greening requires directives and support from the highest levels of executive management. Typically, this is followed by concerted efforts to cultivate leaders within the organization and then to drive and reinforce cultural change through training and/or the inclusion of sustainability components in regular performance evaluations throughout the company.

Retailing, especially in a mass market, is highly competitive and cost driven. For this reason, greening actions often begin with changes that save the retailer money in the short term. Savings are then reinvested into innovative and longer-term projects with even greater impact. For example, by supplying its massive fleet of trucks with auxiliary power generators and instituting a no-idle policy, Wal-Mart was able, in the short term, to save millions of dollars in fuel costs. In the meantime, the corporation's transportation team began working with truck manufacturer Peterbilt to develop a hybrid-electric semi-truck tractor, which may yield even greater long-term savings in both fuel costs and carbon emissions.

Reducing waste is key to reducing costs while becoming sustainable. By eliminating waste to landfills, retailers are able to convert costs (e.g., trash disposal) into revenue streams (e.g., consolidated recyclables). Collaborating with suppliers is key to waste reduction in retail. For example, actively greening retailers may specify to suppliers that shipping and packing materials must be reduced wherever feasible and completely recyclable by stores or end consumers.

Using only renewable energy is a critical goal in the battle against greenhouse gases. Retailing is energy intensive, both in the operation of stores and in the distribution of product. In most of the world the supply of renewable power is inadequate to meet all retailers' needs. Greening retailers use multiple strategies to reduce their reliance on fossil fuels. The first and most immediately successful is energy conservation. Tactics such as recapturing heat waste from in-store cooking, optimizing the use of natural light, and adding glass doors on refrigeration cases can all save enormous amounts of both fossil fuels and money. Retailers with large facilities can also invest in power generation through solar, wind, or geothermal sources as their geographic locations and footprints allow.

Retailers employ a range of strategies for making their product lines more sustainable. Small-scale and specialty retailers, such as Whole Foods Markets, choose products selectively to appeal to particular target markets seeking the tangible and psychic benefits of greener products. Large-scale retailers, such as Wal-Mart, Target, Nike, and Kroger, have the channel power to exert significant pressure on their supply chains to provide greener product alternatives, which in turn can change entire industries at the infrastructural level. For example, Wal-Mart's practice of price guarantees for both organic and transitional cotton is accelerating change in worldwide cotton agriculture. Similarly, Nike's development and specification of water-based, as opposed to solvent-based, adhesives are changing the entire athletic shoe industry.

Because the expertise of many retailers lies more in logistics processes and merchandising than it does in manufacturing and product design, they tend to rely on strategic partnerships and alliances to further their sustainability goals. Since the turn of the millennium, there has been an unprecedented level of collaboration among retailers, their suppliers, and other groups such as environmental nongovernmental organizations, government regulatory agencies, and academic researchers. With diverse networks devoted to specific industries, operations, or product-line objectives, retail firms draw upon synergies to create innovation in sustainable practices. Innovation, in turn, drives competitive advantage by numerous means including cost reductions, new products, and access to new markets.

The other major piece of the green retailing puzzle is influencing and facilitating more sustainable purchasing, consumption, and postconsumption patterns among customers. As the main interface between manufacturers and users, retailers are in unique positions to provide consumer education and enrichment. In their media tool kits, they have external marketing communications as well as in-store merchandising tools such as signage, personal demonstrations, and product displays.

Retailers' sustainability goals and messages must necessarily vary according to customers' values, knowledge, and interests. For example, Portland, Oregon–based companies New Seasons (groceries) and Nau (sportswear) highlight the sources and the ecofriendly natures of their products in their in-store communication with customers who are already socially and environmentally proactive. Nike, on the other hand, stays true to its market position of superior athletic performance in virtually all its customer communication, allowing its green messaging to leak out only in secondary ways. Wal-Mart also maintains its price leader position with its core message: "Save Money. Live Better." In its attempts to encourage greener consumption, Wal-Mart connects its green messaging to deeper customer values, such as family health and thrifty living.

Retailers also are well placed to receive and consolidate certain kinds of postconsumer waste, such as plastic bags, bottles, cardboard, and polyesters. In some cases, retailers even become their suppliers' suppliers, turning postconsumer recyclables, such as plastic shopping bags and cardboard packaging, into raw materials for manufacturers of such products as durable decking materials and corrugated shipping pallets.

See Also: Closed-Loop Supply Chain; Corporate Social Responsibility; Ecoefficiency; Environmentally Preferable Purchasing; Supply Chain Management.

Further Readings

Arnould, Eric J. and E. J. Faulkner. "Animating the Big Middle." *Journal of Retailing*, 81/2 (2005).

Martin, Diane M. and John W. Schouten. "Engineering a Mainstream Market for Sustainability: Insights From Wal-Mart's Perfect Storm." In *Explorations in Consumer Culture Theory*, John F. Sherry, Jr. and Eileen Fisher, eds. London: Routledge, 2009.

Milian, Rudolph E. *The Retail Green Agenda: Sustainable Practices for Retailers and Shopping Centers*. New York: International Council of Shopping Centers, 2008.

Vargo, Stephen L. and Robert F. Lusch. "Evolving to a New Dominant Logic for Marketing." *Journal of Marketing*, 68 (June 2004).

Diane M. Martin
John W. Schouten
University of Portland

GREEN TECHNOLOGY

The extensive use of technology distinguishes the activities of humans and their civilization from other living organisms. Technologies comprise the tools human use to transform their environment, from those that shape their social organization, like markets and financial instruments, to the products they produce and use, such as automobiles and computers. While technology has improved human living standards over time (some more than others), it is also implicated in the deterioration of ecosystems, including many of those on which human civilization is dependent. Whether by adding carbon into the atmosphere, damming a major river, or releasing industrial chemicals into the soil, the evolution of human technologies has the raised the potential for increasingly destructive human relationships to the environment. As the human–environment relationship evolves toward greener values and practices, technologies can potentially be harnessed to make that relationship more sustainable.

Green technology describes those technologies applied to environmental problems or technologies that are more environmentally benign than conventional ones. The term itself is very broad, describing an array of technologies from those dealing with water treatment and purification, to those related to energy efficiency, renewable energy generation and storage. Green technologies are often referred to as clean technologies, which means technologies that strive to utilize or generate renewable energy, reduce resource or raw material use, and eliminate toxic materials and waste streams.

In 2009, investments in green technologies were in the tens of billions of dollars in venture capital and private equity, and in the hundreds of billions of dollars for government spending and tax incentives. The three largest green technology sectors as of 2010 are wind power, solar photovoltaics, and biofuels, all of which would fall into the broader renewable energy sector. As the global economy confronts issues ranging from climate change, aging energy infrastructure, water scarcity, and other environmental problems, green technology will continue to grow in the short term, especially if they are able to capitalize on synergies between government infrastructure spending and green jobs growth.

In the United States green technology investment is centered in Silicon Valley in Northern California. Companies headquartered there produce solar photovoltaics, concentrating solar thermal, LED lights, fuel cells, water filtration, recycling technologies, and electric cars. Many of the start up companies in these sectors utilize emerging nanotechnologies (solar PV, fuel cells, and water filtration, for example) and biotechnologies (biofuels or bioplastics). Other nations—such as Germany, Japan, China, and India—have pursued solar and silicon valleys, seeking to capitalize on synergies of clustering similar sectors regionally.

To explore green technology's key characteristics, it is important to describe an ideal conception of green technology. Barry Commoner's "Four Laws of Ecology" from his 1971 book, *The Closing Circle: Confronting the Environmental Crisis*, can be applied to green technology, especially when trying to make modern industrial systems resemble living ecosystems: (1) everything is connected to everything else; (2) everything must go somewhere; (3) nature knows best; and (4) there is no such thing as a free lunch. This utopian vision of green manufacturing looks to exploit industrial metabolisms and symbioses where the outputs of one metabolic activity become the inputs for others. This approach is often referred to as industrial ecology.

Because of the rising concern over electronic waste and products that pose end-of-life problems, interest in green technology has prompted cradle-to-grave or cradle-to-cradle product stewardship. One version of this is called extended producer responsibility, which requires that product manufacturers take back a product at the end of its useful life. This approach has been used in the electronics and white appliance industries where discarded products have caused problems from potential chemical leaching and fewer available landfills.

Various technologies can be enacted to reach these goals, such as design for environment (DFE)—designing products that incorporate features to make the product more easily recyclable, reusable, or more durable. Green chemistry—designing chemical products and processes that reduce or eliminate the use of hazardous substances—is another approach. Green chemistry seeks to reduce the toxicity of the chemical load on workers and the environment by adhering to the 12 principles embraced by the U.S. Environmental Protection Agency (EPA), including preventing waste, using renewable chemical feedstocks, reducing the energy requirements of chemical manufacture, designing chemicals to degrade more rapidly, using safer solvents, and making chemicals inherently more safe to prevent the possibility and consequences of accidents. Similarly, green engineering embraces reducing pollution at the source and minimizing risk to the environment through the use of new product design and processes. Green engineering utilizes tools such as life cycle assessment to understand a product's impact along the entire lifespan of the product, from raw material acquisition through manufacturing, use, and eventual disposal.

Areas of Green Technology Innovation

Green technology has been incorporated into a number of industrial sectors, including those related to energy, transportation, water, and materials. While many of these areas overlap, the most significant growth of green technology has been in the area of renewable energy generation, energy storage, energy efficiency, and transmission.

Solar photovoltaics create electricity from the sun, utilizing the photovoltaic effect, where incoming photons displace electrons in the top semiconductor layer and migrate to the opposite pole, helping to create an electrical current. This technology receives more investment than any other current green technology. Government policy has stimulated the industry by advancing tax incentives for both consumers and manufacturers. Growth in solar photovoltaics has also been driven by new technologies, novel semiconductor materials and manufacturing processes, and economies of scale, which have significantly reduced their cost.

Wind power technologies are another major green technology sector. Of the renewable energy technologies, large-scale wind power is coming the closest to achieving what is known as grid parity, where the price of the renewable energy is cost competitive with conventional energy sources. Along with issues related to aesthetics and wildlife impacts, wind power also suffers from being an intermittent energy source, meaning that is cannot provide consistent power to the grid without energy storage. Micro-wind turbines have also become popular for homeowners in windy regions.

Solar thermal hot water heating is an older green technology that was used in its current form as early as the 1880s. Solar thermal hot water uses the sun's heat to warm water by heating it directly or through a heat exchanger. It is considered one of the more economically and energy efficient renewable energy technologies.

Solar thermal concentrating electricity generators come in a number of forms, most of which utilize mirrors or heliostats to concentrate the sun's heat onto a surface that can transfer heat to water and generate steam to turn a turbine and produce electricity. A stirling engine, otherwise known as an external combustion engine, uses concentrated solar energy, focused on a parabolic mirror that drives a piston in a chamber that is able to take advantage of the temperature difference.

Energy storage technologies include batteries and fuel cells. Advances in batteries— increased capacity, cheaper, and lighter weights—will be critical to the success of electric and hybrid motor vehicles. Fuel cells can also be used to power homes and commercial buildings. Silicon Valley–based venture capital startup Bloom Energy is piloting small fuel cell power plants at buildings owned by companies like Staples and Wal-Mart. These fuel cells are typically hydrogen-based, and work by electrolyzing hydrogen and oxygen into water and energy. Other energy storage technologies include pumped hydroelectric, where water is pumped uphill by, for example, wind or solar energy, and released when energy from the intermittent source is not available (i.e., when the sun goes down or the wind stops blowing). For solar energy technologies, using molten salts to store energy captured during the day is being explored as a potential solution to the problem of intermittency. Biofuels and bioenergy are also being explored as green technologies to fuel transportation. Biofuels investments typically rank in the top three of the green technology sectors. Although biofuels have recently been criticized for not having the carbon benefits described by proponents, advances in cellulosic biofuel technologies and waste-sourced biofuels will likely advance the industry, but it remains unclear what role these energy crops will play in the future.

There are a number of water-related green technology sectors. New technologies for water treatment and purification enable the reuse of nonpotable water and allow water to be reclaimed where it otherwise would not have been usable, even for agriculture. The oldest of these technologies are rainwater catchment systems, collecting rainfall from rooftops or similar structures and storing the water in a cistern, dam, or other impoundment. Energy efficient desalination technologies are receiving significant attention in light of water scarcity concerns, particularly in desert or coastal areas. Increasing the amount of available drinking water can also be achieved with new technologies like ultraviolet light filtration, reverse osmosis, and other membranes, some of which employ advances in nanotechnology.

Remediation is another area of green technology that emerged, both from growing public interest or from a legal directive to clean up contaminated lands. Conventional remediation technologies include carbon filters, chemical treatment, thermal oxidation, ion exchangers, or soil vapor extraction. Many of these technologies are green only in that they clean contaminated sites, and are actually conventional technologies. Greener forms of remediation include phytoremediation and mycoremediation. Phytoremediation utilizes plant biology to remove heavy metals or destroy other contamination found in contaminated soil. Mycoremediation is similar, but utilizes fungi to accomplish goals of sequestered toxins or breaking down hydrocarbons or other organic chemicals contaminating the soil.

Another area of green technological development involves contruction, which consume a significant portion of the energy in the industrialized world. Standards for green building, such as those set by LEED, encourage architects to incorporate design features and materials conducive for minimizing waste and consumption in buildings. This has also fostered the growth of green construction materials, such as low-CO_2 cement, bioplastic materials, insulated window glass, and building-integrated photovoltaics, such as solar shingles.

Technologies are not just physical objects and practices, but also include concepts and metrics that characterize impact, such as carbon footprint or food miles. Metrics are tools that can be used to collect information about social or environmental dimensions of a technology, or to create markets based on emissions permits, such as those for carbon or sulfur dioxide. Policy instruments, like taxes breaks (or taxes themselves), are also considered technologies.

Challenges Facing Green Technology

As with any technology, green technology often becomes entangled with issues of policy and politics. In the social sciences, the push for green technology is characterized as the central motivation of ecological modernization. Taken literally, the term can basically be defined as making modern industrial society more ecologically minded. However, the term is more commonly used to describe an approach with no inherent conflicts between industry interests and the environment. Because ecological modernization sees cooperation with industry as the best approach, it is often associated with voluntary or market-based environmental policy approaches. The main criticism of ecological modernization is that is fails to effectively deal with the concept of sustainable growth, which according to some is an oxymoron. A second criticism is that it does not effectively deal with the power of industrial interests. These critics assume the impulses of capitalism overwhelm any fundamental transformations towards a sustainable society, and instead only offer reformist, marginal solutions to environmental problems.

A major challenge as we embrace green technology will be gaining knowledge of exactly how "green" touted technologies are, which will involve relying on the emerging practioners of green design, chemistry, engineering, life cycle analysis, environmental justice, environmental policy, and other allied interests.

See Also: Design for Environment; Green Design; Greenwashing; Resource Management; Sustainable Design.

Further Readings

Ayres, R. U. and U. E. Simonis. *Industrial Metabolism: Restructuring for Sustainable Development.* Tokyo: United Nations University Press, 1994.
Benyus, J. *Biomimicry: Innovation Inspired by Nature.* New York: Morrow, 1998.
Commoner, B. *The Closing Circle: Confronting the Environmental Crisis.* New York: Knopf, 1971.
Hajer, M. *The Politics of Environmental Discourse: Ecological Modernization and the Policy Process.* Oxford, UK: Oxford University Press, 1995.
McDonough, W. and M. Braungart. *Cradle to Cradle: Remaking the Way We Make Things.* New York: North Point Press, 2002.

Dustin Mulvaney
University of California, Santa Cruz

GREENWASHING

Greenwashing is a term used to describe businesses, organizations, and individuals who exaggerate or make false claims about the environmental benefits of their products and services. In short, greenwashing is an unjustified claim of environmental virtue. While such accusations are now widespread, qualifying to what degree communications have intentionally distorted environmental harm or benefit can be difficult, especially when judgments are often subject to interpretation. This can include accusations of greenwashing against those who have taken action to improve environmental performance, but are perceived to have not done enough. These difficulties may have delayed legislation on greenwashing issues, although both public and private governance is now emerging.

There are several different forms of greenwashing, including providing false information about environmental attributes, exaggerated claims of sustainability, taking credit for environmental improvements mandated by outside influences, and deceptive claims based on inappropriate comparisons to outdated technologies.

According to some, greenwashing has been confined to instances where environmental claims have not actually been false, but have failed to present the entire environmental picture. However, recent research has uncovered a category of greenwashing where claims of environmental credibility were adjudged to be false and deceitful. These examples vary from claims that certain products have certification from recognized independent bodies when they did not (for example, shampoos that claimed to be "certified organic"), to those where marketing overrepresented the environmental advantages of a given characteristic. A number of these

Some would argue that this Poland Spring company web page is an example of greenwashing. The company touts an improved "ecoshaped" bottle with a smaller paper label to contribute less volume to the waste stream. Nevertheless, an estimated 80 percent of all bottled-water containers still end up in landfills.

Source: polandspring.com

cases have been highlighted in the car industry in the United Kingdom (UK). Suzuki GB released direct mail marketing stating that cars have "great green credentials" and are "guilt-free," featuring graphics of butterflies, blue skies, flowers, and a rainbow. The UK government's Advertising Standards Agency felt that although the advertising implied that the car caused little or no harm to the environment, this fact was not substantively justifiable. In a similar case, the Norwegian government has banned companies from claiming that any vehicle is "green," "clean," or "environmentally friendly."

The growing realization that being "green" can be profitable for a business leads to a second form of greenwashing: exaggerating a product or service's sustainability or green features. Many companies have started to identify, after the fact, features of current products as being environmentally sound while no fundamental or physical change has been made. A subcategory of this type of greenwashing occurs when a company touts that changes to a product or process were environmentally motivated, when they were in fact motivated by financial gain. For example, in what many believe to be the first use of the term *greenwashing*, Jay Westerveld noted that hoteliers' requests for customers to reuse their towels and linens were not motivated out of concern for the environment as they claimed, but out of the desire to cut costs and increase profits. Indeed, this type of activity has been noticeably widespread, as companies have rolled out new "green" programs for energy and material inputs, packaging, and logistics that in fact save them money.

A third form of greenwashing involves implementing mandated environmental standards and improvements while taking credit for green behavior. For example, many companies have been accused of opportunistically marketing their withdrawal of legally prohibited chlorofluorocarbons (CFCs) as a customer benefit. Another example involves the American Coalition for Clean Coal Electricity (ACCCE), which has been criticized for its claim that modern coal power plants are "70 percent cleaner," even though this refers to emissions of sulfur dioxide and nitrogen oxides that has been under legislation for years, and not to the emissions of carbon dioxide associated with climate change. It can also involve business interests that have projected an image of being environmentally proactive, while simultaneously lobbying against further legislation. An example of this would be the car industry promoting clean cars while pressing regulators not to enact stricter emissions laws.

Greenwashing can also occur when a company compares their product or service to outdated or unrelated technologies so that they appear to be the "greener" consumer choice. For example, the benefits of gasoline–electric hybrid cars can be overstated by comparing their emissions to more powerful vehicles, outdated emissions standards, or to more powerful versions of a similar type of car. Another example can be found in the claim made by a commercial airline that they are "flying greener," and that their fleet performs at 55 passenger-miles per gallon, fewer miles per gallon than a hybrid car transporting one person. The misleading portion of this statement is that it would not be feasible for one person to drive a car thousands of miles for a short vacation, a feat only made possible by flying.

The phenomenon of "green marketing" arose on a large scale in the 1980s in response to societal concern over unsustainable business practices. While accusations of greenwashing initially came from individuals and nongovernmental organizations (NGOs) on specific issues, there have also been institutional efforts to expose misleading information. For example, at the 2002 Earth Summit in Johannesburg, a group of NGOs held a Greenwash Academy Awards event to identify the most misleading examples, and there are now various websites enabling users to input and then vote on examples of greenwashing. However, just as the concept of sustainability raises the question of what it to be sustained, by whom, and for how long (given that infinite sustainability of the global economy is obviously physically impossible), accusations of greenwashing often raise contextually specific and subjective issues.

As seen in the previous examples, despite often being grounded in subjective interpretation, the concept of greenwashing has become a widely used tool of protest against companies that are not considered to be representing their relationship with the environmental in an open and honest way: the ultimate goal being to prevent misleading claims and prevent genuine claims from being crowded out. However, there is evidence to suggest that while dishonest activity is being curtailed, the effects are not always positive. For example, while companies that try to promote genuine gains can be criticized for misrepresentation, those that do nothing and remain quiet often attract far less attention. Indeed, some marketers are now avoiding promoting environmental achievements, due to fears that stakeholders might view the activity as self-serving or as an attempt to misrepresent the organization.

To some, this means that the protest against greenwashing has not achieved its goal, and that despite the current emphasis on market solutions to environmental problems, customers will be given less information to differentiate between environmentally friendly and harmful operations. However, now that governments in the UK and the United States are more actively recognizing the problems of greenwashing in monitoring marketing claims, perhaps the green calls for change might still be effective.

See Also: Blended Value; Corporate Social Responsibility; Ecological Footprint; Sustainable Development.

Further Readings

Beder, Sharon. *Global Spin: The Corporate Assault on Environmentalism*. White River Junction, VT: Chelsea Green, 1998.

Greenpeace. "Greenwashing." http://stopgreenwash.org (Accessed January 2010).

Greenwashing Index. http://www.greenwashingindex.com (Accessed January 2010).

Laufer, William. "Social Accountability and Corporate Greenwashing." *Journal of Business Ethics*, 43/1 (2003).

Lotzia, Emerson and Benjamin Sykes. "MisLEEDing: Examples of 'Greenwashing' With LEED." *Florida Real Estate Journal* (October 15, 2009). http://www.frej.net/news/green-report/2009-10-15/misleeding-examples-%E2%80%98greenwashing%E2%80%99-leed (Accessed January 2010).

Lubbers, Eveline. *Battling Big Business: Countering Greenwash, Front Groups and Other Forms of Corporate Deception.* Monroe, ME: Common Courage Press, 2002.

McGinn, Daniel. "The Greenest Big Companies in America." *Newsweek* (September 21, 2009). http://www.newsweek.com/id/215577 (Accessed January 2010).

Peattie, Ken and Andrew Crane. "Green Marketing: Legend, Myth, Farce or Prophesy?" *Qualitative Market Research: An International Journal*, 8/4 (2005).

Tokar, Brian. *Earth for Sale: Reclaiming Ecology in the Age of Corporate Greenwash.* Cambridge, MA: South End Press, 1999.

Walsh, Bryan. "Eco-Buyer Beware: Green Can Be Deceiving." *Time* (September 11, 2008). http://www.time.com/time/magazine/article/0,9171,1840562,00.html (Accessed January 2010).

Alastair M. Smith
ESRC Centre for Business Relationships,
Accountability, Sustainability & Society

GROSS NATIONAL HAPPINESS

Gross National Happiness (GNH) is one of a number of indices that have been devised to measure national quality of life by taking such things as physical well-being, safety, income equality, and access to information into account. Because GNH tries to capture the effects of what are essentially intangible attributes, the major challenge has been determining how best to measure the effects. The effort to refine a GNH model has been criticized because most of the data inputs derive from responses to surveys that are highly subjective. Proponents defend the GNH approach on the basis that the gross national product (GNP), though frequently touted as a measure of national well-being, was never intended for that purpose and is inadequate for the task. The proxies it assumes for well-being (i.e., income, income security, and income equality) have been shown to be unrelated or very weakly related to individual happiness. They maintain that a measure of GNH will prove a powerful tool particularly in examining how public policy will affect the welfare of citizens. Research around GNH is discovering what things contribute to happiness, and that information will be valuable not only to policy makers but to businesses as well in the quest for sustainability.

The term *Gross Domestic Happiness* initially emerged in 1972 as the guiding philosophical principle of the Kingdom of Bhutan in south Asia. In its original context, GDH was not intended as a measure, but was a statement of policy. Rather than focusing on the monetary value of material products, Bhutan indicated that the general well-being of its populace was of the utmost importance.

The effort to find a measure of national well-being predates the coining of the term *Gross Domestic Happiness* by at least a decade, having originated from discussion about limits to growth that accompanied the development of the discipline of ecology during the

1960s. Scholars began to discuss what a postmaterialistic era would look like, and pondered what measures could be devised that would reflect the whole of the economy in a way that GNP does not. The problem, as they saw it, was that GNP only accounts for the value of goods and services that flow between producers and consumers in established markets. Any activity conducted in informal markets (e.g., housework, gardening, or barter exchanges) were excluded, as were accounts of social capital, human capital, and ecosystem services. The search for better measures took different approaches. Some researchers took their cue from ecology. Others started to search for ways to measure quality of life, holding, as Robert F. Kennedy stated in a 1968 speech at the University of Kansas, that GNP "measures everything, in short, except that which makes life worthwhile." The effort to establish a quality-of-life measure has benefited greatly from the advances in survey research methods made during the 1990s and 2000s. Highly sophisticated statistical approaches have allowed researchers to move beyond criticisms that their research rested on subjective responses and vague, self-rated measures of health or satisfaction. Even Bhutan is now developing a measure, having established the Centre for Bhutan Studies in 2008.

The items being included in GNH models can be generally sorted into four categories: (1) living conditions, which includes the ambient conditions measured by climatological and environmental characteristics, location relative to amenities, sanitary conditions, and the like; (2) life prospects, which includes such indicators as health, status considerations, access to resources that facilitate advancement or adaptation to changing conditions; (3) life purpose, which includes attitudes about the meaning of life and one's place in the scheme of things; and (4) emotional status, which encompasses satisfactions and dissatisfactions with life as a whole.

Some have argued that many of the measures used to identify a healthy society or nation are predominantly indicators of Western standards. While trust, fairness, competence, equality, knowledge, shared belief, openness, empowerment, and innovation are arguably all positive values, the degree of engagement with each one and the relative merit assigned may vary according to context. The recognition of the importance of cultural "nuancing" and the context in which these metrics are being applied is important for the validity of the results.

Models of GNH have found wide differences between countries, but contrary to expectations of many, the disparities show up among countries with seemingly similar attitudes regarding income. Data from the World Database of Happiness, which comprises data from several independent studies, reveal no correlation between income and national level of happiness. Though they enjoy approximately the same per capita income, Germans are apparently less happy than their Austrian neighbors.

As the necessity to adopt environmentally sustainable economies has become increasingly apparent and urgent, the development of measures, such as a GNH index, has received wider attention and is moving into the mainstream of economics. The Organisation for Economic Co-operation and Development (OECD) has focused a great deal of attention on the issue, and the European Commission has launched an initiative titled "Beyond GDP." These moves are positive recognition that measuring GHP will in fact help governments shape equitable societies and sustainable economies. It will also help businesses by revealing what really matters to their customers and employees, information that will help them not only provide better goods and services, but also build and maintain a happy, productive workforce.

See Also: Corporate Social Responsibility; National Priorities List; Superfund; Sustainable Development.

Further Readings

Anielski, M. *The Economics of Happiness: Building Genuine Wealth*. Gabriola Island, British Columbia, Canada: New Society Publishers, 2007.

Costanza, R., M. Hart, S. Posner and J. Talberth. "Beyond GDP: The Need for New Measures of Progress." The Pardee Papers, No. 4. Boston, MA: Boston University, The Frederick S. Pardee Centre for the Study of the Longer Range Future, 2009. http://vip2.uvm.edu/~gundiee/publications/Pardee_Paper_4_Beyond_GDP.pdf (Accessed January 2010).

Hofstede, G. *Culture's Consequences: Comparing Values, Behaviors, Institutions, and Organisations Across Nations*. Thousand Oaks, CA: Sage, 2003.

Veenhoven, Rutt. "Measures of Gross National Happiness. Statistics, Knowledge and Policy: Measuring and Fostering the Progress of Societies." Paris: Organisation for Economic Co-operation and Development (OECD), 2007. http://213.253.134.43/oecd/pdfs/browseit/3008081E.PDF (Accessed January 2010).

Elaine D. Kirkham
University of Wolverhampton

HERMAN MILLER

Herman Miller is an office furniture manufacturing company, established in 1923, and headquartered in the town of Zeeland in southwestern Michigan. In addition to manufacturing furniture and office products, it also provides design services to help clients more efficiently organize their office space. While its best-known product is the ergonomic Aeron chair, it has also been consistently recognized as a leader in green manufacturing and design. (This company is not to be confused with Howard Miller, the clock-making company that was spun off in the 1930s from Herman Miller and is also headquartered in Zeeland.)

There are three main reasons for Herman Miller to be highlighted in a compendium on green business. First, most of its products are certified under third-party organizations that designate products with low chemical and particle emissions in order to improve indoor air quality. Second, it is heavily involved in the green building industry, not only by providing products that help meet green building goals, but by having its own facilities certified under the Leadership in Energy and Environmental Design (LEED) program of the U.S. Green Building Council. Third, its holistic approach to environmental management not only includes its buildings and products but its entire production process and has done so for many decades, not only as part of the early 21st-century green wave.

One of the difficulties involved in making manufacturing processes environmentally friendly is the complexity of those processes. Particularly in the age of global assembly lines, many different suppliers contribute to the final product. For example, an office chair incorporates fabrics, plastics, and metals. Herman Miller's Design for the Environment team incorporates all of these items in a database that lists every material used in each of their products. The database was a necessary first step in requiring all of their suppliers to use materials with low emissions and/or recycled materials. This results in products that enable customers to meet their own goals with regard to indoor air quality by using low-emissions furniture. The company was one of the first to have many of its products follow third-party certification schemes such as Green Guard or cradle-to-cradle. By definition, cradle-to-cradle means the use of materials that are not only sustainably harvested or manufactured, but can be recycled at the end of their useful life. Since 2001, all of Herman Miller's new products must meet these standards, and the goal is to have every one of its products meet cradle-to-cradle certification by the year 2020.

One of the common criticisms of such third-party certifications is that they focus on the characteristics of a single manufactured object and only tangentially on the environmental impacts of the manufacturing and distribution processes. In contrast, Herman Miller has devoted as much effort to its facilities as to its products. In particular, it has been among the pioneers of green building in southwestern Michigan (along with competitors Haworth and Steelcase), leading to the Grand Rapids–Holland metropolitan area having the highest number of green buildings per capita anywhere in the United States. While only a few of these buildings are associated with Herman Miller, they have served as a model for other institutions in the region in the same way that universities or local governments have in other parts of the country.

Additionally, Herman Miller was one of the founding members of the U.S. Green Building Council (USGBC), the leading organization in certifying green buildings in the United States. As a founding member, it developed a green manufacturing facility that became one of the pioneer projects for the LEED program. As of 2009, it had 14 certified green buildings and one more pending certification, spread across three different countries. Most of these are not actually new buildings; the most environmentally efficient building is one that is reused, and so most of Herman Miller's LEED-certified buildings are renovations of existing facilities under the Commercial Interiors designation. Most of these buildings (9 of the 14) are certified at the Gold level, the second-highest level of four. In addition, all of Herman Miller's manufacturing facilities are ISO 14001 certified, meaning they meet international standards for environmental management systems.

Beyond their products and buildings, Herman Miller has placed importance on other aspects of environmental management. Early in the company's history, founder D. J. DuPree made environmental stewardship a key part of the company's philosophy. For example, 50 percent of all facility sites consist of green space. While green buildings today often include employee access to daylight as one of their key and innovative features, as early as the 1950s, DuPree declared that no employee would be more than 75 feet from a window. The company's Environmental Steering Committee works on improving internal company processes such as recycling in offices, while the Design for the Environment team focuses on keeping the manufacturing process and products as environmentally friendly as possible.

Herman Miller's environmental innovations have not gone unnoticed by its industry and beyond. *Fortune* magazine's annual list of "Most Admired Companies" has named it as the most admired furniture company for 20 of the past 22 years, and it is in the top 10 for all companies. While this ranking is based on more than environmental stewardship, innovation and social responsibility do contribute to the rankings. Herman Miller has also been recognized by the state and national environmental protection agencies for their products and facilities, and they are the only furniture manufacturer of the 318 firms listed on the Dow Jones Sustainability World Index.

Of course, there are economic as well as social and environmental reasons for Herman Miller to support sustainability. Namely, the LEED certification criteria for green buildings include credits earned for improving indoor air quality. Furniture and other products that are low or zero-emissions can help to certify a building as green, particularly for the Commercial Interiors and Existing Buildings designations. Many furniture manufacturers besides Herman Miller, including Steelcase, Haworth, and Allsteel, have been at the forefront of green manufacturing and green buildings for reasons of marketability. Practicing what they preach in terms of reducing environmental impact, both through their products and their facilities, can be seen as a way for these firms to attract clients.

However, Herman Miller's lofty future goals, as expressed in the Perfect Vision 2020 Initiative, suggest that there is more than self-interest motivating its actions. By 2020, the goal is for all of the company's processes to have net zero environmental impact. First, as mentioned above, all of its products should be cradle-to-cradle certified, which means the manufacturing process and outcomes are sustainable. Second, its goal is to use entirely renewable energy. Currently, about 30 percent of the company's energy is renewable, purchased through the third-party organization Green-e and generated at various locations in the United States. Third, zero net impact includes generating no landfill waste or water or air emissions through production, distribution, and consumption of its products. The company admits these goals are probably not practically achievable; in particular, the emissions created by transporting goods to market are likely to be difficult to overcome. Nevertheless, simply trying to meet the Perfect Vision goals will help the company significantly lower its environmental impact. For example, a 1990 goal to achieve zero landfill by 1995 resulted in a 50 percent reduction in waste production in only four years. In 2008, the company's yearly report claimed a 79 percent reduction in its footprint, which is truly remarkable. Continued financial success for the corporation should demonstrate to other firms that a reduction in environmental footprint can actually lead to financial benefits in the long run.

See Also: Certification; Cradle-to-Cradle; Design for Environment; Green Building; ISO 14000; Voluntary Standards.

Further Readings

Herman Miller. "Environmental Advocacy." http://www.hermanmiller.com/environment (Accessed April 2009).

Kadleck, Chrissy. "Office Product Maker LEEDs Green Charge." *Waste News*, 12/26:13–15 (April 2007).

Zaun, Tim. "Herman Miller, the BHAG, and Internalizing Sustainability." *Greener Design* (June 22, 2007). http://www.greenerdesign.com/news/2007/06/21/herman-miller-bhag-and -internalizing-sustainability (Accessed April 2009).

Julie Cidell
University of Illinois

I

IKEA

Founded in Sweden, IKEA has grown to be the world's largest furniture retailer, specializing in flat-pack kits to be assembled at home. IKEA is also renowned for its prominent efforts to reduce its environmental impact and increase the social good generated throughout its supply chains. Although IKEA has certainly been reactive to wider trends and demands, many argue that responsibility has been at the core of IKEA's operation throughout its history. Having said this, IKEA has not gone without criticism, and perhaps the biggest area of controversy is the company's unusual legal structure that some say fails tests of transparency, and possibly implies tax evasion.

IKEA's progressive tenets include a pledge to minimize their own ecological footprint and to forge partnerships with trade unions to ensure fair and safe working conditions for their own and their supplier's employees.

Source: Christian Koehn/Wikipedia

IKEA was founded as a general trading company in Sweden in 1943 by Ingvar Kamprad. After first introducing furniture in 1947, it was decided that this would constitute the entirety of the company's range from 1951 onward, and in the same year, IKEA launched its first catalog: a marketing approach that has become synonymous with the company's brand. IKEA has subsequently gone on to become the world's largest furniture store, with some 301 franchised stores in 37 countries across the world, and although having only a few factories of its own, contracts some 1,500 third-party suppliers in 55 countries, mostly in Asia.

Some argue that the concept of corporate social responsibility (CSR) has always been ingrained in IKEA's business operation, and perhaps the most innovative aspect of the company is the degree to which a more sustainable approach has not superseded the goal of providing accessible goods. For example, it is argued that the minimal use of materials has both reduced environmental impact and price, and likewise that the flat-packed nature of the product leads to both reduced environmental and financial transport costs. The degree to which such financial savings are passed on to the consumer or accumulated in profit (which would lead some to categorize the above measures as greenwashing), is of course a different issue.

Whether responsible operations are intrinsic or not, IKEA's core business, which involves transforming natural resources into economic commodities at minimal cost, has certainly attracted attention in the era of greater environmental and social awareness. The first of the concerns emerged in the mid-1980s, when formaldehyde emissions from IKEA's widely used particle board designs were found to exceed some recommended standards. Later on, movements for the preservation of the rainforests accused IKEA of making a significant contribution to the loss of important environmental capital, and in the early 1990s, the company was faced with allegations that it profited from child labor.

The first controversy led IKEA to apply the new E-1 German standards on formaldehyde emissions, which were recognized to be the strictest in the world, even in countries where the existing standards were more lax. Indeed, the application of the most rigorous standards across all markets became a standard of IKEA operations. The response to environmental concerns involved engaging both with external and internal stakeholders, and has been seen as a prominent element of IKEA's innovation. For example, although Karl-Henrik Robèrt (the founder of The Natural Step) was initially invited to IKEA by Lennart Dahlgren (vice president) and Russell Johnson (head of quality) as a means of trying to overcome resistance to reform, genuine engagement with wider management yielded a practical way to reduce environmental impact without impacting finances. Instead, the approach was to gradually introduce a range of more sustainable equivalents to standard products, the Gustavian line, but to justify the additional costs not in reference to sustainability, but by the exclusivity of the range. As the marketing strategy promoted new options over conventional products, consumer demand increased, production volumes grew, and unit prices fell. This was also complemented by improvements to other items, and all sectors of the business were reformed by making structural changes in six key categories: (1) management and personnel; (2) products and materials; (3) customers; (4) suppliers; (5) buildings equipment and consumable materials; and (6) transport. What is more, responsibility has not just remained a policy, but has been translated down the corporate structure through a number of educational and interactive schemes, such as "environmental coordinators" and "ideas banks" for employee initiative. IKEA has also partnered with other external organizations, including the Forest Stewardship Council (FSC). As a result, in November 1999 the company banned timber from intact natural forests, except those certified by the FSC. While there has been substantial criticism of such private governance schemes, many argue that IKEA has set a strong example on how voluntary action can reduce the impact of business on the environment.

IKEA is also seen to have made an exemplary effort to reduce the tension between low prices and corporate responsibility through a shift from "market interactions" with third-party suppliers, to a more "relational" form of interactions; in a shift from a "trading" to a "purchasing" model of business. The implication of this strategy is that short-term contracts distributed in line with market prices are replaced with longer term relationships in which the abilities and constraints of the supplier are recognized. Under this approach, IKEA is less seen as a dictator of a required level of quality, price, and corporate responsibility, and instead is seen more as a partner in helping supply companies develop the same aspirations. Alongside this new approach, IKEA launched "The IKEA Way on Purchasing Home Furnishing Products" in 2000: a code of conduct with which all suppliers are expected to comply. IWAY, as it is known, includes 19 different areas, divided into more than 90 specific issues, such as working conditions, child labor, and environmental management.

In order to help suppliers meet the requirements, IKEA has structured expectations around a Staircase Model. This means that while there are basic requirements that any potential supplier must meet before engaging is possible (which include social and working conditions, such as no forced or bonded labor, no child labor, and standards for wooden merchandise that aims to ensure the appropriate management of forests), those seeking a longer term relationship with IKEA are required to go further. Level Two of the code sets requirements relating to the outside environment, social and working conditions, and forestry (where appropriate). However, even if companies are registered as noncompliant during audits carried out by an IKEA Audit team (which is itself audited), this does not necessarily mean an end to the relationship, as long as the supplier shows a willingness to improve. This approach is also argued to be applicable right across the supply chain, as while IKEA has concentrated on bringing as many suppliers as possible to Level 2, the next stage will see the scheme extended to those organizations that supply to IKEA's suppliers. In order to verify the results of audits by IKEA staff, an internal Compliance and Monitoring Group also arranges external verification of compliance by suppliers. According to independent observation, this has resulted in retraining auditing teams that have not been applying a sustainable standard.

As well as seeking to reduce negative implications internal to its business operations, IKEA has also engaged in activities to promote social good. For example, the company works closely with Save the Children and the United Nations Children's Fund (UNICEF), often in cause-related marketing campaigns for products that raise money for a certain cause. In 2005, IKEA launched its Social Initiative program, designed to promote the rights of children to a healthy and secure childhood. Specifically, IKEA has targeted south Asia, and especially India, where it has had long-term business engagement through local suppliers of carpets.

While there are no doubt possibilities for criticizing IKEA's programs, perhaps the strongest ethical critique of the company has been the allegation that it uses legal loopholes to rein back on the requirement for financial transparency and engage in what appears to some as tax evasion. The reason for this is that the parent organization for most of IKEA's companies held by the Stichting Ingka Foundation, a Dutch-registered, tax-exempt, nonprofit legal entity which in turn channels its funds to the Stichting IKEA Foundation. While the aim of this organization is to fund "innovation in the field of architectural and interior design," little information on its activities was encountered by an investigation by the *Economist* in 2006, and there are serious questions concerning how the lightly taxed income of IKEA is actually spent.

Overall, IKEA is considered one of the most progressive companies in developing its CSR agenda, moving beyond low prices to include both environmental and social considerations across the supply chain. Having said this, given the legal structure of the company and the tax and remuneration issues that this promotes, perhaps the best summary might be that IKEA offers a truly excellent case study on what it really means for a business to behave responsibly.

See Also: Corporate Social Responsibility; Greenwashing; Supply Chain Management; Sustainable Development; Voluntary Standards.

Further Readings

Anderson, M. and T. Skjoett-Larsen. "Corporate Social Responsibility in Global Supply Chains." *Supply Chain Management: An International Journal*, 14 (2009).

Brunner, Robert, Stewart Emery and Russ Hall. *Do You Matter? How Great Design Will Make People Love Your Company*. Upper Saddle River, NJ: FT Press, 2008.

Edvardsson, Bo and Bo Enquist. *Values-Based Service for Sustainable Business: Lessons From IKEA*. London: Routledge, 2009.

IKEA. "The IKEA Way on Purchasing Home Furnishing Products." http://www.ikea-group .ikea.com/repository/documents/1293.pdf (Accessed January 2010).

Pedersen, Esben R. and Mette Andersen. "Safeguarding Corporate Social Responsibility (CSR) in Global Supply Chains: How Codes of Conduct Are Managed in Buyer-Supplier Relationships." *Journal of Public Affairs*, 6/3–4 (2009).

Alastair M. Smith
ESRC Centre for Business Relationships,
Accountability, Sustainability & Society

INDUSTRIAL ECOLOGY

Industrial ecology is a field of study concerned with analyzing the interactions between humans and the environment that takes a broad, long-range view to the analysis of industrial systems and views them in a manner analogous to an ecosystem. It draws from both systems engineering and ecological principles, and is particularly useful in addressing concerns such as sustainability at a systems level. Industrial ecology studies the flow of materials and energy through industrial systems, and aims to develop ways to design, produce, use, recycle, and dispose of products in order to optimize the use of resources and energy while minimizing any environmental impacts. In this period of diminishing natural resources and global climate change, and industrial ecology can provide a pathway toward a sustainable society.

One of the current problems faced by modern society is the generation of a wide variety of hazardous by-products in its industrial production. During most of humankind's existence on Earth, small, widely scattered groups of people have had little impact on the planet's service systems for providing food and recycling waste. Simple dwellings, small trails, and small, spatially diverse harvests left virtually no footprint. However, in the last several thousand years, human population has increased to a point of modifying natural environmental systems. Since the industrial revolution, and especially during the 20th century, human society built structures, such as megacities, that demanded extreme levels of product production, increasing the hazardous by-products that are discharged into the atmosphere and water. These changes mean that people can no longer take the relationship with the natural environment for granted, and one hope of industrial ecology is that it can provide a basis for a more sustainable and manageable technological society, and allow humans to live in harmony with other species and with their surroundings, ensuring the habitability of the earth for an indefinite period.

Integration of Human Activities and Systems
With Planetary Ecosystem Services

Modern manufacturing activities have a large influence on the environment, resulting in an imbalance of material and energy flow when integrated into the ecosystem. Raw materials, such as petroleum, coal, and ores, must be extracted from the Earth to be used as inputs to the industrial system, while the outputs include wastes as well as finished products. In the early industrial age, pollution by-products were ignored, and wastes were dumped into the air, water, and soil. Once it became clear that these practices had a deleterious effect on human health, regulations were instituted, generally aimed at "end of pipe" emissions (local, ad hoc efforts, involving mostly emission control technologies) and initiating some remediation of current and legacy hazardous waste problems.

Although these measures have been successful in reducing pollution, they do not reach the root of the problem: rather than eliminating or recycling wastes, they simply transfer the discarded waste from one part of the environment to another. Industrial ecology, in contrast, takes a systems approach and a global perspective, and considers how materials may be reused or recycled to reduce both reliance on virgin raw materials for inputs, and the amount of waste that must be disposed of. In the early stage, U.S. regulations regarding the industrial process focused mainly on monitoring pollution and preventing entry of certain materials into the industrial waste stream: this approach may be seen in legislation such as the Clean Water Act, the Clean Air Act, and the Toxic Substances Act.

The second stage of regulation was oriented toward reducing pollution by modifying industrial processes to encourage less production of toxic material. For instance, the Pollution Prevention Act of 1990 required the release of reports, and compiled them in a Toxics Release Inventory (TRI). Because members of the general public can access the TRI, it has been used to put pressure on companies to reduce TRI emissions of hazardous wastes, and also serves as a type of informational regulation: the mere fact that toxic emissions must be reported publicly provides an incentive for companies to reduce them.

The third stage or regulation, as represented by the U.S. Environmental Protection Agency's (EPA) encouragement of clean production strategies, is more in line with industrial ecology because it takes a systems approach to integrating human industrial activities with the environment. Ultimately, these strategies will prove the most economical, because they are relatively more efficient and effective.

The economies associated with industrial ecology consider natural capital and social capital in addition to the financial aspects of labor and raw materials. Natural capital refers to the Earth's ability to provide resources for human activity, and includes clean air and water, biodiversity, forests, and topsoil. Natural capital may be degraded or lost by resource consumption or environmental pollution, and two major goals of industrial ecology are to reduce resource consumption by recycling, and to reduce environmental degradation by converting hazardous by-products into benign forms, or using them as fuel.

Social capital refers to the connections among individuals and includes shared values, relationships, and behaviors that create bonds between people and make communities cohesive. Although not traditionally included in economic analyses, it is a vital consideration in industrial ecology: basically, the effect of any industrial process or system on both individual people and their communities is considered alongside economic and environmental considerations.

Principles and Goals of Industrial Ecology

Industrial enterprises are designed to handle materials and energy extracted from natural resources and to reassemble materials into components for products in a manner similar to biological organisms and their systems. These enterprises can be assembled into industrial ecosystems. Such systems require a group of large and diverse industrial enterprises acting together, with each enterprise utilizing products, as well as wastes, from other members of the system. Industrial systems are generally considered open, that is, they take in raw materials from outside the system and return them outside the system in the form of products and wastes. Recycling closes this loop so that outputs become inputs: ideally the system would need only materials to start the process and enough energy to maintain throughput. Transforming industrial processes to closed systems does not remove their impact on the environment, but can limit its effects and reduce its harm.

A functioning industrial ecosystem should have (1) renewable energy sources, (2) recycling of materials, and (3) diversity of enterprises for resilience to external stresses. In a typical industrial ecosystem, the flow of energy in a system is considered open because it enters in a relatively concentrated, usable form (such as coal), and leaves in a dispersed, unorganized form (such as waste heat). This system can be reformed in two ways: by making the use of existing materials more efficient (e.g., using more of the energy in the coal and discharging less as waste), and by using more stable or renewable sources of energy, such as wind, water, nuclear power, and biofuels.

A successful industrial ecosystem is one that minimizes both the materials required to enter the system and the unusable wastes leaving the system while maximizing the flow of materials within the system. In principle, with a limited supply of resources and unlimited energy, a complete recycling of material is possible, although in most cases it must be practiced on a larger scale than simply that of a single product or even an individual enterprise. A central need of industrial ecology is to develop efficient technologies for recycling that reduce the need for raw materials to the lowest possible levels. However, in practical terms, not all materials are recyclable, thus a related goal of industrial ecology is to minimize dissipative uses of toxic substances, such as mercury or lead, which are not biodegradable, and find effective, ecologically friendly ways to dissipate other materials that are impracticable to recycle. A third goal is to reuse materials whenever possible, reducing the need for raw materials as input as well as reducing waste.

Diversity for resistance to stress creates the quality of robustness in industrial ecosystems, so that if one component of the system is perturbed, there are other parts that can replace its function. A robust system can recover from changes in the external environment, for instance a plant using renewable energy rather than petroleum products will be less affected if the price of petroleum rises or the supply is shut off. Similarly, a plant that uses its own recycled materials for inputs is less affected by the reduced availability of virgin raw materials.

Natural Ecosystems and Industrial Ecology

A natural ecosystem is driven by solar energy and photosynthesis, and is populated by interacting species, where materials are interchanged in a largely cyclical manner using biochemical processes that build and alter biomolecules. An industrial system, like an ecosystem, also synthesizes and degrades substances. Both natural and industrial ecosystems ideally exist in a steady-state condition without causing any net degradation to the environment.

In both systems, stocks of materials are maintained in separate compartments connected by flows of materials. The quantity of material in each compartment remains relatively stable because material is cycled back into the compartment by one or more, usually complex, subcycles. With the exception of energy and a small amount of nonrecyclable waste, these systems are closed, and quantities of material in the various compartmental reservoirs remain constant. Natural systems are self-regulating, mainly through competition, whereas the self-regulation in an industrial ecosystem must come from the economic system in which the drivers are supply and demand. The current models of industrial ecosystem have raw materials imported into the system, with one major product exported from the system, and associated enterprises coexisting synergistically, utilizing each other's products or services. To be sustainable, an industrial ecosystem requires that as many of its by-products and wastes as possible be used to maintain a steady state. Table 1 compares current industrial systems with natural ecosystems.

Table 1   Characteristics of Industrial Systems and Natural Ecosystems

Characteristic	*Industry*	*Nature*
Basic unit	Enterprise/Firm	Species/Organism
Material pathways	Mostly linear	Closed loops
Process cycles	Few	Many
Recycling	Low	High
Material fate	Dissipative	Concentrative
Sustainability	Low	High

Source: Adapted from Manahan (1999).

Material and Energy Flows

The flow between material reservoirs and energy flow from high-grade to low-grade heat are the two central aspects of industrial ecology. Energy generally enters current systems as high-grade fossil fuels and leaves as waste heat. An industrial ecosystem is designed to extract as much usable energy as possible as it passes through the system by using heat exhaust to produce power or steam for other needs. Similarly, an industrial ecosystem wants to extract as much use as possible from materials flowing through the system through recycling and reuse, thus reducing both the use of raw materials and the production of waste.

In analyzing an industrial system, industrial ecology identifies the flows of materials and energy from their initial sources through the entire system, including various subcycles to the consumer and eventual disposal. The system can be defined as anything from an individual unit operation to a factory to the entire globe, but analysis is most commonly carried out at the regional level. This is because at the regional level, industrial processes are large enough to have a number of industries with a variety of products and wastes that might be recycled or exchanged, but small enough to permit economical transport, because effects on the environment (e.g., the watershed) are often best evaluated at a regional rather than a local level.

Industrial ecology is also concerned with optimizing energy and material flows and assessing the results. Individual processes may be evaluated for their productivity, typically defined as the economic output divided by the material and energy input. Recycling efficiency can be measured by dividing the amount of recycled materials by the sum of recycled and raw materials. The combination of these indicators provides a ratio that can monitor material and energy inputs as well as output of recycled materials.

It is only possible to optimize an industrial system for efficient production, minimum waste, and minimum pollution if the materials cycle is based on internalization.

This means that the materials cycle is closed so that the materials do not need to be transported long distances, for instance, by developing local markets for recycled or upgraded waste materials. For example, upgraded waste material could be used to generate electricity on-site, or sulfur dioxide waste could be used to generate sulfuric acid for use in other production processes. This internalization of the materials cycle leads to systems integration, which is essential for any industrial ecological system. Maximum efficiency and profit requires all materials, that in a linear system would be wastes, are now used within an integrated complex. However, since low-value waste materials cannot be economically transported long distances from an industrial enterprise, local boundaries of the system must be well designed.

Ecoefficiency is a term developed in the 1990s that advocates producing more goods and services with fewer resources, while producing less waste. Sustainability is a key goal, and an ecoefficient assessment of a business includes consideration of their success in reducing environmental impacts and raw materials consumption, as well as in increasing production. Specific concerns include the durability of products, quality control (so fewer goods are discarded), recycling, the use of renewable energy sources, a reduction in the use of toxic materials, and waste reduction. Developing measures of ecoefficiency is key in encouraging green businesses because it broadens the definition of success from a simply accounting of profits and output of products to include the goals of satisfying human needs and enhancing quality of life. Industrial ecology can aid in this process because it looks at industrials systems from a broad perspective that includes the goals of reducing waste and impact on the environment as well as increasing productivity.

Examples of Industrial Ecology in Practice

Analyses informed by the concepts of industrial ecology can help businesses become more green by not considering them in isolation, but as part of a functioning system that allows for a greater variety of uses for waste products, as well as greater sources of inputs. Kalundborg, Denmark, offers a well-known example of a functioning industrial ecosystem in which the waste products of one industry become the inputs of another. This system begins with two energy supplies (a coal-fired electrical generation plant and a petroleum refining complex). The electric power plant sells steam to the oil refinery, and receives both produced gases and cooling water in return. Sulfur removed from the oil by the power plant is sold to the Kemura sulfuric acid plant. Waste heat from the electrical power plant and the refinery is sold for heating homes and commercial establishments, greenhouses, a fish farming enterprise, and a pharmaceutical company. The biological waste (sludge from microorganisms) is sold as fertilizer to regional farmers. Other industries included in this industrial ecosystem include a dry wallboard company that uses the calcium sulfate from the lime scrubbing used by the power plant to reduce greenhouse gases, and a cement company that uses the ash produced by burning coal. Waste from the fish farm is also used as fertilizer. The evolution of the Kalundborg complex began in the 1950s and 1960s, and developed a system of self-organization over time, analogous to that of a natural ecosystem. Kalundborg's self-organization leads to an increase in communication, cooperation, and structured interactions that increase the social capital in the industrial ecosystem.

Another example of a successful industrial ecosystem exists in Burnside Industrial Park in Nova Scotia, Canada. It was built in 1995 and developed under the leadership of Professor Ray Côté, an industrial ecologist from Dalhousie University. Burnside includes a Cleaner Production Centre, devoted to facilitating green practices and cooperation among businesses, and offers various services, including auditing energy conservation and waste reduction, searching for technologies to reduce resource use, and facilitating waste and energy links among firms, including waste exchanges. Current successes include a company devoted to collecting corrugated cardboard and sending it to be recycled into liner board, companies that use the excess polystyrene from a computer company for packing material, and several firms devoted to recycling and reusing materials discarded by other companies in the group, including furniture, tires, and toner cartridges.

Environmental Justice and Ethics

Since its goal is to ensure continued habitability of the planet, industrial ecology poses several critical ethical questions for industry, environmentalism, and society in general. The "consumer society" that has evolved since the 1950s is in conflict with resource conservation and ecosystem health. Nonsustainable use of energy and materials for growth contradicts the principles and goals of industrial ecology, so the adoption of industrial ecology requires that the public adopt an ethic that respects the limits of Earth's ecosystem services. This prospect will require both education about the environment and opportunities to benefit from the industrial ecosystems. Industrial ecology enterprises can provide just solutions to cross-scale environmental problems that require both local mitigation and adaptation as well as strategic political and economic activity, in both national and international arenas, while increasing regional adaptive capacity allows for comprehensive community sustainability.

Environmental justice addresses the issue of inequality, where some people bear disproportionate toxic burdens from industrial by-products, or have inequitable access to goods and services. For example, indigenous people throughout the Arctic, who have an intimate relationship with their local ecosystems for both physical and cultural sustenance, are experiencing significant health impacts from the accumulation of industrial products or their by-products, such as persistent organic pollutants, as well as the impacts of climate change related to industrial gases and carbon dioxide. The case of Arctic climate impacts raises the related issue of transnational equity, as well as intergenerational justice, with regard to benefits and burdens from industrial production and consumption. Industrial ecology, which stresses coalitions and collaboration in addressing sustainable systems, can be a broad-based approach toward productive justice. It allows the collective view that "everyone's backyard is my backyard."

Industrial ecology offers a way of looking at the relationship between humans and the environment that aims to facilitate the production of material goods while also making that production more sustainable and less harmful to the environment. It looks at issues such as pollution and resource use at a systems level, and seeks to reduce waste through reuse and recycling rather than merely moving the waste from one location to another. Industrial ecology offers important tools for green businesses because it offers methods to trace the path of materials and energy through the industrial process, and can help find ways to reduce waste and find new uses for materials that would otherwise be discarded. As such, it can increase industrial efficiency and profitability while also increasing environmental justice by limiting the harm done to the least powerful members of society who often bear a disproportionate burden of industrial pollution.

See Also: Corporate Social Responsibility; Cradle-to-Cradle; Dematerialization; Exposure Assessment; Extended Producer Responsibility; Extended Product Responsibility; Green Chemistry; Industrial Nutrients; Toxics Release Inventory; Waste Reduction.

Further Readings

Ayres, Robert and Leslie Ayres. *Industrial Ecology: Towards Closing the Materials Cycle.* Cheltenham, UK: Edward Elgar, 1996.

DeSimone, Livid and Frank Popoff. *Eco-Efficiency: The Business Link to Sustainable Development.* Cambridge, MA: MIT Press, 1997.

Frosch, Robert and Nicholas Gallopanlos. "Strategies for Manufacturing." *Scientific American*, 261:94–102 (1989).

Manahan, Stanley. *Industrial Ecology: Environmental Chemistry and Hazardous Wastes.* Boca Raton, FL: CRC Press, 1999.

Trainor, Sarah, et al. "Environmental Injustice in the Canadian Far North: Persistent Organic Pollutants and Arctic Climate Impacts." In *Speaking for Ourselves*, J. Agyeman, et al., eds. Vancouver, Canada: University of British Columbia Press, 2009.

Lawrence K. Duffy
University of Alaska Fairbanks

INDUSTRIAL METABOLISM

Industrial metabolism describes, using an analogy of human metabolism, the process by which industry convert raw materials, energy, and labor into finished products and wastes in a (more or less) steady-state condition. This characterization was developed in the 1960s when increased concerns about the environment prompted analyses of industrial operations analogous to the way material and energy balances are described in natural ecological systems.

Applications

The concept of industrial metabolism is useful for ecological economics as it provides a way to analyze the flow of physical resources in industrial and societal systems from an integrated perspective, allowing the study of system-wide effects and problem shifting due to environmental policies. It can be usefully applied at many different levels from global through national, regional, sectoral, company, site, and household. At smaller scales, industrial metabolism addresses how resources are used in the human economy through the development of resource-accounting frameworks for political and economic entities, as well as life cycle analyses of the materials used in the manufacture of industrial and consumer products. At the largest scale, it can highlight anthropogenic contributions to atmospheric concentrations of trace gases and the flow of excess nutrients from agricultural activities to water bodies. The information can be used to construct a materials balance of the whole system and draw conclusions regarding the actions needed to improve the environmental character of its metabolism.

Measures of Industrial Metabolism

A useful measure of industrial metabolic efficiency is the economic output per unit of material input, which can be termed *materials productivity*. In principle, this can be determined for the economy as a whole, as well as for each sector and major nutrient element (e.g., carbon, oxygen, hydrogen, sulfur, or iron).

Various approaches have been developed to measure sustainability, including material flow analysis, physical input–output tables, life cycle assessment, cost minimization models, equilibrium models, and dynamic optimization and system dynamics. Unsustainability can be determined by the amount or degree of dissipative loss (materials that are not reused or recycled) and analyzing the reasons for dissipative loss may help develop systems to improve sustainability. There are three principle reasons why materials are not recycled: recycling may be inherently unfeasible, or technically feasible but not economically viable under current conditions, or both technically and economically feasible, but not occurring for other reasons.

Future Directions

To increase sustainability, the material and energy flows in industrial metabolism must be altered so that they more closely resemble biological metabolism: in particular, by returning more materials currently discarded as waste back into the system through reuse and recycling, rather than simply changing waste from one form to another or moving it from one location to another. This will reduce the amount of virgin materials required as inputs into the system—a change that is necessary as the supply of resources is limited and in some cases close to exhaustion.

Research in the field of industrial metabolism has traditionally been focused on measuring and describing physical flows of economic systems, because economic systems are traditionally viewed as the regulators of industrial metabolism, exerting control through the price mechanism that determines the supply and demand for labor and products. The metabolism of economic systems, however, changes over time, and research on changing industrial metabolism is needed to improve our understanding of how physical materials flow through the economic system.

Far more work needs to be done in environmental science to establish causative relationships between human emissions and damage to the environment. The inescapable influence of specific geographical and cultural features on human–environment interactions requires advanced models of individual and collective behavior to describe and improve environmental quality at all levels.

See Also: Appropriate Technology; Best Available Control Technology; Best Management Practices; Closed-Loop Supply Chain; Cradle-to-Cradle; Environmental Management System; Green Technology; Life Cycle Analysis; Natural Capital; Resource Management; Stewardship; Supply Chain Management; Waste Reduction.

Further Readings

Ayres, Robert U. "Industrial Metabolism: Theory and Policy." In *The Greening of Industrial Ecosystems*. Washington, D.C.: National Academy Press, 1994.

Ayres, Robert U. and Udo Ernst Simonis. *Industrial Metabolism: Restructuring for Sustainable Development*. New York: United Nations University Press, 1994.

Green, Gary Paul and Anna Haines. *Asset Building and Community Development*. Thousand Oaks, CA: Sage Publications, 2001.

van den Bergh, J. C., et al. *Theory and Implementation of Economic Models for Sustainable Development*. New York: Springer, 1998.

Rasmi Patnaik
Gopalsamy Poyyamoli
Pondicherry University

Industrial Nutrients

The term *industrial nutrients* refers to materials used in manufacturing, the use and disposal of products, and emphasizes the fact that materials often discarded as waste can be reused or recycled, thus becoming useful new inputs or nutrients to sustain the industrial system, analogous to the nutrients that humans ingest to sustain their bodies. Additionally it emphasizes the fact that industrial nutrients can cycle through the industrial ecosystem just as nutrients such as water and carbon cycle through the natural ecosystem. The industrial nutrient approach to manufacturing facilitates the design of the manufacturing process in order to maximize sustainability and minimize harm to the ecosystem by reusing materials or returning them to the environment in useful forms.

The industrial nutrient concept is integral to analyses of industrial metabolism, as advocated by Robert Ayres, which trace the life cycle of materials through industrial systems. It is also a vital component of the cradle-to-cradle (C2C) approach to manufacturing, which offers an alternative to the cradle-to-grave system that theorists such as William McDonough and Michael Braungart argue is no longer sustainable. The cradle-to-grave approach is based on continually taking raw materials, such as coal and iron ore, into the industrial system, and disposing of the resulting wastes outside the system, without much consideration about causing harm to the environment or the possibility that the supply of raw materials may not be unlimited. This type of approach to the industrial process is often referred to as a linear flow of materials (because the materials move in only one direction, from their raw state through manufacturing to waste) or an open-loop system (because materials are taken in from and return to locations outside the system). The C2C system, in contrast, is a circular or closed-loop system in which the end products of the manufacturing process, including waste, are seen as potential inputs to the same or other processes, that is, as nutrients to feed the process. In the cradle-to-grave approach, raw materials are used only once and discarded, while in the C2C approach, materials are seen as the source of nutrients that can be reused or recycled in further industrial or biological processes, ideally with no degradation of quality or ecological harm.

Braungart and McDonough classify industrial nutrients as either technical nutrients (synthetic materials such as plastics and metals that can be used over and over again without degradation) or biological nutrients (organic materials that will decompose in the appropriate natural environment). In a C2C system, both kinds of nutrients flow through the manufacturing and use processes in a closed loop that minimizes the amount of raw materials which must be taken in and the amount of waste discarded. Ideally, products made from technical nutrients will be designed so they can be reused or recycled more or less endlessly, while those made from biological nutrients will be designed so they can be completely returned to the environment without harm (meaning, for instance, that they must not contain toxic chemicals).

In Braungart and McDonough's conceptualization, products made from biological nutrients are called products of consumption because they are not intended to last forever, and are designed to break down in the natural environment after use. One example is the Climatex® Lifestyle fabric developed by Rohner Textil and DesignTex (in conjunction with the Environmental Protection Encouragement Agency and McDonough Braungart Design Chemistry), which is made from natural fibers, including wool and ramie, and can safely be returned to the environment as a biological nutrient. Another example is the Trigema Compostable T-shirt, which contains no toxins or heavy metals, and will biodegrade in a household compost heap.

Products such as automobiles and television sets made from technical nutrients are called products of service because they are durable goods that will not biodegrade, but can circulate endlessly in a closed-loop system of manufacture, recover, and reuse without degradation of quality. An example of a product of service is the nylon EcoWorx carpet fiber, which can be recycled into high-quality yarn without degradation, in contrast to most synthetic materials where recycling decreases the polymer length and ultimately limits the usefulness of the recycled material.

Many consumer products combine both biological and technical nutrients: these are known as complex products and require careful design and planning so their nutrients may be separated when the product is no longer useful in its original form, and either recycled (technical nutrients) or returned to the natural environment (biological nutrients). Products and materials that have lost their commercial value but are not suitable for either recycling or biological decomposition are known as unmarketables, and require redesign using biological or technical nutrients. For instance, many computers now fit the category of unmarketables (and, in fact, constitute hazardous waste), but could be redesigned to facilitate recycling of their technical nutrients.

Only a few products on the market currently meet the standard of being either completely biodegradable under realistic conditions or completely recyclable without degradation (and the accuracy of some such claims has been questioned). However, the industrial nutrients approach to analyzing industrial processes is a useful framework that assumes that natural resources are limited, as is the ability of the environment to absorb industrial waste. It is intended to spur design and production of products whose nutrients are either fully recyclable or biodegradable.

See Also: Appropriate Technology; Best Management Practices; Cradle-to-Cradle; Industrial Ecology; Life Cycle Analysis; Waste Reduction.

Further Readings

Aldy, Joseph E. and Robert N. Stavins, eds. *Architectures for Agreement: Addressing Global Climate Change in a Post-Kyoto World*. New York: Cambridge University Press, 2007.

Ayres, Robert and Leslie Ayres, eds. *A Handbook of Industrial Ecology*. New York: Edward Elgar, 2002.

Braungart, Michael. "Cradle to Cradle Production." In *Surviving the Century: Facing Climate Chaos and Other Global Challenges*, Herbert Girardet, ed. London: Earthscan, 2007.

McDonough, William and Michael Braungart. *Cradle to Cradle: Remaking the Way We Make Things*. San Francisco, CA: North Point Press, 2002.

Sarah Boslaugh
Washington University in St. Louis

INFORMATIONAL REGULATION

Informational regulation (IR) refers to regulation that requires that specific information be provided to affected stakeholders, which may include the general public. It is an alternative or supplement to third-party regulatory monitoring and enforcement because the assumption is that interested parties will use the disclosed information to reduce harmful activities, such as releasing hazardous wastes into the environment. IR plays a large role in the management of environmental risks and is an approach that appeals to many because it avoids command and control mechanisms, still requires some degree of corporate accountability, while the provision of information theoretically places power directly in the hands of the people affected by pollution. Examples of informational regulation include the Toxics Release Inventory, a publicly accessible database maintained by the U.S. Environmental Protection Agency (EPA) that includes information on the release of over 650 hazardous chemicals, and food labels mandated by the U.S. Food and Drug Administration (FDA) that contain information about risks (such as mercury contamination in fish).

Informational regulation is intended to achieve two goals: informing individuals of the risks and/or environmental costs associated with a product, and providing an incentive for manufacturers to reduce risks and environmental harm. IR is best suited for situations, such as the example of mercury contamination of fish, where the risk applies only to a small part of the population (e.g., children and pregnant women) or where consumer choice is considered important (for instance, labeling paper products as recycled or not). In such cases, informational regulation may be more efficient than more coercive alternatives, such as outlawing the sale of nonrecycled paper.

IR preserves individual autonomy by allowing individuals to evaluate risks and environmental costs relative to their specific set situation and values, and may strengthen the democratic process by providing all parties with essential knowledge. Finally, IR creates incentives for self-regulation by industry: because many consumers prefer to patronize products from environmentally friendly companies, some corporations have reduced emissions of hazardous chemicals listed in the TRI and stated this fact in their annual reports and advertising campaigns. Although IR is usually considered less coercive than command-and-control legislation, industry often fights the requirement for reporting by claiming that the expense involved is burdensome or that the mere existence of labeling will unduly frighten away customers. From the consumer's point of view, IR is an imperfect solution, in many cases, because the information provided is often incomplete and may be insufficient to truly empower private citizens to make informed choices or to exert pressure on corporations. Consumers may not be able to use the information provided effectively: for instance, studies have found that most consumers say they read nutritional labels on foods, but do not have a comprehensive understanding of their importance. An additional problem is that as the number of warnings increase, each becomes less effective as consumers cease to pay attention to them.

Applications of IR

A well-known example of IR is California's Proposition 65, passed in 1986, which requires businesses to provide warning labels on products that contain carcinogens or reproductive toxins at a level representing significant risk. Businesses that fail to comply are subject to heavy fines, one-quarter of which are awarded to the complainant initiating the enforcement. Proposition 65 is considered to have succeeded in lowering the overall amount of carcinogens in California's food and water (as businesses were strongly motivated to cease using listed substances in their products in order to avoid the warning labels), but less successful in helping consumers make informed choices. Many manufacturers feel that the legislation was alarmist and that consumers overestimate the amount of risk involved in consuming a labeled product.

Another example of IR is the regulation of seafood sold in the United States. Banning the sale of fish containing mercury would be tantamount to banning the product entirely, as almost all fish contain some amount of mercury. Regulations to reduce mercury emissions are not a practical solution, because fish sold in the United States are caught all over the world. In addition, the harm caused by mercury is related to the amount retained in the body, so the amount consumed is an important factor. Only 1–2 percent of the U.S. population eats fish daily, and only about half of them would exceed the recommended daily dose of mercury through fish consumption. Risk is not uniform across populations: pregnant women, nursing mothers, and young children have a much greater risk of harm from mercury poisoning than does the general population. Therefore, the EPA and the FDA chose to require warning labels on fish and shellfish, but not to restrict their sale directly. Spokespeople for the fish industry protest that such labels have depressed sales because they unnecessarily scare people away from eating fish at all, and that most individuals are unable to differentiate between high- and low-risk fish, or to understand that the risks are not equally applicable to all individuals.

There are many other examples of IR in industry, including mandated disclosure requirements for regulated substances under consumer information, product information sheets, material safety data sheets, and hazardous goods transportation regulations. However, often these sources do not include information about technologies and processes employed in production: in large facilities, major technology and equipment may be obvious to an observer, and may even be voluntarily identified in corporate publications and promotional materials.

Additionally, where the law requires operators and owners to have permits, the applications for those permits may include disclosure of pertinent information. For example, the permits for operations with air quality impacts directly address the disclosure of information by regulation. However, information collected in this manner may not be publicly available, and therefore is of limited use to private citizens. Although some technology and process information is publicly available in patent documentation and in intellectual property registries, these seldom include statements of environmental impacts and risks. Information may also be gleaned through securities documents, which are required to disclose certain information to investors and other stakeholders.

For instance, the Exchange Act in the United States and similar legislation in other jurisdictions set up regulatory bodies with authorization to formulate and administer securities regulations, including the establishment of information disclosure requirements. Section 13(a) of the Exchange Act establishes reporting requirements, and the Securities and Exchange Commission (SEC) is given regulatory authority to specify the content and form of the reports. Under the Canadian constitution, securities regulation is under provincial jurisdiction; consequently, the Ontario Securities Commission jurisdiction is provincial, while the SEC in the United States has federal jurisdiction. The Toronto Stock Exchange and the New York Stock Exchange each have their own rules, which are similar. A company listed at a particular exchange must comply with general securities reporting regulations as well as the requirements of each exchange on which it is listed. Although little information about the environmental impacts of corporate activities in securities reporting is currently required in securities reporting, this could be made a requirement in the future.

In summary, IR can be a powerful tool in reducing environmental harm, and in some cases can substitute for or supplement command-and-control legislation. However, IR is not always the best choice, and can have unintended deleterious effects. Successful use of IR requires that the information supplied be complete and accurate, and that the public is able to interpret it meaningfully.

See Also: Corporate Social Responsibility; Ecolabels; Emissions Trading; Exposure Assessment; Transparency; Triple Bottom Line; Voluntary Standards.

Further Readings

Applegate, John S. "Bridging the Data Gap: Balancing the Supply and Demand for Chemical Information." *Texas Law Review*, 86/7:1365–407 (2008).
Broll, Udo and Bernhard Eckwert. "The Competitive Firm Under Price Uncertainty: The Role of Information and Hedging." *Decisions in Economics & Finance*, 31/1:1–11 (2008).
Renshaw, Katherine. "Sounding Alarms: Does Informational Regulation Help or Hinder Environmentalism?" *New York University Environmental Law Journal*, 14:3:654–97 (2008). http://www1.law.nyu.edu/journals/envtllaw/issues/vol14/3/v14_n3_renshaw.pdf (Accessed January 2010).
Sunstein, Cass R. "Informational Regulation and Informational Standing: Akins and Beyond." *University of Pennsylvania Law Review*, 147/3:613 (1999).

Lester de Souza
Independent Scholar

INTEGRATED BOTTOM LINE

The integrated bottom line (IBL) is an extension of the triple bottom line concept, which suggests that the environment, society, and the economy are the three critical entities in any business operation. The IBL goes a step further by suggesting that all three of them should be considered in unison in any undertaking. All measures are combined into one balance sheet instead of three independent ones. The environment and society are critical elements of the financial bottom line and should be considered as such. The coinage of the term is attributable to Theo Fergusson of Sustainable Ventures.

The precursor term *triple bottom line* (TBL, 3BL, or People, Planet, Profit) succinctly describes the goal of sustainability—the phrase was coined by John Elkington in 1994 and later expanded and explained in his book, *Cannibals With Forks: The Triple Bottom Line of 21st Century Business*. The term *sustainability* was first defined by the Brundtland Commission of the United Nations in 1987. The notion of this term mandates that a company's responsibility is to the stakeholders rather than the shareholders, where "stakeholders" refers to anyone who is influenced, either directly or indirectly, by the actions of the firm. The goal of any business entity should be to coordinate stakeholder interests, instead of shareholder (owner) profits.

The Three Bottom Lines

"People" or human capital in the triple bottom line refers to fair and beneficial business practices toward employees, customers, community, and region in which the firm conducts its business. An IBL enterprise seeks to benefit as many stakeholders as possible and not to exploit or endanger any of them. The enterprise pays fair wages to its workers, maintains a safe working environment and reasonable working hours, and does not otherwise exploit the community or its labor force. In addition, the business proactively gives back to the community by offering services such as healthcare and education. Quantifying this bottom line is often subjective and problematic. The Global Reporting Initiative (GRI) has developed guidelines to enable firms and nongovernmental organizations (NGOs) alike to comparably report on the social impact of a business.

The use of the word "planet" in the triple bottom line refers to Earth's natural capital and resources. A company using an IBL approach attempts to minimize the environmental impact by reducing its ecological footprint by, among other things, carefully controlling the consumption of energy, reducing manufacturing waste, and careful disposing of toxic materials in a safe and legal way. Managing all facets of the supply chain by addressing cradle-to-grave issues is critical in firms using the IBL approach. This includes conducting a life cycle assessment of products to ascertain the actual environmental cost from the growth and harvesting of raw materials to the manufacturing to distribution to eventual disposal by the end user. Firms focused on IBL also proactively avoid ecologically destructive practices such as endangering depletion of resources such as overfishing or deforestation.

Finally, "profit" in the triple bottom line is the capital earned by businesses. Within an IBL system, the "profit" issues pertain to not just the firm, but to the society at large. Thus, in an IBL approach, profitability is interpreted as traditional corporate accounting as well as social and environmental impact. In other words, traditional objectives of profitability, efficiency, and economic growth are judged juxtaposed with their synchronization with biodiversity, ecological sustainability, equity, community support, and maximized well-being for all types of stakeholders.

Arguments for the Concept

IBL is supported by many because it has allowed firms to find financially profitable niches and hence reach untapped potential, which was missed when monetary gains were the only driving factor. This includes developing goods and services for the underserved markets, for example, building Sun Ovens for donation and retail sale and creating Fair Trade jobs in the process.

Other businesses have successfully adapted to new opportunities in the realm of social entrepreneurship by being socially beneficial, ecologically sustainable, and yet financially profitable. Fair Trade and ethical trade companies require ethical and sustainable practices from their suppliers as well as their service providers.

Finally, with the Earth's carrying capacity at risk and in order to avoid a catastrophic breakdown of climate and nature's services, there is a call for a comprehensive reform in global accounting principles that incorporate the IBL.

Arguments Against the Concept

While most people agree with the significance of improved social conditions and protection of the environment, others disagree with the idea of integrating the three bottom lines as a way to enhance these conditions. First, some believe that companies can best contribute to the welfare of the society by doing what they do best, and hence balancing three bottom lines is likely to divert their attention away from their core competency. Hence, the Sierra Club is best suited to address environmental but not social issues, while the Red Cross is better suited to address social but not environmental issues.

Second, the IBL concept is a rich-country phenomenon, where wealthier citizens feel an increasing desire for a cleaner environment as well as a just society and are willing to financially contribute to a more compassionate society. Yet, as noted by Adam Smith in *The Wealth of Nations*, businesses should focus solely on the task at hand, while the "invisible hand" will ensure that business contributes most effectively to the improvement of all areas of society—social and environmental as well as economic.

It is difficult, however, to achieve a global agreement on simultaneous policies on all three dimensions and, hence, an integrated bottom line may not be enforceable. For example, the shareholders and management of a company may not be prepared to undergo a sustained loss to replenish lost ecosystems.

See Also: Best Management Practices; Body Shop, The; Bottom of the Pyramid; Corporate Social Responsibility; Environmental Indicators; Fair Trade; Genuine Progress Indicators; Pollution Offsets; Recycling, Business of; Social Entrepreneurship; Social Return on Investment; Triple Bottom Line.

Further Readings

Bryson, J. and R. Lombardi. "Balancing Product and Process Sustainability Against Business Profitability: Sustainability as a Competitive Strategy in the Property Development Process." *Business Strategy and the Environment*, 18/2 (2009).

Chappell, Tom. *The Soul of a Business: Managing for Profit and the Common Good.* New York: Bantam Books, 1996.

Elkington, John. "Towards the Sustainable Corporation: Win-Win-Win Business Strategies for Sustainable Development." *California Management Review*, 36/2:90–100 (1994).

Hart, Stuart L. *Capitalism at the Crossroads: The Unlimited Business Opportunities in Solving the World's Most Difficult Problems.* Philadelphia, PA: Wharton School Publishing, 2007.

Painter-Morland, Mollie. "Triple Bottom-Line Reporting as Social Grammar: Integrating Corporate Social Responsibility and Corporate Codes of Conduct." *Business Ethics*, 15/4:352–64 (2006).

Perrini, Francesco and Antonio Tencati. "Sustainability and Stakeholder Management: The Need for New Corporate Performance Evaluation and Reporting Systems." *Business Strategy and the Environment*, 15/5 (2006).

Pham, D. T., P. T. N. Pham and A. Thomas. "Integrated Production Machines and Systems—Beyond Lean Manufacturing." *Journal of Manufacturing Technology Management*, 19/6:695–711 (2008).

Savitz, Andrew W. and Karl Weber. *The Triple Bottom Line: How Today's Best-Run Companies Are Achieving Economic, Social, and Environmental Success—and How You Can Too*. San Francisco, CA: Jossey-Bass, 2006.

Snee, R. and E. Gardner. "Putting It All Together." *Quality Progress*, 41/1:56–59 (2008).

Willard, Bob. *The Sustainability Advantage: Seven Business Case Benefits of a Triple Bottom Line (Conscientious Commerce)*. Gabriola Island, British Columbia, Canada: New Society Publishers, 2002.

Abhijit Roy
University of Scranton

INTEGRATED PEST MANAGEMENT

Integrated pest management (IPM) is an approach to pest management that relies on a combination of simple and economical practices that can be applied in any agricultural or nonagricultural setting where pest management is necessary. The starting point of any IPM approach is a current and systematic understanding of the pest and its interactions with the environment, and an assumption that full eradication is impossible or unnecessary. Based on this, appropriate monitoring, action thresholds, and evaluation methods are established to control pests. In contrast to organic production methods that share many of the same practices as IPM but prohibit the use of chemical pesticides, IPM can combine limited use of pesticides with other available pest control methods that are most appropriate to the specific environment and pest life cycle. IPM is well established in agriculture, but is just beginning to enter into discussions centering on food ecolabels and certification.

IPM evolved out of supervised insect control, an approach developed after World War II by entomologists at the University of California and in the cotton-belt region of the United States. As opposed to standard scheduled applications of insecticides, supervised insect control involved monitoring of pest life cycles and populations and based insecticide applications on projections of pest and natural enemy populations. In the 1950s, University of California entomologists expanded supervised control by further articulating principles governing the compatibility of biological and chemical controls: integrated control meant that chemical controls should disrupt biological control as little as possible and should only be used if the economic loss caused by the pest in question was projected to exceed the cost of chemical control. Over time, other measures such as cultural controls—the manipulation of the plant's environment—became part of IPM as a multidisciplinary cast worked to develop the approach and legitimize IPM even into national policy in the 1970s. Recent debates surrounding IPM have focused on the question of whether IPM should be geared toward reduced pesticide use or maintaining pesticides as a viable tool.

IPM can be applied to any type of agriculture as well as to commercial and residential sites, lawns, and gardens. Within agriculture, IPM is highly compatible with smaller-scale organic agriculture, given its reliance on a labor-intensive cycle of observation, monitoring, and combinations of control techniques. In larger-scale agriculture, IPM can reduce negative impacts on human and environmental health by possibly lowering costs with reduced pesticide application. An IPM system works with a set of basic tools that generally include proper identification of the pest, the establishment of an action threshold, preventative cultural practices, monitoring, mechanical controls, biological controls, and chemical controls. The idea of an IPM system is to control the pest population rather than eradicate it. Eradication measures that are chosen must be less costly and less environmentally disruptive than nonaction. To lower impact and costs, preference is also given to prevention measures rather than to intervention.

How IPM Works

To start with, it is important to properly identify the pest and the host plant, as misidentification can result in costly and ineffective actions. Observation and monitoring of the pest should begin upon identification before it actually becomes a problem, to establish the presence or nonpresence of the pest, the distribution of the pest within the target field or site, and whether the pest population is increasing or decreasing. Acceptable pest levels are then established, and these, known as action thresholds, are site specific, meaning that the action threshold will differ from one field or crop to the next, based on economic, health, or aesthetic requirements as well as the crop's resistance to or tolerance of a specific pest. The economic threshold is the point at which the cost of pest damage exceeds the cost of control. The health threshold is the point at which they become a health hazard, generally a situation of low tolerance. The aesthetic threshold is where the pest causes cosmetic damage, which is more tolerated than economic or health damage. Varying the action threshold between fields assures that genetic diversity is maintained in the pest population and nonresistant pests will survive chemical pesticide treatment alongside resistant pests to reproduce, maintaining lower resistance. For instance, if a farmer sprays insecticide on her soybeans four times yearly regardless of the caterpillar pest's life cycle, it is likely that the genetically resistant individuals will survive the spraying to reproduce, repopulating the field with a resistant population. Approaching the control of caterpillars with different action thresholds in each field retains the genetic variety of the caterpillar population and prevents the development of a superpopulation of resistant individuals.

Extensive knowledge of local growing conditions is necessary for selecting appropriate varieties and maintaining healthy crops. This and other preventive cultural practices, including crop sanitation and quarantine techniques, proper waste disposal, and mulching to prevent weed growth, are major tools in preventing infestation and avoiding costly interventions later. Constant monitoring and record keeping of pest behaviors and reproductive cycles is key to projecting outbreaks and planning preventative action measures. If pest levels become unacceptable, mechanical controls such as traps, barriers, or tilling to disrupt breeding cycles are first utilized. Biological controls, which include promoting natural predators to control pest populations, as well as the use of biological insecticides are also an inexpensive and low-impact option. Chemical controls, including the application of insecticides, herbicides, and horticultural oils, are used as a last resort when deemed necessary and appropriate to the pest's life cycle. Plant-based pesticides are also available.

Pest management tactics in an IPM system generally include a combination of the mechanical and physical, cultural, biological, and chemical controls outlined above, determined by taking into account local growing conditions and action thresholds. Evaluation of the effectiveness of the combination of strategies utilized is the next step. Determining if the actions taken had the desired impact, if the pest was prevented or controlled satisfactorily, and if there were any unforeseen negative economic, environmental, or health impacts is key to understanding the overall effectiveness of the IPM program. With this understanding, the IPM manager can then modify the set of strategies to be used in the future, to work further toward keeping pest populations below the action threshold.

What IPM Is Not

Integrated pest management is not a management strategy with the express goal of reducing pesticide use; indeed, pesticides are one tool in the overall toolbox of IPM and are to be used where their application is less economically and environmentally costly than the damage or injury caused by not controlling the pest in question. The implication is that an IPM plan can just as easily lead to increased pesticide use as to decreased use, depending on specific context and decision-making processes of the IPM manager. For this reason, IPM's place within policies to reduce or regulate chemical pesticide use is questionable. Besides IPM's often mistaken identity as a "green" strategy within alternative agriculture circles, it does, in fact, promote interactive control strategies, like mechanical and biological controls, that are often associated with "green" agriculture. The challenge that promoters of IPM face is the formation and implementation of a system of value-added products and services that would make IPM a financially viable industry in and of itself. The argument that IPM will never be viable without a financially viable pest management industry and as long as it is guided by subjective "moral" arguments rather than value-driven decisions, it faces a challenge as IPM moves to backyard gardening and small-scale agricultural settings where managers are well exposed to the bad publicity surrounding chemical pesticides in the marketplace.

One hurdle that IPM faces in achieving a distinct and recognizable identity in the market place—in terms of both products and services used in IPM places of production as well as in products grown or produced within IPM systems—is the lack of branding of IPM. Unlike organic foods certified under the U.S. Department of Agriculture, there is no national certification for foods grown using IPM. The major obstacle facing promoters of IPM foods is that IPM processes and practices differ not only from crop to crop, but from field to field. It is impossible, therefore, to formulate a universal set of standards that can be used to certify an agricultural field as IPM. However, there are efforts by individual commodity growers to define what IPM means for their individual crops and regions, and the market is beginning to see IPM-labeled foods in limited areas. Another challenge is that IPM deals only with pest management and does not integrate other facets of sustainable agriculture like soil fertility, water management, and waste management. The final challenge also remains of how to communicate the benefits of IPM-grown foods to consumers, when IPM is not so clearly "green" as other labeled foods, like organic.

See Also: Environmental Marketing; Newman's Own Organics; Organic; Stonyfield Farm; Sustainability.

Further Readings

Dent, David. *Integrated Pest Management*. London: Chapman & Hall, 1995.

Kogan, Marcos. "Integrated Pest Management: Historical Perspectives and Contemporary Developments." *Annual Review of Entomology*, 43 (1998).

Radcliffe, Edward B., William D. Hutchison and Rafael E. Cancelado. *Integrated Pest Management: Concepts, Tactics, Strategies and Case Studies*. Cambridge, UK: Cambridge University Press, 2009.

van den Berg, Henk and Janice Jiggins. "Investing in Farmers—The Impacts of Farmer Field Schools in Relation to Integrated Pest Management." *World Development*, 35/4 (2007).

Heather R. Putnam
University of Kansas

International Organization for Standardization

In an era of rapid globalization, technological innovation, population boom, and environmental change, the need for international collaboration and standardization exists across disciplines in order to decrease duplication and the reinvention of competing standardizing platforms and processes for business, government, and society. Accordingly, the International Organization for Standardization (ISO) was founded in 1947 to address these needs and to facilitate knowledge transfer for the sake of collaborative standardization within a global civil society. Today, the ISO is a network, composed of the national standards institutes of 162 countries, and is headquartered in Geneva, Switzerland. It runs under the auspices of the United Nations. It is a nongovernmental organization (NGO) intended to form a bridge between the public and private sectors, and which strives to find solutions that meet the needs of both business and the general public. Each country has one representative, and some are part of their country's government, while others have their roots in business and industry.

The mission of the ISO is to partner with a global network of standardizing bodies and stakeholders in order to identify, synthesize, arbitrate, and bring forth consensus on voluntary standards for operational platforms and practices in areas such as business, technology, environmental management, government, and public policy. Many of these standards, particularly the ISO 9000 and ISO 14000 family, are relevant to sustainability and other environmental concerns, and are therefore of interest to companies and organizations who want to "go green," and use that choice as part of their public image. The ISO certification process provides an important means for businesses to organize and implement their environmental concerns while also assuring stakeholders that the business is adhering to sustainable management practices.

Although the ISO standards are voluntary, they have become increasingly relevant for business and governmental bodies, as consumers demand that organizational processes, policies, and products meet a globally held standard as a threshold of competence and quality. In line with these global market trends, consumers and governments are demanding the private sector, as responsible community actors, meet a level of transparency and organizational competence. As such, ISO standards have greatly influenced the field and the practices of corporate social responsibility (CSR), global organizational development, and environmental management.

ISO 9000 is a family of quality-management standards intended to provide a framework for companies and organizations to ensure customer satisfaction and a best-practice level of organizational competence and is intended to be both applicable and achievable by all organizations, regardless of size, location, or industry. For instance, ISO 9001:2008 (meaning that it was passed in 2008) includes requirements that the company regularly monitor processes for effectiveness, keep adequate records, check output for defects and make corrections as necessary, monitor the overall quality system for effectiveness, and facilitate continuous improvement. If an independent auditor certifies an organization to be in conformance with the requirements of ISO 9001, the company can advertise that it is "ISO 9001 certified," a distinction that can increase its competitiveness in the global marketplace. Companies often prefer to do business with other companies who take operational excellence seriously, and ISO 9001 certification can be an important credential in establishing credibility in this field. The certificate must be renewed regularly so that it provides an ongoing indication that quality assurance procedures are being used within a company. It should be noted, however, that ISO 9001 applies to business processes and does not directly certify the quality of the product or services produced by a firm, only that they are following a set of procedures that are associated with quality.

The core organizational and managerial requirements of the ISO 14000 body of standards have much in common with the practices of ISO 9000. The ISO 14000 standards focus on environmental management systems (EMS) and provide internal and external guidelines on implementing, managing, auditing, and reporting on organizational practices that impact the environment. The ISO 14000 standards also set expectations that an organization's EMS program can provide assurance to external stakeholders that there is a level of quality environmental stewardship occurring at the organization in question.

No certification process is perfect, and the ISO process has been criticized on several grounds. The first is that large amounts of time and money must be devoted to the paperwork required for certification, and there is some question as to whether that time and effort is well spent. It has also been noted that companies that choose to apply for certification are often well on the way to adopting more environmentally sustainable practices, and that the certification process is primarily a matter of documenting what has already taken place. Additionally, the 9000 and 14000 standards apply to management procedures rather than to documenting, for instance, that the products manufactured actually are of the specified quality or that pollution has been reduced (which are the kind of questions of interest to stakeholders), but rather simply documents that the companies follow specified procedures. This, at worst, may represent the latest fads in management rather than procedures that truly help achieve the stated ends. In addition, because auditors receive their fees from the companies they audit, there may be temptation to certify a firm that is not truly qualified—a danger exacerbated when there are multiple auditing organizations competing for business.

ISO Functioning

The ISO is a democratic-style assembly whereby members of the ISO participate in a voting process. Each member body has one vote with equal influence. Standard setting is ultimately a market-driven process. As a result, ISO standards often influence laws and policy created by member countries by virtue of the ISO members' eliciting feedback from various subject matter experts, the public, and targeted consumer groups. The ISO does not, however, hold authority or jurisdiction over each of the member countries.

ISO members propose new standards and participate in the research and development process of new standards. They typically appoint delegation committees that work with a variety of field experts to engage in the vetting of new standards and processes. Annually, the ISO reports, over 50,000 experts participate in the ISO's standard-setting process.

ISO standards apply to many fields besides quality management and environmental management, including agriculture, engineering, construction, environmental management, product interoperability, medical safety, and technical specifications and terminology. The ISO reports that since its inception, it has established 17,000 standards that impact nearly every aspect of the public and private sectors. The ISO also has a strategic partnership with the World Trade Organization (WTO) in order to promote practices that reduce barriers to global trade.

As an example of the ubiquity of ISO standardization, within the field of information technology, computer files and related products are often referred to as ISOs, as they are produced under the standard filing system created by the ISO. Accordingly, this ISO standardization has influenced the development of many well-known and widely used products, such as the ISO 9660 computer disk image standard that has led to the creation of the CD-ROM and the DVD. Through this use of international standard setting, consumers and manufacturers alike can rely on the knowledge that products like the CD-ROM can be run on all computer systems that have a CD-ROM drive.

Overall, by working to set standards across boundaries, reduce barriers to trade, and provide standards for environmental practices, the ISO is a key player in supporting a globally sustainable civil society. Although the ISO certification process has been criticized on several grounds, it still provides an important tool for businesses who wish to adopt more green practices, and provides evidence to other businesses, governments, and customers that a company has met an accepted standard for environmental quality and sustainability.

See Also: Global Reporting Initiative; ISO 14000; ISO 19011; Leadership in Green Business; Supply Chain Management; Voluntary Standards.

Further Readings

International Organization for Standardization. "International Standards for Government, Business and Society." http://www.iso.org/iso/standards_development.htm (Accessed April 2009).

Murphy, Craig and Joanne Yates. *ISO, The International Organization for Standardization: Global Governance Through Voluntary Consensus*. London: Routledge, Taylor & Francis, 2009.

Jeff Leinaweaver
Fielding Graduate University

ISO 14000

The International Organization for Standardization (ISO), headquartered in Geneva, Switzerland, develops international standards for products, services, processes, materials, and systems. These standards carry the ISO logo—the name derived from the Greek word *isos*, meaning "equal." Currently, there are 162 member organizations, generally representing countries, involved in the standard-setting process. This organization does not carry out certification of conformity to its own standards.

ISO 14000 is a series of international standards designed to meet the needs of business, industry, governments, and consumers in the environmental field. These standards are developed by ISO Technical Committee 207. They have produced consensus work products such as the basic standard for environmental management systems—ISO 14001. There are a number of other environmental standards in the ISO 14000 family, including ISO 14015 (environmental assessment of sites and organizations), ISO 14020 (environmental labels and declarations), ISO 14040–14049 (life cycle assessment), and ISO 14063 (environmental communication). A complete listing of accepted standards and developing standards can be found on the ISO website.

The United States participates in the development and review of these standards with a Technical Advisory Group (TAG) that reports to TC 207. The TAG is an American National Standards Institute (ANSI) committee with the American Society for Quality (ASQ) serving as the secretariat. Other countries have similar arrangements through their own organizations.

The first two standards in the ISO 14000 series are ISO14001:2004 and ISO 14004:2004. (Please note that the date of the most recent version is provided following the colon after the standard's number.) ISO 14001:2004 provides the requirements of the management system and ISO 14004:2004 provides some guidelines for organizations seeking to plan and to implement an environmental management system. There is another standard, ISO 19011, that specifies how this conformance should be conducted and the competencies required by the environmental management system auditors.

An organization begins the ISO 14001 process by determining the scope of coverage. Then it must create an environmental policy that meets a number of important criteria including three commitments:

- Compliance with legal requirements and with other requirements to which the organization subscribes
- Dedication to the prevention of pollution (e.g., pollution prevention or cleaner production)
- Dedication to the demonstration of continual improvement

The rest of the ISO 14001 management system standard is based on the well-accepted methodology known as "plan-do-check-act" (PDCA).

The planning involves the creation of an environmental footprint for the organization's activities, products, and services. Included are the environmental aspects (i.e., how each activity can interact with the environment) and the environmental impacts (i.e., any changes, adverse or beneficial, to the environment resulting from the environmental aspects). The organization must use risk management to determine the significant environmental aspects. Part of the determination of significance includes the legal and other requirements related to the activity associated with the environmental aspect. The organization then establishes objectives and targets, taking the significant aspects and legal and other requirements into account. An environmental management program is established to guide the projects designed to achieve the objectives and targets. All the projects have definitive action plans.

The establishment, implementation, and maintenance of the environmental management system include provisions for each of the following components:

- Resources, roles, responsibility, and authority
- Competence, training, and awareness
- Communication—internal and external to the organization
- Documentation
- Control of documents and records
- Operational control
- Emergency preparedness and response

Checking the environmental management system includes the following provisions:

- Monitoring and measurement
- Evaluation of compliance
- Corrective actions and preventive actions

Internal Audits

Finally, the act provision involves the review of all of these components by top management at planned intervals. These reviews ensure the continuing suitability of the management system and its adequacy and effectiveness. Opportunities to improve the management system are used to drive the organization's commitment to continual improvement. A company may gain several advantages by obtaining ISO 14001 certification. The first is a more efficient and productive business through reduced use of energy, raw materials, and pollution, and lowered risk of accidents. Two more are increased market access and improved corporate image: meeting an internationally recognized environmental standard demonstrates commitment to quality as well as environmental responsibility. Finally, obtaining the certification can increase confidence of shareholders, insurers, and investors that the company is well run and will continue to be competitive in the marketplace.

Of course, some organizations obtain ISO 14001:2004 certification only because a customer requires it, and thus there may be a temptation to institute a program that is minimally conforming to the standard. However, other organizations take to heart the ISO 14001 definition of environmental management system: "part of an organization's management system used to develop and implement its environmental policy and manage its environmental aspects." This means that the environmental aspects of the activities, products, and services become part of the way the organization itself is managed—part of the work of every employee and part of every management decision. Additionally, the improved efficiency and competitiveness that are assumed to follow conformity to the ISO 14001 standard should help make the process self-sustaining.

For instance, ALcontrol Laboratories, a European environmental and food laboratory testing organization operating in the United Kingdom and Europe, claims that the money invested in going through the ISO 14001 certification process has resulted in operational savings more than 10 times the certification cost. Other internal benefits cited by

ALcontrol include assurance that they are in compliance with environmental legislation, and improving the company's "green" image. Environmental results include substantial reduction in electricity use (over 20 percent at some of the company's locations) and in liquid waste sent for recycling (over 40 percent at some locations).

ISO 14001:2004 provides four options for an organization to demonstrate conformity to this international standard:

1. Self-determination and self-declaration

2. Seeking confirmation of conformance by parties with an interest in the organization, such as customers

3. Seeking confirmation of conformance by parties external to the organization

4. Seeking certification/registration by a certified external organization

The fourth option includes certification (registration) and accreditation. The rules for accreditation and registration are developed by ISO's Committee on Conformity Assessment (CASCO). These CASCO rules are Guide 61 (the rules governing accreditation bodies), Guides 62 and 66 (the rules governing registrars), and ISO 17024 (the rules for organizations that certify individuals). There is currently an ISO standard-setting effort under way to replace Guides 62 and 66 with a new ISO standard, ISO 17021, Conformity Assessment; Requirements for Bodies Providing Audit and Certification of Management Systems.

In order to be certified, an organization must follow the rules established by the registrar who issues it a certificate, and, in turn, the registrar must follow the rules established by the accreditation body that accredits that particular registrar. These rules may be more stringent than the requirements established within ISO 14001 and ISO 19011.

In the United States, the accreditation body that certifies ISO 14001 registrars is the ANSI/ASQ National Accreditation Board (ANAB). There are several other accrediting bodies around the world. These accreditation bodies have entered into a mutual recognition agreement to recognize each other's accreditations as members of the International Accreditation Forum (IAF).

In order to be accredited, each registrar must develop written procedures to govern its ISO 14001 registration process. Registrars are subject to periodic surveillance audits by their accreditation bodies to ensure audit reliability and impartiality. One of the requirements registrars must meet is to use competent auditors. RABQSA is an organization that provides a program for certifying that individual environmental and quality auditors have met auditor competency requirements. Registrars may, but do not have to, use RABQSA-certified auditors.

See Also: Certification; Environmental Audit; Environmental Management System; Environmental Risk Assessment; International Organization for Standardization; ISO 19011.

Further Readings

Icon Group International. *International Organization for Standardization: Webster's Timeline History, 1900–2007.* San Diego, CA: ICON Group International, 2009.

International Organization for Standardization. http://www.iso.org (Accessed November 2009).
Miller, Frederic P., Agnes F. Vandome and John McBrewster. *International Organization for Standardization*. Beau Bassin, Mauritius: Alphascript Publishing, 2009.

Robert B. Pojasek
Harvard University

ISO 19011

ISO 19011 is an international standard of the International Organization for Standardization (ISO) that sets detailed guidelines for quality management systems auditing and environmental management systems auditing and provides guidance on the competence of quality and environmental management systems auditors.

Founded in 1946, the ISO is a global standard-setting organization headquartered in Geneva, Switzerland, that promulgates international, industrial, and commercial ISO standards. The ISO 19011: 2002 standard was prepared and published in October 2002 by ISO in accordance with the rules given in the ISO/IEC (International Electrotechnical Commission) Directives, part 3. The standard was produced in response to a growing demand from the commercial sector for guidance on combined management systems audits. The cost of failing an accreditation/reaccreditation audit by an awarding body, such as the British Standards Institute, referred to officially as the "third-party audit," can affect an organization in a number of ways. These include suspension or withdrawal of certification leading to loss of clients, costly improvements to policy or procedures and working instructions, and hidden costs associated with inefficiencies of internal reviews.

Through its adoption, ISO 19011 provides a "best practice template" by clarifying the objectives of the environmental or quality audit program, seeking commitment to the audit program and individual audit goals, reducing duplication when instigating dual environmental and quality audits, ensuring that audit reports conform to the ideal template and record all essential data, and benchmarking the competence of the audit team against suitable criteria. ISO 19011 can also be adapted for auditing any management system.

ISO 19011 provides four instrumental decision/support resources to successfully scope the audit plan, audit execution, and critical evaluation of environmental and/or quality audit:

- A clear explanation of the principles of management systems auditing
- Guidance on the management of audit programs
- Guidance on the conduct of internal or external audits
- Advice on the competence and evaluation of auditors

The fundamental role of an audit is to record a sample of evidence to confirm the validity and suitability of an organization's data and to "independently" assess an organization's internal control procedures. The audit report compares planned conformance to actual conformance and provides a mechanism for continuous improvement. In order to maintain its credibility and acceptance auditing is dependent on five fundamental principles: ethical conduct, fair presentation, due professional care, independence, and evidence-based approach. Ethical conduct is essential to maintain the reputation of the auditing profession. Fair presentation of the audit exercise must be truthful and accurate and based purely on the evidence discovered. Due professional care must be maintained throughout by the auditor/auditing team, so it is essential that the auditors are qualified and competent. Independence of auditors is fundamental to maintain objectivity and impartiality while carrying out the on-site audit, auditing report findings, and presenting these findings to the management team. An evidence-based approach means that results must be verifiable in that samples are recorded on the observation sheets and documented in the final audit report competently, reflecting the variance between actual performance and the applicable ISO standard.

The audit program may consist of one or more audits that are dependent on the size and operations of the organization. Each audit may be carried out by a single auditor, or an audit team coordinated by a lead auditor. Senior management must take an active role in the audit program, a requirement particularly relevant during a third-party accreditation audit. All participants must be comfortable with the auditing principles and have the relevant management skills and technical/operational competence. The senior management team must also ensure that the audit program is adequately resourced. Clarification of the objectives of the audit program is paramount to all parties; examples of objectives include to satisfy the criteria for certification, to meet contractual requirements, to demonstrate competence in the supply chain, or to act as a catalyst for continuous improvement. In addition, an audit schedule must reflect the scope of the operations of the organization that is seeking or has ISO certification. Over a set period of time, normally 12 months, each process must be audited. The audit observation sheets record evidence to support conformance, nonconformance, or observations, and normally three examples of evidence are recorded to support each finding. The lead auditor then documents the findings/conclusions in an audit report that is presented to management, which may result in continued certification, suspension of certification pending a follow-up audit, or a formal withdrawal of certification.

Prior to the on-site audit, the auditee's documentation undergoes a desktop review to assess conformance against the relevant ISO standard. An audit plan is then designed, which may include a checklist that will vary depending upon internal and external audits. However, all nonconformances and observations from the previous audit will also be checked during the initial on-site audit to ensure corrective action has been taken. If this is not the case, then the organization's certification or application could be withdrawn.

The on-site audit commences with an opening meeting between the auditor and the management team to confirm the audit plan, audit activity schedule, lines of communication, guides and observers, and a consultation and feedback meeting with senior management. The auditee should not attempt to influence or interfere with the conduct of the audit. The auditor should be sensitive to the confidential nature of the business operations and systematically record samples of verifiable evidence. It is understood that an audit cannot obtain a complete holistic picture but rather a snapshot of observable evidence. There is, therefore, an element of uncertainty in auditing and this needs to be made explicit to the auditee. The audit evidence is evaluated against the audit criteria and documented in an audit findings report. The audit findings may indicate the need for corrective, preventive, or improvement actions. The report is then presented to senior management to highlight the extent of conformity. If agreement cannot be reached between both parties all opinions should be recorded. In the case of third-party audits, the auditee has the right to appeal to the accrediting body and request a second audit. If the initial audit findings are upheld, then the auditee must pay the costs of the second audit.

Overall, the reliability and credibility of the audit process rests upon the competence of those conducting the audit. The auditors needs to have the essential personal attributes and generic knowledge and skills to allow him to perform his duties in accordance with the principles of auditing and to operate competently and professionally within diverse operational environments. It is, therefore, paramount that auditors are not only suitably qualified but have relevant experience in conducting audits and a career profile reflecting credible work experience. In addition, auditors should be able to produce a generic personal development plan highlighting their continuing professional development.

The ISO 19011 standard has an indirect but important effect on reducing the environmental impact of organizations. The purpose of the standard is to improve the auditing of environmental management systems by standardizing the auditing process and making it more reliable and efficient. This standard should encourage more companies to institute environmental management systems (processes and practices which help reduce the environmental impact of an organization) and help to make preexisting systems operate more efficiently by ensuring that they will be evaluated regularly and fairly, by competent auditors, and with a resulting audit report which is objective and verifiable.

See Also: Certification; Environmental Assessment; Environmental Audit; Environmental Impact Statement; Environmental Risk Assessment; International Organization for Standardization; ISO 14000.

Further Readings

International Organization for Standardization (ISO). *Quality Systems—Model for Quality Assurance in Design, Development, Production, Installation and Servicing, International Standards ISO 9001:2008(E)*. Geneva: ISO, 2008.
ISO Directives. "Part 1: ISO/IEC Procedures for the Technical Work," 5th Ed. Geneva: ISO, 2004.
ISO Directives. "Part 2: ISO/IEC Rules for the Structure and Drafting of International Standards," 5th Ed. Geneva: ISO, 2004.
ISO 19011. *Guidelines for Quality and/or Environmental Management Systems Auditing*. Geneva: ISO, 2002.

Derek Watson
Mitchell Andrews
University of Sunderland

LEADERSHIP IN GREEN BUSINESS

Traditionally, many mainstream business leaders have argued that environmental protection, social responsibility, and economic success are fundamentally incompatible goals. The concept of green business challenges this assumption. Also known as "sustainable business," green business ideas demand that managers make choices that meet contemporary economic needs without compromising the ability of future generations to meet their own needs.

Because green business is an emerging concept, the implementation of green business practices faces many obstacles. Going green must be a clear company priority endorsed at the highest levels of leadership, who must then communicate with lower-level managers and employees and deal with their concerns. Reducing environmental impact may increase costs, at least at first, and may seem less efficient than the practices the company is familiar with. Suppliers, who have their own concerns about profitability, may be unwilling to comply with requests to green their own processes unless sufficient incentives are offered. And even after the decision is made for a company to go green, this may conflict with deeply entrenched practices and market regulations based on economic theories which do not consider externalities such as climate change and environmental degradation relevant to an economic analysis.

Given this context, strong leadership is necessary for any business attempting to adopt more environmentally friendly practices. Green business leadership requires that management think broadly and proactively and devise business models that deliver financial, social, and environmental benefits.

Sustainability and the Triple Bottom Line

One of the most valuable tools for businesses to become more sustainable is the concept of the "triple bottom line," a measurement system that examines the impacts of business actions on three core areas: people, profits, and planet. For business practices to be considered sustainable, they must take into account three areas of impact: society, the environment, and the financial health of the company.

A company can be considered environmentally sustainable if it replaces 100 percent of the natural resources it consumes and has no negative impact on the natural environment.

It can be considered socially responsible if the company's decision makers act ethically, respect human rights, maintain health and safety, compensate employees and suppliers fairly, and ensure a high quality of life. Yet, if a company is not economically sustainable, that is, if it does not maintain a healthy balance between income and expenses, then it will not survive.

Balancing these three demands is not easy, but green business leaders argue that doing so is a moral imperative. Earth's biosphere is in steep decline and our industrial economy is the primary cause of this degradation. As Ray Anderson, founder and chairman of Interface stated, "The only institution on Earth that is large enough, wealthy enough, persuasive enough and influential enough to really lead humankind out of the mess it's making for itself is the same one doing the greatest damage: the institution of business and industry."

Going green need not be motivated entirely by altruism, however. Many major businesses have chosen to reduce their environmental impact, yet have retained or even enhanced their profitability. For instance, the furniture manufacturer IKEA has incorporated environmental awareness into their corporate image, and their website touts their many initiatives, from requiring documentation of the origins of wood used in their products to reduced packaging that reduces waste and lowers the amount of fuel required to transport products. Similarly, the outdoor clothing manufacturer Patagonia widely publicizes its concrete actions to reduce the environmental impacts of its products and safeguard the health and safety of those who work in its factories, as well as donates money to environmental causes and encourages other companies to do likewise through the alliance "1% for the Planet," founded by Patagonia.

What Is Green Business Leadership?

Green business leadership can be defined as the ability to think broadly and proactively to devise business models that deliver financial, social, and environmental benefits. According to a 2009 study by the Massachusetts Institute of Technology (MIT), green business leaders respond to environmental concerns first and consider social pressures, like consumer demand and regulation, second. Green business leaders acknowledge that economic values are not "the only values in the world" and do not automatically favor cost-based arguments yet are always looking for ways to improve the environment while also improving their bottom line.

At the same time, green business leaders are able to make a strong business case for sustainability: not only are consumers demanding greener practices and more transparency around these practices, but shareholders and powerful stakeholders such as nongovernmental organizations (NGOs) are also pressuring businesses to go green and address their own roles in environmental degradation. Green leaders are able to envision how sustainability can become the driver of innovation, which, in turn, is said to drive competitive advantage.

Despite the emergence of compelling reasons for businesses to go green, green business practices remain nonexistent or underdeveloped in the majority of North American companies. Less than a dozen Fortune 500 companies consider sustainability a key business strategy, and small businesses (which make up 98 percent of all companies in the United States) are even more reluctant to adopt a green strategy, largely because the financial incentives are not sufficient.

For this reason, green business leadership requires an entrepreneurial spirit, and a willingness to take risks and forge a new path. Green business leaders are game changers according to Joel Makower, people "who can envision a future that combines elements of the present in unexpected, creative, and productive ways."

There is no one method for green business leadership, but there are a number of relevant principles that have been proven successful thus far:

- Think holistically.
- Demonstrate courage, personal commitment to a vision, and strong analytic and communication skills.
- Articulate the company's green vision and repeat it often.
- When communicating green business changes, be sincere. Both consumers and investors need to see a company's commitment to the environment, not just hear about it. Consumers can detect greenwashing (the process of overinflating a company's environmental achievements) and will reject a greenwashed product when they see it.
- When analyzing problems, be careful to define them correctly. Most environmental problems have technical, emotional, economic, and social dimensions that must be addressed.
- Look for solutions that benefit all stakeholders. This includes suppliers, distributors, consumers, shareholders, regulatory bodies, and also the environment.
- Find ways to make benefits of green business tangible to employees, shareholders, consumers, suppliers, distributors, and other stakeholders.

Although some traditional business leaders still maintain that there is a conflict between environmental concerns and business profitability, many companies are finding that going green can actually improve the company's bottom line. However, pressure for a company to change its environmental practices is unlikely to succeed unless the owners and top management buy into the process and make it a clear priority. Green leadership at the highest levels of a company is necessary in order to successfully adopt practices that allow the company to remain economically competitive while also reducing the negative impact it has on the environment.

See Also: Balanced Scorecard; Body Shop, The; Ceres Principles; Ecological Footprint; Environmental Accounting; General Electric (GE); Herman Miller; IKEA; Newman's Own Organics; Patagonia; Sainsbury's; Seventh Generation; Social Return on Investment; Stonyfield Farm; SunEdison; Systems Thinking; Triple Bottom Line.

Further Readings

Estes, Jonathan. *Smart Green: How to Implement Sustainable Business Practices in Any Industry and Make Money.* Hoboken, NJ: John Wiley & Sons, 2009.

Hawken, Paul, Amory Lovins and L. Hunter Lovins. *Natural Capitalism: Creating the Next Industrial Revolution.* New York: Back Bay Books, 2008.

Makower, Joel. *Strategies for the Green Economy: Opportunities and Challenges in the New World of Business.* New York: McGraw-Hill, 2009.

Winston, Andrew S. *Green to Gold: How Smart Companies Use Environmental Strategy to Innovate, Create Value, and Build Competitive Advantage.* Hoboken, NJ: John Wiley & Sons, 2009.

Maura Troester Nuñez
University of Colorado, Boulder

LIFE CYCLE ANALYSIS

Life cycle analysis (or assessment) or LCA, is a method for evaluating the total environmental impact of a product or service. The major work involved in conducting an LCA is to compile all the inputs and outputs of a product or a system of products, and to evaluate the potential environmental impacts of a product or system throughout its life cycle, from extraction of raw materials through manufacturing, use, and disposal.

LCA is a methodological tool that helps calculate the total environmental costs (including externalities such as pollution) of a particular product or service over its lifetime or to compare the costs of different options by applying a life cycle perspective of activities related to processes or products. The life cycle perspective focuses on the entire life cycle of a good or service, starting from the extraction and processing of raw materials, followed by manufacturing, transportation and use, and ending with waste management as well as recycling and final disposal. Each of these life cycle stages consumes energy and resources, and generates emissions and wastes, which result in a number of environmental impacts. Without LCA, many of these impacts may be overlooked.

A full LCA is sometimes called "cradle-to-grave" because it covers all phases from manufacture to disposal. This type of LCA can provide information to help consumers, governments, and other purchasers evaluate the full environmental costs associated with existing consumption and production strategies, preventing a piecemeal approach. An LCA can help avoid shifting the environmental burden from one stage to another, from one geographic area to another, and from one environmental domain or protection target to another. For example, focusing on making improvements in one part of the life cycle (for example, in production) may lead to even higher impacts in other parts of the same life cycle (for example, the product use), and without LCA, the added impacts might not be considered. To take another example, without LCA, it would be very easy to overlook the fact that a reform that reduced one type of environmental impact associated with a product (for instance, air pollution) might increase the impact in another part of the environment (for instance, solid waste in a landfill).

In an LCA, the use of resources, raw materials, parts and products, energy carriers, electricity, and so on, are documented as inputs for each single step in a process. Emissions to air, water, and land, as well as waste and other by-products, are recorded as outputs. For inputs not coming directly from raw materials (for example, other manufactured products), their life cycle, which is often referred to as their environmental history, has to be included in the analysis as well. For outputs, the subsequent processes they undergo need to be included in the analysis also.

For example, in the case of a car, raw materials are first extracted from the Earth, such as mineral ores, water, and oil. Second, raw materials are processed into finished materials, such as bauxite ore, that is processed into aluminium, and oil that is refined into plastics. Third, the materials are manufactured or assembled into a final product—in this case, a car. Often, this stage can be split into two parts. In the first, materials are manufactured into parts (for example, an aluminium sheet is manufactured into an automobile body panel). The parts are then assembled into a final product (for example, the body panel, along with the windows, engine, and the multitude of parts that are assembled into a car). Fourth is the use stage, when a consumer has control of the product: the car, for example, will mainly consume oil refined into gasoline and will emit pollutants into the air. Finally comes the end-of-life stage, when the product is broken down into component materials for remanufacturing or recycling, or is discarded.

A sixth stage of distribution, when the materials and products are transported between stages, should be also taken into account.

As can be seen in the example mentioned above, carrying out an LCA is a lengthy and very detailed exercise: the data collection stage is relatively uncomplicated, provided the boundary of the study is clearly defined, the methodology is rigorously applied, and reliable, high-quality data are available.

Once all the inputs and outputs are taken into account, the sum of inputs from and outputs to nature is the basis for the analysis and assessment of the environmental effects related to the product. This phase is often referred to as life cycle impact assessment.

There are a number of ways to conduct life cycle impact assessment. While the methods are typically scientifically based, they are heavily dependent on the quality of data provided and the accuracy of assumptions used, while the complexity of environmental systems often make the accuracy of the studies difficult to evaluate. As mentioned earlier, an LCA can help identify potential environmental trade-offs. However, converting the impact results to a single score, and thus in simple policy measures, requires the use of value judgments, which must be applied by the commissioner of the study or the analyst. This can be done in different ways, such as through the use of a panel of experts or of stakeholders, but it cannot be done based solely on natural science.

LCA has found widespread applications in environmental analysis, research, and policy making, as well as in product development and marketing of products. For instance, information from an LCA is used to differentiate the impacts of two comparable products. Perhaps a university cafeteria is trying to decide if they should use plastic or paper cups instead of glassware, and wishes to choose the alternative with the least impact on the environment. To conduct the analysis, they would have to collect information on the raw materials required for manufacture (for example, petroleum, trees, or sand) and the different types and amounts of energy to produce either type of cup, as well as the wastes and emissions from the manufacturing process.

They would also gather information about the environmental costs of use; for instance, the plastic or paper cups would create more solid waste because they are used only once, while glass would require more water and energy resources because they have to be washed and sterilized. Factors such as the environmental impact of transportation would be considered. Perhaps the glass cups must be purchased from a plant 1,000 miles away, while the paper cups are available from a local plant, and glass cups are much heavier, so would require more fuel per cup to transport. Packaging would also be included in the analysis.

LCA is also used to assess design options for the same product. For example, automobiles use a wide variety of materials in their various parts. Steel has typically been used, but plastics and composite materials have been replacing its use. Steel is heavier than the plastics or composites, adding weight to the car, which increases the fuel needed to operate the car, but steel parts are easily recycled at the end of the vehicle's life.

LCA is a powerful tool for evaluating the environmental impacts of products and services. The International Organization for Standardization (ISO) has created a framework for LCA in the 14040 series that includes descriptions of four distinct phases (goal and scope definition, inventory analysis, impact assessment, and interpretation). However, the complexity of the process (including the boundaries of the life cycle), the quantity of data, and the amount of interpretation that goes into the process means that conducting an LCA remains an art as well as a science.

See Also: Ecological Economics; Environmental Assessment; Environmental Risk Assessment; International Organization for Standardization; Quantitative Risk Assessment.

Further Readings

European Commission's Information Hub on Life Cycle Thinking–Based Data, Tools and
    Services. http://lca.jrc.ec.europa.eu/lcainfohub/index.vm (Accessed July 2009).
Guinée, Jeroen B. *Handbook on Life Cycle Assessment: Operational Guide to the ISO
    Standards (Eco-Efficiency in Industry and Science)*. New York: Springer, 2002.
Hawken, Paul. *The Ecology of Commerce: A Declaration of Sustainability*. New York:
    HarperBusiness, 1994.
Horne, Ralph, Tim Grant and Karli Verghese. *Life Cycle Assessment: Principles, Practice and
    Prospects*. Collingwood, Australia: CSIRO Publishing, 2009.

Marco Orsini
Institut d'Études et Conseils
en Développement Durable (ICEDD)

LIFESTYLE OF HEALTH AND SUSTAINABILITY (LOHAS)

LOHAS is an acronym for lifestyles of health and sustainability. It describes three increasingly significant cultural and business phenomena: (1) a lifestyle-based social movement; (2) a growing market of consumers and producers who adhere to this lifestyle; and (3) the LOHAS Institute, an organization founded in 1999 to support the growth of this market. In 2008, the U.S. LOHAS market was estimated at 41 million people, or about 15 percent of the U.S. adult population, and was worth $209 billion. Although the term *LOHAS* was coined in the United States, LOHAS consumers can be found throughout North America, Europe, and Asia, and LOHAS marketers can be found following right behind them.

LOHAS consumers can also be understood as cultural creatives, a term coined in 2000 by social scientists Paul Ray and Sherry Anderson in their book of the same name. In their book, Ray and Anderson argue that—although largely under the radar of the mainstream media—about 50 million people, or 26 percent of the American adult population, care deeply about so-called hippie concerns of the 1960s. These include personal spiritual growth, human rights, and ecology and serve as guiding principles for how cultural creatives conduct their lives and respond to the world around them.

Cultural creatives were initially unaware of themselves as part of a larger social force. They were a loosely knit subculture of (primarily) college-educated, middle-class people who simply shared a distinct way of thinking about the world. Ray and Anderson describe this way of thinking as synoptic, fueled by a desire to understand even the most basic things by looking at the bigger picture and figuring out how all the parts in a system influence one another. Cultural creatives tend to see the world as an interconnected web of living things. As a result, they have developed a moral code that demands that they examine the impact of their actions on the environment, the lives of others, and their own personal well-being.

From a Subculture to a Market

The term *LOHAS* was coined shortly after Ray and Anderson published their book *Cultural Creatives*. That year, two journalists, Frank Lampe and Steven Hoffman, came to

In a lifestyles of health and sustainability consumer's home, one might find Fair Trade coffee, such as these green and roasted coffees undergoing quality testing.

Source: Courtesy of TransFair USA

the conclusion that there was some connection between the growing number of cultural creatives, on the one hand, and the booming market for natural and organic products on the other. They founded a small media company, Natural Business Communications. With Paul Ray, they began conducting research into the possible connections between cultural creatives values and the natural products market. According to Monica Emerich, research director for Natural Business Communications, they found that cultural creatives were basically frustrated consumers who struggled to find products that supported their values and goals. Natural Business Communications saw this as a market opportunity that needed a name. They developed the LOHAS concept and popularized it by launching the *LOHAS Journal* and the LOHAS forum as media for connecting LOHAS-minded producers with a like-minded target market.

LOHAS consumers can be considered a distinct market segment because they make purchasing decisions based on a set of values markedly different from consumers in the mainstream market. In addition to concerns about quality and performance, LOHAS consumers share a commitment to wellness, broadly defined. These consumers care not only about their own wellness, but also the wellness of their family, their community, the environment, and all creatures living on this planet. They understand the connection between what they buy and how it impacts the lives of other living things. Thus they seek to change the world by buying products that lessen their negative impact on the environment and on the lives of the people who produce the product. In their homes, one may find Fair Trade coffee, organic food, bisphenol A (BPA)–free baby bottles, solar-powered water heaters, hybrid cars, homeopathic medicine, original art, yoga mats, and magazines from environmental and social justice organizations to which they donate money.

LOHAS-oriented businesses share these goals to varying degrees. The most dedicated LOHAS companies consciously use their business model as a tool for improving social and environmental conditions around the globe. For example, companies like Pangea Organics, Green & Black's, and Patagonia constantly seek to restructure their sourcing, manufacturing, and marketing processes to lessen their ecological impact; all three also ensure that the companies they source from provide a fair wage and safe and human working conditions to all employees. Large companies such as Sony Corp., Wal-Mart, and the Ford Motor Company have been called LOHAS companies based on their attempts to employ ecological principles in the design of new products and through developing more sustainable business practices.

From a consumer perspective, however, there is a dichotomy here. The size of these companies means that even a small positive change can have a large impact on the environment, so while LOHAS consumers may applaud, for example, Wal-Mart's environmental actions, they remain skeptical about the company's overall integrity. These consumers argue that Wal-Mart's labor practices don't fit their values, and that the presence of Wal-Mart stores has devastated economies in many small towns. As more and more mainstream companies are attempting to target the LOHAS market, LOHAS consumers have begun to demand that corporations walk their talk by buying only products that have been third-party certified as meeting high standards for health and wellness. Primary among these are U.S. Department of Agriculture (USDA) and other organic standards for food and clothing, Leadership in Energy and Environmental Design (LEED) green building standards, and Fair Trade standards developed by TransFair USA.

Global LOHAS

Cultural creatives can be found in industrialized countries around the globe and the term *LOHAS* has gained global popularity in recent years. LOHAS markets have been identified in countries as diverse as Australia, South Korea, Japan, Germany, France, and Taiwan, and research is under way to establish the similarities and differences between these markets. For example, a report by the research firm Porter Novelli found that European LOHAS consumers are 32 percent more likely than Americans to purchase products that have been certified green, organic, or Fair Trade. Taiwan, on the other hand is strongly influenced by trends in Japan and Korea. To varying degrees, all three countries draw upon their own, well-developed traditions of Buddhism and holistic medicine—aspects of LOHAS culture that would be considered "alternative" within Western mainstream cultures.

In 2003, Natural Business Communications was sold to Conscious Media. Conscious Media has become the largest promoter of the LOHAS market and is also a large purveyor of LOHAS-related marketing communication. Conscious Media continues to grow and publishes the *LOHAS Journal* both online and in print. It sponsors the annual LOHAS Forum, a trade show that attracts business leaders from over 360 corporations around the world.

See Also: Cause-Related Marketing; Corporate Social Responsibility; Environmental Marketing; Fair Trade; Social Entrepreneurship.

Further Readings

Florida, Richard. *The Rise of the Creative Class: And How It's Transforming Work, Leisure, Community and Everyday Life.* New York: Basic Books, 2003.

Horn, Greg. *Living Green: A Practical Guide to Simple Sustainability.* Evanston, IL: Freedom Publishing, 2006.

The LOHAS Journal. http://www.lohas.com/journal (Accessed April 2009).

Ray, Paul and Sherry Ruth Anderson. *The Cultural Creatives: How 50 Million People Are Changing the World.* New York: Three Rivers Press, 2000.

Maura Troester Nuñez
University of Colorado, Boulder

M

MATERIAL INPUT PER SERVICE UNIT (MIPS)

Material input per service unit is a measure of ecoefficiency, developed at the Wuppertal Institute for Climate, Environment, and Energy, a German sustainability research institution that focuses on the intersection of ecology, economy, and society. The basic idea behind the concept is that, at some point down the line, every material input (natural resources, energy) becomes an output in the form of waste or emissions. The subsequent argument is that measuring the inputs thus allows for a rough approximation of the overall environmental burden. The Wuppertal Institute defines MIPS as "an elementary measure to estimate the environmental impacts caused by a product or service. The whole life cycle from cradle to grave (extraction, production, use, waste/recycling) is considered. MIPS can be applied in all cases where the environmental implications of products, processes, and services need to be assessed and compared."

As a concept, in short, the method can be used to estimate the environmental burden caused by a product, service, or even a lifestyle, and can be performed on a macroeconomic scale (examining national economies) or on a microeconomic level (evaluating a single product, industry, or service).

MIPS is the reciprocal of resource productivity, which would be expressed as service units per material input (analogous to the familiar formulation of an automobile's mileage). A lower MIPS means a greater degree of ecoefficiency. MIPS values are calculated for the entire life cycle of the product or service—production, use, recycling, or disposal—with the service units varying according to the nature of the product. A common example given by Wuppertal is an automobile, the service unit of which would be passenger miles traveled over the course of its life. The fewer material inputs required per passenger mile across that life span, the greater the environmental efficiency of the car. The material input includes the raw materials and energy used to produce the product or service, to operate it, and to dispose of it. Reducing any of those obviously increases the efficiency and MIPS. Using recycled raw materials, for instance, brings the number down, as does a more energy-efficient factory or production process. Efficiency gains can also result from reducing the waste in the production process—whether caused by errors (products discarded when they fail quality inspection), an unnecessarily excessive use of material (such as vehicles larger than they need to be), or material that is wasted in the production process but is neither recycled nor part of the final product (such as the scraps of material left after the process).

Wuppertal and other sustainability advocacy groups promote the MIPS figure because they seek to encourage reducing materials consumption—a necessity in order to achieve sustainable manufacturing sectors. One of the concepts promoted by Wuppertal is the "Factor Four" goal (from the book of the same title by Ernst Ulrich von Weizsacker et al.), which calls for "doubling wealth, while halving resource use." This is sometimes called "dematerializing" the economy. MIPS considers the use of a resource—a material input—to begin at the point it is extracted from nature. Such material inputs include biotic resources (living and renewable, such as plant materials), raw materials, abiotic resources (nonliving and nonrenewable, such as mineral ore), water, air (including the air consumed by combustion and chemical reactions), and earth (i.e., the soil used in such things as agriculture). Energy used for manufacture and transport is also measured. The focus on inputs reflects the institute's desire to reflect the fact that all inputs eventually become outputs, in the form of waste or emissions, and that measuring inputs thus provides as accurate an understanding of eventual environmental impact as measuring outputs does. An automobile's environmental impact is not limited to its exhaust emissions—a car that is never driven still represents a use of energy to manufacture it, the materials taken out of the Earth to build it, and the scrap it becomes when it is no longer usable. The use of a car certainly leads to its greatest environmental impact, and more gains are made by focusing on fuel efficiency and lower emissions than on dematerialization; but that does not negate the impact of the manufacturing phase. Likewise, though putting clothes through the laundry has obvious environmental impact and consumers have awakened to the desirability of environmentally friendly washing machines, dryers, and the environmental gains of returning to air-drying, the consumer is largely unaware that the resource consumption required to manufacture clothes can be as high—or even greater—as the cumulative resource consumption of laundering over their life span.

Similarly, MIPS includes the idea of an "ecological backpack" or "ecological rucksack," which is derived by subtracting the weight of the product from its total material input. This has become a fairly popular concept and can be accessed online, where users can estimate the environmental burden their own individual lifestyle generates. The estimates include housing, energy and material consumption at home, mobility, leisure time activities, food and beverages, and, of course, waste produced. The "ecological backpack" is a practical application of the MIPS concept and can be employed by a variety of users, from individuals to businesses.

The weight of the materials used to create a product is not usually evident through a simple examination of the product. Comparatively massive amounts of displaced earth go into the procurement of diamonds, for instance.

Calculating the MIPS for a product or service consists of seven steps:

1. The object of the MIPS analysis is defined, as is its service unit. A service unit can be "one passenger mile in the automobile," "one 12-ounce can of cola consumed," "one day of wearing an article of clothing, and its subsequent laundering," and so on.

2. The process chain is constructed, showing the life cycle of the product and the individual steps used to create it and to bring it to market, in relation to one another.

3. Inputs and outputs are represented in a "process picture."

4. Material input total is calculated for the creation of the product ("from cradle to product"), by gathering specifics on individual material inputs (such as the raw materials for various components).

5. Material inputs are calculated "from the cradle to the grave," the end point of the product's life span.

6. Data from the "use phase" (when the car is owned and driven) and recycling/disposal stage (when the product is discarded in some fashion) is collected. Though the manufacturer does not have total control over this, it does have control over how easy the product is to recycle and how efficient it is at consuming resources.

7. The results are interpreted and calculated for the life span of the product.

Once a MIPS analysis is done of one product—an automobile by Ford, for instance—it is easier to construct one for a relatively similar product, such as an automobile by General Motors. Many of the types of data will remain the same, with only the specifics varying.

While the need for sustainable economies is increasingly apparent, how to define, much less how to realize, sustainable forms of production, consumption, and disposal are still far from clear. One important, albeit ill-defined, criterion in the sustainability debate has been "constant natural capital." The MIPS approach clarifies this concept and provides relatively clear and operational applications for a wide variety of users. Above all, it provides clear and tangible results that can help individuals and businesses alike to "dematerialize" their operations, and it helps illuminate how using fewer resources does not have to translate into a decline in utility.

See Also: Cradle-to-Cradle; Ecoefficiency; Sustainability; Sustainable Development.

Further Readings

Ashby, M. F. *Materials and the Environment: Eco-Informed Material Choice.* Oxford, UK: Butterworth-Heinemann, 2009.
Bleischwitz, Raimund. *Corporate Governance of Sustainability: A Co-Evolutionary View on Resource Management (Esri Studies on the Environment).* Northampton, MA: Edward Elgar, 2007.
Ritthoff, Michael, Holger Rohn and Christa Liedtke. *Calculating MIPS: Resource Productivity of Products and Services.* Wuppertal, Germany: Wuppertal Institute for Climate, Environment, and Energy, 2002.
von Weizsacker, Ernst Ulrich, Amory Lovins and Hunter Lovins. *Factor Four.* Denver, CO: Rocky Mountain Institute, 1998.

Bill Kte'pi
Independent Scholar

MAXIMUM ACHIEVABLE CONTROL TECHNOLOGY

In the United States, the maximum achievable control technology (MACT) and the state-of-the-art (SOTA) control technologies are the highest standards of technology required by governmental regulation to minimize risks and to manage the effects of human actions. More specifically, MACT is the technology capable of reducing pollution to the greatest extent possible. The implications of MACT for green enterprises are significant as such considerations are important to environmental management systems. In the context of air quality requirements, the MACT is specified to obtain a reduction in residual risk beyond a base operating standard that may have harmful effects.

Specific or fixed standards in legislation or administrative regulations may be higher than the MACT standard. For instance, health concerns may require standards, like that of zero emissions, regardless of the technology used or if the outcome is achievable. In such circumstances, the specified standard may anticipate and promote the MACT or SOTA technology. Standards requiring SOTA control technologies, such as those in the state of New Jersey, may include the federal MACT standards as well as other higher standards. For comparison, under a nonregulatory common law standard in the United States and other jurisdictions, it is not permitted to externally impose risk on individuals or communities. Therefore, the relevant common law threshold comparable to the MACT or SOTA regulatory requirements would be one with the lowest measurable risk.

Arguably, the MACT, SOTA, and common law requirements are similar insofar as they all rely on evolving technology for measurement. Determining the specifications and standards requires an understanding of the causes and results of human actions that may have harmful consequences. These determinations contribute to the development of standards that may be incorporated either by regulation or by agreement—the violation of which can then be enforced through criminal and civil penalties. Conceptually, the MACT standard may be intended as a minimization principle requiring reduction of adverse impacts from human actions and events toward zero. In practice, under the Clean Air Act (CAA), the MACT standard, like the best available control technology (BACT) and other technology standards, is derived from and established through industry experience in implementing the precautionary principle. At times, such standards may be used as proxies for this principle in regulations or in agreements.

Organizations may demonstrate their cultural and societal positions, as well as their leadership in green issues, by voluntarily including explicit reference to the MACT in their own management and operations. Where such an organization is involved in activities that cause adverse environmental impacts, the installation and operation of the MACT as intended in its design may not only be in compliance with regulatory requirements, but may also serve to demonstrate due diligence, to limit liability, and to retain regulatory compliance as a complete defense. The implementation of a cap-and-trade system may serve as an additional or an alternative to the regulatory control technology approach in limiting adverse environmental impacts.

The MACT is not a singular standard. For instance, MACT standards to control hazardous atmospheric pollutants in the oil industry in Houston, Texas, include standards called Refinery MACT1, Refinery MACT2, Organic Liquids Distribution MACT, Boiler and Process Heater MACT, and Turbines and Engine MACT. Other MACT standards exist for hazardous waste combustors and site remediation activities. In some instances, the SOTA technology could be specified as equivalent to the MACT. In other situations where the maximum is not required, more emission tolerant technology standards such as the BACT may be recognized or specified. More emission tolerant standards in the available range include the best environmental technology, best practical technology, best available control technology economically achievable, the reasonably available control technology (RACT), the lowest achievable emissions rate (LAER), and those required to keep adverse impacts and risks to a level that is "as low as reasonably achievable" (ALARA), which take into account relevant economic and social factors. The selected standard may be implemented by name in relation to the desired emission standards that are to be met. The MACT and other technology standards are deployed to achieve regulatory compliance through the application of measures, processes, methods, systems, or techniques that affect the quantity and quality of pollutants to reduce or eliminate their emissions at a facility.

These factors are important as companies and consumers increasingly seek sustainable products and services.

Technology-based standards include command-and-control regimes that may specify permitted emission limits as well as how the emissions sources must be operated. Agencies responsible for implementing the regulatory regime may set specific standards based on demonstrated achievable standards for environmental impacts after the application of existing or new technology. Depending on the source, the required standard for the technology may be the LAER, the RACT, or the BACT. For instance, since 1990, the U.S. Environmental Protection Agency (EPA) established a permit system by the authority granted in the CAA and in the Resource Conservation and Recovery Act. The permits prescribe limits on emissions into the atmosphere after implementation of the MACT and based on the National Ambient Air Quality Standards (NAAQS) as required in Phase I of the amended CAA. In the CAA, MACT is defined as the technology that provides the "maximum degree of reduction in emissions" of the pollutant. The legislation requires the MACT standard to be revised every eight years to incorporate technological developments.

Under both the Clean Water Act and the CAA, there is an interest in protecting public health. As such, it is mandatory to achieve the applicable quality standards that specifically refer to state implementation plans. In such circumstances, the imperative to achieve the specified standard may require equipment and processes that exceed the MACT standard. Further, where there is a difference in regulatory requirements in the same geographical area, the higher water- or air-quality standards will usually apply, whether it is the MACT standard or a higher standard. In the United States, through the judicial process, courts have determined that for new sources of pollution (1) costs may not be considered in setting the MACT "floor"; and (2) the minimum standard required by law is to be not less the emission control that is achieved in practice by the similar best-controlled source. For existing major sources, the standard is to be not less than the best-performing 12 percent of existing sources. These requirements have substantial influence on business practices especially for companies trying to incorporate ecoeffectiveness into their strategic plans.

The EPA has been slow to issue MACT standards and has often failed to meet statutory deadlines. Since the MACT requirement is mandated in legislation for the purpose of reducing residual risk and requiring a risk analysis involves probabilities rather than certainty, there is inherent and unavoidable uncertainty in assessing and measuring residual risk. In the context of air pollution, the risk determination in eliminating carcinogenic emissions in an intermediate range is a more difficult task. This range of residual risk extends from the determination of lifetime excess cancer risk to the most exposed individual at or above the threshold risk standard of one in 10,000 persons and the higher standard of one in a million persons. In this context, if more than one in 10,000 persons develops cancer caused by air pollutants, it would be a problem. However, if less than one in a million persons develops cancer for the same reason, for purposes of this regime, it would be less than the residual risk and negligible. Green companies must consider these factors as they develop products that meet regulations and market demand.

The risk approach to implementing air quality standards has been recognized as applicable to the EPA when establishing the standards under the CAA. The additional question of whether an agency must use a risk assessment approach to enforce the standards remains problematic. If a risk approach is used as a general basis for enforcement in large geographic areas or in an industry, variations within the area or industry cannot be recognized, resulting in uneven enforcement of the regulations. Uneven enforcement imposes an unfair burden on the sources that have the best performance and cleanest operations. This also leads to the free rider problem in that the sources that cause the most pollution get a free ride on the performance of the least-polluting sources.

Another related issue was raised in the case of *Natural Resources Defense Council v. Environmental Protection Agency* (529 F.3d 1077 [D.C. Cir. 2008]), where the EPA opposed the position that it must eliminate all risks above the one in a million trigger in fulfilling its obligations under the CAA Section 112(f)(2)(A). The court agreed that the EPA had provided an ample margin of safety in its technology-based standards and, consequently, that it had satisfied its statutory obligations. This solution does not address the situation where an individual incurs damage to their health, including cancer caused by regulated pollutants. In providing an ample margin of safety in a risk-based approach, the established standard accepts that beyond the one in a million persons, the regulated carcinogen will cause cancer for which the victim will then have no legal recourse. This interpretation of the regulatory standard permits harm and is different from the common law principle that prohibits one from causing harm to another. These provisions are vitally important as green businesses try to compete against other companies that may be able to underprice their products, even if they later prove to be harmful.

Achievement of the standards implies operational management and enforcement provisions have been fulfilled. Regulations may permit some discretion in the enforcement of the requirements to achieve applicable toxicity standards. Monitoring and enforcement of MACT standards are identified as means of achieving the EPA priority objective of compliance and environmental stewardship. In particular, the EPA has publicized its objectives in seeking to reduce toxic air emissions as required by the CAA through 2010. The identification of a priority is distinct from the core activities of the EPA and reflects short-term concerns and particular interests based on public awareness.

An example of a particular interest is the regulation of mercury emissions. The presence of mercury in the environment has been subject to attention for some time, and since mercury is a hazardous air pollutant, the implementation of the MACT is relevant as mandated by legislation. The health effects database for mercury is extensive, and its persistence through multiple pathways has been well documented. Reduction of exposure to elemental, organic, and inorganic mercury has recognized health benefits. Mercury emissions as a component of hazardous air pollutant emissions from coal- and oil-fired electric utility units were recognized as a particular concern by the EPA in December 2000. As a result, these types of units were included in units subject to the provisions of the CAA Section 112 and MACT standards.

The qualifying term maximum in MACT specifications and the eight-year update requirement should reasonably result in a regime with progressively higher standards and greater scope with the continuing dominant consideration being the availability of technology rather than its cost. Since technology changes are not regular, the rate of change to the standard will vary accordingly, resulting in some unpredictability and cost consequences in addition to the implementation of continuous evaluation, decision making, and risk management. The competence of management and operations, as well as the processes and tools for risk assessment, should have a direct bearing on the effectiveness of the implementation of technological standards such as the MACT. The implementation of MACT standards, therefore, also requires the monitoring of its effectiveness and enforcement. The benefit of a focus on technology is the ability to implement measures and equipment in a systemic manner that could also improve overall facility efficiency and sustainability. In circumstances where no adverse impact on the environment is the only acceptable result, installation of technology, regardless of the MACT or other installed and operating capability, may be insufficient. Permits, as required by the CAA, may also be instrumental in effectively enforcing applicable standards. In addition to regulatory command-and-control regimes, specified technology and technology standards may be components in fee-for-use strategies as well as in market systems, including commercial valuations in tradable permit markets.

In conclusion, the MACT framework provides an opportunity for increased levels of environmental protection to be ensured. Green companies should continue to provide leadership by exceeding the existing standards, developing new technologies to reduce pollutants, and by lobbying state and federal officials to pass stricter environmental standards.

See Also: Appropriate Technology; Best Available Control Technology; Best Management Practices; Clean Technology; Emissions Trading; Environmental Risk Assessment; Voluntary Standards.

Further Readings

Flatt, Victor B. "Gasping for Breath: The Administrative Flaws of Federal Hazardous Air Pollution Regulation and What We Can Learn From the States." *Ecology Law Quarterly*, 34/1:107–73 (2007).
Harvey, Pamela D. and C. Mark Smith. "The Mercury's Falling: The Massachusetts Approach to Reducing Mercury in the Environment." *American Journal of Law & Medicine*, 30/2–3:245–81 (2004).
"RCRA Regulatory Scorecard." *Hazardous Waste Consultant*, 27/2 (2009).

Lester de Souza
Independent Scholar

NATIONAL INCOME AND PRODUCT ACCOUNTS

The National Income and Product Accounts (NIPAs) measure the totality of goods and services produced, as well as the income generated, as valued by current market prices (though, for comparative purposes, results are adjusted for inflation). NIPAs are generally put together on a quarterly and annual basis by the federal government's Bureau of Economic Analysis (BEA). They represent the nation's largest and most significant single economic scorecard. While NIPAs are little known outside the circle of economists and accountants, a key component of NIPAs has become famous: the gross domestic product (GDP).

Unique among indicators, the GDP often is used as shorthand to define national economic success or failure. It captures, in one figure, all the output generated in a given country during a specified period, regardless of the type of services and goods that make up that output. As today's preeminent economic indicator, the GDP single-handedly determines whether a national economy is growing, stagnating, or shrinking. A recession, for instance, is officially defined as two quarters of negative GDP growth. Equally, reporters, economists, and politicians alike speak of economic recovery when the GDP moves back into plus territory, irrespective of any other important indicators. Indeed, as the period following the last deep recession demonstrated, unemployment can remain high (what pundits called a "jobless recovery") the environment can continue to deteriorate, increasing numbers of Americans can continue to lose healthcare, high-tech jobs can continue to be exported, the gap between rich and poor can continue to widen: As long as the output and income of the total economy is growing, according to GDP, the economy appears healthy.

History and Background

Scholars largely agree that the main impetus for the creation of national economic accounts has always been taxation. In more recent history, national governments are especially in need of reliable information about the productive capacities of their respective economies during times of real or impending war, and during times of economic depression. It is not surprising, therefore, to discover that the first official national income statistics were published in the United States in early 1934 (at the height of the Great Depression), and in the United Kingdom in 1941 (in the midst of depression and war). This was done in direct response, in the words of one participant, to the "almost complete lack of reliable data upon which to make major policy decisions regarding what the government could and should do to improve the economic outlook." By the end of World War II, NIPAs in both the United States and the UK had become an indispensable instrument of their respective national governments for planning, mobilizing, and directing resources.

In the aftermath of World War II, based in large part on the desire to monitor and control Marshall Plan aid, the Organisation for European Economic Co-operation (OEEC, the forerunner of the Organisation for Economic Co-operation and Development [OECD]), under the leadership of the United States and the UK, developed an international system of national accounts. Officially approved in 1953 by the United Nations (UN), the System of National Accounts (SNA) was soon adopted by all capitalist nations worldwide.

Following several, albeit minor, alterations over the years, it today represents the basic economic scorecard for every major economy, and provides the foundation for virtually all transnational and cross-historical economic comparisons. As observers have pointed out, it would be an understatement to call NIPAs, and the GDP in particular, the "800-pound gorilla of economic indicators." NIPAs provide the very foundation for economic policy making—interest and tax rates, subsidies, regulations, foreign aid, tariffs, and investments.

As a political reality, the GDP has become the one-stop indicator for economic health. And as a social reality, though not by design, and despite explicit warnings of the creator of the national income accounts, Simon Kuznets, that the "welfare of a nation can scarcely be inferred from a measurement of national income," the GDP has also become the most important marker of social well-being. As any listener of news programs can attest, well-being is routinely portrayed as some function of "per capita income." For instance, the progress of a nation or a particular group of people is measured as "per capita" growth. By the same token, per capita income or per capita GDP is used to make comparisons between different groups (black/white, male/female), between today and years past, or between so-called advanced and so-called poor nations. Indeed, among observers and theorists across the ideological spectrum, a widely believed assumption is that economic growth (as measured by GDP) and social well-being generally moves in the same direction. The idea that a community or society with a lower per capita GDP may possibly be better off—perhaps because people work less, are sick less, have a cleaner environment, have depleted their resources less, or enjoy more equality—is as culturally alien as it runs counter to basic assumptions of neoclassical economics and politics on both sides of the aisle alike.

Role of NIPAs/GDP in Society

Activities throughout U.S. society are geared toward increasing the numbers of the GDP scorecard. Americans work for economic "growth" without asking "What, exactly, are we growing?" "Is what we are counting as 'success' really promoting welfare, or is it actually undermining it?" As an indicator, NIPAs are blind to all such questions, providing no qualitative data distinguishing between growth that increases well-being versus growth that diminishes well-being. Nor do they distinguish between growth that is sustainable versus growth that is not sustainable. To the extent that we value what has a price, and work for what pays, Americans—as producers, consumers, and citizens—pursue (virtually indiscriminately) GDP growth as the preeminent goal. Like it or not, in other words, national economies tend to become what they measure.

The social and environmental implications of a performance indicator like the NIPA or GDP are difficult to exaggerate. By attaching value to certain things (the car ride but not the walk; the pollution cleanup but not the clean river; the burning of fossil fuel but not the oil in the ground; the war campaign but not the stable peace), NIPAs literally affect all aspects of life in modern economies. Even people who never think about or know anything about NIPAs or the GDP are profoundly impacted. One follows the course of the road without having to be cognizant of road-building techniques, the technical specifications of the vehicle used, or even a clear awareness of the destination. As all green businesses that have engaged in resource or waste reduction measures not readily profitable can attest, deviating from the course of the road can be costly.

According to predominant economic indicators, here are some common examples of what barely, if at all, contributes to well-being (either economic or social, as measured by GDP): reducing hours worked to be with one's family; the child raised at home; the home-made meal; the tree in the forest; clean water; walking to work; stable marriages; relative equality of incomes/wealth between chief executive officers and employees, men and women, or whites and blacks; health; and peace. And here are the same examples of things that contribute a great deal to our economic and social well-being (as measured by GDP): the breadwinner working overtime; the child raised by daycare facilities and nannies; the takeout meal; the tree cut and turned into lumber; contaminated water that requires filtration and allows a blossoming bottled-water market; many miles commuting by car/SUV; divorce; burgeoning corporate incomes and bonuses; job insecurity; sickness; and war.

To be clear, while the primary motive of individual businesses, large and small, is to make a profit (rather than feed the GDP), the course and direction of how that profit can be realized directly follows what we measure. Fundamentally, by determining what is (and is not) considered "final output" or "investment" or "consumption," and by establishing the ground rules for what is valued and priced, the GDP defines the parameters of what can or cannot be profitable. If resource depletion or pollution were measured as a cost, rather than an income, for instance, prospects for profitability would change dramatically.

By today's (quite indiscriminate) GDP standards, the more goods and services being produced, marketed, sold, and consumed, the better. What is coveted by the priced values included in GDP is not only more of the same, however. An increasingly crucial component to growing the amount and value of goods and services is to find new ways to monetize things that previously did not have a price. The more natural goods can be turned into monetized commodities (clean water, uncontaminated soil, oil), the better. The more social goods can be turned into monetized commodities (advice, friendship, play, safety) the better. This includes the burgeoning industry of "defensive costs": everything from locks and security systems to maintain safety, to increased mortgage payments in order to be in the vicinity of functioning schools, or Superfund cleanups to reestablish some semblance of a clean environment.

As a steadily growing number of critics of the NIPA system began to point out as early as the 1960s, in short, our premier economic gauge is not only broken, but badly misleading. As far as the environment is concerned, continuing the values and measures as defined by the GDP may be catastrophic.

Originally intended as a rational and objective measure of volume of economic output, today's NIPA and GDP accounts are far from either. While increasingly sophisticated tabulations of values and numbers that flow into GDP appear straightforward and neutral, in reality they have come to be used in ways that promote a system that is both frivolous and reckless. To summarize: the more goods and services are consumed, the larger the contribution to presumed "well-being." As Robert F. Kennedy concluded in 1968, our current indicators "measure everything except that which makes life worthwhile."

Difficult Alternatives

The GDPs fundamental flaws as a welfare measure are well recognized by now. Even the GDPs validity as an economic indicator is increasingly questioned. As of today, there are few trans-national organizations concerned with global economic matters that have not produced reports and summoned commissions addressing the many flaws of GDP accounts. The OECD, World Bank, World Resources Institute, European Union, French government, think tanks, and dozens of nonprofit and educational institutions around the world have criticized the nature and uses of GDP. Many have also generated possible alternatives: broadly defined, they either reject the "one-indicator" model and provide a spectrum of social, economic, and environmental measures (e.g., Sustainable Development Indicators), or they reconceptualize the one-indicator model to include social and environmental considerations (such as the Genuine Progress Indicator, or the Happy Planet Index).

Most economists have notably stayed away from such efforts, either arguing that NIPAs were never intended to measure well-being (but presumably do a fine job measuring economic output), or claiming that NIPAs are not, in terms of daily economic decision making, very important. Essentially unnoticed by the public, Congress has held two hearings on the validity of the GDP in the last 20 years, acknowledging major problems but providing no legislative policy follow-up. Those directly involved in tabulating the massive volume of data that flows into measures like the GDP point to its global preeminence as an economic indicator, and its success in becoming the premier measure for national and transnational comparisons and evaluations. Changing the measure, the argument goes, would "sharply diminish its usefulness," in the words of the director of BEA, by robbing policy makers of a much-needed comparative indicator.

To this day, consequently, none of the critical evaluations, and none of the attempts to generate alternative models, have gained much political traction. Widely recognized and supported, the need for change is rich in analysis and good intentions, but very short in concrete implementation. With the exception of Bhutan, no country or transnational organization has yet adopted an alternative set of indicators as the official basis for economic policy making. Circumscribed only by national policies subsidizing or regulating a few select activities (e.g., corn subsidies, fuel-efficiency standards, and pollution caps), the primary goal of virtually every economy worldwide remains indiscriminate growth. Increasingly, the question is whether either the environment or human communities can afford it.

No "Green" in National Accounting

As currently configured, the environment appears in the national income accounts only in one of two ways: either when natural resources are exploited and turned into priced commodities (turning trees into lumber, well-water into bottled water, crude oil into gasoline), or when environmental pollution and degradation require compensatory market activity (installing filters, cleaning toxic waste sites, purchasing cap-and-trade permits). An oddity even by neoclassical economic standards, the environment, as the very foundation of all economic activities, does not represent a capital asset. While the Net National Product, for instance, is specifically calculated by subtracting depreciation from GDP, this depreciation concerns only manufactured capital, but not natural capital. As any accountant understands, equipment suffers from wear and tear and gets used up, and therefore depreciates in value. Inconsistent with this basic logic, depletion of natural resources is not counted, any more than the degradation of millions of other vital services nature provides to the economy, and human welfare in general.

Again, many observers understand how such accounting standards can send fundamentally flawed, and potentially catastrophic, signals to national and international markets. Countries around the world are beginning to confront the reality that the short-term gain of natural resource exploitation and depletion is followed by long-term degradation and poverty. A World Resource Institute study documented, for instance, that the impressive GDP growth rates of Indonesia in the 1970s and 1980s were essentially bought with clear-cutting its forests and exhausting its topsoil with intensive cash-crop farming. Degrading, depleting, and despoiling the very plot of land one stands on, while perhaps improving one's economic income accounts in the short term, inevitably fails to sustain well-being in the long term. In response, a few countries have implemented so-called environmental satellite accounts ("satellite" because they are add-ons that circle around the main account, which is GDP). But since such satellite accounts are recorded outside the income and production accounts, they have remained without any effect on the conventional indicators of cost, income, product, and capital formation, and thus largely without consequence on markets and economic policy making. The UN issued in 1993, and revised in 2003, its call for national accounts that include environmental depletion and degradation. This, too, has yielded no substantive results to date.

Given predominant national accounting standards, green businesses face a fundamental dilemma: what is good for the environment and the community is frequently not good for the bottom line. The problem is that depletion and pollution make multiple contributions to what is tabulated in national accounts, while environmental protection and conservation come at a cost to potential economic activity. Absurdly, the consequence is not only that environmental problems are promoted by the national accounting system; worse, economic and ecological goals are brought into inevitable conflict. Rather than recognizing the existential significance of a stable and healthy environment to long-term economic success, national accounts instead reward only consumption, waste, and degradation. As a result, despite increasing public awareness and subsequent growing consumer demand for production and consumption patterns that are less wasteful and destructive, green or sustainable business practices still occupy a relatively marginal existence.

Today, good environmental stewardship is profitable only to the extent to which it either can create cost savings (less waste and resource use) or successfully respond to niche demands of consumers willing to pay higher prices for things like organic food or sustainable energy. This would instantaneously and radically change through any number of logical economic course corrections. To name but two that have been widely discussed: (1) as described above, allow for natural capital depreciation and (2) include "externalized" costs in the pricing of goods and services. Though it is difficult to determine with precision, the results would undoubtedly produce a tectonic economic shift. One relevant example that has attracted some attention: according to several studies, gasoline in either scenario would immediately cost somewhere in the range of $8 to $14 a gallon. This, in turn, would radically alter consumer choices (smaller cars, moving back to urban cores, less waste, green construction with emphasis on conservation and renewables, etc.), free up billions of investment dollars for everything from sustainable energy resources to conservation efforts to a wide variety of green technologies, and fundamentally revise economic cost-benefit analyses. Businesses with a focus on sustainable operations and products, in turn, would suddenly face virtually endless opportunities to accomplish both: make money and protect the environment.

There is nothing natural or free about what we measure, or how we measure it. All national economic systems in history have been a product of political decisions, of rules and laws regulating what players in the marketplace can and cannot do. The creation of

NIPAs, and its subsequent spread around the globe, is a perfect example for a design that, according to its founders, is full of subjective choices and value judgments. Since the 1950s, these choices and values have had a profound impact on the world, generating unprecedented growth rates while pushing the globe to the brink of environmental collapse. It has been a process fundamentally guided by the goals and performance indicators embedded in the GDP. Creating alternative goals and performance indicators that account for environmental impact is not only possible, but has become necessary. To be sure, businesses with a focus on environmental sustainability would greatly profit; but so would future generations of market participants.

See Also: Ecological Economics; Environmental Economics; Externalities; Genuine Progress Indicator; Gross National Happiness; Sustainability.

Further Readings

Cobb, Clifford, Ted Halstead and Jonathan Rowe. "If the GDP Is Up, Why Is America Down?" *Atlantic Monthly* (October 1995).

Daly, Herman E. *Beyond Growth: The Economics of Sustainable Development.* Boston, MA: Beacon Press, 1997.

Fleurbaey, Marc. "Beyond GDP: The Quest for a Measure of Social Welfare." *Journal of Economic Literature*, 47/4:1029–75 (2009).

Foss, Murray F. *The U.S. National Income and Product Accounts: Selected Topics (National Bureau of Economic Research Studies in Income and Wealth).* Chicago, IL: University of Chicago Press, 1983.

Jorgensen, Dale W., J. Steven Landefield and William D. Nordhaus. *A New Architecture for the U.S. National Accounts: Studies in Income and Wealth.* Chicago, IL: University of Chicago Press, 2006.

Lequiller, François and Derek Blades. *Understanding National Accounts.* Paris: Organisation for Economic Co-operation and Development (OECD), 2006.

Kuznets, Simon. *National Income: A Summary of Findings.* New York: National Bureau of Economic Research, 1946.

McKibben, Bill. *Deep Economy: The Wealth of Communities and the Durable Future.* New York: Times Books, 2007.

Vanoli, André. *A History of National Accounts.* Amsterdam, the Netherlands: IOS Press, 2005.

Dirk Peter Philipsen
Virginia State University

NATIONAL PRIORITIES LIST

The National Priorities List, also known as the NPL and the Superfund List, is a federally managed list of severely polluted, hazardous waste sites targeted for long-term cleanup action by the U.S. Environmental Protection Agency (EPA) under the Superfund Program.

Although wind is an economical alternative to fossil fuels, a modern wind turbine is a major industrial investment. A substantial contribution to national energy supplies will require many thousands of wind turbines.

Source: GE

Hazardous waste sites that require long-term or remedial cleanup action are placed on the NPL after a review of the site by the EPA, following its Hazard Ranking System and after public comment. These sites then undergo extensive hazardous waste cleanup efforts, either by the EPA or under its direction, with funding coming from responsible parties (the companies or groups who originally polluted the area) and from federal and state funds.

The NPL has been criticized for its chilling effect on private investment and development of a site; for the long amount of time sites stay on the NPL until cleanup efforts are completed; for the disproportionately low inclusion of minority and low-income areas on the list; and for the high cost associated with some Superfund projects. At the same time, sites on the NPL have shown significant reduction in the amount of hazardous waste, to the extent that hundreds have been removed from the list after being fully rehabilitated, and more than a thousand have been classified as "construction complete," no longer requiring active cleanup efforts.

History and Background

The Superfund Program was created in response to the environmental disasters of 1970s, most prominently the chemical exposure at Love Canal, near Niagara Falls, New York; the dioxin exposure at Times Beach, Missouri; and the toxic waste at the "Valley of the Drums," near Louisville, Kentucky. The political consensus of the time was that it was necessary to establish a program to safely dispose of this toxic material and to return hazardous sites to a safe condition, protecting the health of members of the community as well as the environment.

To this end, the U.S. Congress passed the Comprehensive Environmental Response, Compensation, and Liability Act (CERCLA) in 1980, authorizing the EPA to remove hazardous waste and clean up toxic waste sites that threaten public health and the environment. The act provides for removal actions, where immediate and localized dangers are removed from public exposure, and remedial actions, where long-term response is required to return a polluted site to a safe condition. Remedial actions include, for instance, capping and monitoring landfills; excavating and disposing of river sediments; pumping and treating groundwater; and incinerating or biologically treating soils. The EPA is required to give preferences to treatments that permanently reduce the potency of hazardous substances, rather than those that merely move the waste to other locations.

The act also provided for a tax on polluting industries, such as chemical and oil companies, to fund these cleanup efforts. When this tax expired in 1995, however, it was not reinstated by the Republican Congress of the time. Superfund has since experienced a chronic shortage of funds for cleanup efforts.

CERCLA requires the EPA to maintain a list of the most severe sites that threaten to release hazardous substances. This is the basis for the NPL, which the EPA maintains and which is published and updated in the Federal Register. The sites placed on the NPL are those that the EPA considers too complex for basic removal efforts—generally, those efforts that cost over $2 million.

Relevance to Green Business

The NPL has had a significant impact on businesses in three related respects: (1) as the responsibility for paying for cleanup programs lies with the responsible parties, when a particular site is placed on the NPL, the business responsible for the toxic contamination finds itself liable for significant hazardous cleanup costs under CERCLA; (2) NPL projects offer opportunities for businesses specializing in environmental cleanup and the creation of technologies for hazardous material filtering and removal; and (3) the inclusion of a site on the NPL often attaches a toxic stigma to the site, particularly for green businesses, decreasing the viability of the site for investment and development, at least until it has achieved rehabilitated status.

CERCLA specifically authorizes the EPA to find and contact responsible parties through the EPA's Office of Enforcement and Compliance Assurance (OECA). Responsible parties are required to either clean up sites themselves, under the EPA's direction and supervision, or reimburse the EPA for cleanup expenses. The EPA generally attempts to reach a settlement dictating which parties will pay for what cleanup costs; this is often complicated when multiple responsible parties or potentially responsible parties exist who have contributed to the hazardous waste at the site. The Department of Justice assists the EPA when, as often happens, these cases require legal action to settle. In fact, legal expenses make up from 28 percent (for large cleanup efforts) to 46 percent (for smaller cleanup efforts) of the total cost at Superfund sites on average.

The EPA, through its Technology Innovation Program (TIP), supports the development, both through government and private enterprises, of more effective and less costly approaches to assess and clean up contaminated waste sites. According to TIP, federal, state, and local governments, as well as private companies, spend billions of dollars yearly to clean up hazardous waste contaminated sites. To encourage private research and development, the EPA sponsors programs such as the Superfund Innovative Technology Evaluation program, and the Office of Research and Development Land Research Program. One recent innovation has been the development of nanoparticles, such as microscopic iron particles, designed to remove underground chromium-6, carbon tetrachloride, trichloroethylene, and other pollutants at Superfund sites, reducing the cost by as much as 75 percent compared to current technologies.

Inclusion on the NPL

To have sites placed on the NPL, individuals and interest groups must first petition the EPA to review the site, which is then evaluated according to the criteria of the EPA's Hazard Ranking System (HRS), a screening system with structured criteria that returns a numerical result. Generally, a site is included on the NPL if its HRS score is at least 28.5.

The factors considered by the HRS include the possibility that a site either has already released or has the potential to release hazardous material into the environment; the toxicity and quantity of the waste itself; and the communities and environments affected by release of the hazardous waste.

In evaluating the release of toxic material, the HRS considers contamination and exposure through ground and surface water, soil, and air. The scoring system is designed to return a high numerical value of the hazard of a site even if only one source of contamination is prevalent, since a single, large source of toxic waste contamination is still dangerous. However, it is more likely to return a high ("hazardous") score when multiple waste exposure sources are active.

As of January 2010, 1,270 sites have been finalized on the NPL, and a further 340 have been deleted from the list as all cleanup goals have been reached, for a total of 1,610 NPL sites to date. Of all of these sites, 1,082 have reached construction completion status and 52 sites have been partially deleted. The number of sites added to the NPL per year has, after 1993, stabilized to an average of 28 sites per year.

Estimates of the total number of sites that will be placed on the NPL vary, from as low as 2,100 to as high as 10,000, according to the Congressional Budget Office (CBO). The CBO's best estimate for total cost of the Superfund Program is $74 billion in the base case, but, depending on variables, the estimate can range from $42 billion (low case) to $120 billion (high case).

Once sites have been proposed for inclusion on the NPL, a proposal notice is placed on the Federal Register. The proposal for inclusion on the NPL is then open to comment from the public. The EPA can then choose to place the site on the NPL or not; if so, another notice is placed on the Federal Register. Cleanup action then begins, under the direction of the EPA's Office of Solid Waste and Emergency Response (OSWER), passing through several stages: investigation and study, selection and design of cleanup method, and implementation of cleanup ("remedial action").

Site Cleanup and Funding

Sites may be included on the NPL for containing a variety of hazardous material. The most common hazardous waste causing a site to be listed on the NPL is asbestos, lead, mercury, and radioactive material. CERCLA's priority list of hazardous substances includes, in order, arsenic, lead, mercury, vinyl chloride, polychlorinated biphenyls (PCBs), benzene, polycyclic aromatic hydrocarbons, cadmium, benzo(a)pyrene, and benzo(b)fluoranthene. In total, 83 percent of sites exhibit contaminated groundwater, 78 percent soil, 32 percent sediment, and 11 percent sludge.

Hazardous sites on the NPL are removed from the list only when the EPA declares that the pollution at the site has been completely remedied. Generally, sites are first declared "construction complete," meaning that active efforts to remove waste have been completed and passive remediation technologies, such as treatment pumps that remove pollutants from the groundwater, have been installed. However, as a site may remain hazardous even after active efforts are complete, requiring the installation of passive cleanup technologies, many remain on the NPL until they are considered completely remedied; it is only when sites are considered to no longer pose a threat to human health or the environment that they are deleted from the NPL with a deletion notice in the Federal Register.

The cost of cleanup for every site varies depending on the size and scope of the project, including the type and amount of hazardous material present. Under CERCLA, the EPA is authorized to try and recover these costs from responsible parties, those entities found to be responsible for the hazardous waste. Generally, responsible parties and state governments pay for more than half the total cost, while the federal government initially pays 38 percent.

However, a statute of limitations exists for recovering these costs. For remedial actions (long-term cleanup), this statute of limitations is six years from when the cleanup construction begins on the site. A three-year statute of limitations exists for the costs associated with removal actions (short-term cleanup), remedial investigation/feasibility studies, and remedial design projects. When responsible parties or potentially responsible parties fail to cooperate on recouping costs, the EPA works with the Department of Justice to collect the debt. To date, the EPA has recovered approximately $30 billion or equivalent work from responsible parties for Superfund cleanup activities.

From 1994 through 2007, Superfund litigation to recover cleanup costs from responsible parties has decreased 50 percent. At the same time, federal Superfund funding has decreased significantly. Up until 1995, a federal tax on polluting companies, such as those in the chemical and oil industry, feed a trust fund that helped pay for Superfund cleanup efforts. This tax provided for roughly 67 percent of federal funding of Superfund cleanup efforts, but expired in 1995, and was not reinstated under a Republican-led Congress. Democratic President Barack Obama has signaled that he plans to reinstitute the tax in 2011.

Effectiveness and Criticism

The overall effectiveness of the Superfund Program in general and the NPL specifically is difficult to gauge, in particular because of a lack of knowledge of the risks of toxic substances. Gauging effectiveness also relies heavily on the results one prioritizes. The program has many beneficial results, from improving public health, to rehabilitating real estate, to protecting natural resources, to improving quality of life and the environment in an area with a toxic waste site. Quantifying these benefits in order to facilitate a cost comparison is difficult, if not impossible. Still, it is important to note that, as of 2006, 39,000 sites have been evaluated for possible hazardous waste, 82 percent of the sites on the NPL had human exposure under control, 68 percent had groundwater migration under control, and 62 percent had achieved construction complete status.

The NPL has been effective in drawing attention to hazardous waste sites and dramatically increasing funding at these sites for cleanup and rehabilitation efforts. It has brought together private groups, state governments, and the federal government in addressing concerns about hazardous sites across the country. It has also allowed the EPA to prioritize those sites that pose the highest risk to the community in order to most efficiently allocate resources to these efforts.

However, the NPL has encountered criticism for some of its unintended consequences. Designation of a site on the NPL can have a chilling effect on private development and remediation of an area, as it attaches a toxic stigma to the site. This problem is exasperated by the long cleanup time: on average, it takes more than a decade for a site to reach the status of "completely remedied." Designation as a Superfund site can have a counterproductive effect on the viability of an area, at least in the short run, as it drives away investment and development. For this reason, community groups frequently oppose including a local waste site on the NPL. On the other hand, after the site is cleaned up and designated safe, it is often easier to attract investment to the area.

Another criticism of the NPL, and the Superfund Program in general, is that sites often experience long delays in rehabilitation. The CBO estimated the average time between proposed listing on the NPL and a designation of "construction completion" (completion of the active stages of rehabilitation) to be between 13 and 15 years. There is, however, much variability in this estimation. Some sites, particularly those where state action was taken before federal Superfund designation, are often completed shortly after listing on the NPL. Other sites, however, are estimated to take as much as 30–40 years. The cause for these delays are generally both intrinsic site difficulties, such as size and extent of contamination, and enforcement difficulties, such as negotiations with potentially responsible parties who, if found responsible, would be compelled to pay for the rehabilitation effort. The EPA has also been criticized for the high costs associated with remedial action of sites on the NPL. Superfund remedial actions have cost approximately $134 million to date, while the CBO estimates that a few "mega-sites" can cost as much as $169 million per site. However, the average for the cost of all Superfund sites is $24 million.

Environmental Justice

Ensuring equal representation of hazardous waste sites on the NPL across socioeconomic status is critical to ensuring environmental justice. To that end, President Bill Clinton signed Executive Order 12898 in February 1994, requiring that federal agencies address disproportionately high health and environmental hazards in low-income and minority populations. Indeed, evidence has shown that minority and poor populations are disproportionately affected by environmental burdens.

The EPA has been criticized by many, including the U.S. General Accountability Office and the U.S. Office of Inspector General, for falling short of its obligations, for it failed even to develop criteria for determining populations that have been disproportionately affected by hazardous waste, much less to have developed ways to remedy these differences. In particular, despite regional environmental justice programs implemented by EPA field offices, no environmental justice programs have been implemented at the federal level and no federal funding exists for these efforts. In fact, inclusion on the National Priorities Site is not as likely for sites in areas with high minority and poor populations. Increases in minority population, poverty rates, and rates of people without high school diplomas in an area all lower the chances of NPL listing. A 10 percent rise in minority population lowers the chance of NPL listing by 2 percent; a 10 percent rise in poverty levels lowers the chance of NPL listing by 13 percent. In short, not only are poor communities much more likely to be the sites of toxic contamination, but they are also much less likely to be the recipient of federal cleanup efforts.

Since the creation of the Superfund in 1980, the NPL has served as a collection of the worst hazardous waste sites in the country, those requiring long-term remedial cleanup efforts. The effectiveness of the list is difficult to judge, but it has allowed the EPA to focus on the most important cleanup efforts under Superfund. Nevertheless, the NPL has come under criticism for its chilling effect on investment, the length of work and cost associated with cleanup efforts at sites on the NPL, and the unequal representation of socioeconomic communities on the NPL.

It is clear that the work required of the Superfund is far from over, and the NPL is likely to have several thousand more sites listed as requiring remedial, long-term attention. Given political and industrial pressures, it is not clear where the funding for these projects will come from, and the litigation between the EPA and responsible parties as well as the debate over federal funding from the general fund and taxes on polluting industries is likely to continue into the foreseeable future.

See Also: Environmental Assessment; Environmental Justice; Restoration; Superfund.

Further Readings

Beider, Perry. *Analyzing the Duration of Cleanup at Sites on Superfund's National Priorities List.* Washington, D.C.: U.S. Congressional Budget Office, 1994.

Beider, Perry. *The Total Costs of Cleaning Up Nonfederal Superfund Sites.* Washington, D.C.: U.S. Congressional Budget Office, 1994.

Hogue, Cheryl. "Superfund Slowdown." *Chemical and Engineering News,* 85/46 (2007).

O'Neil, Sandra G. "Superfund: Evaluating the Impact of Executive Order 12898." *Environmental Health Perspectives,* 115/7 (2007).

U.S. Environmental Protection Agency (EPA). "National Priorities List." http://www.epa.gov/superfund/sites/npl/ (Accessed January 2010).

U.S. EPA Office of Solid Waste and Emergency Response. "Cleaning Up the Nation's Waste Sites: Markets and Technology Trends, 2004 Edition." Cincinnati, OH: National Service Center for Environmental Publications, 2004. http://www.clu-in.org/download/market/2004market.pdf (Accessed January 2010).

U.S. Government Accountability Office (GAO). *Superfund: Litigation Has Decreased and EPA Needs Better Information on Site Cleanup and Cost Issues to Estimate Future Program Funding Requirements.* Washington, D.C.: U.S. GAO, 2009.

Dirk Peter Philipsen
Virginia State University

NATURAL CAPITAL

Natural capital is a term utilized to identify the economic value of natural resource stocks and environmental processes that yield goods and services. Maintaining natural capital through wise, productive, and efficient use of resources is considered essential to ecological sustainability, but is also argued to be able to positively contribute to an individual, firm, or state's income earning potential. Natural capital has become a core green business theme, as demonstrated through the use of the term to unify ecologically responsible consumers, nonprofit organizations, green firms, and sustainability consultants within various online communities and interactive web spaces.

Capital is considered to be a stock of materials or information that exists at a period of time and that can yield a flow of valuable goods and services into the future. The use of natural capital as a concept complements the definition of other types of capital, like human capital, social capital, or manufactured capital. Ecological economists argue that quantifying natural capital helps internalize potentially external costs, such as pollution or the depletion of stocks of nonrenewable resources that are not regularly counted in economic transactions. They even suggest that the value of the vital services that nature provides should be calculated in national accounts such as the gross domestic product (GDP), with subtractions occurring if natural stocks, cycles, and balances are not maintained. While the accurate valuation of ecosystems services, such as nutrient cycling, pollination, waste treatment, climate regulation, and freshwater supply, can be difficult, it is important to estimate prices in order to demonstrate the important and often indispensible role of such processes. Assigning values can also be used to clarify additional cost-benefit equations. For example, the cost of maintaining wetlands as storm buffers may be less expensive than paying for hurricane damage when wetlands are lost, or the price of rehabilitation to re-create natural storm buffers.

The influential book *Natural Capitalism* published in 1997 predicted that a new industrial revolution would inspire the greening of industry. The book's authors make an economic case for environmental protection and also create biological and social framework to transform commerce. The authors essentially argue for a revolution that would emerge from within the business sector, rather than in response to pressure from outside. Nonetheless, independent green seals of approval that assuage consumer sustainability concerns, such as the Forest Stewardship Council and the Marine Stewardship Council, have also been effective in influencing corporate shifts to green behavior. Another factor that has played a role in some instances has been the potential threat of consumer boycotts of unsustainable or immoral practices.

A number of interesting applications of the concept of natural capital have emerged in the past decade in business and nonprofit sectors alike. For example, the Natural Capital Project is a joint venture between the Woods Institute for the Environment at Stanford University, the Nature Conservancy, and World Wildlife Fund. The organization's goal is to align economic forces with conservation and to educate the public about how destroying nature ultimately also sacrifices human life-support systems. The project aims to improve human use of the world's lands and waters by making clear the economic and life-sustaining services they provide.

The nonprofit organization Natural Capitalism Solutions (NCS) was founded by L. Hunter Lovins, one of the coauthors of *Natural Capitalism*, to promote economic advantages that sustainability generates. NCS uses the three core principles listed below to identify opportunities to reduce capital investment, lower operating costs, and generate market leadership advantages.

Natural capitalism principles are as follows:

1. Using resources more productively: this slows resource depletion, lessens pollution, lowers costs, halts the degradation of the biosphere, makes it more profitable to employ people, and preserves vital living systems and social cohesion.

2. Redesigning industrial processes and the delivery of products and services to do business as nature does: biomimicry enables materials to be produced with low energy flows or powered with renewable energy. It promotes the reuse or recycling of materials and reduction or elimination of toxicity.

3. Managing all institutions to be restorative of natural and human capital: such approaches enhance human well-being and enable the biosphere to produce more wealth from its intact communities and abundant ecosystem services and natural resources.

Another coauthor of *Natural Capitalism*, Paul Hawken, founded the Natural Capital Institute (NCI) in 2002, with a similar goal to support socially and ecologically responsible investment and action. NCI's main project, WiserEarth, is an online community space connecting people, nonprofit organizations, and businesses. It offers social networking tools and groupware for people to connect and collaborate. The site was officially launched to the public in 2007 and since has become the world's largest directory of environmental, social justice, and indigenous community–focused organizations. Three years of extensive project planning began in 2004. NCI brought in software engineers and researchers and consulted with movement leaders to build the database. NCI partnered with open source software experts and trained data-collecting teams in India and Ecuador, as well as dozens of volunteers with expertise in a variety of languages, sustainability topics, and geographic regions. The international NCI team added publicly available data from the Internet and other sources, established content standards, and developed the database infrastructure. The success of this social networking—focused around the theme of natural capital—demonstrates the concept's international applicability and relevance to all regions of the globe.

The concept of natural capital has been productively applied to specific business sectors. For example, the Natural Capital Group provides finance to create competitive options for green commercial real estate. A primary focus is Leadership in Energy and Environmental Design (LEED)–certified buildings. Green buildings have been demonstrated to lower operating costs, create healthier work environments, increase employee productivity, and lead to additional business opportunities with like-minded investors and entrepreneurs.

Green improvements that benefit business owners include the following:

1. Reduced energy cost: green building reduces present and future energy costs through efficiency.

2. Reduced water use: water conservation saves this valuable resource for future generations while lowering expenses.

3. Reduced waste: when you use less, you spend less. Streamlining business operations and creating a sustainability plan encourages tenants and employees to consume fewer resources.

4. Increased productivity: green systems and nontoxic work environments keep employees healthy and more productive, resulting in a stronger bottom line for business.

5. "Halo" effect: green businesses are leaders in their industry. They demonstrate that they care about tenants, employees, and the planet.

6. Higher demand: potential tenants recognize the value of green space. Green buildings often receive higher rent contracts and experience lower vacancy rates.

7. Increased value: green buildings are simply worth more than traditional buildings. Improvements will recapture a greater portion of initial investments.

By documenting the economic value of natural capital, sustainability experts highlight how green practices can also foment profit. The adoption of green production tactics and plans for business sustainability has increasingly been mainstreamed to form a foundation of corporate responsibility.

See Also: Biomimicry; Corporate Social Responsibility; Cost-Benefit Analysis; Ecological Economics; Ecosystem Services; Externalities; Green Building; Leadership in Green Business; Resource Management; Waste Reduction.

Further Readings

Costanza, Robert, et al. "The Value of the World's Ecosystem Services and Natural Capital." *Nature,* 387 (1997).

Hawken, Paul, Amory Lovins and L. Hunter Lovins. *Natural Capitalism: Creating the Next Industrial Revolution.* Boston, MA: Little, Brown, 1999.

Natural Capital Group. "Why Go Green?" http://www.naturalcapitalgroup.com/whygogreen
    .html (Accessed April 2009).
Natural Capitalism Solutions. "Our Principles." http://www.natcapsolutions.org/
    index.php?option=com_content&view=article&id=48&Itemid=60 (Accessed
    April 2009).

Mary Finley-Brook
University of Richmond

Natural Step

The Natural Step (TNS) is a process of first considering and then engaging in improvement of environmental performance, associated with an organization of the same name. While most commonly used in organizations, especially private sector companies, the process can also be applied to other settings, such as communities. The Swedish nonprofit organization was founded in 1989 by Dr. Karl-Henrik Robèrt (b. 1947), a leading cancer researcher whose work led him to consider the role of preventable environmental factors in cancer. He concluded that since the late 19th century, human civilization had been "disrupting the cyclical processes of nature at an accelerating rate," consuming raw materials at unprecedented levels and producing waste in unheard-of quantities. Robèrt spent 21 drafts composing a paper soliciting consensus from the Swedish scientific community, reasoning that there were certain realities that would hold true regardless of political or social views. If a framework of that common ground could be established, he felt, resources could be spent addressing the issues brought into focus rather than arguing about areas of dissent. This iterative, interactive, consensus-seeking approach forms the basis for TNS.

Robèrt's framework for sustainability is based on the laws of thermodynamics: energy can only be modified, not created or destroyed; and in a process known as entropy, matter and energy disperse over time. While the Earth is an open system with respect to energy from the sun, the principles of thermodynamics reflect the fact that it is a closed system with respect to matter and waste. The essential challenge of sustainable development is to meet human needs and wants while preserving the capacity to continue to do so. This is difficult, given that humans remove material from the Earth faster than it can be returned through degradation, deplete nonrenewable resources, consume renewable ones faster than they are renewed, and continually manufacture plastics and chemical compounds, including many used in pesticides and industrial applications, faster than natural processes break them down. Robèrt identified the following four system conditions as essential to ensure a sustainable society:

1. Concentrations of substances extracted from the Earth do not systematically increase

2. Concentrations of man-made substances do not systematically increase

3. Physical degradation of nature does not systematically increase

4. The capacity of people to meet their needs is not systematically undermined

Applicability of TNS

In using TNS, leaders of an organization or community are introduced to the four system conditions by a TNS consultant. They are then asked to reflect on and discuss them until they reach consensus that the conditions are in fact accurate, and that humanity's (and their organization's) activities are in fact unsustainable. If and when this consensus is achieved, the leaders are encouraged to consider a vision of the company as a sustainable organization, imagine that this has been achieved, and through a process of "back-casting," explain how the organization moved to sustainability from its current condition. A similar process of education and discussion then occurs with others in the organization or community, and action plans for moving toward sustainability are eventually developed. This approach helps develop shared mental models, thus increasing the likelihood that leaders and members are fully committed to sustainability and to putting energy into action planning. Thus, the likelihood of action plan implementation and long-term change increases.

TNS has been successful in moving many businesses toward improved environmental performance. The retailer IKEA is a prime example of a business applying TNS framework. An environmental awareness already existed within IKEA in 1991 when Robèrt was invited to speak with the board of directors. Instead of simply describing environmental concerns facing IKEA—specifically the use of tropical rainforest wood in furniture—TNS helped create an awareness of solutions, both from a strategic and an operational viewpoint. A consensus was reached within IKEA for strategies in the areas of management and personnel, products, customers, suppliers, buildings and equipment, and transport. This use of TNS has resulted not only in environmentally beneficial change, but in adoption of the continual process of reevaluation and consensus seeking.

Interface, an Atlanta-based commercial floor covering manufacturer, was the first company in North America to adopt the use of TNS to rework its business practices. Its goal is to move not only toward sustainability but toward doing no harm to the Earth or its inhabitants. Nike partnered with TNS beginning in 1998. As a result, Nike's response to protests against conditions at its international contract factories evolved into one of proactivity and stakeholder engagement. Rohm and Haas, an international organization dealing in technology solutions, was challenged by one of its customers, Hydro Polymers, to develop sustainable polyvinyl chloride (PVC) technology solutions using TNS. It received the 2007 Hydro Sustainability Award and, in collaboration with Hydro Polymers, has developed a course on sustainability for the chemical industry. Rio Tinto Alcan, one of the largest global suppliers of aluminum, worked with TNS to cement its sustainability commitment despite increasing reliance on projects in developing nations.

TNS networks exist in 10 nations. At www.naturalstep.org, links are available to request a speaker for an event, course materials, or a customization of the course. Regular events are hosted and a newsletter is offered. Associated educational programs are outlined, including online courses, master's-level courses, and several master's degree programs.

TNS framework has also been adopted by communities around the world, 12 of which are located in the United States. In Sweden, a quarter of the country's municipal areas have become "ecomunicipalities," adopting a town or city charter that includes ecological justice, sustainable development, and the explicit goal to meet human needs more efficiently while reducing the use of nonrenewable resources and synthetic chemicals.

See Also: Ecological Economics; Ecological Footprint; Green Design; IKEA; Sustainability.

Further Readings

Nattrass, Brian and Mary Altomare. *The Natural Step for Business: Wealth, Ecology & the Evolutionary Corporation (Conscientious Commerce)*. Gabriola Island, British Columbia, Canada: New Society Publishers, 1999.

The Natural Step. http://www.naturalstep.org (Accessed January 2010).

Robèrt, Karl-Henrik. *Natural Step: A Framework*. Berkeley, CA: Critical Press, 1997.

Robèrt, Karl-Henrik. *The Natural Step Story: Seeding a Quiet Revolution*. Gabriola Island, British Columbia, Canada: New Catalyst Books, 2008.

Bill Kte'pi
Independent Scholar

NEWMAN'S OWN ORGANICS

Actor Paul Newman (1925–2008) started a philanthropic food business in 1982, donating all after-tax profits from his line of natural salad dressings, popcorn, lemonade, and pasta sauces to charitable organizations. His daughter, Nell Newman (1959–), was concerned about how agricultural pesticides affected humans and wildlife. Nell Newman and her partner, Peter Meehan, devised a business plan and established Newman's Own Organics as a subsidiary of Newman's Own in 1993. The offshoot was designed to support the environment and reduce pesticide use by encouraging the growth of organic agriculture.

The first offering from Newman's Own Organics—pretzels—became the nation's best-selling organic pretzel. The company focused on cultivating wide consumer appeal by using familiar ingredients. Its slogan is "Great tasting food that happens to be organic." Kraft granted permission for them to produce fig cookies under the name Fig Newmans™, which led to Newman's Own Organics becoming the top purchaser in the organic fig market. The line expanded to contain 80 items, including popcorn, chocolate chip cookies, chocolate bars, peanut butter cups, dried fruit, mints, coffee, tea, and pet food. In *American Gothic* fashion, a folksy "Pa" Newman and his daughter Nell posed on the packaging of Newman's Own Organics foods. Newman's Own Organics became a separate, independent business in 2001. It is based in northern California, with Nell Newman as president and Meehan as chief executive officer.

Committed to sustainable organic farming, Nell Newman had long tried to convince her father that organic foods were healthier and also delicious; however, he believed that organic food were unpalatable concoctions for "hippies." After secretly cooking a delicious organic Thanksgiving dinner, Nell finally persuaded him to cover the initial costs for exploring organic products. As her company succeeded and the organic movement gained momentum, Nell was quickly able to return her father's seed money.

Accredited third-party certification agency Oregon Tilth verifies that all organic suppliers to Newman's Own Organics are certified under the U.S. Department of Agriculture National Organic Standards Program. Organic ingredients for Newman's Own Organics come from a variety of sources, including corn from Texas, cacao from Costa Rica and Panama, and coffee from Peru. Some critics disapprove that Newman's Own Organics has a partnership with Green Mountain Coffee that provides a blend of its Fair Trade organic coffee to McDonald's restaurants. Nell Newman has argued that one of her goals was to introduce organics to people who might not normally consume them, and she contends that coffee farmers in economically challenged countries would be pleased to see sales of organic coffee rise.

Since all consumer product packaging contributes to waste, and national distribution uses energy resources, Newman and Meehan acknowledge that their business entails contradictions. Still, they have made the case that Newman's Own Organics helps expand the scope of ecologically beneficial organic farming methods. Furthermore, its donations fund many environmentally responsible groups. After the first decade of operation, the enterprise had already donated more than $2 million to small groups involved with wildlife preservation, organic agriculture research, education, and other socially and environmentally progressive causes. Newman has stated that, though the positive impact of Newman's Own Organics may be small, it is "better than nothing."

Nell Newman has said that her environmental consciousness arose from her love of birds. As a child, she was alarmed to learn that the synthetic pesticide DDT was putting peregrine falcons at risk of extinction. She became a licensed falconer and worked with the Ventana Wilderness Sanctuary in California to reestablish populations of falcons, bald eagles, and condors. When she and the other scientists had their own blood analyzed, they all tested positive for contaminants like DDT, PCBs, and chlordane. Though she was only 30 years old and eating mainly organic foods at the time, omnipresent background levels of these toxic chemicals had seeped into the food chain and thus entered her body.

Nell Newman also grew up with an appreciation for garden-fresh foods, which later inspired her to maintain her own organic vegetable gardens, keep hens in the yard, and regularly patronize farmers markets. Newman firmly believes that organic farming is safer for humans and better for the environment, and this conviction continues to inspire her dedication to the success of Newman's Own Organics. If nothing else, the company can be seen as a successful inspiration to businesses that are attempting to pursue a more sustainable path of doing business.

See Also: Corporate Social Responsibility; Ecolabels; Leadership in Green Business; Organic; Stonyfield Farm.

Further Readings

DeVault, George. "Using Money to Make Change." *Mother Earth News* (February/March 2004).

Mendelsohn, Lotte. "Nell Newman's Own." *Vegetarian Times* (October 2002).

Newman, Nell. *The Newman's Own Organics Guide to a Good Life: Simple Measures That Benefit You and the Place You Live.* New York: Villard, 2003.

Newman's Own Organics. http://www.newmansownorganics.com (Accessed April 2009).

Tracy, Kathy. "Nell Newman's Organic Thanksgiving." *Eating Well* (November–December 2007).

Robin O'Sullivan
University of Texas at Austin

NEW SOURCE REVIEW

New Source Review (NSR) is a major air pollution program in the United States instituted under amendments to the Clean Air Act (CAA) (42 U.S.C. § 7401 and following), which is the successor to the Air Quality Act of 1955. The 1977 amendments provided comprehensive regulations to control air pollution and introduced the National Ambient Air Quality Standards (NAAQS) and the New Source Performance Standards (NSPS). The CAA also introduced requirements to install and operate the best available control technology (BACT) using a permit system within the CAA regime to implement the NSPS (42 U.S.C. § 7407 & 7411) and NSR provisions. Specifically, NSR and the legislation that implements this program arose out of an important green policy debate. Essentially, advocates for NSR intended to create a regulatory

Idling traffic contributes to air pollution, which new source review seeks to address.

Source: U.S. Environmental Protection Agency

scheme where businesses are required to use BACT when building new facilities. Furthermore, this regulatory framework grew out of a desire for increased sustainable outcomes, whereby adding substantial improvements or upgrades to existing facilities that would significantly increase air pollutants is considered the equivalent of building new facilities, which, therefore, requires commensurate BACT.

Under the NSR provisions of the CAA, facility owners and operators are obligated to inform the U.S. Environmental Protection Agency (EPA) and seek permits from it, or other state or local agencies, prior to modification or construction of facilities that emit regulated pollutants. When issued, these permits are legally binding documents that specify the required installation of the BACT, the emission limits, how the emissions source must be operated, and the other terms under which the approval has been granted. There are three types of NSR permits. The first two types of permits are related to the attainment and nonattainment of the NAAQS (42 U.S.C. § 7409). Specifically, Prevention of Significant Deterioration (PSD) permits, as outlined in Part C of Title I of the CAA (42 U.S.C. § 7470), apply to geographic attainment areas where the air is relatively clean and has attained the applicable quality standards. Nonattainment NSR permits, as outlined in Part D of Title I of the CAA (42 U.S.C. § 7501), apply to geographic nonattainment areas where the air quality is below the standards. Finally, the third type of permit is the Minor NSR permit, which is applicable to stationary sources that do not require either a PSD or NSR permit to comply with the air quality standards. The three sets of requirements apply to new sources of air pollution and to existing sources of air pollution that are making a major modification, which is reflective of the intended policy environment in which NSR was implemented. (In current literature, reference to NSR permits tends to include both the first and second sets of permits; earlier discussions focused on the nonattainment NSR permits, which is primarily due to the fact that the PSD requirements were only added in the 1977 amendments to the CAA.)

Title V of the CAA amendments requires facilities to obtain an operating permit in addition to the construction and modification permits. As a self-reporting permit regime, the determination of whether or not to report and apply for a permit rests with the owners and operators of a facility. Once granted a permit, these facilities are subject to inspection and permit term enforcement by the EPA. Limitations of the EPA's enforcement capacity likely influence the scrupulousness with which permits are sought, which has obvious consequences for the integrity and effectiveness of the regime on air quality. Noncompliance with NSR leads to continuing degraded air quality with the introduction of unaccounted pollution estimated to be in the thousands of tons each year. These pollutants include nitrogen oxides, volatile organic compounds, and particulate matter.

The EPA may delegate authorization to issue permits to state or local agencies. Permits under NSR, including PSD, minor NSR, and operating permits, provide legal authorization for a facility to act within their stated terms. The terms may specify permitted construction, emission limits, and frequency of operations. Permits may impose conditions and may outline parameters and dimensions of equipment similar to those indicated as conditions self-reported by the applicant facility. Such conditions may be imposed unilaterally by the local, state, or federal agency or, more likely, by agreement with the facility. Oftentimes, a facility would prefer the conditions of accepting a minor NSR permit rather than agreeing to the more rigorous procedures entailed in obtaining a PSD permit. Permits have self-reporting obligations that must be supported by the continual monitoring and maintenance of business records.

Under certain circumstances, existing sources of pollutants may continue to operate without triggering the requirement for a permit. The regulations allow routine maintenance and repairs to equipment that do not constitute a major modification. Consequently, existing sources of pollutants may continue to operate indefinitely without a CAA-mandated PSD or NSR permit by conducting "routine maintenance" that does not trigger the "modification" threshold, or by not making changes that result in a "significant increase" in environmental impacts. They can thereby continue to cause degradation of the air quality until other legal provisions apply. The definition of "modification," including the provision for "routine maintenance," was incorporated from the NSPS provisions of the CAA (42 U.S.C. § 7411). Judicial interpretation of the NSPS and the PSD provisions have established that a modification is "any physical change" if it results in significantly increased emissions of any regulated pollutant from the facility source by taking into account the nature, extent, purpose, frequency, and cost of the work within the context of reasonable industry practices.

The effective exemption from PSD permits may give existing major sources an advantage in costs over new facilities for an indefinite term. As beneficiaries of the provisions, existing sources also have a greater incentive to perpetuate the scheme rather than participating in air quality solutions. The uncertainty in the scope of the terms *routine maintenance* and *modification*, or the threshold between the two, has led to litigation. Additionally, to the extent the persistence of existing pollution sources is favored over facilities that can achieve actual reductions in the environmental loading of pollutants, the objectives of the regulated entities and the enforcement agencies could be frustrated. Attempts to avoid the NSR provisions may actually operate to block investments to increase efficiency and, hence, be counterproductive to the intended objectives.

Still, it should be noted that installation of improved technology has cost consequences for everyone. For instance, the EPA has taken particular note of the contribution of electricity utility facilities and their effect on air quality. A 1997 report to Congress noted that coal-fired utility and industrial boilers accounted for more than half of the 158 tons of mercury released into the atmosphere in the United States. The demand for energy and its costs flow through to the consumer, systemically affect the productive capacity, and add costs to the economy. Notwithstanding EPA and other agency efforts to balance implementation of the CAA, industry and environmental groups have been critical of the implementation and operation of the NSR provisions.

The common public interest and stated CAA purpose is to reduce existing air pollution. Perpetuating existing operations as provided under current practice may not serve to achieve the intended result. It is conceivable that the PSD provisions could be considered transition provisions and eventually eliminated. Should that occur, all facilities could be subject to compliance with the NSR requirement for BACT installations but may still not require existing operations to upgrade to the BACT standards. An alternative to PSD is an emissions trading system. A further alternative is an improved statutory definition of routine maintenance. These measures may also be used to complement each other to provide redundancy in the air quality regime.

The EPA set standards for permitted emissions by determining the demonstrated emission levels that are achievable following the application of technology. For new major stationary sources in nonattainment areas, the technology standard was the lowest achievable emissions rate (LAER). For existing sources in nonattainment areas, the standard was reasonably available control technology (RACT). Finally, emitting facilities in attainment areas were required to have the BACT. Under the authority of the CAA and the Resource Conservation and Recovery Act, the EPA promulgated standards based on the maximum available control technology (MACT). However, under the Clean Water Act and the CAA, neither the BACT nor the MACT may be sufficient if the applicable quality standards, such as State Implementation Plan specifications, are not achieved. Where there is a difference, the higher of the BACT or the other governing water or air quality standards will apply. The same regulations permit some discretion in the enforcement of the requirements to achieve applicable toxicity standards.

Whether it is required under the NSR provisions of the CAA or voluntarily by the owners and operators of a facility, the installation of the BACT operates as a function of the precautionary principle and can preempt reasonably foreseeable adverse impacts or events. From a risk management approach, absolute certainty is not necessary, only a level of risk that the standard is intended to address. Additionally, as a dynamic standard, implementation of the BACT and related measures can contribute positively to industries where competition is based on reducing input and operational costs and improving quality of deliverables. Improved technology can contribute positively to efficiency, productivity, and quality and to managing risks with implications for calculating the balance between the marginal costs of pollution control and social benefit. The impetus for improving technology itself can drive innovation that could produce competitive advantages.

The external imperatives to consider the NAAQS and NSR compliance can arise from evolving industry standard practices, escalating efficiency requirements, competition, customer requirements, and escalating input costs. The development of cap-and-trade or emissions trading markets can provide an alternative to the NSR requirements or an additional incentive to implement the BACT in order to generate tradable credits. Internally, installation of the BACT pursuant to a permit may be cost effective if it reduces private costs to the organization and attends to environmental and social interests. For instance, the major modifications that trigger an NSR/PSD permit requirement may be implemented as part of an environmental management system to improve total quality, reduce waste, and obtain market advantages and social acceptance. Where command-and-control regulatory or industry market requirements or other circumstances require, obtaining an NSR permit may simply be a cost of doing business.

When regulatory requirements, such as compliance with air quality standards, focus on outcomes rather than technologies, obtaining the applicable NSR permits may be an implicit requirement as well. To the extent that the NSR provisions involve equipment, technology, and management, the onus remains with the operator to achieve the intended effects. Where a business implements technology solely to meet the legal requirements, it may not necessarily be operating from either an interest in the social impacts of its operations or from a sustainability perspective. In a context of voluntary continuous improvement strategy, obtaining an NSR permit and implementation of the BACT could be an acceptable risk that contributes to the intentional reduction of adverse impacts on the environment from human activities and events. In the emergent paradigm, it is possible that a business may be required to comply with NSR type of requirements as a means of including the cost of ecoefficiency and environmental quality, as well as social impacts that are essential to continuity and the cost of doing business. A complementary social and regulatory acceptance of air quality standards and implementation of ongoing reviews may then become routine matters.

See Also: Best Available Control Technology; Corporate Social Responsibility; Emissions Trading; Leadership in Green Business; Precautionary Principle; Sustainability; Triple Bottom Line; Voluntary Standards.

Further Readings

Rankin, Katherine K. "Big Win for Environmentalists in *New York v. EPA* May Have Limited Impact on Air Quality." *Ecology Law Quarterly*, 34/3:837–60 (2007).

Reitze, Arnold W., Jr. "Federal Control of Carbon Dioxide Emissions: What Are the Options?" *Boston College Environmental Affairs Law Review*, 36/1:1–77 (2009).

U.S. Environmental Protection Agency. "Clean Air Act." http://www.epa.gov/air/caa/title1 .html#ic (Accessed October 2009).

U.S. Environmental Protection Agency. "New Source Review." http://www.epa.gov/NSR/ (Accessed October 2009).

Zorn, Graham. "Prevention of Significant Deterioration and Its Routine Maintenance Exception: The Definition of Routine, Past, Present, and Future." *Vermont Law Review*, 33/4:783–804 (2009).

Lester de Souza
Independent Scholar

ORGANIC

Organic food and farming developed through the 20th century in response to crises in the farmed environment and technological innovations in farming and food technologies. Organic food and farming is based on not using nitrogen fertilizers, synthetic pesticides or insecticides, prophylactic antibiotics, artificial hormones, or genetically engineered plants or animals. Instead of using these technologies, the farm is managed to ensure that fertility is gained with compost, manures and leguminous plants, companion planting, and encouragement of natural predators. Plant-based insecticides are used to control insects, with homeopathy and less-intensive production techniques used to secure animal welfare. To ensure that organic food is produced in accordance with the principles and ethos of the organic movement, farms and food processors are inspected and certified, allowing the product to display a logo indicating its organic status.

Instead of using chemical fertilizers and pesticides, organic products, such as these cauliflowers, are managed to ensure that fertility is gained with compost, manures, and leguminous plants; companion planting; and encouragement of natural predators.

Source: Doug Wilson/U.S. Department of Agriculture

Organic farming, according to its proponents, is better for the environment because it encourages a complex ecosystem on the farm and uses fewer natural resources. Organic food, it is argued, is healthier because of the absence of chemical residues and the presence of trace minerals and chemicals missing from nonorganic products.

The origins of the organic movement lie in two separate, and initially independent, developments in Germany and the British Empire in the first half of the 20th century.

During the 1920s, the Life Reform movement in Germany brought into question the use of nitrogen fertilizers in agriculture and sought to find a method of producing agriculture more in tune with nature. Simultaneously, Rudolf Steiner, the founder of anthroposophy, delivered a series of lectures on agriculture that were developed into biodynamic agriculture. This perspective on organic farming requires the farm to be as closed a system as possible, with wastes and nutrients returned to the farm, but adds a cosmological dimension whereby various preparations are used to optimize the soil and crops of the farm. In the British Empire, the encounter of Western agricultural science with traditional Indian farming practices led to the development of a vision of agriculture that sought to nurture the ecosystems of the soil and valued traditional agriculture. This perspective argued that nitrogen fertilizers destroy the ecosystem of the soil, and therefore undermine the health of the plants, animals, and humans dependent on it. Albert Howard, then a prominent British agricultural scientist, argued that proper soil management with composts based on animal manure and plant matter would maintain the health of the soil and those reliant on it. Howard had little time for the mystical ideas of Steiner, labeling them "muck and magic," while promoting his own ideas as based in science.

Despite some interest within the leadership of Nazi Germany, the organic movement there foundered, leaving the British movement as the fulcrum of the heart of the movement. The most prominent proponent of organic agriculture was Lady Evelyn Balfour, who after hearing the arguments of Balfour and his associates, undertook to start a comparative experiment on her own farm at Haughley, Suffolk, England. Balfour's plan was a farm-scale trial managing one set of fields with "artificials," one left fallow, and the other organically. The outbreak of World War II led to many of those who had been experimenting with organic farming and food in the previous decade to record their experiences. It was at this time that the term *organic* began to be used by the movement, rather than terms such as compost or humus farming. Balfour's account of her experiment, *The Living Soil* (1943), catalyzed the networks into thinking about an organization that they could form at the end of the war. The Soil Association was founded in 1946, with the direct purpose of promoting organic farming and food, as well as the evidence necessary to underpin it. The association sought in the following years to coordinate the organic movement across the planet and manage Balfour's research at Haughley. At this time, the organic movement began to gain a foothold in North America with a number of experiments being undertaken, the most prominent of which were those of Jerome Rodale. Rodale had read the work of, and then entered into a correspondence with, Albert Howard in the early 1940s. He began to publish a magazine, *Organic Farming*, and to offer scientific research funding to those prepared to investigate organic agriculture.

Agriculture and the New Technologies

The war led to the introduction a new range of technologies to agriculture. A powerful new insecticide—DDT—had been developed to protect Allied troops from the threat of lice and mosquitoes; this was transferred to agricultural use in the postwar period. In the postwar period, the Rockefeller Foundation promoted combination of hybrid seeds, farming machines, chemical fertilizers, and insecticides that became known as the "green revolution." The organic movement launched a trenchant critique of the damage that these new farming technologies did to the farmed environment and the health of those eating its produce. The publication of Rachel Carson's *Silent Spring* (1962) highlighted the damage that the new insecticides were doing to the natural environment and how these chemicals could find their way into the human body. That agricultural pollution was damaging the health of consumers introduced a new generation to the organic movement, as did the increasing argument that food technologies were debasing the nutritional value of food. Organic increasingly became a way of critiquing not just farming but the entire food system. In the late 1960s and early 1970s, influential environmentalists such as Barry Commoner, Fritz Schumacher, and Teddy Goldsmith were associated with the organic movement.

It was a joint initiative by the Soil Association and a project led by the Rodale Institute in California in 1972 that saw the introduction of the first organic standard scheme. Some experiments in Germany in the 1920s had seen prototypes of organic standards, but this new scheme was far more ambitious. In order for organic farming to increase in scale, consumers had to be certain that the produce was grown according to the principles of the organic movement. After nearly 30 years of experimentation, by the early 1970s, the organic movement had not been able to produce scientific evidence of the value of organic food. By turning to a certification scheme, the movement would grow through giving people the opportunity to buy organic food and so, in turn, create a form of market that would encourage farms to produce organic food. Farmers would agree not to use a range of artificial fertilizers, insecticides, synthetic animal hormones, and routine use of antibiotics, submitting to an annual inspection to ensure that they were not; in return, they could use the organic logo on their produce. Exactly how farmers would manage their land and crops was left to them as long as they followed the standards. Initially this was a private agreement without legal force, but by 1979, California had enacted legislation that would allow prosecutions in the case of fraud. In Europe, parallel legislation came from the European Union, but along with it came proposals of far greater ambition.

The organic movement was divided by the possibility of working in tandem with the state. Some were libertarians opposed to working with capitalism, while others were in favor of a free market and saw the state as an infringement of the market, while an older group still hoped for scientific proof. During the 1980s, ethical consumerism, as championed by operations such as The Body Shop, increasingly became a way for consumers to reconcile their environmental aspirations with their daily needs and organic food became an important part of that development. Multiple retailers saw opportunities in selling organic foods as part of a turn to marketing health and whole foods. This was particularly apparent in some nations where very small markets for organic produce were growing very quickly, for example, in the United Kingdom in the late 1980s, demand outstripped supply to the point that 95 percent of some goods had to be imported.

Growth of the Organic Movement

The strategy of growing the organic movement through selling organic produce was boosted by the arrival of legislation that spanned all the European Union in 1992. This required all member states to initiate legal frameworks for inspecting organic produce to a broadly common set of organic standards. One of the most important global markets for food products was united by common standards that had to be met by those wishing to sell produce as "organic." Agricultural exporting nations such as New Zealand and Australia with strong cultural links to Europe were well positioned to exploit this market, which in turn boosted their domestic organic movements.

With the introduction of the common standards, member states of the European Union could provide various forms of financial support to encourage farmers to convert to organic production or support organic farms. From being a movement outside the networks of agricultural policy and politics, the organic movement found itself increasingly intertwined with it.

Federal legislation in the United States was a far more contested process than in Europe, with opponents of organic farming seeking to have the very technologies the movement opposed approved under the proposed regulations. Only a massive protest by organic consumers in 1997 ensured that the new standard did not include genetically engineered (GE) plants under the new standards. The eventual U.S. standards, the National Organic Program (NOP), have been relatively weakly enforced, with ongoing disputes about what is organic and attempts by corporate interests to change the standards to reflect their financial goals.

Until the late 1990s, the organic movement had been one of the less-aggressive parts of the wider environmental movement, preferring polite lobbying to public protests. This changed with the mobilization of the movement against the introduction of GE plants. These plants had been introduced with little controversy in the early 1990s in North America, but efforts to grow them in Europe were fiercely resisted and the organic movement was at the forefront of that resistance. In alliance with more radical environmental groups, the organic movement argued that GE crops threatened organic crops and the freedom of choice for consumers as well as their health. These protests saw test crops destroyed, ships suspected of carrying seeds being boarded, and hundreds of naked protesters occupying legislative chambers discussing the plants. European supermarket chains, some of the largest in the world, were pressured to remove GE products from their shelves and supply chains. The result of the protests was that GE plants are not grown in Europe and the influence of the organic movement has increased.

As a diverse cultural movement, the organic movement has been subject to criticism from a number of perspectives. Many supporters of the new agricultural technologies see the movement's positions as being romantic and not based on scientific evidence. From the other direction, the movement has increasingly been criticized as becoming too similar to the food system that it was supposed to oppose; many in North America argue that it has become "conventionalized." Certainly in areas where agriculture and food are tied closely to conceptions of rural life and protected by a broadly social democratic state, organic farming and food appears to be secure. Where the fate of the organic sector is left to the market, its fortunes appear to be more uncertain, where either weak regulation as in the United States or faltering sales such as in the UK weaken the movement. Certainly its influence in pointing to the damage caused to the wider environment by intensive agriculture, the dangers of foodborne contamination by agrochemicals, and the damage done by highly processed foods is far wider than the sales of foods certified as organic alone would suggest.

See Also: Bio-Based Material; Body Shop, The; Certification; Integrated Pest Management; Newman's Own Organics; Stonyfield Farm.

Further Readings

Conford, P. *The Origins of the Organic Movement.* Edinburgh, UK: Floris Books, 2001.
Guthman, J. *Agrarian Dreams: The Paradox of Organic Farming in California.* Berkeley: University of California Press, 2004.
Howard, Sir Albert. *The Soil and Health: A Study of Organic Agriculture (Culture of the Land).* Lexington: University Press of Kentucky, 2007.
Reed, M. *Rebels for the Soil: The Rise of the Global Organic Food and Farming Movement.* London: Earthscan, 2010.

Matt Reed
Countryside and Community Research Institute

PATAGONIA

Patagonia is an outdoor clothing company based in Ventura, California. It is best known for its environmental and social consciousness, which manifests itself from its marketing to its business practices to its support of environmental charities.

Patagonia was founded in 1972 by southern California rock climber Yvon Chouinard (b. 1938). The environmentally conscious son of a French-Canadian blacksmith—he joined the Sierra Club as a young man and founded the Southern California Falconry Club—he first became interested in rock climbing through falconry. It was a necessary skill to investigate falcon aeries in craggy peaks in places like Yosemite Valley. After a while, he decided to make his own climbing tools, using the blacksmithing skills he had learned from his father and a secondhand coal-fired forge. Chouinard was a serious rock and mountain climber: he was the first to ascend the North Face of Mount Sir David in the Canadian Rockies, and with legendary climber Fred Beckey, he discovered the Beckey-Chouinard Route on South Howser Tower in the Bugaboos, a range in the British Columbian Purcell Mountains.

Dividing his time between climbing expeditions and surfing, he supported himself selling his homemade hard-steel pitons out of the back of his car. Specially made for the climbing challenges of the Yosemite Valley, the pitons were instrumental in stirring up interest in "big-wall climbing" in Yosemite, which in turn led to more piton sales, leading to the founding of Chouinard Equipment Limited. Working with fellow climber Tom Frost, he redesigned the basic tools of ice climbing for use on steeper ice, using what they had learned from rock wall climbing. By the late 1970s, this innovation led to the modern sport of recreational ice climbing.

Chouinard's commitment to the environment was demonstrated when he discovered that the hard-steel pitons he had introduced to Yosemite Valley climbing were damaging the rock. Beginning the following year, in 1971, Chouinard Equipment began selling more "rock-friendly" aluminum chockstones to replace pitons—which had accounted for more than two-thirds of the company's business—and championed the cause of "clean climbing." While traditional rock climbing implements like pitons, bolts, and copperheads do permanent damage to rock, "clean" gear like chockstones, nuts, slings, and camming devices allow for climbing without needing to hammer or drill into the surface of the rock. The introduction of clean climbing tools by Chouinard and other companies changed the sport of rock climbing again; today, the average free climber never uses a hammer or a drill. Chouinard Equipment eventually went into bankruptcy in 1989 as a result of liability lawsuits from amateur and beginner climbers who alleged that the equipment did not bear an adequate warning about the dangers of rock climbing. It was purchased by its employees and reemerged as Black Diamond Equipment.

Around the same time that he was introducing clean climbing, Chouinard founded Patagonia as a separate company to sell outdoor clothing. He had previously had unexpected success selling rugby shirts he had picked up on a trip to Scotland, but Patagonia was a drastic step forward, making specialized clothing for a variety of outdoor activities popular in southern California. Over time, the company had specialized in apparel for climbing, camping, skiing, surfing, and hiking. Today, Patagonia is one of the most popular American brands of such outdoor apparel.

Beginning in the 1980s, Chouinard's environmental and social consciousness became part of official Patagonia company policy. The company began providing on-site healthcare and a cafeteria providing healthy fresh-made meals—mostly, but not exclusively, vegetarian. Since 1986, the company has "tithed" a portion of its revenue to environmental activism—10 percent of profits or 1 percent of sales, whichever is greater. Over 1,000 charities have received a total of over $25 million since Patagonia began the practice. In 2001, Chouinard cofounded One Percent for the Planet, an international group that encourages other companies to do likewise. Over 1,200 companies have joined, including Sony, SIGG, and Mountain Equipment Co-op.

An environmental audit of Patagonia revealed that its most environmentally harmful practice was, counterintuitively, the cotton used in much of its apparel. After investigating the state of the cotton industry, the company committed in 1994 to using only organically grown, pesticide-free cotton—it was largely the pesticide use that had led to cotton, one of the country's most important cash crops, having such an environmental downside. The influx of demand for organic cotton as a result of Patagonia's switch has been a major boost to the industry.

More recently, Patagonia has focused on recycling challenges. In 2005, its Capilene Performance Base Layer line of clothing was introduced, made of recyclable materials, along with the Common Threads Garment Recycling Program, through which customers could return used clothing to be recycled. Almost 15,000 pounds of clothes have been recycled through the program, and several times that amount has been warehoused, awaiting recycling. Concurrent with the inception of the program was the announcement of a goal to recycle 100 percent of its products by 2010, a goal the company is on track to meet, according to the 2009 annual report.

As of 2009, 80 percent of Patagonia apparel was made of recyclable material: spandex, small items with disproportionate amounts of trim, down, and polyurethane account for most of the remaining fifth. Still, the real difficulty lies in encouraging customers to do the recycling, since, unlike paper, aluminum, cardboard, glass, and plastic, few customers have access to, or awareness of, textiles recycling centers. Judging from the figures of the Common Threads program, most consumers are not taking advantage of available recycling opportunities. Patagonia, in short, represents an example of a company that, rather than being pushed into a more environmentally friendly direction by the public, is ahead of most of its customers in its attempts to develop a sustainable business model.

See Also: Corporate Social Responsibility; Environmental Audit; Green Retailing; IKEA; Recycling, Business of; Social Entrepreneurship.

Further Readings

Chouinard, Yvon. *Let My People Go Surfing: The Education of a Reluctant Businessman.* New York: Penguin, 2006.

Chouinard, Yvon. "Patagonia: The Next Hundred Years." In *Sacred Trusts: Essays on Stewardship and Responsibility*, Michael Katakis, ed. San Francisco, CA: Mercury House, 1993.

Rarick, Charles A. and Lori S. Feldman. "Patagonia: Climbing to New Highs With a Smaller Carbon Footprint." *Journal of the International Academy for Case Studies*, 14/7:121–25 (2008).

Bill Kte'pi
Independent Scholar

PERSISTENT POLLUTANTS

Persistent pollutants are substances that contaminate the environment, causing harm to humans, living organisms, and ecosystems over extended periods of time. Defined broadly, persistent pollutants could include substances as diverse as greenhouse gases, radioactive waste, heavy metals, and even transgenic gene sequences because all can cause harms over extended periods of time.

Radioactive waste is one type of persistent pollutant that is carcinogenic for thousands of years.

Source: iStockphoto

Chlorofluorocarbons (CFCs) and hydrochlorofluorcarbons (HCFCs), for example, are greenhouse gases that contribute to destruction of the ozone layer. While CFCs have largely been banned and use of HCFCs is slated to end by 2030, their effects continue. The chemical properties of HCFCs, and most especially of CFCs, are such that it takes several decades for their negative effects to subside; in the interim, ozone continues to be destroyed in the stratosphere. While the ozone layer will likely recover by 2065, global warming produced by CFCs, HCFCs, and other greenhouse gases will likely extend for millennia. Similarly, radioactive wastes are carcinogenic for millennia. Mercury in laptops and flat-screen televisions is another persistent pollutant. While in use there is little harm, but when these products break down, heavy metals are released. Small quantities of mercury can damage the brain, nervous system, and reproductive systems; mercury is also highly toxic to aquatic ecosystems and moves through the food web. Yet another example of a persistent pollutant could be a transgenic gene sequence. Transgenic genes from Bt maize escaped into native crops in Oaxaca, Mexico, where maize originates. If the transgenic sequence proves harmful, genetic pollution could cause harm indefinitely. Thus, harm can persist for the duration of the existence of the substance in question or, as in the case of climate change, long after.

Persistent Organic Pollutants

Persistent organic pollutants (POPs) are a class of chemicals that are especially persistent because they resist degradation by biological, photolytic, or chemical means. Over time, they bioaccumulate in organisms that are exposed to them, increasing in concentration in organisms higher in the food chain. Animals that ingest POPs store them in fatty tissues. As these animals are eaten by predators, the concentrations of POPs increase up the food chain. POPs can also travel easily to remote locations via water or migratory species. POPs are semi-volatile. In hotter locations, POPs travel by air, but in cooler climates, deposit in large concentrations; high levels of POPs have been found in Norwegian polar bears.

The Stockholm Convention was formed to protect human health and the environment from POPs. Signatories pledged to cease or to limit the production and emission of 12 POPs by 2004: aldrin, chlordane, dieldrin, endrin, heptachlor, hexachlorobenzene, mirex, toxaphene, polychlorinated biphenyls (PCBs), DDT, PCDD (dioxin), and PDCF (furans). Aldrin, dieldrin, chlordane, and endrin are pesticides phased out in the 1980s in the United States. Heptachlor is an insecticide phased out in the United States in 2000. Hexachlorobenzene, a fungicide and by-product of combustion, was phased out in the United States in 1985. Mirex, phased out in 1977 in the United States, was originally used as pest control and as a fire retardant in plastics, rubber, and electronics. Toxaphene, an agricultural insecticide, was phased out in the United States in 1990. PCBs had various industrial uses in electronics and as additives in plastics; PCB is also a by-product of combustion. The use of PCBs in the United States ended in 1978. DDT, another pesticide, was phased out in 1989 in the United States, in large part because of publicity generated by Rachel Carson's *Silent Spring*. The United States continued DDT export to developing nations into the 1990s; DDT is still used globally in combating mosquitoes carrying malaria. Dioxins and furans are by-products of industrial processes and of burning municipal waste.

In May 2009, the Stockholm Convention considered adding to its list commercial octabromodiphenyl ether and pentabromodiphenyl ether (both polybrominated diphenyl ethers or PBDEs), chlordecone, hexabromobyphenyl (HBB), lindane, alpha and beta hexachlorocyclohexane, pentachlorbenzene (PeCB), perfluorooctane sulfonic acid, its salts, and perfluorooctane sulfonyl fluoride (PFOs). PBDEs are brominated organic substances used as fire retardants in electronics. Since the 1970s, levels of PBDEs in human blood, milk, and tissue have increased by 100-fold. Younger people have a higher concentration of PBDE for their years, suggesting bioaccumulation. Chlordecone, an agricultural pesticide, has no current production. HBB, a flame retardant, ceased production in the 1970s. Lindane is an insecticide used on wood and trees, by veterinarians, and in products treating head lice. Intentional production of beta hexachlorocyclohexane ceased, but it can be released in the production of lindane. PeCB was created for intentional use in PCB products, dyes, fungicides, and flame retardants; it is unintentionally produced by combustion in industrial processes. PeCB is an impurity in several solvents and pesticides. PFOs are used as flame retardants in electronics, firefighting foam, and textiles. POPs are used in the production of polyvinyl chloride, solvents, and pharmaceuticals.

POPs are found in the depths of the oceans, in remote deserts, and in rain and snow. POPs are known to cause cancer and birth defects, may disrupt immune and reproductive functions, and may diminish intelligence. Increased POPs are correlated with species declines in marine animals such as bottle-nosed dolphins. Nearly every person on Earth likely has POPs in their bodies. Obese persons with higher concentrations of six key POPs had a much higher incidence of Type II diabetes. Most POPs enter food supplies because of pesticides, chemical manufacturing, and incinerated waste. PBDEs, for instance were found in shrimp, chicken, cheese, and soy formula. Like PCB, PBDEs may be endocrine and thyroid disruptors. Several companies including Intel, Sony, Apple, and IBM have opted to cut some or all use of PBDEs.

Equity Issues

Kristen Shrader-Frechette has argued that the production of toxic substances whose victims are by and large not its beneficiaries violates moral requirements of distributional equality. Inuit suffering health problems from POPs, for example, receive few benefits and suffer disproportionate harms. Future generations are also subject to greater harms caused by persistent pollutants and few, if any, of their benefits, resulting in intergenerational inequity. Because of the persistent harmful properties of POPs, the only viable and effective way to restrict their widespread and long-term damaging effects is to restrict their production and use in the first place. Yet dealing with persistent pollutants can be difficult because companies often create new chemicals faster than regulators can document problems. Developing countries often lack infrastructure to handle health risks and the regulatory strength to prevent the use of POPs. Even in developed nations that restrict POPs, safe disposal of toxic items, particularly electronic goods, is often lacking.

See Also: Brownfield Redevelopment; E-Waste Management; Pollution Prevention; Superfund; Toxics Release Inventory.

Further Readings

Brown, Phylilda. "Trouble in Store." *New Scientist*, 199/2673 (September 10, 2008).
Grossman, Elizabeth. *High Tech Trash*. Washington, D.C.: Island Press, 2006.
National Toxicology Program. "Public Health." http://ntp-server.niehs.nih.gov/?objectid= 720161D5-BDB7-CEBA-F912171282926F28 (Accessed April 2009).
Stockholm Convention on Persistent Organic Pollutants. "POPS Review Committee: Listing New Chemicals." http://chm.pops.int/Portals/0/docs/publications/sc_factsheet_002.pdf (Accessed April 2009).
U.S. Environmental Protection Agency. "Pollutants/Toxics." http://www.epa.gov/ebtpages/ pollutants.html (Accessed April 2009).

Mary Lyn Stoll
University of Southern Indiana

POLLUTION OFFSETS

For air pollutants of regional concern, such as the acid-forming gases of sulfur dioxide (SO_2) and nitrous oxides (NO_x), air quality depends more upon the net amount of a pollutant emitted within a region than on the specific quantity coming from an individual source. For pollutants of global concern, such as those that contribute to global warming, including methane (CH_4), carbon dioxide (CO_2), and other gases, the global emissions matter more than the local emissions. Within watersheds, key pollutants such as biological oxygen demand (BOD) or total phosphorus (TP) affect overall water quality. A unique approach to limiting the net quantities of specific pollutants in air and water is discussed in this article.

A variety of pollution control technologies exist for reducing the amount of pollutants released into air. Within the United States, industry-specific standards are established for the maximum achievable control technology (MACT). But sometimes current technology is not adequate at removing a specific pollutant, from a specific source, down to an acceptable level. When this occurs for a pollutant of regional or global concern, an equal or greater reduction in the emissions of the same pollutant from a different source, known as a pollution offset, may be an acceptable alternative for compliance with emissions standards. Offsets may also be bought by organizations or individuals interested in reducing their carbon footprint. To ensure that the actual pollution reduction occurred, third-party verification of the offset is required by all reputable offset registries.

One of the most common types of offsets available for sale or trade is a carbon offset. Carbon offsets are typically quantified in units of metric tons of carbon dioxide equivalents (CO_2e). Other global warming gases may be included in carbon dioxide equivalents, such as methane, but their quantities are converted to the equivalent amount of carbon dioxide that possesses a similar potential to cause global warming. Offsets may be sold on the open market, through existing trading platforms. Offsets may also be traded and sold directly between two parties.

Offsets may be traded or sold as means of meeting emissions commitments made by signatory members of the Kyoto Protocol. The United Nations' Clean Development Mechanism (CDM) verifies projects to ensure that the actual emission reductions occurred and requires that the projects be "additional" to normal modes of operation.

Carbon offset projects must reduce the amount of carbon dioxide, or other global warming gases, released or present in the atmosphere. A wide variety of projects may qualify for carbon offsets, some examples of which are as follows:

- The replacement of fossil fuels with noncarbon-emitting energy sources, such as wind, solar, and hydropower
- The capture and use or destruction of methane generated by the following sources, among others:
  - Municipal wastewater treatment
  - Landfill component decomposition
  - Oil and coal extraction
- The sequestration of carbon by planting trees, which utilize carbon dioxide from the air to produce biomass

- The utilization of carbon neutral fuels. Fossil fuels take carbon that is sequestered underground and release it into the air when burned, increasing the net amount of atmospheric carbon dioxide. In contrast, burning biomass for power or heat is considered carbon neutral because biomass, such as trees, sequesters carbon from the atmosphere as it grows and releases carbon dioxide back into the atmosphere when it is burned.

These are just a few among the many projects that can create carbon offsets to help reduce the net global emissions of global warming gases.

An atmospheric issue of regional concern is that of acid rain. Sulfur dioxide and nitrous oxides are the primary gases that cause acid rain. The goal of the U.S. Environmental Protection Agency's (EPA) Acid Rain Program is to reduce the amount of sulfur dioxide and nitrous oxide emitted into the atmosphere, much of which is generated by coal-fired electric generating plants. As part of this program, the EPA issues sulfur dioxide allowances to electric generators based on historical fuel consumption and emission rates. The allowances are in units of 1 ton of sulfur dioxide and linked to a specific year. Similar to carbon credits, these allowances may be traded or sold among utilities.

As a part of the sulfur dioxide allowance trading program, at the end of each year, each utility must account for its annual emissions and may buy, sell, or trade allowances to meet its emissions requirements. Coal-fired electric generating plants that emitted less sulfur dioxide than allotted for the year may sell, trade, or bank the difference between their allotted and emitted amounts. As the total amount of sulfur dioxide that may be emitted is capped and does not change, this is referred to as a cap-and-trade system. Sulfur dioxide emissions may be reduced through the use of low-sulfur coal, the commingling of biomass-based pellets or cubes for combustion with coal, and various other pollution control technologies. This program effectively promotes sulfur dioxide offsets, as reductions in sulfur dioxide at one source can be utilized to offset emissions at another source.

The Acid Rain Program's nitrous oxides reduction efforts focus on coal-fired electric utility boilers but do not include allowance trading as a compliance mechanism. However, nitrous oxides emissions have been capped for large combustion sources in the eastern United States through other programs.

Offset programs are not limited to air emissions but may also be used within a water body. When the amount of a specific pollutant within a watershed limits further industrial growth, unique opportunities may emerge for offsetting a new source of the pollutant through a reduction in an existing source. As aqueous emissions are often referred to as wastewater effluents, this practice is also referred to as effluent trading. Within the United States, effluent trading may not be used to avoid any applicable Clean Water Act technology-based requirements. Effluent trades must meet all water quality standards and appropriate regulations, including the prevention of backsliding of the quality of water. Trades must meet the established limit of the total maximum daily load (TMDL) of the pollutant in the water body. Trade boundaries must coincide with the boundaries of the affected watershed.

Perhaps the most famous example of effluent trading occurred on the Fox River in Green Bay, Wisconsin. In 1995, a proposed marina would have increased the biological oxygen demand (BOD) of the Fox River beyond the amount allowed by the Wisconsin Department of Natural Resources. As a result, the Wisconsin Department of Natural Resources could not issue a permit and the marina could not be built. The marina entered into an agreement with an upstream paper mill, purchasing the required amount of biological oxygen demand, which allowed the marina to be constructed, providing additional recreation and economic benefits to the town. Through this example, it can be seen how offsetting pollution can allow for business development in an environment limited by the pollutant load it can receive.

See Also: Carbon Footprint; Carbon Neutral; Carbon Sequestration; Carbon Trading; Compliance; Maximum Achievable Control Technology.

Further Readings

Jarvie, M. and B. Solomon. "Point-Nonpoint Effluent Trading in Watersheds: A Review and Critique." *Environmental Impact Assessment Review*, 18/2:135–57 (1998).
Napolitano, Sam, Melanie LaCount and Daniel Chartier. "SO_2 and NO_X Trading Markets: Providing Flexibility and Results." *EM Magazine* (June 2007).
United Nations Clean Development Mechanism. http://cdm.unfccc.int/index.html (Accessed April 2009).

Michelle E. Jarvie
Independent Scholar

POLLUTION PREVENTION

According to the Pollution Prevention Act (PPA) of 1990 the term *pollution prevention* (P2) means source reduction and other practices that reduce/eliminate pollution creation through more efficient use of resources (raw materials, energy, water) or conservation of natural resources. This definition addresses both resource use (inputs) and resource loss (outputs) from various processes.

PPA specified a hierarchy in reference to pollution:

- Prevent or reduce at the source whenever feasible
- If unpreventable, recycle in an environmentally safe manner whenever feasible
- If unpreventable or unrecyclable, treat in an environmentally safe manner whenever feasible
- Use disposal or other release only as a last resort in an environmentally safe manner

Practices that can result in source reduction include equipment/technology modifications, process/procedure modifications, product reformulation/redesign, raw materials substitution, leak and spill prevention, and improvements in housekeeping, maintenance, training, and inventory control. Although the PPA definition does not mention recycling, energy recovery, treatment, and disposal, some practices commonly described as "in-process reuse" may qualify as pollution prevention.

Prior to the passage of the PPA, industry in general and businesses in specific tended to focus on complying with environmental laws requiring the control of pollution already created. Such an approach resulted in an emphasis on technological fixes that could be applied at "end of pipe." For example, adding an electrostatic precipitator to remove particulates from flue gases is a pollution control approach. Attempting to reduce the amount of particulates created—by changing combustion times, temperatures, fuel mix, and/or maintenance schedules—is a pollution prevention approach.

The focus on pollution control rather than prevention was not solely a result of industry preference, but reflected both the nature of environmental laws and regulations and the training of environmental regulators. Laws such as the Clean Air Act and Clean Water Act, and their predecessors at both the federal and state levels were all based on pollution control logic. They relied on "command and control" approaches, such as (1) establishing limits on the amount of pollution that could be emitted and as a result regulators processing, issuing, and monitoring compliance with emission permits; and (2) requiring the installation and use of specific pollution control technologies—typically the "best available technology" at the time of legislative or regulatory enactment—which required that regulators process reports of purchase and installation of such equipment. In either case, both regulators and environmental compliance staff of regulated companies had well-defined duties.

The logic of pollution prevention focuses instead on examining each individual process and attempting to find the most appropriate means to reduce the creation of pollution in the first place. This is far less amenable to application of "one size fits all" solutions, and is far less amenable to regulatory oversight. It is much more akin to the total quality management (TQM) logic of continuous improvement, and therefore represented a challenge for both regulators and company staff trained in and used to operating under the pollution control logic. It is probably not coincidental that the growth of interest in and acceptance of pollution prevention in the United States followed the growth in understanding and application of TQM by American companies in the 1980s.

During this period, companies, regulators and environmentalists all began to recognize that the traditional pollution control approach, while still necessary, had not solved environmental problems, and that the nature of the laws actually retarded environmental improvement beyond a certain point. Under these laws, once firms were in compliance, they had no legal incentive to reduce pollution further. A few firms had begun to experiment with going "beyond compliance" by voluntarily reducing pollution more than they were legally required to, primarily through process improvements rather than by adding more stringent pollution control technologies. As they did so, many were surprised to discover that they not only reduced pollution, but saved money, largely due to reduced expenditures for raw material inputs and for disposal of pollutants captured by control equipment. In the late 1970s, 3M introduced its Pollution Prevention Pays (3P) program to encourage scientists and managers to consider pollution prevention activities, and by the mid-1980s, reported a total of over $500 million in savings during first-year implementation alone. Publicity regarding the success of 3P and similar efforts, such as Dow Chemical's Waste Reduction Always Pays (WRAP), began to spark widespread interest in P2 among business, regulators, and environmental groups. After the eight years of the Reagan administration, which was generally hostile to environmental regulation, all parties found they could agree upon the PPA, which emphasized providing technical assistance to companies willing to engage in voluntary activities that promised both cost savings and environmental performance improvements.

Since the passage of PPA, thousands of companies, both large and small, have applied pollution prevention methods, frequently with the assistance of the U.S. Environmental Protection Agency (EPA) or state regulatory agencies. Many state environmental agencies maintain websites devoted to presenting examples of pollution prevention success stories. The success and logic of pollution prevention have given rise to several other "ecoefficiency" approaches that emphasize continuous improvement and prevention rather than control. These include Factor 4 and Factor 10 (referring to the proposed magnitude of reduction of pollution and environmental impact) and Zero Emissions.

A Traditional Approach to Pollution Prevention

The traditional approach to pollution prevention originated in a series of industry sector–specific waste minimization reports developed in the state of California in 1987. The information outlined in these reports was compiled in an EPA guide known as the *Waste Minimization Opportunity Assessment Manual* (1988), revised and republished in 1992 as the *Facility Pollution Prevention Guide* and still widely used in the P2 community. An *Organizational Guide to Pollution Prevention* recently published by the EPA outlines four main activities involved in the traditional approach to P2—establishing the program, preparing the plan, implementing the program, and maintaining it.

Establishing a P2 Program

Traditionally, pollution prevention has a "top down" focus, beginning with management approval and goal setting. A P2 policy statement communicates management decisions to the workforce, and consensus-building efforts promote general acceptance. The principles reflected in this approach are very typical of management-by-objectives programs—very popular at the time of the first waste minimization manual publication.

Management organizes a P2 task force under the direction of a "champion," which conducts a preliminary P2 assessment to collect data, review program sites, and help establish program priorities. This information will later help the task force prepare a written P2 plan.

Organizations using the traditional approach generally view this assessment as a "walk-through" activity to be performed by the task force itself or an outside service provider (such as a state P2 technical assistance provider or consultant). A term commonly used to describe this activity is pollution prevention opportunity assessment (PPOA).

Preparing the P2 Program Plan

Focus now turns to preparing a written plan. Input should be sought from all stakeholders, for example, employees, regulators, environmental interest groups, community organizations, and the investment community. To define clear program objectives and targets, it may be necessary to conduct a "baseline review" identifying the following:

- Process/material flow profiles
- Process inputs and outputs, together with a mass balance
- Total costs and benefits
- Relevant legal requirements
- Related organizational policies

The P2 plan should anticipate implementation obstacles and suggest approaches to overcoming them. A schedule is also required, which can be challenging to set based solely on the information gathered up to this point. However, it is essential for tracking the plan's progress.

Implementing the P2 Program

A detailed P2 assessment is the starting point for program implementation. An assessment team—generally a higher-level, multidisciplinary team that may include some workers—is assembled to review the data and visit prospective P2 sites. Outside technical assistance may be sought for these visits. The purpose of this detailed assessment is to help the team specify alternatives (options) for pollution prevention. For each target P2 opportunity, the team brainstorms to identify potential alternatives and, once they are selected, examines the environmental, technical, and economic feasibility of each one. The alternatives are ranked based on environmental benefits, technical feasibility, and cost. This evaluation also considers strategic issues important to the organization. At this point, a formal written assessment report is prepared detailing the analysis of the P2 assessment team and transmitting that information to management.

Once specific P2 projects have been selected, the P2 team works with management to assign responsibilities, resources, and time lines for implementing them. During implementation, projects are further reviewed and adjusted if necessary to make sure they meet the company's objectives. Finally, progress is measured by comparing data from the baseline review to data obtained from the implemented P2 projects. Benefits can be stated in terms of both total cost savings and reductions in resource flows.

Maintaining the P2 Program

The EPA's *Facility Pollution Prevention Guide* details several maintenance-related activities that can be employed at this point. Program maintenance requires integrating the P2 program into the organization's other activities. As part of this process, the company assigns accountability for wastes, discharges, emissions, spills, and leaks to the processes responsible for them. These wastes need to be carefully tracked and formally reported. This tracking, in addition to project results, should be evaluated on an annual basis to maintain proper program documentation.

Employee P2 training should be updated every year and must include both orientation for new employees and advanced training for those who have been involved in P2 program implementation. Considering the importance of communication in the success of any program, employees are routinely given information regarding P2. Their suggestions are solicited and receive follow-up. An employee reward program is the traditional incentive to encourage employee participation, but acknowledgments in employee performance reviews, recognition among peers, and financial benefits might also come into play. The final step in program maintenance is also communication based: public outreach and educational programs inform the community about the organization's pollution prevention activities.

Systems Approach to Pollution Prevention

The systems approach to P2 relies much more on employee involvement, and less on "top-down" management. The various steps described here rely on quality management tools used in the systems approach. Fundamentally, P2 requires an understanding of the various elements of any organizational process that creates pollution, in order to identify possible means of reducing or eliminating that pollution. This requires understanding, or mapping, these processes.

Process Mapping

All programs with a process focus incorporate a process-mapping tool, helping to ensure that everyone understands the process they are seeking to improve. Typically, the process under consideration contributes to the product or service offered by the organization. Process mapping may be facilitated by outside contractors, but is always verified by employees involved in the process. This means that employees play a direct role in ensuring that the process maps accurately and adequately portray the process as it currently operates. During the verification activity, each employee is asked how he or she would improve the process. In many cases, employees have never been asked this question before.

Process Improvement Ideas

Once this stage is complete, a list of potential process improvement ideas—specific recommendations from employees—is prepared and presented to management, along with the verified process maps, promoting ease of management visualization of P2 opportunities. When ideas are presented this way, managers are typically surprised at the number and quality of ideas generated.

Root-Cause Analysis

The next step involves appointing employee teams to work on each of the specific problems identified in the process. The teams generally begin by analyzing root causes in order to better understand these problems. This careful analysis prevents blindly implementing solutions.

Employee Involvement

All systems approach efforts note the importance of involving employees in both the planning and the implementation of the process improvement program. Combining a strong bottom-up component (employee involvement) with a traditional top-down management strategy always works better than a one-sided program.

The employee involvement feature helps ensure that employees will want to improve their processes. There is an adage that says "Employees never resist their own ideas." Because ideas come from the bottom up in the systems approach, organizations do not need to inculcate the kind of behavior modification that is required in initiative-driven programs.

Typically, employee teams formulate more (and better) potential solutions to problems than do outside experts who often restrict their thinking to tried-and-true ideas, gravitating to their favorite "right" answer. By contrast, employee teams rarely search for a single right answer. Instead, the systems approach brings out creativity and builds on desires for process improvement. The systems approach builds in "provocation," prodding participants to generate creative, seemingly outrageous, solutions through discussion of a wide range of alternatives.

Draft Action Plans

Once a team has decided on a particular solution to a process problem, the team members draft an implementation action plan. At this stage, performance standards must be defined for each step of implementation. The combined performance standards for all the steps involved become the overall project goal. Management can also set a "stretch" goal by negotiating with the team on ways to enhance or accelerate the project.

In the systems approach, there often is a great deal of astonishment when employees present their draft action plans to management, despite the fact that the quality management tools used in the systems approach are designed to yield comprehensive and far-reaching solutions. They help employees address the underlying causes of process problems, rather than merely fixing the symptoms.

Goal Setting

The systems approach method offers a superior approach to setting goals. Many companies still set goals at the start of their improvement programs, before they have gathered much information. Employees are then expected to write action plans for delivering on the goals. While this approach works in some instances, in most cases employees are much more comfortable setting goals after the work has been properly studied and defined.

Resource Requirements

Using systems approach tools, employees discuss all the alternative solutions that are generated, choose a solution, and decide how best to approach it. Each team thus becomes aware of the resources needed to complete the desired work. By contrast, the outside expert typically has a particular monetary figure in mind when providing a service and will try to negotiate the fee in his or her own favor. This means that the dynamics of the improvement process and the resource equation are very different.

When managers approve the resources asked for in an employee-drafted team action plan, the employees will be held accountable for delivering the promised solutions. This is generally not problematic since the action plan, instead of being imposed by outside entities, is conceived and written by the employees themselves.

Bringing Out the Best in Each Approach

Both the traditional and the systems approaches to P2 have discernable strengths. Companies often can benefit from combining features of both:

- Although it is possible to learn from others' success stories, source reduction comes from the persistent application of a systematic approach that is focused clearly on the individual process. The improvement effort must operate within the vision, mission, and core values of the organization. Sustainable source reduction cannot be achieved simply by copying what others have done.
- When a company decides to utilize outside expertise, it is preferable to retain more than one expert. Having alternative points of view might help challenge the in-house team to come up with better solutions or to work more closely with one of the experts.
- No matter what approach is chosen, improvement efforts work best if employees are at least given an opportunity to study the problem to be addressed before having an expert's solution imposed on them. This also allows the expert to get a much better idea of what employees are thinking. The expert can then at least consider this information in conjunction with his or her "best practice."
- When relying on an employee-driven approach, the organization should provide employee teams with outside professional facilitation until they are able to develop the

skills needed to use the systems approach tools effectively. If necessary, the facilitator can have outside experts present ideas to the employee team. Employees can then interactively question the expert and gather the information they need to prepare their draft action plans.

- Employees can benefit from visiting other companies that have tackled the process improvement issues they face. This again allows them to ask questions and gain more information to complete the draft action plans. A significant question for such an opportunity would be, "If you had it to do over, would you have selected the same approach to the problem?"

- It is always a good idea to seek multiple points of view when trying to resolve difficult problems. Soliciting these views can be done before team members work through the various systems approach tools; however, they will benefit more from these exchanges if they first utilize the tools to analyze the problem and complete a preliminary draft action plan. At this point, they will have a much better sense of the problem.

Process improvement is crucial to the sustainability of every organization. Using the optimal combination of approaches can put an organization on the path to continual process improvement.

Where P2 Falls Short

The typical pollution prevention program faces gaps in various performance areas, significantly the areas of leadership, strategic planning, stakeholder involvement, and information and analysis.

Adequate leadership cannot be provided simply by signing a pollution prevention or environmental management system policy statement and attending a few briefings about the program's progress each year. Leaders need to see the value of the program, and should not view pollution prevention as another costly compliance program.

Strategic planning defines how the organization identifies, develops, and implements long-term and short-term goals and objectives for continuous improvement in environmental quality. A good environmental management system uses the plan-do-check-act cycle. However, P2 takes the process a step further by asking how environmental goals and objectives relate to the organization's overall business objectives, seeking alignment between its programs and the organization's core business practices. Pollution prevention needs to become part of every business decision from the design of a new product or service to the process improvements made by the organization every day.

Stakeholders are very important to a P2 program. The organization must involve all interested parties in the development and implementation of its continuous environmental improvement program. It must also focus on customer and market requirements and expectations.

The information and analysis component looks at how the organization selects, collects, and uses information to assess the overall effectiveness of its continuous improvement program. P2 practitioners often seek to use existing management information systems to do this work. By contrast, managers adopting environmental management systems may seek to create new and expensive information systems dedicated solely to environmental management.

In addition, with few exceptions, pollution prevention has been seen as an internal organizational matter. This may change now that large companies are beginning to require their suppliers to use (and often certify to) environmental management systems such as ISO 14000. P2 may become a key part of these requirements. To have P2 become part of the way an organization operates, the following performance characteristics must be addressed:

- Employee involvement considers how the organization prepares and involves employees in the development and implementation of its program approaches. This component also involves looking at employee value and employee well-being. By contrast, environmental managers often bring in outside experts to help identify pollution prevention opportunities and alternatives. Employee involvement is not a strong point for many environmental management systems.
- Process management asks how the organization identifies processes that impact the environment, and how it analyzes those processes to understand their impacts and underlying causes. This component also looks at how the organization manages processes to achieve environmental excellence. Environmental management systems are a very important contribution to the last item. However, these systems rarely emphasize the "process view" of the organization, upon which TQM insists.
- Results are essential to both quality and pollution prevention programs. Pollution prevention practitioners typically measure results in terms of the weight and volume of pollutants avoided.

But from a business point of view, it is very important to see these regulatory-based metrics in financial terms, allowing management to understand their value. Quality practitioners have created the concept of "cost of quality," which allows them to express results in ways that management understands. Quality programs also look beyond environmental and financial results to understand how the program impacts other interested parties and meets the measures of performance for each of the six major categories named above. Quality practitioners compare anticipated results to actual results. They also track and trend the results, looking for demonstrated continuous improvement.

In short, pollution prevention should be a key operational component for all companies with serious sustainability programs.

See Also: Clean Production; Ecoefficiency; Environmental Management System; ISO 14000; Sustainability; Waste Reduction.

Further Readings

Batty, Lesley C., Kevin B. Hallberg and Adam Jarvis. *Ecology of Industrial Pollution (Ecological Reviews)*. New York: Cambridge University Press, 2010.

Bishop, Paul L. *Pollution Prevention: Fundamentals and Practice*. Long Grove, IL: Waveland Press, 2004.

Pojasek, R. B. "Selecting Your Own Approach to P2." *Environmental Quality Management*, 12/4 (2003).

U.S. Environmental Protection Agency (EPA). "Pollution Prevention Act of 1990." http://www.epa.gov/p2/pubs/p2policy/act1990.htm (Accessed September 2009).

Robert B. Pojasek
Harvard University

PRECAUTIONARY PRINCIPLE

Achieving sustainable development relies on the proportionate exercise of the precautionary principle in the regulation of risks. The basic sentiment of the precautionary principle (hereafter, the Principle) is "better a little caution than a great regret" or "better safe than sorry." The Principle is a regulatory mechanism that enables institutions and decision makers to take into account scientific uncertainty and potential risks of activities before a causal link has been established. In this way, the Principle is a proactive tool used to prevent possibly negative effects on human health and the environment. Embedded in the Principle are controversial issues such as science, perceptions of risk, public involvement, cost and benefits, moral values, and policy making. All of these should be considered within the regulation of risk and are all dependent upon the institutional and administrative arrangements of the state. While the Principle originates as a means of preventing potential irreversible environmental harm from activities, substances, new technologies, and innovations, it is capable of much broader application. In fact, it is in the area of food safety and health where the precautionary principle appears to have found a firmer footing in regulatory decision making.

Precaution is defined by the *Oxford English Dictionary* as a mode of action where caution exercised beforehand provides against mischief, otherwise viewed as "prudent foresight." However, over the past few decades, precaution has become a matter of legal, scientific, political, economic, philosophical, and academic debate. The *Vorsorgeprinzip*, or Principle of Foresight, established in German environmental law in the 1970s, was the precursor to the precautionary principle. Its objective was to balance the avoidance of environmental damage against economic viability.

The 1987 Ministerial Declaration of the Second International Conference on the Protection of the North Sea was the first international agreement introducing a precautionary approach toward the protection of the North Sea from possibly damaging effects of dangerous substances. In the same year, the Convention for the Protection of the Marine Environment of the North-East Atlantic (the OSPAR Convention) considered the use of the Principle in circumstances where there is no conclusive evidence of a causal relationship between inputs and effects. Possibly the most commonly used definition of the Principle is that found under Principle 15 of the 1992 Rio Declaration on Environment and Development in which states set out their collective intention to apply the precautionary approach according to their capabilities in order to protect the environment. Principle 15 provides: "[w]here there are threats of serious or irreversible damage, lack of full scientific certainty shall not be used as a reason for postponing cost effective measures to prevent environmental degradation." Other international conventions integrating the Principle include the Convention on Biological Diversity and the Framework Convention on Climate Change.

Australia was one of the first countries in the world to adopt the Principle as part of its national environmental policy as well as within many of its environmental statutes. In the United States, the Principle has a short history. The 1998 Wingspread conference called for government, corporations, communities, and scientists to implement the Principle in making decisions. Nevertheless, it wasn't until 2003 that a U.S. government body (the City and County of San Francisco) established the Principle as the basis for all environmental policy in its jurisdiction.

At the European level, the Principle was enshrined under Article 174(2) of the Treaty of the European Community (EC) as an overarching policy aspiration upon which EC environmental policy should be formulated and that member states should endeavor to incorporate into their regulatory systems. While challenges to applying the Principle, such as the Commission's export ban on British beef, have led the European Courts to consider the scope of the Principle, the Courts have been reluctant to interfere with institutional regulatory discretion or to provide anything akin to a transparent definition, preferring to leave it to the member states and their competent authorities.

The ways in which the Principle has been phrased in law and policy varies significantly. Cass R. Sunstein recognized in excess of 20 definitions. However, it is generally accepted that there are both weak and strong forms of the Principle. Weak versions of the Principle, for example, Principle 15 of the Rio Declaration, highlight that a lack of scientific evidence or certainty of a particular consequence should not prevent mitigating action to the extent that this avoidance is cost effective, whereas stronger versions of the Principle require rethinking regulatory policy in order to prevent undesirable consequences even where there is scientific uncertainty regarding the potential damage (OSPAR Convention), and even when the cost of preventing a speculative detrimental effect is disproportionately high.

These scientific uncertainties and knowledge limitations to the possible later impacts of scientific advances and technological innovations generate risk that the Principle can respond to. Ironically, the Principle is reliant on science to protect against science. An activity can also manifest itself as a risk through the psychology of public risk perception (varying in acceptability of risk between societies and grounded in intuition, perceived causal links, and heuristics). In legal challenges, the onus has been on the party bringing the case to prove the causal link between the activity or substance and the harm or risk of harm. The application of the Principle cannot be based on fear alone, but is considered to involve a shift in the burden of proof onto those proposing an activity, for example, a manufacturer of market drugs who must provide evidence that the drug is safe and that it does not constitute a risk.

Following a series of hazards resulting from scientific and industrial advance, such as benzene and polychlorinated biphenyls (PCBs), it is now generally accepted that science can never be conclusive, although a risk assessment helps to determine the likelihood, magnitude, frequency, and type of potential negative effects. Incorporating public concerns alongside science in the regulation of risk and the application of the Principle has been coined by Ulrich Beck as "reflexive modernization." When science has not identified a risk, the benefits of taking preventative action outweigh the costs associated with potential negative human health or environmental impacts. The difficulty with this is that the benefits yielded from such protection do not easily lend themselves to a materialistic or economic valuation.

The most important aspect of the Principle is its ability to be applied in practice. It is likely that through international, European, national, and domestic developments in the field of sustainable development, the regulatory and policy application of the precautionary principle will both intensify and be clarified.

See Also: Abatement; Cost-Benefit Analysis; Environmental Risk Assessment; Quantitative Risk Assessment; Sustainable Development.

Further Readings

Beck, Ulrich. *Risk Society: Towards a New Modernity*. Newbury Park, CA: Sage, 1992.

Peel, Jacqueline. *The Precautionary Principle in Practice: Environmental Decision-Making and Scientific Uncertainty*. Sydney, Australia: The Federation Press, 2005.

Raffensperger, Carolyn and Joel Tickner. *Protecting Public Health and the Environment*. Washington, D.C.: Island Press, 1999.

Sunstein, Cass R. *Laws of Fear: Beyond the Precautionary Principle*. Cambridge, UK: Cambridge University Press, 2005.

Hazel Nash
Cardiff University

QUANTITATIVE RISK ASSESSMENT

Identifying, understanding, and managing risks is an imperative tool in preventing accidents during the course of business operations. Engaging in emergency preparedness and preventing harm to human health and the environment can be significantly informed by undertaking a risk assessment.

Risk assessment is a scientific process that defines, identifies, and analyzes the nature and magnitude of risks, uncertainties of the risk, and its impacts. It can involve either quantitative or qualitative analysis. Risk assessment is used to establish priorities in which the decision maker is able to target the allocation of resources according to the immediacy and acuteness of the threats in order to make unwanted events less likely to happen, according to the U.S. Department of Health and Human Services. In this way, it contributes toward decision making through proactive risk management and mitigation of potentially detrimental impacts.

Probability should be understood as the likelihood that a particular hazard will result in damage at a particular location. Hazards are defined as any situation that has the potential to cause damage: for example, chemicals, electricity, or working while standing on a ladder. Allocating probabilities to specific outcomes or risks is a way of assessing the likelihood of particular negative eventualities and is a means of informing risk management decisions.

Quantitative Risk Assessment (QRA), also referred to as Probabilistic Risk Assessment, is a systematic scientific process that identifies the possible causes of harm (hazards) and their effects. It then attributes probabilities to each scenario based on the likelihood, magnitude, and severity of the occurrence. The emphasis in QRA is on formalized, quantifiable data, which inform an organization's regulatory systems and risk management operations and allow preventative action to be taken. QRA requires calculations of two components of risk (R): the magnitude of the potential loss (L), and the probability that the loss will occur (p). In essence, QRA seeks to answer three questions. What can go wrong? What is the likelihood of X going wrong? What are the consequences of X going wrong?

Broadly, the QRA procedure is to

1. identify and define undesirable end states;

2. develop a set of disturbances to the normal state for each end state (initiating events);

3. identify sequences of events using event or fault trees are (this generates accident scenarios);

4. evaluate probabilities of these scenarios using all available evidence, past experience, and expert judgment; and

5. rank scenarios according to their expected frequency of occurrence.

The resulting hierarchical probability scale from a QRA can be useful in evaluating risk for decision making. This enables a more informed and precautionary decision to be reached. QRA has emerged in environmental toxicology as the predominant tool for describing the public health consequences of human exposure to environmental contaminants. Existing situations are measured and compared according to a measure of population health risk. Proposed preventative measures are then compared according to the reduction in population health risk in order to determine the best course of action to mitigate potential detrimental environmental consequences. It is this modeling of quantitative relationships between cause and effects and action and consequence that characterizes QRA.

QRA can be based on a deterministic or stochastic modeling approach. In deterministic modeling, analysts assign values for discrete scenarios to see what the outcome might be in each. This method recognizes only those elements capable of being measured through the assignment of probability of occurrence that is taken into account. Those factors that remain uncertain and thus unable to quantify are ignored. The main problems with the deterministic approach are (1) it oversimplifies complicated interdependent and natural relationships; (2) each outcome is given equal weight; and (3) not all possible outcomes are considered. Stochastic modeling, on the hand, takes uncertainties as well as risks into account. The Monte Carlo simulation enables uncertain inputs in a model to be represented using ranges of possible values known as probability distributions. The result is a probability distribution of possible outcomes.

QRA is intended to be an objective evaluation of risk in which assumptions and uncertainties are clearly considered and presented. K. Shrader-Frechette et al. argue that all forms of risk assessment employ value judgments. David Santillo and colleagues explain that "risk-based approaches extend from the view that environmental risks can be quantified and managed at sustainable and acceptable levels . . . it assumes it is possible to know enough." A central element in Environmental QRA is the formulation of acceptance criteria for damage. The formulation of acceptance criteria is a matter of controversy requiring caution and careful consideration regarding nature's ability to withstand external disturbances, according to E. Wiig and colleagues.

Probabilities are scientifically informed guesses that shed some light on the possible or likely affects of a particular action or inaction. Qualitative information and differences among risks is also valuable and should not be overlooked by overreliance upon QRA. Although conclusions drawn are dependent upon subjectivities and the limitations of research techniques, probabilities seeks to inform and preempt future events, facilitating the opportunity for society to take a proactive role in attempting to prevent detrimental effects, be these environmental, political, economic, social, or cultural.

See Also: Best Management Practices; Cost-Benefit Analysis; Environmental Risk Assessment; Precautionary Principle.

Further Readings

Helliar, Christine, et al. *Attitudes of UK Managers to Risk and Uncertainty*. Glasgow, UK: Bell and Bain, 2001.

Humber, James M. and Robert F. Almeder. "Quantitative Risk Assessment." *Biomedical Ethics Reviews* (1986).

Knight, Frank, H. *Risk, Uncertainty, & Profit II*. Boston, MA: Houghton Mifflin, 1921.

Megill, R. E. *An Introduction to Exploration Economics*. Kingwood, TX: Petroleum Publishing Company, 1973.

Santillo, David, Paul Johnston and Ruth Stringer. "Management of Chemical Exposure: The Limitations of a Risk-Based Approach." *International Journal of Risk Assessment and Management*, 1/1–2:160–80 (2000).

Shrader-Frechette, Kristin. *Environmental Justice: Creating Equity, Reclaiming Democracy (Environmental Ethics and Science Policy)*. New York: Oxford University Press, USA, 2002.

U.S. Department of Health and Human Services. "Focus on Prevention: Conducting a Hazard Risk Assessment." Pittsburgh, PA: U.S. Department of Health and Human Services, 2003. http://www.msha.gov/training/trainingtips/risk%20analysis.pdf (Accessed January 2010).

Wiig, Erik, Ellinor Nesse and Anita Kittelsen. "Environmental Quantitative Risk Assessment (EQRA)." SPE Health, Safety and Environment in Oil and Gas Exploration and Production Conference, June 1996. New Orleans, LA: Society of Petroleum Engineers, 1996.

Hazel Nash
Cardiff University

RECYCLING, BUSINESS OF

Numerous products support our professional and private lifestyles. Absent intervention, they will all end up in the garbage at the end of their useful lives. Compounding the situation is society's growing consumptive appetite both for more products, and for different varieties of the same product. A global population that has quadrupled in the last 50 years, possessing increasing material needs and desires, further increases the pressures that will be exerted on Earth's natural resources. Recycling of materials is one piece—along with reduction (of consumption) and reuse (including refurbishment) of products—of the overall approach needed to effectively utilize these resources in a sustainable manner. Because of the greater energy requirements of recycling, it is somewhat less desirable than either reduction or reuse.

Recycling can be defined as the entire chain of activities involved in the collection and utilization of all or part of old products in the manufacture of new products. Therefore, the business of recycling involves all concerns along the recycling value chain, including product design, recycling processes, reliability of secondary raw material supplies, markets for recycled goods, regulatory mandates/legislation, and the overall economics of recycling.

The recycling of materials has long been practiced by individuals out of economic necessity, since the cost of using secondary raw material (scrap) was typically less than the cost of using virgin material. As economies industrialized and manufacturers pursued technological innovation and ever-increasing economies of scale, this relationship frequently reversed. As collection and sale of scrap materials became less profitable, fewer scrap collectors existed. The economic incentive for individuals to recycle therefore declined, rendering it more convenient to simply dispose of end-of-life products, further reducing the viability of the recycling business. Although recycling had been part of the framework of daily life, the habit thus diminished. Recycling received a significant boost during World War II for patriotic reasons—as a means of providing materials for the war effort—but again declined in the postwar period. Interest revived with the growth of the environmental movement in the 1970s. Concerns about declining availability of landfill space, water and air pollution from landfills and incinerators, and resource depletion led to the initiation of municipal recycling efforts. These in turn increased the supply of postconsumer materials, necessitating the need for more use of

Approximately 120 million kilograms of Hewlett-Packard products are recycled annually. Here can be seen a bin full of green envelopes that contain used printer cartridges.

Source: Hewett-Packard

such materials in products, thereby "closing the loop." Increased material complexity of products has complicated recycling and affected its economics, however.

Customer Requirements

In any business, satisfying the requirements of the customer is paramount. The size, appearance, and functionality of products made from recycled materials need to be as competitive as their traditional counterparts. A product, green or otherwise, must be designed to meet the preferences of consumers if it is to perform well in the marketplace.

When the customer is a manufacturing company, it is imperative that the supplier employs rigorous quality assurance practices to ensure homogeneously high quality of its secondary raw material supply. When these secondary raw materials are obtained from end-of-life products originating in areas with diverse environmental regulations, this poses greater challenges.

Product Design

Advocates of sustainable product design and design for the environment (DFE) assert that design is the key to product sustainability. Design for disassembly, design for remanufacturing, and design for recycling are each DFE strategies that can enhance the ease and economic feasibility of subsequent reuse and recycling efforts. Incorporating DFE into product design policies should form part of the overall objectives, policies, strategies, and vision of erstwhile green businesses. This could effectively boost a company's image, improve its competitive advantage, and encourage investor interest.

Design for recycling can be influenced by environmental legislation in the country of export, as well as by environmental regulations in the manufacturer's location country. Even absent legislation, product designers can adopt William McDonough's dictum "waste equals food," and attempt to ensure that all products can either be recycled or composted at end of life.

Dismantling Analysis

Dismantling analysis—information about nondestructively separating components of a product for recycling or for reuse—is essential in the business of recycling. A product designed without dismantling in mind will require much more time and expense to recycle. Dismantling analysis helps companies establish the real cost of recycling as well as forecasting potential problem areas. Some useful questions to be answered are as follows:

1. Were hazardous materials used in the product?

2. Are the connections in the product's components and parts easily accessible and separable?

3. Are material bonds separable?

4. Are plastic types and other materials identified and characterized?

5. Which product components can be reused?

6. Would upgrading or increased maintenance be more cost effective than recycling for this product?

7. How much time is required for dismantling?

Photos taken of the dismantling process can be useful for later evaluation and in future product design.

This analysis is very useful in estimating the cost of the entire recycling process from dismantling, separation, and processing to environmental compliance, logistics, and marketing. A business example is the establishment of the International Dismantling Information System (IDIS) to provide data for environmentally friendly management and treatment of vehicles at end of life.

Recycling Process

The logistics of recycling includes collection, separation, packaging, warehousing, and transportation. In addition, various marketing activities, including finding customers, must be completed before recycling of materials can be completed. Municipalities frequently collect materials through drop centers and curbside collection. Responsibility for commercial collection of specialized materials can involve either collection by recyclers or delivery by the generator of the scrap. The amount of energy used in collection negatively affects the economics of recycling—particularly of lightweight, low-value materials.

Once collected/dismantled, secondary raw materials are then broken down into smaller particles. When these particles are composed of several materials, such as mixed plastic and metal types, they must be separated. Different methods include screening, gravitational-based, magnetic-based, and electric-based separation. In screening, large screens are utilized to recover materials based on their sizes and shapes. Gravitational-based separation makes use of the force of gravity to separate heavier particles from lighter ones. Magnetic-based methods employ the use of magnets to separate ferrous from nonferrous materials, and electric based separation relies on the electrostatic attraction, repulsion, or neutrality properties of various particles. In nations with weak environmental laws, separation can occur using unsophisticated and hazardous means such as burning. Such practices can outweigh environmental benefits of recycling and the resulting publicity can discourage individuals from recycling.

Techniques for packaging recyclables for delivery to customers vary according to the material being recycled. Since warehousing entails costs, both suppliers and users of recycled materials attempt to minimize the space devoted to warehousing of scrap.

However, as scrap prices can vary widely and rapidly, warehousing will sometimes be chosen by either party in the pursuit of more favorable prices. Decisions dealing with these logistical issues must occur within the context of both micro- and macro-level marketing considerations.

Supply of Secondary Raw Material

Every household and business generates waste. These wastes can be common, such as paper, newsprint, food scraps, or highly specialized—specialty chemicals, metals, components. Nearly all of these have the potential to serve as raw materials for some company. The security of the raw material supply is critical to the success of any business activity. Any limitation to security of the secondary raw material supply is a disincentive to entrepreneurs who are considering a recycling business. For example, as a result of the End-of-Life Vehicle (ELV) Act (discussed below) and the additional fees it imposed, the supply of end-of-life vehicles available for dismantling in Germany decreased substantially. Only 60 percent of ELVs that were supposed to be recycled actually made it to the dismantling and shredding companies. The remaining 40 percent were sold to used car export dealers, who sold them in the huge secondhand car and parts market in developing countries, thus enabling higher earnings.

Prices and availability of scrap supplies are influenced by national and global industry and macroeconomic conditions. During the 2008 recession, various scrap material prices dropped by more than 60 percent before rebounding to or above previous levels within a matter of months. As a result of price volatility, collection of certain materials may at times be halted, discouraging individuals and companies from separating recyclables from the waste stream. Such disruption in recycling behavior can have long-term negative impacts on recycling rates and security of supply.

Corporations and Recycling

Businesses outside the recycling industry have several choices with regard to recycling. Brand owners must, as noted above, decide whether to design products with recycling in mind. They must also decide whether to promote the take-back and recycling of their own products. Examples of recycling activities by brand owners include initiatives by Hewlett-Packard increasing the number of its products recycled by 6 percent globally. Approximately 120 million kilograms of Hewlett-Packard products are currently recycled annually. Other consumer electronic companies with active recycling programs include Xerox, Siemens, AEG, Toshiba, Dell, Panasonic, and Sharp. Most automotive manufacturing companies have recycling rates ranging from 75 to 95 percent of vehicular weight. Examples of car companies with very high recycling rates include Toyota, Honda, BMW, Volkswagen, and Volvo.

Companies must also decide how transparently to communicate their recycling activities. This includes reporting of recycling rates as well as accurate labeling. For example, some companies state that plastic containers are recyclable when few opportunities for recycling that specific plastic exist. Companies may also fail to distinguish between pre- and post-consumer content. Such practices can add to individual confusion and disillusionment with recycling. Finally, all companies must decide how actively to promote diversion of various materials, such as office paper, from their waste streams.

Environmental Legislation and Business Recycling

Environmental legislation that promotes recycling efforts is essential to safeguarding our health by preventing toxic substance dumping in landfills. Forty percent of all lead and 70 percent of all mercury and cadmium in U.S. landfills comes from consumer electronic products. Cadmium, lead, and mercury in landfills can leak into groundwater sources, polluting drinking water and causing health implications, ranging from allergic reactions to brain damage to cancer.

In an attempt to halt addition of hazardous materials to landfills and encourage greater recycling and reuse of consumer electronic products, the European Commission passed the Waste Electrical and Electronic Equipment (WEEE) directive, put into force in February 2003. WEEE bans the use of hazardous material such as lead, mercury, cadmium, chromium, and bromine compounds that serve as flame retardants—polybrominated biphenyls (PBB) and polybrominated diphenyl ethers (PBDE)—and stipulates the use of environmentally friendly alternatives. Member countries in the European Union (EU) are expected to achieve a collection target of 4 kg per person per year. However, challenges to implementation show that approximately 70 percent of such material still show up in landfills, having been handled at uncertified treatment sites both within and outside the EU. An unintended consequence has been a growing trend in illegal export of electrical and electronic equipment waste to non-EU regions with less stringent environmental legislation, specifically Asia, Africa, and South America where obsolete electrical and electronic equipment is often burned in open fires to reclaim the copper. Burning this equipment results in the release of dioxins, which are highly carcinogenic. The European Commission is currently considering amendments to the WEEE directive to address these issues.

Another example of environmental legislation enacted to promote recycling is the *Altautoverordnung* or End-of-Life Vehicle (ELV) Act, enacted in 1998. Approximately 75 percent of all ELV scrap by weight are effectively recycled, the remaining 25 percent being landfilled. The primary goal of the ELV Act was to establish basic guidelines for the removal of hazardous materials and fluids, promote reuse and recycling, and encourage the recovery of at least 15 percent of end-of-life vehicles by weight. When the German government instituted the end-of-life vehicle take-back act, owners were required to pay the equivalent of about 150 euros (about US$211) to have their automobiles scrapped. As with the WEEE directive, there were implementation problems with the ELV Act. Unfortunately, the environmental problems the ELV Act tried to solve were actually compounded by a sudden change in societal behavior—people illegally dumping unwanted cars in order to avoid paying the 150 euros. This behavior led to a subsequent environmental challenge resulting from oil and gas leaking directly into the soil. The German government amended the ELV Act by adding 150 euros to the initial cost of purchasing a new car, transferring the cost to the first buyer. The rationale supporting this decision was that the purchaser of a new car would be less likely than the end-of-life owner to mind the additional cost.

A third example of recycling-oriented legislation is Germany's Packaging Ordinance, which requires manufacturers to be accountable for recycling or disposal of their own packaging materials. This legislation prompted German industry to create the Duales System Deutschland. This system allows participating companies to place a "green dot" logo on their packages, making them eligible for free collection from consumers. The desire to avoid government legislation can also lead to voluntary industry initiatives, such as the creation in the United States and Canada of the Rechargeable Battery Recycling Corporation, which provides consumers free recycling of rechargeable batteries.

Governments can also adopt policies that give purchasing preference to products containing recycled content. Such action can stimulate market development by increasing sales, encouraging expanded production, and lowering prices, thereby increasing purchases by other organizations and individuals and further continuing the process.

Economics of Recycling

The economics of recycling varies for different material types. The total costs of plastic material recycling, for example, range from 1–1.50 euro/kg ($0.64–$0.96/lb). The total cost of producing high-density plastic material from virgin sources varies from 1 euro/kg for producing high-density polyethylene (64/lb), to 6 euro/kg ($3.84/lb) for polyamide (PA). From this information, it is apparent that it makes greater economic sense for a manufacturing industry to recycle its high-density plastic materials than to obtain them from virgin sources. In such a case, environmental legislation is not required to support the recycling of high-density plastic materials. On the other hand, the total cost of producing low-density plastic materials such as polyvinylchloride (PVC) from virgin sources—estimated at 0.50 euro/kg ($0.32/lb)—is cheaper than obtaining it from recycled materials. In this case, there is no financial incentive for entrepreneurs and manufacturing industries to engage in the business of recycling low-density plastic materials, thus requiring legislative support and policy incentives to encourage and sustain recycling initiatives.

The economics of metal recycling are predominantly dictated by the difference in energy usage between obtaining the metal from secondary raw materials or from virgin sources (i.e., ore extraction, processing, and refining). For example, the energy intensity to produce aluminum from virgin sources is estimated to be 200–300 megajoule/kg while that obtained from recycled materials is 5 megajoule/kg representing a total energy savings of 195–295 megajoule/kg and a 3,900–5,900 percent reduction in energy intensity. Recycling of aluminum is therefore a very viable business venture, even in the absence of any support from environmental legislation.

In conclusion, recycling is necessary, although insufficient, for responsible stewardship of Earth's natural resources. Good product design and appropriate government policy are essential to facilitating recycling, providing appropriate incentives and thereby achieving a viable recycling industry.

See Also: Clean Production; Clean Technology; Closed-Loop Supply Chain; Compliance; Dematerialization; Design for Environment; Ecoefficiency; E-Waste Management; Extended Producer Responsibility; Extended Product Responsibility; Green Design; Industrial Ecology; Life Cycle Analysis; Reverse Logistics; Sustainable Design.

Further Readings

Cui, Jirang and Eric Forssberg. "Mechanical Recycling of Waste Electric and Electronic Equipment: A Review." *Journal of Hazardous Materials*, B99:243–63 (2003).

Curlee, Randall. *The Economic Feasibility of Recycling: A Case Study of Plastic Wastes.* Westport, CT: Praeger, 1986.

European Commission. "Recast of the WEEE and RoHS Directives Proposed in 2008." http://ec.europa.eu/environment/waste/weee/index_en.htm (Accessed October 2006).

Lucas, Rainer and David Schwartze. "End-of-Life Vehicle Regulation in Germany and Europe: Problems and Perspective." Wuppertal Papers. Wuppertal, Germany: Wuppertal Institute for Climate, Environment and Energy, 2001.

VDI 2243. Recycling-Oriented Product Development. Verein Deutscher Ingenieure, 2002. http://www.wupperinst.org/uploads/tx_wibeitrag/WP113.pdf (Accessed January 2010).

Emmanuel Kofi Ackom
University of British Columbia

Juergen Ertel
Brandenburg University of Technology

REMANUFACTURING

Remanufacturing is a process whereby used items are disassembled, and their components are inspected, cleaned, and then used in the manufacture of new products. A product is considered remanufactured if its primary components come from a used product, although it should not be inferior to, or less durable than, comparable new products that solely use virgin materials, which are materials that haven't been used, consumed, or subject to processing before. Remanufacturing can reduce both production costs and demand for increasingly scarce virgin natural resources, although other options such as reuse or recycling may be preferable depending on product type and the value of the materials. Remanufacturing differs from recycling in that recycling takes a material (newspapers, plastic bottles, cans) that often feeds into the production of different goods, unrelated to the original raw material.

The Freitag company makes bags from old truck tarpaulins, used seat belts, and cycle inner tubes, which could be considered an example of remanufacturing.

Source: Peter Würmli/Freitag

In their report "Remanufacturing and Product Design: Designing for the 7th Generation," Casper Gray and Martin Charter have tried to clarify the meaning of remanufacturing and note that it can be easily confused with similar terms beginning with "re," such as repair, refurbishment, and reconditioning. The only way to distinguish its correct definition is to consider the process adopted. There are different types of remanufacturing, and they offer the following synonyms that also represent remanufacturing across different industries: rebuilt (remanufacturing in motor vehicles and systems), recharged (remanufactured imaging products, such as toner cartridges), retread or remolded (for the tire industry), rewound (remanufacturing certain electrical equipment), overhaul (most notably in the aerospace industry).

The feasibility of remanufacturing is often dependent upon design for environment, or more specifically, design for remanufacture (DFR); so that the original product design ensures that core remanufacturable components are easily accessible for cleaning and reinserting into new products. DFR is, at the time of writing, yet to be adopted widely. In fact, due to concerns over intellectual property, some manufacturers may design their products in such a way that third-party remanufacturers cannot recover core components at all. Alternatively, products may also be designed with obsolescence or disposal in mind, so as to increase revenue for the manufacturer as consumers have to buy new products at the end of a product's useful life.

Most manufactured products could be remanufactured, although the decreasing costs of brand-new products coupled with low consumer demand for remanufactured goods will impact the economic incentives for investing widely in business-to-consumer (B2C) remanufacturing capacity. Research has revealed some unfavorable consumer attitudes toward remanufactured goods when they are perceived to be of lesser quality, perhaps inferred by a cheaper price. For example, an often-cited example is retreaded tires and the belief that that these are inferior in performance and safety to new tires, even though nowadays there are equivalent safety assurance standards for both to pass. There is a marketing challenge here in overcoming and redressing these consumer preconceptions.

A growing number of original equipment manufacturers (OEMs) are engaging in remanufacturing, in particular in a business-to-business (B2B) environment, in light of heightened awareness of sustainable development, with Xerox seen to be one of the pioneers in this field. Xerox has adapted its Product Service System (PSS) so that its products can only be rented instead of purchased, meaning that every three years, it can recover and remanufacture the core components. Such activities divert waste that would otherwise be sent to landfill, and Xerox reaps both economic and environmental benefits.

There can be financial benefits for manufacturers opting to create zero waste or closed-loop supply chains, where manufacturers are involved in the take-back of items. In addition, increasing regulatory pressures on the premise of extended producer responsibility (EPR) may also drive such decision making. The concept of EPR underpins directives that aim to promote more efficient use of natural resources, cleaner products and production methods, more integrated environmental management approaches, and greater responsibility through the supply chain. Some examples of EPR directives are Waste Electronics Electrical Equipment (WEEE), End-of-Life Vehicles (ELV), and Energy Using Products (EUP). However, third-party remanufacturers may lose out on their ability to access product components with the increased proliferation of EPR and resultant obligations on OEMs to take back goods. Conversely, opportunities may arise for OEMs and third-party remanufacturers to work in partnership, as opposed to competing against each other, in particular where OEMs have no existing in-house capacity or experience in dealing with such products.

See Also: Closed-Loop Supply Chain; Corporate Social Responsibility; E-Waste Management; Extended Producer Responsibility; Extended Product Responsibility; Green Design; Industrial Ecology; Life Cycle Analysis; Recycling, Business of; Resource Management; Service Design; Sustainable Design; Take Back; Waste Reduction.

Further Readings

Centre for Remanufacturing & Reuse (UK). http://www.remanufacturing.org.uk (Accessed July 2009).

Ferrer, Geraldo and D. Clay Whybark. "From Garbage to Goods: Successful Remanufacturing Systems and Skills." *Business Horizons,* 43/6 (2000).

Guide, V. Daniel R., Jr., and Luk N. Van Wassenhove. "Managing Product Returns for Remanufacturing." *Production and Operations Management,* 10/2 (2001).

Hormozi, Amir M. "Parts Remanufacturing in the Automotive Industry." *Production and Inventory Management Journal,* 38/1 (1997).

Hutchens, Stephen P. "Consumer Attitudes Toward Remanufactured Products as a Form of Recycling." *Dissertation Abstracts International Part A: Humanities and Social Sciences,* 43 (1983).

Hutchens, Stephen P. and Jon M. Hawes. "Consumer Interest in Remanufactured Products: A Segmentation Study." *Akron Business and Economic Review,* 16/1 (1985).

King, Andrew M., et al. "Reducing Waste: Repair, Recondition, Remanufacture or Recycle?" *Journal of Sustainable Development,* 14/4 (2006).

Lund, Robert T. *Remanufacturing: The Experience of the United States and Implications for Developing Countries.* Washington, D.C.: World Bank, 1984.

Mondal, Sandeep and Kampan Mukherjee. "An Empirical Investigation on the Feasibility of Remanufacturing Activities in the Indian Economy." *International Journal of Business Environment,* 1 (2006).

Peattie, Ken and Margarete Seitz. "Meeting the Closed-Loop Challenge: The Case of Remanufacturing." *California Management Review,* 46/2 (2004).

The Remanufacturing Institute. http://www.reman.org (Accessed July 2009).

Smith, Vanessa M. and Gregory A. Keoleian. "The Value of Remanufactured Engines: Life-Cycle Environmental and Economic Perspectives." *Journal of Industrial Ecology,* 8 (2004).

Steinhilper, Rolf. *Remanufacturing: The Ultimate Form of Recycling.* Stuttgart, Germany: Fraunhofer IRB Verlag, 1998.

Cerys Anne Ponting
Cardiff University

RESOURCE MANAGEMENT

There are many different types of resources, including human, social, information, and natural resources. This entry focuses specifically on natural resource management. While community organizations and state policies are very important for defining equitable and sustainable use of the Earth's resources (discussed below), this entry highlights private sector practices, with particular attention to resource management benefits and costs associated with the rapid growth of green businesses. A discussion of participatory natural resource management models linked to private sector and state initiatives follows an introduction to the characteristics of resources, ownership rights and responsibilities, and parameters for sustainable management.

Natural Resource Classification and Status

Natural resources are generally characterized as renewable or nonrenewable. Renewable resources, such as forests, can regenerate if they are not overharvested. Soil and water are both renewable, but can become depleted if not sustainably managed so that the rate at which they can become replenished is equal to or higher than the rate at which they are used. Once renewable resources are consumed at a rate that exceeds their natural rate of replacement, the standing stock will diminish and may eventually run out. Severe resource degradation can occur prior to total depletion, and often occurs at accelerated rates in specific at-risk areas.

Fossil fuels, like coal, petroleum, and natural gas, take thousands of years to form, and are not created at the same speed as they are being consumed. These energy sources are considered nonrenewable and are at risk for depletion. Nuclear energy is also nonrenewable because it requires uranium, a finite resource. A natural resource's value rests in the amount and extractability of the material available and the demand for it. If we do not transition away from gasoline-powered vehicles, the price for oil may reach surprising levels.

Natural resources are currently under increasing pressure, threatening environmental and human health and development. Air, soil and water pollution, water shortages, soil exhaustion, loss deforestation, and degradation of coastlines are some of the many existing problems reflecting gaps and weakness in environmental policy and natural resource management. Most developed countries currently consume natural resources at a pace much faster than they can regenerate. Seventy-five percent of energy resources are consumed by 25 percent of the world's population, living in developed countries. The same population also consumes the majority of the world's extracted and marketed mineral and forest resources in any given year. On average, a child raised in the United States utilizes more than 30 times as many resources as a child reared in India.

Sustainable natural resource management requires addressing global inequity in resource use, since resource extraction may increase as low-income countries seek to develop and prosper. While many people in the industrialized world overconsume, their counterparts in many developing countries do not have access to enough resources to meet basic needs. A significant portion of the poorest families live in rural areas and depend directly on natural resources for their survival, a situation that often makes their future insecure as stocks of resources become diminished and the health of the environment deteriorates. Even where natural resource stocks remain, natural resource endowment may contribute to destabilization and conflict in many developing countries. In places such as Angola, the Democratic Republic of Congo, Nigeria, and Sierra Leone, diamonds and fossil resources have fueled conflicts by challenging livelihoods, threatening the environment, raising disputes over rights to control the resources, or by overusing revenue to cover the cost of war. In the Middle East and China, water access is a source of conflict. Managing natural resources is particularly challenging in unstable and conflict-affected areas because of the vulnerability of the economy to illicit trade and lack of transparency.

The *resource curse* is the term used to describe why the existence of abundant natural resources may not lead to economic growth in some developing countries, including such reasons as the following:

- Reliance on exports of raw resources limits potential "value added" and labor generation advantages.
- Excessive borrowing based on the expectation of future resource revenue.
- Revenue volatility, due to the fact that prices for some natural resources are subject to wide fluctuation; for example, crude oil prices rose from around $10 a barrel in 1998 to over $140 a barrel in 2008.

- Conflict over resource access and between resource users.
- Corruption and abuse of power, whereby corrupt leaders in resource-rich countries may maintain authority through allocating benefits to favored constituents. Money from natural resources sometimes fuels political corruption or may finance military forces or armed groups that are used to control internal populations attempting to denounce state injustice or abuses of power.

Resource Policies

Many countries are characterized by legal pluralism—the operation of different bodies of formal and informal laws and legal procedures within the same sociopolitical space. These legal orders may be rooted in the nation-state, religion, ethnic group, local custom, international agreements, or other entities. They often overlap, resulting in different legal bodies that can be complementary, competitive, or contradictory. Resource conflicts often emerge because there is a lack of harmony and coordination among these different legal orders, particularly when policies, programs, and projects fail to consider local situations.

In developing countries, resource policies and interventions are often formulated without the active and sustained participation of communities and local resource users. For example, many governments have long relied on centralized resource management strategies, and make decisions based solely on the expertise of technical experts. These policies and practices frequently fail to take into account local rights to, and practices regarding, natural resources. For example, the introduction of new policies and interventions without local input may end up supplanting, undermining, or eroding community institutions governing resource use. Nevertheless, comanagement, involving the sharing of responsibilities between governmental institutions and groups of resource users, is rapidly increasing in popularity. In many countries, environmental management has been reformulated from exclusive state control to various kinds of joint management in which local communities, indigenous peoples, and nongovernmental organizations (NGOs) share differing degrees of authority with governmental institutions. Community-based comanagement is a promising approach where local populations play a key role in the management of natural resources and ecosystems. However, not all programs delegate true authority to local actors.

Adaptive comanagement is an emerging approach for flexible governance of socioecological systems, combining the iterative learning dimension of adaptive management and the linkage dimension of collaborative management, in which rights and responsibilities for managing resources are jointly shared. These features can promote an evolving, place-specific governance approach for resources in which strategies are sensitive to feedback (both social and ecological) and oriented toward system resilience and sustainability. Such strategies include dialogue among interested groups and actors and the development of complex, redundant, and layered institutions to facilitate experimentation and learning through change.

Integrated natural resource management (INRM) is an approach aimed at improving livelihoods, ecosystem resilience, productivity, and environmental services. In other words, it aims to augment social, physical, human, natural, and financial capital. Its efficiency in dealing with these problems comes from its ability to do the following:

- Empower relevant stakeholders—people and institutions who possess an economic, cultural, or political interest in or influence over a resource
- Resolve conflicting interests of stakeholders

- Foster adaptive management capacity
- Integrate levels of analysis
- Merge disciplinary perspectives
- Make use of a wide range of available technologies
- Generate policy, technological, and institutional alternatives

Yet, despite growing recognition of the need for integrated approaches to natural resource management, many governmental and other agencies still rely on sectoral approaches with limited interagency cooperation. Overlapping and competing jurisdictions and activities among state agencies and with other development organizations, such as foreign donors and NGOs, may result in their inability to reconcile the needs and priorities of various stakeholders. Many participatory and adaptive models for natural resource management have been developed at a theoretical level, but implementation on the ground remains highly challenging.

Business Approaches to Resource Management

Market-based approaches to resource management are the dominant model in foreign aid agencies and have been integrated into many state, nongovernmental, and private sector development projects and policies throughout the globe. These approaches are often promoted as "win-win" situations that bring income to local populations and increase the degree and efficacy of resource management. Nevertheless, one should proceed with caution when introducing market environmentalism into developing areas, as cultural changes are likely, and there are often many unexpected consequences. Market environmentalism is often uncritically formulated around the idea of "selling resources to save them": in other words by simply recognizing or increasing economic value, one will encourage more sustainable management. However, since many natural resource-based projects link to imperfect and highly competitive export markets inundated with uncounted negative social and ecological consequences, known as externalities, these projects may still deplete the targeted commodity or negatively impact other nontarget resources (e.g., groundwater) and potentially cause a subsequent devaluation of resources customarily used for subsistence practices. Prices often do not reflect true costs of extraction or loss of resources, and thus market-based management may not always achieve the triple bottom line of ecological, social, and economic sustainability. Successful resource management projects generally proceed very slowly, are highly participatory, and respect and value local differences. Market-based natural resource management cases designed with narrow expectations, like the rapid economic returns on investment, are unlikely to achieve long-term sustainable resource use in conjunction with human and social development.

Important lessons have been learned from natural resource management experiences in recent decades. Projects with local processing that add value, in contrast to the mere extraction of raw materials, increase employment and community revenue. It is important to build participation and avenues for representation at the local level: empowering rural communities and urban neighborhoods through cooperatives and other associations is often successful. Grassroots organizations often need financial support, ideally provided as revolving microloans, for start-up investment capital. Support from private sector can be instrumental, but research suggests there are often problems with corporate donors, including paternalism. Nonetheless, there have also been mixed results from bilateral and multilateral aid agencies, with a major concern being condition-setting so that outside organization have undue influence over domestic or local spheres because they control the purse strings. In some instances, there are also problems of unequal power and access to resources when local groups partner with domestic and international NGOs. Fortunately, in many locations, collaboration between local communities and outside NGOs is improving. Many NGOs, although not all, aim to empower local partners. Furthermore, there are growing expectations for transparency in NGO operations. Ratings and statistics, like the percent of total expenditures utilized in administrative costs as compared to programs and projects, are posted regularly on the Internet sites of independent charity evaluators, like Give Well or Charity Navigator.

Looking Forward: Decarbonization, Green Technology, and Renewable Energy

There are a growing number of green business models based on the premises of the widely used slogan "reduce, reuse, recycle." These initiatives are important to goals of natural resource management as they limit the pressure placed on existing resource stands or deposits. Solid waste experts in Europe suggest that the region has enough aluminum and steel already in use to supply Europeans for hundreds of years into the future. With effective programs to recycle existing components, there is no arguable need to extract these metals from untapped deposits. The efficient reuse of these resources, however, requires manufacturing products so that they can be efficiently disassembled and reused. Automobiles in Europe, and other products, are currently being constructed with cost-effectiveness and ease of reuse in mind.

Business models and community enterprises are emerging around green technologies and a decarbonization or dematerialization of resource use. While decarbonization aims broadly at the reduction of emissions of carbon dioxide and other greenhouse gases (GHGs), specific goals often focus on diminishing energy use and decreases in energy intensity of activities and processes. Decarbonization can be achieved through energy efficiency improvements, as well as from the substitution of fuel sources with high carbon content by those with low carbon content. Dematerialization roughly translates into doing more with less, but specific initiatives often focus on absolute or relative reduction in the quantity of materials, including natural resources, required to complete an economic or social transaction. A commonplace example of dematerialization is reading newspapers, magazines, and books in digital rather than printed format.

Momentum for decarbonization and dematerialization projects has quite rightly begun in industrial countries, but global applications are growing and possibilities are nearly endless. In some instances, developing countries can leapfrog to new green technologies that reduce natural resource input and carbon output. In other instances, the existing low efficiency systems need to be replaced. One of the goals of the Kyoto Protocol and the United Nations Framework Convention on Climate Change (UNFCCC) was to encourage the expansion of sources of renewable energy in developing countries in order to decarbonize processes of economic development and service provision.

Unfortunately, many verified Clean Development Mechanism (CDM) projects focus narrowly on utilizing infrastructure and investments already in place rather than promoting innovative green technologies with the highest potential for energy intensity and GHG emission reductions. One consequence is a limited transition away from polluting technologies and industries in many countries. In some instances, the CDM provides supplemental income to agribusiness and energy companies, such as hog farms, sugar and palm plantations, or fossil fuel–burning electrical plants, which continue to have high resource inputs and high carbon outputs even when their processing efficiency is improved or new waste products are captured to be flared or utilized. In general, there should be more critical, holistic assessments of social and ecological implications of proposed GHG reduction solutions. For example, the use of landfill gas has the potential to be positive by lowering the release of GHG emissions. However, projects that restrict access to local populations who depend on solid waste resources for survival, or initiatives that contradict broader efforts to reduce overall waste or recycle materials before landfilling can simultaneously have negative social and ecological repercussions that need to be addressed to assure overall sustainability and long-term economic and social development.

Decarbonization requires fundamental shifts in international and national policies, in addition to dramatic technological advancements and major changes in consumption patterns. If not, subsidies and other financial incentives for green industries support inadequate small steps, such as movements away from conventional fossil fuels toward transitional energy sources, like nuclear power or "clean" coal, instead of investing now to develop and utilize long-term renewable energy sources like solar, hydrogen, or cellulosic biofuel. Since corporate lobby groups representing particular industries, such as oil, mining, transportation, and biotechnology, have a strong influence on national and international development policy, currently, the solutions to climate change that are pursued are the ones with the largest profit margin. This is a particularly short-sighted economic agenda, given the predictions in the Stern Report and other rigorous economic studies that climate change will likely cause significant losses in future gross domestic product earnings around the globe in the absence of aggressive policies to rapidly curb the release of GHGs.

Community enterprises based on green building and renewable energy from localized sources hold exceptional promise for decarbonization and dematerialization goals, but are also particularly important in developing employment in areas of economic distress. Jobs are increasingly being created in industries that reuse or recycle part of the waste stream. An example of a successful nonprofit project is the Green Institute, a decade-long green building industry leader located in Minneapolis, Minnesota. The institute's Clean Energy Resource Teams (CERTs) provide information about and support community-based sources of renewable energy, while a massive ReUse Center promotes the exchange and use of salvaged construction materials. The Green Institute's integrated and symbiotic programs have impressive cost and natural resources savings.

The Green Institute is one of many international organizations spearheading a shift toward industries based on recycling, reusing, and remanufacturing. Remanufacturing involves the repair or replacement of worn out components to extend the life span of machinery. For example, an airplane or a piece of medical equipment can be completely refurbished so that there would be limited, if any, difference from a new product. These reuse concepts, and exact stipulations for product and service delivery, can be verified through Cradle-to-Cradle Certification: focused on using environmentally safe and healthy materials; design for material reutilization, such as recycling or composting; the use of renewable energy and energy efficiency; and efficient, responsible use of water. A final requirement in this certification system is instituting strategies for social responsibility.

Hybrid partnerships between the private sector, state agencies, universities, and non-governmental or citizens organizations increasingly promote green energy transitions, for example through the sale of Renewable Energy Credits (RECs). Renewable energy sources, including solar, geothermal, biomass, tidal, and wind power, can be expensive to establish. Sales of RECs help share those investment costs across various stakeholder groups that will benefit in the long-term from a transition to renewable energy. This is just one example of the multitude of ways that governmental, nongovernmental, and private-sector groups are forging new alliances and partnerships to improve conventional energy production and use resources more efficiently.

Climate Action Plans (CAPs), an increasingly prominent planning tool used by businesses, schools, and state governments, promote ongoing implementation of reductions. CAPs have encouraged state, educational institution, and industry leaders to plan for energy and resource use transitions, find new and additional methods to conserve resources, and initiate annual budgeting to pay for the transition to green production and consumption, with the goal of saving money and resources in the long term. As institutions and individuals work to lower GHG emissions and dematerialize economic pathways, we may also need to redefine social expectations for ongoing and rapid economic growth in industrialized countries such as the United States, as well as in urban areas of rapidly transitioning economies such as China, India, and Brazil.

See Also: Appropriate Technology; Best Management Practices; Biological Resource Management; Closed-Loop Supply Chain; Cradle-to-Cradle; Environmental Management System; Factor Four and Factor Ten; Green Technology; Material Input per Service Unit (MIPS); Maximum Achievable Control Technology; Natural Capital; Remanufacturing; Supply Chain Management; Sustainability; Triple Bottom Line; Waste Reduction.

Further Readings

Food and Agriculture Organization of the United Nations (FAO). "Conflict and Natural Resource Management." Rome: FAO, 2000. http://www.fao.org/forestry/foris/pdf/conflict/conf-e.pdf (Accessed January 2010).

McDonough, William and Michael Braungart. *Cradle to Cradle: Remaking the Way We Make Things*. New York: North Point Press, 2002.

Meinzen-Dick, Ruth, et al. *Innovation in Natural Resource Management: The Role of Property Rights and Collective Action in Developing Countries*. Baltimore, MD: Johns Hopkins University Press, 2002.

Natcher, D. and C. Hickey. "Putting the Community Back into Community-Based Resource Management: A Criteria and Indicators Approach to Sustainability." *Human Organization*, 61:350–63 (2002).

World Resources Institute. "World Resources 2008: Roots of Resilience—Growing the Wealth of the Poor." United Nations Development Programme, United Nations Environment Programme, World Bank, and World Resources Institute, 2008. http://www.wri.org/publication/world-resources-2008-roots-of-resilience (Accessed January 2010).

Gopalsamy Poyyamoli
Pondicherry University

RESPONSIBLE SOURCING

Every product has health and environmental impacts as a result of manufacture, use, and ultimate disposal. Today's consumers increasingly demand high-quality as well as safer, ethical, "green" products. This cannot be achieved by a company's processes alone, but depends on the processes of its suppliers. Businesses increasingly recognize that focusing on price and quality of inputs is one-dimensional, and sourcing is not only about unit cost and just-in-time. Responsible sourcing integrates the three pillars of sustainability—economic, environmental, and social—into the procurement process. It is an integral part of effective supply chain management, bundling health, environmental, social, and ethical considerations with the routine purchasing processes of price, product performance, and delivery schedule.

Responsible sourcing addresses upstream activities from raw material extraction to final inputs and should be complemented by sustainable operations and by extended producer responsibility addressing downstream activities such as use and recycling of products. Responsible sourcing builds on a variety of related concepts from traditional purchasing practices and must involve effective communication between buyer and supplier.

Responsible sourcing can stimulate demand for more sustainable raw materials and products leading to improved supplier efficiencies, thereby assisting in developing and strengthening markets for such products. Companies and governments have thus begun an attempt to foster sustainable behavior using their purchasing power, simultaneously reducing risks associated with their purchasing decisions.

Globalization has facilitated outsourcing to developing/emerging countries in search of a low-wage advantage. Through outsourcing, companies attempt to distance themselves from responsibility for the impact of their own operations. However, they are ultimately responsible for activities in their supply networks—from working conditions to corruption to environmental impacts—and should be held accountable. As the supply chain becomes longer and more complex, there is a greater likelihood of environmental, as well as health and safety, impacts. Corporations often don't have necessary information about supply points in a given country, and the scope of any monitoring often does not extended to raw material suppliers.

Effective supply chain management can allow businesses a competitive advantage, especially in sectors such as clothing, footwear, electronics, or food products where production is largely outsourced. For some companies, improving social and environmental standards in the supply chain has become a natural extension of their commitment to corporate responsibility, forming part of their overall business model. Other companies have continued to pay close attention to current trends in the fields of corporate social responsibility and responsible sourcing as part of their risk management strategy. Companies can demonstrate leadership by clearly stating ethical and responsible values and visions in their sourcing policy and strategy, outlining goals and implementation of the supply chain social responsibility program, and using sustainability reporting to communicate actions and their impacts to stakeholders.

Companies develop their own approaches demonstrating voluntary commitment to responsible sourcing. Such approaches include the use of codes of conduct, supplier questionnaires, social compliance auditing, and supplier capacity training as sources of supply chain monitoring. Corporate codes of conduct are often regarded as part of supply chain policy and are one of the most widely used formats expressing corporate social responsibility commitment. They typically state ethical standards the company claims to uphold—focusing mainly on local labor laws, working conditions, and the environment—as well as communicating company expectations to suppliers regarding wages, child labor, health and safety, and workers rights.

Examples of Responsible Sourcing Practices

Levi-Strauss embedded its social compliance program, the Terms of Engagement (TOE), in its sourcing division in the mid-1990s, and was among the first to do so. For The Gap, responsible sourcing means going beyond ethical business basics to embrace responsibility to people and the planet, thus striving to "make a difference" by improving factory conditions, caring for the environment, investing in communities, and supporting employees.

At UK grocer Waitrose, responsible sourcing means providing the best possible conditions for workers, protecting the natural environment, and promoting high standards of animal welfare. Waitrose launched its Responsible Sourcing Code of Practice and Supplier Audit Programs in 1999, specifying acceptable working conditions and measuring suppliers' compliance. It has implemented improvements to supplier workplace conditions and increased the commercial, environmental, and ethical sustainability of its supply base. The Waitrose Foundation, partnering with South African citrus farmers, provides investment to improve community welfare, while its Locally Produced initiative aims to find the best local suppliers.

The Body Shop has long been recognized as a compassionate company, selling natural, ethically sourced products and preventing animal suffering while protecting the environment and indigenous peoples. Adidas emphasizes improvement of global suppliers' workplace conditions, intensifying business partnerships with suppliers who are independently motivated and able to implement fair factory conditions and governance systems.

IKEA, the Swedish home furnishings company, has partnered with the World Wildlife Fund since 2002 to address climate change while promoting responsible forestry and more sustainable cotton production. L'Oréal has taken steps toward responsible sourcing by insisting on Forest Stewardship Council (FSC) certification for paper companies and printers, specifying the use of recycled materials, and building a reference guide for raw materials.

British retailer Marks and Spencer's "Plan A" is a five-year initiative tackling a wide range of environmental and sustainability issues, including sustainable and ethical sourcing. Starbuck's Coffee and Farmer Equity (C.A.F.E.), a supplier practice program, promotes product quality improvements, economic transparency, social responsibility, and producer-level environmental leadership.

The UK-based Co-operative Group (the world's largest consumer-owned business) pursues a rigorous sourcing policy, using only suppliers sourcing from well-managed fisheries, and actively avoiding vulnerable species. Mountain Equipment Co-op's Ethical Sourcing Program entails a factory audit program, corrective action, and community involvement (transparency). No suppliers are contracted before auditing, and all suppliers are audited on a periodic basis.

Wal-Mart is working to implement vendor scorecards that provide performance summaries on core issues, including price, quality, and labor standards. This step represents an effort to integrate buying and social compliance activities, potentially providing a tool for sourcing from vendors and suppliers demonstrating commitment to improved labor standards.

Multiparty Approaches: Industry Codes and Certification Schemes

In addition to developing its own responsible sourcing approach, a company can incorporate responsible sourcing practices by following industry codes of conduct, and/or by only using suppliers whose practices meet certification scheme requirements.

The toy and electronic sectors exhibit other responsible sourcing initiatives. The Brussels-based Business Social Compliance Initiative is a business-driven platform encouraging social compliance improvement for all consumer goods in all supplier countries. ICTI-CARE, an international toy industry ethical manufacturing program, aims to ensure safe and humane workplace environments in toy factories worldwide. The Electronic Industry Code of Conduct has been adopted and implemented by major electronics brands and their suppliers, improving the electronics supply chain.

Many certification schemes require third-party assessment, transparent processes with built-in stakeholder consultation and objection procedures, and standards based on target issue and management practices, giving suppliers opportunities to demonstrate commitment to buyers' sustainability objectives. The FSC established early responsible sourcing standards in the 1990s through work on forestry management standards, firmly establishing it within the timber and construction industry. The Marine Stewardship Council, the only widely recognized environmental certification and ecolabeling program for wild-capture fisheries, provides the only seafood ecolabel consistent with the ISEAL Code of Good Practice for Setting Social and Environmental Standards and United Nations Food and Agriculture Organization guidelines for fisheries certification. The Rainforest Alliance certifies coffees produced by farms that meet economic, social, and environmental sustainability criteria. The Bonn, Germany-based Fairtrade Labeling Organizations certify products that meet fair trade criteria.

Challenges of Responsible Sourcing

Responsible sourcing practice challenges are found in the following areas: justifying responsible sourcing costs; establishing criteria; communicating standards to suppliers; supplier collaboration, auditing, and assessment; supplier violations response; transparency; and the potential for stakeholder skepticism relative to each of these challenges.

Buyers considering adopting responsible sourcing should realize that such programs represent a serious commitment, and are costly in both time and financial resources. Hastily designed or haphazardly implemented programs will likely encounter difficulties, bringing the company significant criticism. However, inaction has its cost as well. One negative incident with one supplier can lead to a disproportionate amount of adverse publicity, damaging a company's reputation and brand image. Examples include Nike and worker rights; the Gap's link to bonded child labor; the use of illegal workers by Primark's suppliers; and Wal-Mart paying employees less than minimum wage. Following consumer backlash, a number of manufacturers accused of allowing subcontractor sweatshop labor requested social auditing by the Fair Labor Association, Social Accountability International 8000, or the Ethical Trading Initiative.

Companies should develop business policies enabling them to prevent such practices and to respond proactively and responsibly when they occur. Suppliers should be required to implement environmentally sound management practices, strengthen community social conditions, and improve economic returns by producing higher-quality products. Companies must establish sourcing criteria carefully, however, as adherence failure will likely be discovered and publicized to the company's detriment. Choosing an industry code or certification scheme should also be done carefully, as competing codes and schemes often exist. A company choosing less stringent ones is likely to be criticized by stakeholders.

Successful responsible supply chain members share risks, gains, costs, and have a partnership mentality. At the buying end of the chain, internal compliance teams and their buying department should influence, team up, communicate, and standardize suppliers' requirements. Sourcing decision makers must tackle the inherent conflict between social compliance efforts that may push suppliers to make improvements requiring capital expenditures, and buying practices that demand the lowest price and fastest turnaround time. Existing and potential suppliers must understand the details of the criteria for which they will be held accountable. If suppliers' knowledge or criteria-meeting skills are questionable, buyers must either help suppliers develop these skills, or not award the contract. Long-term buyer commitments can enable suppliers to make capital expenditures often required to meet buyer-imposed codes of conduct.

Because of the costs and difficulty of auditing supplier sourcing criteria compliance, this essential element is one of the most difficult in a responsible sourcing initiative. Companies must decide the frequency of and procedure for audits, whether to use their own or with third-party staff, and decide on policies regarding audit findings transparency. When audits unearth violations, buyers must follow through on established policies, including terminating supplier contracts. Failure to release audit results and terminate contracts can severely impact a buyer's stakeholder reputation.

Consumer influence is fundamental to change. They must continually question a firm's eradication of unethical practices, and companies must therefore educate consumers about their responsible sourcing policies, programs, and outcomes. Doing so can help companies committed to responsible sourcing reap the reputational and economic benefits they deserve for making improvements to the social and environmental outcomes of their purchasing decisions.

See Also: Compliance; Corporate Social Responsibility; Ecolabels; Supply Chain Management; Sustainability; Voluntary Standards.

Further Readings

Business Social Compliance Initiative. http://www.bsci-eu.org (Accessed May 2009).
Casey, Roseann. "Meaningful Change: Raising the Bar in Supply Chain Working Standards." Prepared for John Ruggie, UN Secretary-General Special Representative on Business and Human Rights, November 2006. http://www.hks.harvard.edu/m-rcbg/CSRI/publications/workingpaper_29_casey.pdf (Accessed May 2009).
Electronic Industry Code of Conduct. http://www.eicc.info (Accessed April 2009).
Ethical Trading Initiative. http://www.ethicaltrade.org (Accessed May 2009).
Fair Labor Association. http://www.fairlabor.org (Accessed May 2009).
Social Accountability International. "SA8000 Resources and Documents." http://www.sa-intl.org/index.cfm?fuseaction=Page.viewPage&pageID=710 (Accessed May 2009).

World Bank Group. "Strengthening Implementation of Corporate Social Responsibility in Global Supply Chains." Washington, D.C., October 2003. http://siteresources.worldbank .org/INTPSD/Resources/CSR/Strengthening_Implementatio.pdf (Accessed May 2009).
Worldwide Responsible Apparel Production. http://www.wrapcompliance.org (Accessed May 2009).

Joyce Tsoi
Brunel University

RESTORATION

Restoration involves actively working to bring something back to a previous condition, often because it has suffered injury or its integrity has degraded over time. Many things—artwork, architectural structures, or political institutions—can be the subject of restoration efforts.

Ecosystem Restoration

Ecosystem restoration entails returning an ecosystem to its previous healthy state. Humankind has historically disrupted ecosystems by logging, clearing land for agricultural purposes, and altering water levels and flows by draining swamps and constructing dams and dikes. There has been a recent increased appreciation for the "free services" that ecosystems provide, with a resulting increase in laws facilitating restoration. An early realization involved lost diversity through native prairie destruction. The first major prairie restoration was in 1934 at University of Wisconsin–Madison's arboretum.

Site of a Staten Island Salt Marsh Restoration project. In 1990, nearly 600,000 gallons of fuel oil discharged from a ruptured Exxon pipeline on the bottom of the Arthur Kill waterway, killing much of the marsh vegetation and a variety of intertidal organisms.

Source: National Oceanic and Atmospheric Administration Restoration Center, Northeast Region

Federal aid to states was provided as early as 1937 by the Federal Aid in Wildlife Restoration Act. However, it and other laws were aimed at restoring wildlife, sometimes actually leading to habitat modification rather than ecosystem restoration. The Estuary Protection Act of 1968 provided means for restoration, but also required balancing restoration with development for economic purposes—an underlying theme of several other acts. The policy goal of "no overall net loss" of wetlands, established by the Water Resources Development Act of 1990, served to dramatically expand implementation of restoration activities mitigating the destruction of wetlands through development and highway construction.

The Coastal Wetlands Planning Protection and Restoration Act from the same year provided for federal grant funding. Finally, The Estuaries and Clean Water Act of 2000 established an Estuary Habitat and Restoration Council, charged with producing a national restoration strategy.

Not all restoration has been based on good scientific knowledge, and outcomes are difficult to predict. Consequently, there have been increasing calls for ecologically grounded ecosystem restoration. Proponents suggest that successful ecosystem restoration should entail clearly stated goals and a design stemming from ecological principles. Restorationists should gather and utilize quantitative data about the way the ecosystem functions and its reaction to change, and an analysis of results must guide future actions.

An ecosystem might need to be restored because of radical alteration due to either intentional or unintentional introduction of nonnative species. For example, saltcedar—an invasive deciduous small tree—was introduced to the United States in the 1800s as an ornamental species. As early as 1850, it rapidly spread through riparian habitat and replaced native plant life. The plant has a long natural life (50–100 years), absorbs an excess of water (a larger tree is capable of absorbing 200 gallons of water per day), can withstand cutting, burning, cold temperatures, droughts, and floods, and changes the salinity of the surrounding soil by excreting salt from its leaves. Thus, restoration efforts could aim for the eradication of the saltcedar. However, questions are raised when considering species eradication from an ecosystem: is it right to eliminate a species once it has settled into an ecosystem? What constitutes an invasive species and after how long, if ever, would it be considered native? In relation to wildlife, these ethical questions are often more poignant. Another question, more difficult to answer, is what a specific ecosystem should be like. In a warming world, should attempts be made to restore an area to its historical condition? Or should the character of the ecosystem be changed to one that is sustainable in a warmer climate regime? How and to what state we restore ecosystems is largely a reflection of our communal values.

Restoration and Business

Restoration relates to business in several ways. An industry can engage in mitigation for harm it caused—environmental improvements can be undertaken even when the business undertaking them did not cause the degradation—and green businesses can carry out the work of restoration.

Ecosystem restoration and business connect when an industry promises to restore an ecosystem after extracting natural resources. For example, the logging industry promises to replant forests, intimating that it will replace logged trees. Whether such activities will restore preexisting species diversity is an open question, however. Even when such an intent is present, there is often the concern that businesses may not have the financial resources to adequately restore an ecosystem. The ultimate fear is that humans may not possess the necessary knowledge or technological capability to fully restore an ecosystem to a healthy state. A basic question is whether industry should be able to degrade ecosystems, even with the promise of restoration and the ability to deliver on that promise.

Ecosystem restoration can involve brownfields—sites formerly used by business or industry—that may contain low levels of hazardous waste. (The term *brownfield* usually does not apply to areas with high levels of contamination or to superfund sites.) Usually located in industrial areas within a municipality, small brownfields can be found in older residential neighborhoods where, for example, a gas station or dry cleaner had been located. Brownfield remediation often results in redevelopment—improving air and water quality and reducing urban sprawl—but can also result in ecological restoration. Examples include pocket parks in the Chicago area; a former Gulf refining site in Hooven, Ohio, restored to a wetland and natural habitat area; and areas along the Tuscawaras River in northeastern Ohio.

The Wildlife Habitat Council not only facilitates the restoration of brownfields in general, but encourages individual corporations to donate lands, including parts of their corporate campuses, as restoration sites. These sites may have been purchased specifically for the purpose of restoration. The list of corporations originally partnering with Wildlife Habitat Council include Anheuser-Busch, DuPont, ExxonMobil, General Electric, Tenneco Oil, and U.S. Steel. Corporate involvement in restoration has expanded internationally. For example, CEMEX, the Mexico-based multinational cement manufacturer, has funded several restoration projects in the Mexico–U.S. border region.

The business of ecosystem restoration is thriving and expanding into many areas. Substantial business opportunities exist for firms involved in restoration projects on state and federal government lands, and carbon offsetting and sequestration efforts may help provide new funding. Networks connect those involved in restoration projects, and the Society for Ecological Restoration (SER) promotes and provides forums for dialogue in this area, in addition to public support and awareness.

Its website links to networks such as Global Restoration Network (GRN). The GRN is, in turn, a portal to information including a restoration toolkit, a directory of experts, various databases, restoration theory, guidelines, methods, and restoration case studies. SER provides an online marketplace where vendors can advertise services and products, and a career center, connecting job seekers with employers. Types of businesses that are part of this growth in environmental restoration include consulting firms, contractors, media services, academic programs, and nonprofit organizations. Businesses also offer assistance in training and workshops, erosion control, reforestation, mining reclamation, plants and seeds, fertilizers, equipment, and soil and root enhancement.

See Also: Bioremediation; Brownfield Redevelopment; Conservation; Ecosystem Services; Sustainable Design.

Further Readings

Elliot, Robert. *Faking Nature*. New York: Routledge, 1997.

Falk, Donald A., Margaret A. Palmer and Joy B. Zedler. *Foundations of Restoration Ecology*. Washington, D.C.: Island Press, 2006.

Global Restoration Network. http://www.globalrestorationnetwork.org (Accessed January 2010).

Society for Ecological Restoration (SER). http://www.ser.org (Accessed January 2010).

van Andel, Jelte and James Aronson, eds. *Restoration Ecology: The New Frontier*. Hoboken, NJ: Wiley-Blackwell, 2005.

Wildlife Habitat Council. http://www.wildlifehc.org (Accessed January 2010).

Gordon P. Rands
Pamela J. Rands
Western Illinois University

Jonathan Parker
University of North Texas

REVERSE LOGISTICS

Reverse logistics, an important principle in green, or sustainable, business, refers to any systematic reversal of the traditional flows in a value chain. A value chain is the chain of activities and institutions, such as wholesalers, agents, brokers, shippers, and retailers that add value to a product on its way from the manufacture to its end consumer.

Value chains, or marketing channels, are typically depicted as linear throughput systems with goods and services flowing downstream from manufacturers to consumers, and with money flowing upstream from consumers, back through the channels toward manufacturers, with each member of the chain benefiting along the way. From the standpoint of sustainable business, such linear systems are problematic because, although they work well for managing throughput flows of materials, they fail to account for the ultimate origination or destination of those materials. For example, the raw materials that make up a television or a mobile phone originate from the Earth, and the process of extracting them inevitably reduces the Earth's ability to provide ecosystem services. Similarly, the television or cell phone that ends up in a landfill or an incinerator doesn't disappear there. It degrades and cycles back into the Earth's ecosystems, along with all of its toxic components.

Business becomes sustainable only to the extent that value chains can be converted into value circles. A value circle, in contrast to a value chain, is a system in which all waste is reclaimed and converted back into resources. Reverse logistics are the processes of reclamation.

Value circles are the result of biomimicry, meaning that they are modeled after natural systems, all of which are inherently cyclic and sustainable. In nature there is no waste, because the waste from every process or organism functions as food or fuel for some other process or organism. The concept that waste equals fuel is central to the philosophy, developed by Michael Braungart and William McDonough, of cradle-to-cradle design and manufacturing, which maintains that all commercial waste can be eliminated by ensuring that all product materials either serve as biodegradable nutrients for natural systems or are reclaimed for reuse in technological systems. Again, reverse logistics is the mechanism by which reusable or recyclable materials reenter a value circle.

There are a number of forms of reverse logistics that function to close the loop from value chain to value circle. They include the following:

- Refurbishing and remanufacturing
- Product take-back
- Collection and consolidation of recyclables
- Secondary markets
- Deconstruction and design for disassembly
- Leasing, renting, or substituting products with services

Refurbishing or remanufacturing converts end-of-life products back into usable and, therefore, marketable products. The process requires product take-back, a form of reverse logistics, to move a used product or components from users back into manufacturing cycles. For example, Caterpillar achieves this through parts and engine exchanges, wherein customers can exchange worn out machinery for like-new remanufactured machinery at a fraction of new prices. Retailers Staples and OfficeMax collect ink and toner cartridges for remanufacturing by offering customers store credit for returned ones.

Organizations can also achieve product take-back by offering customers free or convenient handling of recyclables, such as electronics, compact fluorescent light bulbs or appliances, and toxic substances, such as paint, motor oil, or batteries. For example, Metro Paint of Portland, Oregon, collects unused house paint from consumers or contractors and recycles it into new, high-quality, low-cost paint in several colors. Dell's Asset Recovery System provides its small-business customers with free pickup of old computers, scrubs the drives of all data, and dismantles the machines to recover usable parts and materials.

Some retailers provide reverse logistics for other recyclable materials, such as cardboard, metals, and different kinds of plastics by acting as consolidators. The economics of recycling are more favorable if recyclable materials can be easily and cheaply sorted and amassed. Retailers like Wal-Mart have learned that they can convert waste-hauling expenses into revenue streams by handling their own recyclable waste. Rocky Mountain Recycling has made that easier with an innovative "Super Sandwich Bale," in which various kinds of waste materials are collected and layered into standard cardboard compactors. These materials are then easily separated by type—cardboard, metal, or plastic—when the "sandwich" arrives at a recycling facility.

Secondary markets are another form of reverse logistics. Businesses such as liquidators and secondhand stores and community sales and flea markets deal in used or surplus goods at prices that typically are well below initial retail prices. Organizations like thrift shops collect and resell used clothing and household goods, while providing jobs for disadvantaged or special-needs people. Most garments can still be recycled at the ends of their useful lives. The majority of textile wastes can be recovered as usable fiber, and fabrics like polyester can be produced more cheaply from recycled materials than from virgin materials.

Secondary markets for building materials support yet another form of reverse logistics: building deconstruction. The ReBuilding Center in Portland, Oregon, can deconstruct many buildings for about the same cost or less than traditional demolition. Traditional building demolition involves heavy equipment and removes debris to a landfill. Alternatively, the ReBuilding Center uses skilled work crews to dismantle or "unbuild" buildings by hand, salvaging up to 85 percent of a buildings' materials for reuse.

Deconstruction is easier and more economically feasible when buildings, or any other durable products, are designed with end-of-life dismantling in mind. This technique, called Design for Disassembly (DfD), is used by numerous businesses on the forefront of cradle-to-cradle manufacturing. Herman Miller and Steelcase, both manufacturers of office furniture, employ DfD to make furniture that can be broken down easily and quickly into reusable and recyclable components. In the electronics industry, companies like Philips and Panasonic are embracing DfD with features like modular and snap-together components, and even provide end-of-life disassembly instructions.

Businesses also facilitate reverse logistics by providing customers with the product benefits while maintaining control and responsibility over the products themselves. This is achieved through leasing, renting, or substituting products with services. Tuxedo rental is familiar example. A customer can have the benefits of a tuxedo without the expenses of purchase, storage, cleaning, or obsolescence. Car sharing programs like Zipcar achieve the same ends with a much more complex and expensive product. Janitorial services can store, maintain, and replace equipment and chemicals as needed, leaving their clients only the benefit of clean facilities.

See Also: Biomimicry; Cradle-to-Cradle; Herman Miller; Remanufacturing; Recycling, Business of; Take Back; Value Chain.

Further Readings

Blumberg, Donald F. *Introduction to Management of Reverse Logistics and Closed Loop Supply Chain Processes*. Boca Raton, FL: CRC Press, 2004.

McDonough, William and Michael Braungart. *Cradle to Cradle: Remaking the Way We Make Things*. New York: North Point Press, 2002.

Vargo, Stephen L. and Robert F. Lusch. "Evolving to a New Dominant Logic for Marketing." *Journal of Marketing*, 68 (June 2004).

Zikmund, William G. and William J. Stanton. "Recycling Solid Wastes: A Channels-of-Distribution Problem." *Journal of Marketing*, 35/3:34–9 (July 1971).

John W. Schouten
Diane M. Martin
University of Portland

RIGHT TO KNOW

Although it can simply refer to public accessibility of facts from any organization, "The Right to Know" (RtK) most commonly refers to citizens' right to information regarding the extent to which they have been/might be/are being exposed to hazardous chemicals, identifying those chemicals, and the potential hazards associated with exposure to them.

A case of citizen misinformation occurred in Niagara, New York, in 1978, at the infamous Love Canal, where residents discovered that they were living close to or on a hazardous waste dump site. Part of the site's cleanup is shown here.

Source: U.S. Environmental Protection Agency

This right is guaranteed in the Emergency Planning and Community Right-to-Know Act (EPCRA) of 1986—part of the Superfund Amendments and Reauthorization Act (SARA)—which continued the policy agenda of the Comprehensive Environmental Response, Compensation, and Liability Act (CERCLA), also known as Superfund. EPCRA required state governments to oversee both the formation of local and statewide committees for emergency planning related to human exposure to hazardous waste materials. These provisions sought to increase the public's access to information about specific chemicals, how those chemicals are used, and what chemicals, if any, are released into the environment by individual facilities. EPCRA also requires individual businesses to track and report hazardous materials. Occupational Safety and Health laws also require businesses that utilize chemicals that potentially contain toxic substances to make Material Safety Data Sheets (MSDS) available to its employees and consumers.

Prior to the passage of EPCRA, there were no clear guidelines or requirements for informing the public about potential risks associated with hazardous waste production and storage. In 1984, many thousands of people in Bhopal, India, died as a result of a release of the toxic chemical methyl isocyanate from a pesticide plant owned by Union Carbide. No information had been given to those living near the facility regarding potential risks associated with this chemical, although scientists within Union Carbide had attempted to warn senior management of the possibility of such an accident. A nonfatal incident at a Union Carbide plant in West Virginia nine months later indicated that similar accidents and lack of information could occur in the United States as well. Six years earlier, residents in the Love Canal neighborhood of Niagara, New York, discovered they were living close to, and in some cases nearly on top of, a hazardous waste dump site. The city had knowingly bought this land on which to build a school and develop a neighborhood. The chemical company that the city purchased the land from admitted to dumping hazardous by-products on the land for years. However, the citizens who bought land and built houses in this neighborhood were not given any information about the potential hazards associated with the dump site. Residents maintain that they were exposed to dangerous chemicals that caused cancer, miscarriages, and birth defects. The RtK movement and associated legislation were an outgrowth of public demand for more information about toxic chemical exposure.

Under SARA, companies and federal facilities are also required to report information about toxic chemical releases to the U.S. Environmental Protection Agency (EPA), which publishes the Toxics Release Inventory (TRI). The TRI database is available to and searchable by the public. The website contains information on topics ranging from a list of chemicals included in the inventory to an analysis of the latest TRI. In 1990, the Pollution Prevention Act expanded the list of TRI chemicals, and the data now required to be gathered include waste management and source reduction activities.

Failure to thoroughly inform consumers of potential chemical hazards is an issue faced not only by chemical manufacturers, but also by green business. After research regarding the potential toxicity of bisphenol-A (BPA) came to light in 2007, Swiss manufacturer SIGG publicized information indicating that the linings of its water bottles did not leach BPA into liquids. What it failed to disclose until 2009, however, was that they did contain BPA. As a result, consumers who had viewed SIGG as a green business felt betrayed and wondered whether they could trust subsequent statements by SIGG indicating its new bottle liners did not contain BPA.

Although extensive business lobbying for reduction of toxic release information requirements is common, the public availability of this information has also encouraged voluntary self-regulation among organizations that do not want the negative publicity resulting from producing large TRI emissions. Indeed, following the publication of TRI data, many reporting firms implemented strategies to reduce the quantity of reportable chemicals emitted to the environment. Community Advisory Panels (CAPs) have been part of an outreach effort by chemical companies in a response to public demand for information and a bid to refurbish their public image. CAPs have been successful in this regard but have been criticized as merely being cooptation mechanisms. Thus, green businesses need to be sincere in their community advisory efforts. RtK and similar disclosure provisions thus provide examples of how overcoming one market failure, such as imperfect information, by requiring full disclosure can be an alternate regulatory strategy to traditional command and control regulation aimed directly at reducing another market failure—negative externalities.

The same desires that undergird the RtK movement have served to promote the broader movement of transparency. This principle was promoted by the Ceres Principles, which in turn led to the creation of the Global Reporting Initiative and is reflected in corporate practice in countless sustainability reports.

See Also: Brownfield Redevelopment; Ceres Principles; Corporate Social Responsibility; Externalities; Global Reporting Initiative; Toxics Release Inventory; Transparency.

Further Readings

Ashford, N. A. and C. C. Caldart. *Environmental Law, Policy and Economics: Reclaiming the Environmental Agenda.* Cambridge, MA: MIT Press, 2008.

Burns, R. G. and M. J. Lynch. *Environmental Crime: A Sourcebook.* New York: LFB Scholarly, 2004.

Goldsteen, J. B. *The ABCs of Environmental Regulation.* Lanham, MD: Government Institutes, 2005.

Hadden, S. G. *A Citizen's Right to Know: Risk Communication and Public Policy.* Boulder, CO: Westview Press, 1989.

U.S. Environmental Protection Agency (EPA). "Protect the Environment." http://www.epa.gov/epahome/r2k.htm (Accessed April 2009).

U.S. Environmental Protection Agency (EPA). "Toxics Release Inventory (TRI) Program." http://www.epa.gov/epahome/r2k.htm (Accessed April 2009).

Jo A. Arney
University of Wisconsin–La Crosse

S

SAINSBURY'S

J. Sainsbury, also known as Sainsbury's, is the third-largest supermarket chain in the United Kingdom and is regarded by many to be its greenest. In its consistent efforts to improve the environmental and social sustainability of its own operations, of its supply chain, and of its customer's lives and behaviors, Sainsbury's serves as a case study for green retailing.

History

The company was founded by the husband-and-wife partnership of John James and Mary Ann Sainsbury in 1869. While growth was initially confined to London and the surrounding area, by 1922, Sainsbury's was the largest UK grocery retailer. At this point, incorporation as a private company (J. Sainsbury Limited) occurred, and by the time the founder died in 1928, there were 128 shops. When the company went public in 1973, at which time all shares were held by various members of the Sainsbury family, it was the largest ever London Stock Exchange flotation. By the mid-1990s, a string of largely ineffective CEOs, beginning with the last Sainsbury family member to hold the reins, had driven the chain into decline. From its position as the United Kingdom's largest grocer, it was overtaken by Tesco in 1995 and, subsequently (upon its acquisition by Wal-Mart at the very end of the previous millennium), by Asda in 2003.

Under the guidance of Justin King, appointed CEO in 2004, a recovery program appears to be paying dividends. Meanwhile, however, takeover speculation is rife, with a number of initiatives having been tabled over the last two years. However, in a situation where 15 percent of the company's shares are still held by family members, they would need to be convinced of the rationale regarding any bid for it to be successful.

Sustainability

Despite its trials and tribulations, Sainsbury's has developed a dedicated attitude toward environmental and social issues, with the company's stated intention that it be environmentally responsible in the way it conducts its business and enable its customers to become environmentally responsible.

A focus on customer health and well-being is central to Sainsbury's efforts to becoming sustainable. This is evident in its commitment to "make the most popular items in our customers' baskets healthier, focusing on products that contribute the most saturated fat, salt or sugar to the UK diet, to deliver a real impact on the nation's health." The goal of providing customers with healthier, more environmentally sustainable products drives several initiatives:

- Stakeholder engagement—active dialogue with customers and nongovernmental organizations (NGOs) leads to the identification and prioritization of key areas of interest and constant improvement.
- Food labeling and reformulation—efforts to enhance food healthfulness include reducing sugar, salt, saturated fats, and chemical additives in processed foods; and communicating improvements to customers through a simple "Multiple Traffic Light" labeling system.
- Customer access and education—the company focuses on providing affordable versions of more sustainable foods and disseminating usage information through health-oriented "Tip Cards."
- Child nutrition and education—a "Kids" line of prepared foods focuses on complete nutrition and introduces children to "Multiple Traffic Light" nutritional labeling.
- Sourcing locally and responsibly—food products are sourced within the United Kingdom in season, and through socially and environmentally responsible sources outside the United Kingdom. Fair Trade, Marine Stewardship Council, Forest Stewardship Council, and other third-party certifications help to guide purchasing decisions.
- Supplier relations—collaboration with the supply chain includes adherence to strict codes of conduct, extensive supplier training, and regular performance monitoring.
- Sainsbury's presently remains committed to the principle of reducing the company's operational footprint and continuing to develop a better understanding of the environmental impact of its products and its buildings. The company's environmental goals for more sustainable business operations include reductions in energy use and carbon emissions, packaging, food waste, and nonfood waste.

The company's energy goal is to reduce carbon dioxide emissions 25 percent by 2012, against a 2005–06 baseline. Efforts include increasing the energy efficiency of stores and other facilities, and increasing the use of renewable power, in particular by purchasing directly from renewable energy generators. It is reducing the carbon footprint of transportation by increasing operational efficiencies and converting to alternative fuel vehicles, such as electric vans and trucks that burn biomethane from landfill capture.

Its packaging goal is to reduce the weight of its own brand packaging, relative to sales, by 33 percent by 2015 against a 2009 baseline. In addition, the company strives to make all packaging reusable, recyclable, or home compostable and, where possible, constructed from recycled materials. Specific initiatives include overall materials reduction, a dramatic reduction in single-use shopping bags, incentives and reminders for using reusable bags, and extensive recycling centers on store properties. The recycling centers include innovations like mixed-plastics collection.

Sainsbury's food waste goal is "zero food waste to landfill." Excess food that is still consumable is donated to charities. The remainder is processed, either into fertilizer, or into biomethane fuel through anaerobic digestion facilities. Other waste, with a similar "zero to landfill" objective, is reused or recycled if possible. Construction waste has achieved over 80 percent diversion from landfills. Other company waste has currently achieved a diversion rate of about 64 percent.

In areas of social responsibility, Sainsbury's has initiated numerous programs to support local communities and charities. Key initiatives include childhood obesity, food security, and community economic development. Programs are funded in part through reinvestment of funds from the sales of reusable shopping bags.

See Also: Biofuels; Clean Fuels; Corporate Social Responsibility; Green Retailing; Responsible Sourcing; Tesco; Waitrose.

Further Readings

J. Sainsbury plc. "Reports." London: J. Sainsbury, 2009. http://www.j-sainsbury.co.uk/index .asp?pageid=20 (Accessed August 2009).

Leroux, Marcus. "J. Sainsbury Pares Down Head Office Amid More UK Job Losses." *Times Online* (January 22, 2009). http://business.timesonline.co.uk/tol/business/industry_sectors/ retailing/article5563065.ece (Accessed December 2009).

McAlexander, James A. and Eric Hansen. *The Business of Sustainable Forestry Case Study: J. Sainsbury Plc and the Home Depot Retailers' Impact on Sustainability*. Washington, D.C.: Island Press, 1999.

David G. Woodward
University of Southampton School of Management

SERVICE DESIGN

A green service is one that meets the needs of consumers by providing the benefits of product use without the necessity of ownership. Rather than purchasing tangible goods to meet every need, consumers can rent or share the use of physical products, or simply receive the benefits the products would normally provide. Green service design follows a number of principles, including service-dominant logic (SDL) and dematerialization.

Scholars Stephen Vargo and Robert Lusch describe service-dominant logic as a way of focusing strategy on meeting people's needs, that is, providing them service, rather than on trying to sell them physical products. Whereas most companies take a product sales perspective, delivering benefits by selling products, green businesses may use SDL to gain competitive advantage in a variety of ways. Digital products are a prime example of SDL in that they don't need to be made into physical forms, such as compact discs, but can be downloaded directly from the source. Consumers get the benefits of product use, without the costs of purchase maintenance, repair, storage, and eventual disposal.

Dematerialization

Green service design commonly seeks to dematerialize the customer experience. Dematerialization refers to reducing the amount of material resources required to deliver a given benefit. Services are generally thought to be intangible; however, most service delivery requires significant tangible components. For example, a hotel stay, from a customer perspective, leaves few tangible traces, but the manufactured capital required to provide that stay is substantial.

Green service design accomplishes dematerialization in a couple of ways. One is simply to find ways to reduce the total amount of material designed into a product. For example, television providers may offer on-demand movies and entertainment as part of their service. Rather than hooking up separate pieces of hardware, consumers can choose to receive streaming digital content through their existing hardware without additional devices, physical media, or postage costs.

The other methods of dematerialization entail moving the locus of product usage and control farther up the value chain, centralizing and consolidating the materials required to deliver a service. This centralization may be achieved through strategies of renting or sharing, or through a complete replacement of products with services. In the cases of sharing and renting—also called use-oriented product service systems—consumers have the use of products as needed, but without the added burdens of ownership. In the latter case—referred to as result-oriented product service systems—consumers are relieved not only of the ownership and possession of products, but also of the labor needed to use them.

Renting allows consumers to save the costs, in terms of money, time, and space, of owning certain products. Financial savings may come from eliminating such costs as product replacement due to obsolescence and depreciation; maintenance, repair, and storage; and insurance, all of which are carried more efficiently by the business. Businesses benefit from efficiencies and from control over product maintenance and longevity, reducing their own costs and allowing them to add value for customers. To be successful, rental service businesses must assure that customers can use their services as easily and conveniently as if they owned the products themselves. Familiar examples of rental services include rental car companies, self-move rental truck companies, men's formalwear, and heavy home and garden equipment rentals, such as stump removers, air compressors, and generators. Innovators are taking the rental idea a bit further. For example, several communities now operate tool lending libraries, which allow patrons to borrow common household tools for a small fee. High fashion accessories, such as designer handbags, can be rented by the week or month.

Sharing durable products is increasing in popularity. Sharing programs are based on monthly or annual membership fees that allow consumers easy and frequent access to goods for a fraction of the cost of ownership. For example, Zipcar in the United States and Mobility Car Sharing in Switzerland work according to this model. Zipcar is popular with urban dwellers, who find the cost of car ownership prohibitive but enjoy the ease of occasional usage.

Results-oriented systems are designed to replace consumer goods completely, providing only the benefit of their usage. Laundry services and janitorial services, for example, provide the benefits of professional cleaning while eliminating the need for customers to own and stock related supplies, chemicals, or machinery. Such services can achieve efficiencies of resource use that are impossible at the household or small-business scale. Moreover, they can design in the use of latest developments in energy- and water-saving equipment and processes, and environmentally benign chemicals.

Green dry cleaners have designed out the use of perchloroethylene, a chemical associated with organ damage, mental deterioration, and irritation of the skin, eyes, and mucous membranes. Ecomat-Cleaners claims a 100 percent reduction of hazardous-waste emissions through multiprocess wet cleaning that uses water and steam, natural soaps and oils, and the manual targeting of stains. "This service performs as well or better than traditional dry-cleaning."

Designing Sustainability Into Services

Services have both front stage and back stage components, and green service design requires attention to both. The front stage is where the customer experiences the service delivery, such as receiving a manicure or picking up the clothing from a cleaner. At the point of delivery, green benefits may be virtually invisible to the customer. In the back stage, however, which involves all the people, processes, and products required to produce and deliver the service, ample opportunities exists to design in greater social and environmental sustainability. For example, a UPS customer may take delivery of a package and remain oblivious to UPS's sustainability initiatives, but behind the scenes the company is making real triple-bottom-line gains. UPS focuses on building human capital, both within the firm and in communities where it operates. On the environmental front, UPS has become a world leader in the use of alternative fuel trucks, efficiency routing systems, and paperless systems for tracking and billing.

Many large corporations and other institutions, such as schools, hospitals, and government agencies, provide on-site cafeteria services for internal and external stakeholders. Institutional food services are another common site for green service design. Sodexo, for example, has increasingly designed its services around foods that are organically grown, locally sourced, humanely raised, fished sustainably, and purchased through Fair Trade agreements. Green food services also design sustainability into basic processes such as dishwashing, with emphasis on reduced use of water, energy, and chemicals, and on waste controls, incorporating recycling and composting into their materials flows.

In the realm of business-to-business (B2B), many services are designed to provide backstage sustainability support for consumer-facing services or other B2B firms. For example, Shorebank Pacific provides financing specifically for sustainable business ventures. Shorebank helps clients increase triple-bottom-line benefits by bundling sustainability consulting with its financing services. The legal firm Tonkon Torp LLP offers specialized legal services to firms in the renewable energy industry. They deal with providers of wind power, biofuels, solar energy, and wave energy throughout the western United States.

Consumers and businesses are becoming increasingly aware of and concerned about the back stage processes behind basic services such as electricity. Many power companies offer their customers green energy packages that aid the development of renewable sources such as wind, solar, and biomass. For example, Portland General Electric allows customers to purchase 100 percent of their power from renewable sources, for which customers are willing to pay a premium.

See Also: Clean Fuels; Ecological Footprint; Sustainable Design; Sustainable Development; Triple Bottom Line.

Further Readings

Ballantyne, David and Richard J. Varey. "The Service-Dominant Logic and the Future of Marketing." *Journal of the Academy of Marketing Science*, 36/1 (March 2008).
Fuller, Donald A. *Sustainable Marketing: Managerial-Ecological Issues*. Thousand Oaks, CA: Sage, 1999.

Halme, Minna. *Sustainable Consumer Services: Business Solutions for Household Markets.* London: Earthscan, 2008.

Vargo, Stephen L. and Robert F. Lusch. "Evolving to a New Dominant Logic for Marketing." *Journal of Marketing*, 68:1–17 (January 2004).

John W. Schouten
Diane M. Martin
University of Portland

SEVENTH GENERATION

Seventh Generation is a leading line of natural, nontoxic household products. The brand name was drawn from the Iroquois saying, "In our every deliberation, we must consider the impact of our decisions on the next seven generations." Seventh Generation has won numerous awards for socially responsible, progressive business practices. Adhering to a model of environmental consciousness, the company aims to save natural resources, reduce pollution, and keep toxic chemicals out of the natural world. Based in Burlington, Vermont, Seventh Generation donates 10 percent of profits to nonprofit environmental, community, and health organizations.

Environmental activist Alan Newman, who founded Seventh Generation in 1988, had lofty goals. The company's first catalog touted "Products for a Healthy Planet." New York businessman Jeffrey Hollender met Newman while researching his first book, *How to Make the World a Better Place*. Hollender joined Seventh Generation in 1989 and helped create a business plan. With increased publicity for the ecopioneers and rising consumer interest in recycled and biodegradable products, the company's monthly catalog orders jumped from $1 million to over $7 million in 1990. The ensuing recession tempered growth, though, and the partnership simultaneously dissolved over conflicting visions. Newman left, and eventually founded the Magic Hat Brewing Co. in Burlington. Hollender took charge of Seventh Generation, planning to extend the enterprise's reach into mass-produced popular retail items. He felt that, instead of being limited in a niche, the business could still make a positive difference while exerting power in the mainstream marketplace. Like other expansion-minded leaders of companies that started small, Hollender believed that commercial success and environmental integrity were not mutually exclusive. He maintained that burgeoning demand for healthy products would unequivocally make the world better for everyone, especially future generations.

To avoid taking money from venture capitalists who may not have appreciated its ethical objectives, Seventh Generation offered shares of its stock to the public in 1993; however, the company bought back all outstanding stock after its stock price dropped significantly. Hollender sold the catalog business to Gaiam, and began pushing into retail. The company was committed to manufacturing safe, effective cleaning options. Hollender made plans to sell the products in conventional supermarkets, in addition to national natural grocery chains. Seventh Generation developed moral, ethical, and health-oriented rationales for appealing to consumers. "Healthier for You and the Planet" became the new tagline.

Chemicals in household cleaning products have been suspected of contributing to cancer, allergies, asthma, and other serious health problems. Seventh Generation caters to people concerned about using chemicals in their homes and willing to pay a premium to avoid these. Its brand-name products include paper towels, toilet paper, facial tissues, household cleaners, dish detergent, laundry products, diapers, baby wipes, and feminine care products. Many of these are made from vegetable-based ingredients, recycled paper, recycled plastic, or organic cotton; and they eschew chlorine bleaches, dyes, artificial fragrances, and phosphates.

A distinctive green leaf graces each Seventh Generation product. Packages feature messages from Hollender that thank consumers for purchasing from Seventh Generation and "caring enough to help make the world a healthier and safer place for this and the next seven generations." The current tagline You Are Making A Difference™ is accompanied by information on how the safe product ingredients "do not pose any chronic health risks and are safe for the environment." Dishwashing detergents explain that buying phosphate- and chlorine-free products keeps these toxins out of lakes and ponds. Boxes and bottles are seen as an appropriate venue for educational communication, and labels contain full ingredient lists with explanations of each component. Seventh Generation strives to use packaging that is recyclable and made from the highest recycled content possible.

Hollender is now known as the "Chief Inspired Protagonist" and executive chairperson of Seventh Generation. He believes strongly in harnessing the power of industry to improve the world. The civic-minded entrepreneur addresses issues pertaining to sustainable business on his blog, www.inspiredprotagonist.com. Hollender serves on the board of directors of Greenpeace USA. He has been named Social Entrepreneur CEO of the Year by *CRO* (Corporate Responsibility Officer) magazine. He has also written the books *What Matters Most: How a Small Group of Pioneers Is Teaching Social Responsibility to Big Business, and Why Big Business Is Listening*, and *Naturally Clean: The Seventh Generation Guide to Safe & Healthy, Non-Toxic Cleaning*. He lectures widely on corporate social responsibility, which, he insists, can coexist with profitability. He sees sustainable business as the standard to which all businesses should be held.

Hollender is adamant that he has adhered to Seventh Generation's original values. The company has been completely privately held since 1999, buttressed by a small pool of investors who support its long-term intentions. It audits the environmental and social performance of its manufacturing partners, and evaluates itself in an annual corporate responsibility report. Seventh Generation uses an Intranet system to help its employees track their carbon footprints and find ways to reduce their carbon impact. Workers who improve the energy efficiency of their home appliances or buy hybrid vehicles earn financial incentives. Green IT initiatives have included switching from desk PCs to laptops that meet certain environmental standards, and replacing personal printers with more efficient printers in a centralized room with vents that remove toxins from the air. Seventh Generation expected to obtain Leadership in Energy and Environmental Design (LEED) Gold certification for the interior space of its Burlington office in 2009. The company is a member of the Social Venture Network and Vermont Businesses for Social Responsibility. The *Better World Shopping Guide* has ranked Seventh Generation as the "Number One Company on Earth." Whole Foods Market invited Seventh Generation to collaborate with them as a founding member of the Whole Planet Foundation's Supplier Alliance for Microcredit, dedicated to alleviating poverty in the developing countries where products are sourced. Seventh Generation has also been given Microsoft's Pinnacle Award for Environmental Sustainability.

Seventh Generation was one of the first makers of green consumer products. Hollender sought to demonstrate that shaping a conscientious and flourishing company would create an incentive for rivals to follow suit. While challenging cleaning-industry giants like Procter & Gamble, Seventh Generation concurrently spawned competition from independent contenders in the natural cleaning products realm, such as Method; and from newly designed lines by conventional companies, such as Clorox's Green Works, that responded to mounting consumer demand. Seventh Generation products are now highly accessible and sold in megachains like Safeway and Target. After Wal-Mart embraced sustainability initiatives, Hollender agreed to begin selling Seventh Generation products to the giant retailer as well. Revenues had reached at least $120 million by 2008, when sales grew by 51 percent over the previous year. Cleaning products account for the majority of sales, followed by baby care and paper products. In an effort to build shopper allegiance, Seventh Generation Nation launched as a customer loyalty program in 2008, providing special offers, the chance to participate in research panels, and the opportunity to swap ecoconscious ideas with other members by posting on forums.

Advocates of consumer sovereignty postulate that shoppers bear the power to choose, and even control what is offered in the marketplace. However, the potential of recycled-content toilet paper to serve as a serious vehicle for social change might be limited. Deeper efforts directed at convincing all businesses and individuals to produce and consume less of everything could be more effectual. Furthermore, since "natural" products like those made by Seventh Generation usually carry higher price tags than conventional ones, this invites charges that they are elitist luxury items.

Proponents of green consumerism argue that it supplies individuals who may abstain from more radical political actions with a space for modest environmental activism. Seventh Generation also believes that achieving incremental changes in its own practices and leveraging the clout it enjoys as a high-profile business will raise people's consciousness about the impact of their purchasing decisions. It hopes to serve as a role model by being a good corporate citizen, demonstrating that it can be successful while still maintaining high environmental standards and swaying other companies to move in a greener direction.

See Also: Body Shop, The; Corporate Social Responsibility; Ecological Footprint; Extended Product Responsibility; Green Design; Green Retailing; Social Entrepreneurship; Sustainable Design.

Further Readings

BusinessWeek. "A Q&A With Seventh Generation's Jeffrey Hollender." *BusinessWeek Online* (August 10, 2009). http://www.businessweek.com/smallbiz/content/aug2009/sb2009085_100107.htm (Accessed April 2011).

Catling, Linda and Jeffrey Hollender. *How to Make the World a Better Place: 116 Ways You Can Make a Difference.* New York: W. W. Norton & Company, 1995.

Hollender, Jeffrey. *What Matters Most: How a Small Group of Pioneers Is Teaching Social Responsibility to Big Business, and Why Big Business Is Listening.* New York: Basic Books, 2004.

Hollender, Jeffrey, Geoff Davis and Meika Hollender. *Naturally Clean: The Seventh Generation Guide to Safe & Healthy, Non-Toxic Cleaning.* Gabriola Island, British Columbia, Canada: New Society Publishers, 2006.

Inc. "The Green 50: The Pioneers." *Inc.* (November 2004).

McCuan, Jess. "It's Not Easy Being Green." *Inc.* (November 2004).

Seventh Generation. "Crossroads: Reinventing the Purpose and Possibility of Business. Seventh Generation 2008 Corporate Consciousness Report." http://www.seventh

generation.com/files/assets/pdf/2008_SevGen_Corporate-Consciousness.pdf (Accessed January 2010).

Werbach, Adam. "When Sustainability Means More Than Green." *McKinsey Quarterly*, 4 (2009).

Robin O'Sullivan
University of Texas at Austin

Six Sigma

Six Sigma is a strategy used within business to manage quality. It is applied across many sectors. It aims to reduce the amount of defects that occur in products. By reducing this variability in the output of produced items, cost can be reduced or profit increased. Defect elimination is an inherently sustainable goal, as defective products result in wasted resources, wasted energy, and wasted labor.

While Six Sigma has now entered the common language in certain business and manufacturing circles, "Six Sigma" is registered by Motorola, Inc., as a trademark. Originally implemented by Motorola as a statistically based way of improving the defect rate in the manufacturing of electronic devices, Six Sigma has now become widely recognized throughout a variety of industry sectors.

What Does Six Sigma Mean?

Manufactured products have specifications that must be met in order for the product to be considered functional and useful. A product needs to fulfill a number of criteria that can often be defined numerically; for example, a nut needs a certain internal diameter and thread in order to securely fix onto a bolt. In statistics, if the "mean" represents the desired specification and the nearest specification limit can be defined as six standard deviations from the mean, then virtually all manufactured products will fit within this spectrum of quality and be considered acceptable. Any process that is said to operate with Six Sigma quality will produce a defect level of less than 3.4 defects per million opportunities.

Any quality management approach assumes that a cultural change can be achieved within the organization and commitment to achieving quality management can be gained from employees at all levels of the organization including top management.

Six Sigma and Sustainability

In order to work toward improving the sustainability of manufacturing operations, it is critical that energy and materials be conserved. When a manufacturing defect occurs, the product is unsuitable for consumption and waste is generated. While this waste may be recycled within the manufacturing system, there will be some embodied energy loss and additionally there may be some degradation of the material. By eliminating defects in manufacturing, and Six Sigma is one of the methods that can be used to accomplish this goal, materials and energy can be used more effectively.

History and Culture of Six Sigma

There is a rich history of quality management methods, and the culture of Six Sigma has grown out of approaches such as total quality management (TQM), zero defect, and quality control. What differentiates Six Sigma from previous quality management approaches is that it focuses on defined projects with specific objectives, with an emphasis on creating a human resource infrastructure within the organization of Six Sigma "black belts" and "green belts" to lead these projects. Six Sigma also seeks to integrate statistics and quantitative data into the process to ensure that the outputs are measureable, quantifiable, and verifiable, rather than making assumptions or guessing in relation to the gains that can be achieved.

One approach, which could be considered "traditional" for quality management in manufacturing operations, is for the workers on the production line to be responsible for ensuring that a consistent quality of product is produced, while analysis is performed, possibly by in-house statisticians after goods are produced, to analyze the "defect rate." Six Sigma to a large degree "professionalizes" the role of quality management, and makes this a more overt role to be played within the organization. Quality management, rather than being an afterthought, becomes integral to all operations, with quality being managed by multiple employees at all levels of the organization, rather than as an afterthought.

Six Sigma Hierarchy

As an employee's knowledge of Six Sigma practice develops, they are able to fulfill different quality management roles within their organization. The Six Sigma technique borrows from the parlance of martial arts, to define a hierarchy of skills:

- *Yellow Belt:* the lowest level of Six Sigma knowledge and applies to an employee who has a basic working knowledge of Six Sigma and who may manage small process improvements but does not lead team projects.
- *Green Belt:* a Six Sigma–trained individual who, although not working exclusively on Six Sigma projects, has duties that include implementing Six Sigma methodologies in team projects.
- *Black Belt:* an employee who has been trained and certified in its methods. As a result, all of the duties of this individual include implementing Six Sigma at a variety of levels within the business. They will commonly lead Six Sigma project teams and provide mentoring to lesser belts.
- *Master Black Belt:* the highest level of expertise, where all duties revolve around the application of Six Sigma and also training black belts.

Six Sigma Methodologies

In approaching a Six Sigma project, one of two methodologies are used: DMAIC (Define-Measure-Analyze-Improve-Control) or DMADV (Define-Measure-Analyze-Design-Verify), also known as Design for Six Sigma (DFSS).

DMAIC stands for:

- Define the goals of the project.
- Measure the current process and collect relevant data on critical control points.
- Analyze the data for causality and relationships between different variables.

- Improve or optimize the process.
- Control the process to ensure that any deviation is corrected before it results in a defect.

DMADV stands for:

- Define the goals of the project.
- Measure the current process and identify things that are CTQs (Critical To Quality).
- Analyze these CTQs to develop alternatives. Create a top-level design and evaluate design capability in order to select the best.
- Design the small details, configure the design in an optimum manner, and simulate the design where appropriate. Plan for design verification.
- Verify that the design works as expected, that the setup pilot runs, and then implement the production processes to produce the design.

Criticism of Six Sigma

Six Sigma has been criticized for effectively being a "rebranding" of existing quality management techniques and methodologies and lacking in originality. Indeed, Six Sigma has been criticized for being oversold as a concept. Furthermore, statistically, Six Sigma methods only apply to products whose specification outputs conform to "normal" distribution curves. If there is a statistical reason why the output specifications of a process might not fit a normal curve, then Six Sigma may not be a suitable process for quality management.

See Also: Leadership in Green Business; Sustainability.

Further Readings

Brussee, Warren. *Statistics for Six Sigma Made Easy*. New York: McGraw-Hill 2004.
Eckes, George. *Six Sigma For Everyone*. Hoboken, NJ: John Wiley & Sons, 2003.
Pande, Peter S., Robert P. Neuman and Roland R. Cavanagh. *The Six Sigma Way: How GE, Motorola, and Other Top Companies Are Honing Their Performance*. New York: McGraw-Hill, 2000.
Pyzdek, Thomas. *The Six Sigma Handbook: The Complete Guide for Greenbelts, Blackbelts, and Managers at All Levels*. New York: McGraw-Hill, 2003.
Yang, Kai and Basem El-Haik. *Design for Six Sigma: A Roadmap for Product Development*. New York: McGraw-Hill Professional, 2008.

Gavin Harper
Cardiff University

Smart Energy

Smart energy refers to the application of a broad range of new information and communications technologies to provide a more reliable and more efficient electric grid. The product of these efforts is commonly referred to as the "smart grid." As currently envisioned, the transition to a smart grid system will require a dramatic shift from our

Strategically placed commercial wind generators use wind to supplement or substitute for generated power.

Source: U.S. Environmental Protection Agency

long-standing "centralized, producer-controlled network to one that is less centralized and more consumer-interactive," according to the U.S. Department of Energy (DOE). The process is likely to draw from many of the same philosophies, concepts, and technologies that enabled the Internet.

The development of the smart grid will be a vastly complex undertaking, as stated by K. Fehrenbacher, requiring intelligent and efficient communications over a shared, interoperable network that, in its current form, includes 14,000 transmission substations, 4,500 large distribution substations, and 3,000 public and private owners. In the future, the electric grid is likely to become even more complex as households and businesses increasingly invest in small-scale, renewable energy technologies that allow for on-site energy generation. In order to successfully manage such a highly complex system, the DOE reports that the smart grid will rely heavily on five fundamental technologies:

1. *Integrated Communications Technologies:* a communications system that will connect components to open architecture for real-time information and control, allowing every part of the grid to both "talk" and "listen"

2. *Sensing and Measurement Technologies:* an automated system that will support faster and more accurate responses including remote monitoring, time-of-use pricing, and demand-side management

3. *Advanced Components:* a system of transmission and storage components that apply the latest research in superconductivity, storage, power electronics, and diagnostics

4. *Advanced Control Methods:* a monitoring system that will assess the status of essential components, enabling rapid diagnosis and precise solutions appropriate to any event

5. *Improved Interfaces and Decision Support:* a system of tools to facilitate human decision making and provide grid operators and managers the ability to see into their systems

Smart metering technology and the broader advanced metering infrastructure will provide the basic framework on which the smart grid will rely for timely information generation, information sharing, and energy management. Smart meters typically employ digital technologies and real-time sensors to measure energy consumption, monitor power quality, and provide two-way communications between the utility and energy consumers (and/or their appliances). According to the U.S. Federal Energy Regulatory Commission (FERC), advanced metering systems include systems that

- provide measurements of customer energy consumption (and potentially other data) on an hourly basis or with greater frequency and

- transmit that information over a communications network to a central collection point on a daily basis or with greater frequency.

Most automatic meter reading (AMR) devices have time-of-use (TOU) and other flexible pricing capability, as well as remote theft and tamper detection. Although not often deployed, AMR systems can also broadcast whole-home Radio Frequency (RF) signals every 30 seconds to two minutes. Compared to AMR devices, more recent solid-state Advanced Metering Infrastructure (AMI) technologies can record and broadcast data as frequently as every 7 seconds. In addition, AMI technologies provide two-way communication between the meter and the utility. AMI also provides the potential for engaging consumers directly in energy management by providing people with timely and meaningful information about their energy consumption patterns and practices. AMI data can be shared with residential and commercial energy customers in one of two ways: the information can be communicated directly to energy consumers through in-home feedback devices (or via the web), or by sharing the data and information with third-party service providers. In some limited cases, current AMI systems are providing communication to the consumer via short-distance wireless, broadband, cell phone, short-range radio, and power line networks.

According to survey data collected by FERC, there were, as of the end of 2008, approximately 130 million electric, gas, and water meters in the United States, including 27 million smart meters (both AMR and AMI). Roughly 28 percent of these smart meters are being used to their full potential. And even though a growing number of utilities around the world are committing to the use of smart meters, only an estimated 6 percent of U.S. utilities and 6 percent European utilities reported using AMI meters. Although current penetration levels are relatively small, the use of AMI meters in the United States is up significantly from 2006, when penetration was less than 1 percent. In the next five years, the number of AMI meters is expected to expand significantly—to nearly 90 percent in North America and 41 percent in Europe. In fact, the European Union has targeted 80 percent smart meter coverage by 2020 as part of its "20-20-20" climate change initiative.

AMI Functions and Benefits

One of the largest potential benefits of the smart grid is its ability to reduce large fluctuations in electricity demand. Of particular concern are those periods during the day, week, and year when electricity demand is particularly high. During these periods, electricity demand may threaten to exceed actual power generating capacity and could potentially result in blackouts. The complexities of ensuring an adequate supply of electricity have become ever more challenging for utilities as a greater number of households and businesses have come to rely more heavily on air conditioning which places a large, weather-specific demand on utilities. The smart grid can help utilities to better manage electricity demand in a number of ways:

- By facilitating the establishment of time-of-use electricity rates that better reflect the actual costs of energy generation and allow utilities to charge consumers more during times of heightened demand
- By communicating peak electricity rates to electricity consumers so as to encourage consumers to shift their electricity consumption to nonpeak periods

- By enabling the automation of peak load demand reductions through the use of new "smart appliances" and utility-controlled switching devices that enable direct load control

In addition to peak load shifting, advanced meters can also allow utilities to collect electricity consumption data more frequently. Utilities have also reported plans to use the new meters to enhance customer service and improve overall system management (e.g., to detect outages, reduce peak demand, and detect theft). Nevertheless, few utilities have taken the necessary actions to extend the benefits of AMI to consumers in a way that could allow for more direct control over their energy consumption practices. To date, the implementation of AMI technologies has strongly benefited utilities. Nevertheless, a few pilot programs are currently exploring the use of AMI technologies in conjunction with Programmable Communicating Thermostats (PCTs), Home Area Network (HAN) components, and/or In-Home Displays (IHD) of energy consumption. These types of programs offer the potential of reducing overall levels of electricity demand as opposed to simply reducing peak demand.

Home Energy Reports, In-Home Displays, and Feedback Devices

Despite the commitment of many (if not most) of the more-developed countries of the world to the development of a new smart grid infrastructure, most initiatives have focused almost exclusively on the large-scale, physical, and technological aspects of the undertaking, and have largely ignored the smaller-scale and social dimensions. In-home energy displays and feedback devices provide the potential of extending the benefits of smart grid technologies either directly or indirectly to energy consumers.

In order to maximize both energy savings and consumer benefits, new smart meter technologies must be accompanied by in-home feedback devices or well-designed home energy reports. A growing number of businesses have begun to focus on developing the appropriate means of extending the benefits of the smart grid to consumers. Providing feedback to energy consumers has taken several different forms, including (1) third-party presentation of energy use data; (2) third-party installation of home networking systems; and (3) do-it-yourself feedback devices. For example, businesses like OPOWER, in the United States, are working with utilities to provide consumers with user-friendly, targeted information about household energy consumption practices and patterns and to provide useful suggestions regarding effective means of reducing energy consumption. Other initiatives by Google, Efficiency 2.0, and others seek to use Internet-based technologies to provide consumers with similar information online. In addition, an increasing number of businesses are developing home networking systems and do-it-yourself feedback devices that allow consumers to view real-time or near-real-time measures of household energy consumption. Moreover, newer feedback technologies are moving toward providing consumers with more detailed information showing energy consumption by end use (such as heating or cooling) or even by so-called smart appliances.

Smart Appliances

In addition to smart meters, smart infrastructure, and feedback devices, appliance manufacturers have been working to develop new appliance technologies that will be able to communicate with the smart grid. Clothes washers, dryers, and dishwashers are among the major appliances that currently consume a large share of household energy. Companies such as General Electric and Whirlpool are working to develop smarter appliances that will be able to receive and respond to electricity pricing information. These new appliances will contain computer chips that can sense when the transmission system is stressed and will be able to cycle off for brief periods of time. According to one estimate, "energy saving appliances could translate into savings of about $70 billion in new power plant construction and power distribution costs over 20 years," as stated by B. Nelson.

Vehicle-to-Grid (V2G) Technologies

The redundancy, resilience, and flexibility of the smart grid may also provide the means of enabling the development and use of vehicle-to-grid (V2G) technologies in which plug-in vehicles will both draw energy from, and contribute energy to, the electricity grid. New electric and hybrid electric vehicles can be plugged in at night (when electricity is cheap) to draw electricity from the grid and recharge their batteries. During the day (when electricity is expensive), the same vehicles can be plugged in to provide extra power to the grid during times of high demand. As such, V2G technologies help benefit utilities by helping to overcome electricity supply constraints associated with times of peak demand that occur during the day when industrial plants, commercial enterprises, and air conditioner use result in a maximum demand for electricity. In fact, new electric passenger vehicles may store enough energy to power several homes for hours, and utilities may be interested in paying vehicle owners to plug-in their vehicles in order to supply needed energy. When combined with microgrid technologies, the potential benefits are likely to be enormous, including the possibility of transportation options that are emission free.

Distributed Generation, Microgrids, and Renewable Energy

As opposed to the current electricity system where energy flows in one direction (from the utility to the consumer) and information exchange is limited to consumption and billing, the Smart Grid provides the means for multidirectional energy flows, as well as the development of more complex energy information systems. As a greater number of homes, offices, and other buildings become connected to the Smart Grid, everyday buildings could potentially be transformed into energy power plants. Whether through the use of photovoltaic systems, mini wind turbines, or other mechanisms, houses and commercial buildings will have the opportunity to produce as well as consume electricity.

Such a transformation would have enormous implications for energy efficiency, carbon emissions, and the role of households and utilities in the energy system. Large efficiency gains would result from a microgrid system in which small-scale electricity production and consumption occur in close proximity, significantly reducing transmission-related energy losses. Of course, this type of system would also greatly expand renewable sources of electricity production and reduce the level of carbon emissions associated with electricity generation. Moreover, the implementation of microgrid generation systems offers households the possibility of becoming energy suppliers by selling surplus energy back to the grid. According to a recent article in *Fast Company*, "The microgrid is all about consumer control—aligning monetary incentives, with the help of information technology, to make renewables and efficiency pay off for the average homeowner, commercial developer, or even a town." Ultimately, microgrid systems provide the opportunity for an electricity system that is redundant, resilient, and secure.

See Also: Carbon Footprint; Clean Fuels; Distributed Energy; Energy Performance Contracting; Energy Service Company; General Electric (GE).

Further Readings

Ehrhardt-Martinez, Karen, Kat Donnelly, John A. "Skip" Laitner, Dan York, Jacob Talbot and Katherine Friedrich. "Advanced Metering Initiatives and Residential Feedback Programs: A Meta-Review for Economy-Wide Electricity-Saving Opportunities." Washington, D.C.: American Council for an Energy-Efficient Economy, 2010.

Federal Energy Regulatory Commission. "2008 Assessment of Demand Response and Advanced Metering." Washington, D.C.: Federal Energy Regulatory Commission, 2008. http://www.ferc.gov/legal/staff-reports/12-08-demand-response.pdf (Accessed January 2010).

Fehrenbacher, Katie. "FAQ: Smart Grid [Electronic Version]." Earth2Tech, 2009. http://earth2tech.com/2009/01/26/faq-smart-grid (Accessed January 2010).

Nelson, Bryn. "Smart Appliances Learning to Save Power Grid." November 26, 2007. http://www.msnbc.msn.com/id/21760974 (Accessed January 2010).

U.S. Department of Energy. "The Smart Grid: An Introduction." Prepared for the U.S. Department of Energy by Litos Strategic Communications. http://www.oe.energy.gov/DocumentsandMedia/DOE_SG_Book_Single_Pages(1).pdf (Accessed January 2010).

Karen Ehrhardt-Martinez
Human Dimensions Research

SMART GROWTH

Smart growth is a holistic urban planning philosophy that aims to reshape cities in a more environmentally sustainable and community-driven manner. Often considered the antidote to sprawling new suburban development, advocates of smart growth favor high-density development within a city's existing footprint.

Policies supporting suburban development and single family home ownership has led to the sprawling growth of urban, suburban, and exurban communities. Each new development requires further expansion of public services and longer commutes, proportionately expanding the environmental degradation that results from such activities. Smart growth has evolved over the last 30 years to combat this trend by promoting the following elements:

- Building a unique sense of place and community around the existing city
- Involving citizens in transparent, consistent development decisions
- Filling in neighborhoods to create mixed-use communities
- Providing a range of integrated living options for households of mixed incomes
- Promoting a broader range of transportation options, including walking and public mass transit
- Preserving open space and allowing controlled citizen access

These elements can be built into a community's master planning process, and then regulated by policy mechanisms, such as zoning laws, environmental impact assessment requirements, and tax credits. Proponents of this school of thought believe that their communities can significantly reduce the environmental impact of growth while fostering more physically healthy citizens who have deeper ties with their neighbors. Opposing arguments vary, but generally include at least one of the following: that smart growth elements do not actually reduce costs or pollution from public services; that it is simply not a viable option for most American cities given the desires of their residents; or that any economic growth will only continue to degrade the natural environment and ultimately cause much social suffering.

As with many broad philosophies or schools of thought, the concepts within smart growth cannot be attributed to a single person. Controls of growth were introduced in the 1960s, and broader growth management strategies developed throughout the 1970s and 1980s. These changes, and the smart growth movement that resulted, signified a more nuanced understanding of growth, as well as a growing environmental ethic and concern of the cost of public services. Suburbs were becoming exurbs. Cities were not just growing outward, but were abandoning their historic, center neighborhoods and re-creating the structures of communities on the borders of town. Driving a personal automobile became necessary for every daily task, and citizens began having less interaction with their neighbors. Smart growth ideas developed to promote long-term thinking, including the three pillars of sustainability—society, environment, and finances—over the short-term economic decisions that had led to these problems.

In the United States, architects Peter Calthorpe and Andres Duany, founding members of the Congress for the New Urbanism in 1993, were two of the primary advocates of redesigning urban planning thinking. Calthorpe spoke of "urban villages," and Duany supported using design codes to encourage a sense of community, while both emphasized alternative transportation options. Urban planner Travis McPhee is thought to have coined the term *smart growth*.

By 1997, smart growth had become well favored in urban planning circles. The U.S. Environmental Protection Agency (EPA) was funding advocacy groups and two major reports were released: the American Planning Association's *Growing Smart Legislative Guidebook* and the Natural Resources Defense Council's "Tool Kit for Smart Growth." Later that year, the state of Maryland adopted the Smart Growth and Neighborhood Conservation Act, a model of state legislation encouraging changes in development practices.

Elements of Smart Growth

There are many different lists of principles or elements of the smart growth philosophy, but the six broad concepts covered here incorporate the most common features of each of these lists. While many of the elements prescribe very specific traits, the first is a fundamental, overarching goal: to create and maintain special places with vibrant communities. Following this edict requires starting with the existing community as a building block ready for improvement or enhancement. Historic buildings should be preserved and actively utilized as a characteristic seed for future growth that would enhance rather than ignore the locale's past.

Maintaining and featuring historic buildings can contribute to the development of a local community identity. For example, a community could be known for great antique shops, a vibrant artists' community, or an active sports following. Those towns and cities have developed a brand identity benefit for their citizens' pride and interest in the future of their hometown. Community members are more likely to attend master planning forums and to think about the long-term interests of the community as a whole, which are fundamental features of a fully functioning smart growth community.

Starting with the existing city footprint as a basis for future development requires an increase in density of population and civic activity to allow future growth. Adding increased residential, office, and retail capacity to existing, often historically neglected, neighborhoods as a means of increasing density and revitalizing the community has been termed *urban infill*, and is a common theme of smart growth development plans. In other words, smart growth is to "grow up, not out." This reduces the environmental footprint of new growth by avoiding the otherwise required clearing of new land and expansion of utilities and other public services. Additionally, denser development allows for shorter commutes and thus decreased vehicular pollution.

The second overarching element of smart growth is public involvement in community planning and development of transparent, predictable development rules. Public involvement can take the form of community visioning forums followed by draft master plans and series of public feedback forums before final plans are issued. While the process can be onerous, urban planning consultancies have been built around guiding communities through the process. Deeper public understanding and support means that smart growth plans will be more likely to achieve their goals, since many of those goals are dependent on individual choices—choosing to live in dense development and to drive personal vehicles less, for example. Furthermore, broadly supported plans are more likely to last over time.

After the community plan has been created, smart growth advocates argue that development processes and regulations must be laid down in a predictable and fair method. Guidelines must be spelled out very specifically to allow standardized application for all proposals. Predictable rulings allow developers and private citizens to make decisions with a longer time horizon, encouraging more sustainable outcomes. Also, predictable systems are more cost effective for the municipal authorities and allow more equitable access to the decision making process, just as collaborative public involvement in creating a city's plan can break down barriers and bring the community together. Access and collaboration are critical goals of towns that seek more integrated and inclusive sense of community.

The third element of smart growth planning is the development of mixed-use neighborhoods. The traditional development pattern created large tracts of single-family detached homes in one area, job sites such as office buildings in another area, and amenities such as grocery stores, banks, and post offices in a third area. This required driving several places throughout the day, contributing to traffic and air pollution. Smart growth prescribes integrating these areas. For example, a family's condominium could be just down the street from the parents' office building, and the retail stores that they visit on a daily or weekly basis could be on the first floor of each building. The children's school could be just around the corner, so the family could walk there together each morning. On that walk, they would meet their neighbors and develop a sense of community. They would get a little more exercise from walking around the area. They can avoid the cost of putting a 20-mile commute on the car each day, and the parents would not commute long distances, gaining an hour a day with their children.

The fourth element of smart growth, dovetailing with mixed-use developments, is mixed-income housing developments. Developing a variety of housing options in a single area allows for a more viable workforce to live within a short distance of their jobs, whatever type of job it may be. For example, a neighborhood could have a gas station, a grocery store, a variety of retail options, and several office buildings. Creating mixed-use and mixed-housing developments within the existing city footprint also helps alleviate gentrification, a problem resulting when wealthier families move into a previously declining neighborhood. Over time, as the neighborhood becomes more desirable and property values rise, the previous tenants are forced out of the area. The historic culture of the area is lost and there is a great sense of injustice. As a frequent problem, often occurring with downtown revitalization and urban infill efforts, it is important for smart growth advocates to carefully plan strategies to avoid gentrification.

Proponents of smart growth strategies also argue that mixed-income housing contributes to breaking down the persistent racial segregation in many communities. Building centralized, diverse neighborhoods avoids the current problem of some neighborhoods having scarce access to public services and amenities, such as affordable and fresh groceries. While these new neighborhoods make daily tasks more convenient for some community members, they may provide first-time access for other citizens. Proponents argue that this integration is fundamental to the long-term social sustainability of a community.

The fifth element of smart growth is planning transit-oriented development: maximizing access to public transportation while maintaining communities that are safe to walk and bike around. There are many tools available to city planners to achieve this goal:

- Ample bus and train service
- Increased parking fees
- Road use fees for congested areas, such as those implemented in London
- Bike lanes, trail systems, and parking
- Sufficient and safe pedestrian crossings
- Bike rental programs
- Enforcement of driving codes to keep streets safe for pedestrians and cyclists

Incorporating these tools allows for more dense development by alleviating the need for expansive parking lots and six-lane roads. Additionally, the city avoids the cost of road maintenance; the consumer avoids the cost of car maintenance due to a long daily commute; and the many citizens will get more exercise.

The sixth and final element of smart growth is to preserve existing natural elements. This could be in the form of greenbelts, farmland, critical habitats, or urban parks. Redeveloping the already-built areas of a community rather than expanding into virgin land will directly help preserve existing green space, but intentionally planning dense neighborhoods to allow for adequate park space is also important. Community members who are aware of nature's unique beauty are more likely to support smart growth principles and to feel pride in their local area. Furthermore, with more dense development, parks become even more important, as fewer families have their own yards and outdoor space.

Policy Tools

Community planners have a wide array of policies available to promote the elements of smart growth, ranging from incentives to regulations. The most successful plans are comprehensive mixtures of policy tools, based on the local political climate, and combined with in-depth public engagement. Potential tools include the following:

- Zoning ordinances specifying what level of development is allowed in an area, but to be successful planning tools ordinances must be maintained in a predictable manner. In many communities, so many exceptions are made that it makes planning moot.
- Grants, loans, and tax incentives provide government funds to support specific projects that contribute to the smart growth plan. These are particularly helpful options at the beginning of a revitalization process, during which it can be difficult to find financially viable development projects to foster future growth.
- Environmental impact statement requirements ensure that new growth meets the community's standards.
- Urban development boundaries and development zones specify that only development within the central zone will receive government-funded public services, such as sewer and trash pickup, thereby creating a disincentive to develop on virgin land outside the zone.
- Convening state and regional planning commissions to ensure that one city's smart growth planning is not simply shifting new growth to another location nearby, often directly to the virgin land that the policies are meant to protect.
- Land acquisition, either by the government, or private trusts. While costly, land banking is the firmest control of growth for a sensitive piece of land. Purchasing development rights or easements may be a more cost effective option.
- Allowing density transfers gives landowners more control over their land while maintaining the overall development goals. If a landowner held a parcel that was allowed 50 lots in one area and 50 in another area, the landowner could be allowed to transfer the density—resulting in one area with 100 developed lots and the second area left entirely undeveloped.
- Some communities have decided to simply specify a growth rate. For example, they may only approve 100 new residential development projects each year.

Arlington, Virginia

Arlington, Virginia, has been recognized by the EPA and a variety of advocacy groups for the success of their smart growth–style planning over the last 30 years. As a bustling suburb of Washington, D.C., Arlington's plans are centered around five Metro (subway) stations. This area is referred to as the "Rosslyn-Ballston corridor." Arlington's General Land Use Plan (GLUP) helps concentrate dense, mixed-use development within a walkable radius of the each station—then tapers out to traditional residential neighborhoods. Sector plans are developed to maintain a distinct sense of place at each station, including preserving the feel of historic neighborhoods and existing natural areas.

Smart growth planning has been a part of Arlington for many years. Having already debated the merits of growth and coinciding use of public transportation in the 1970s, Arlington's county leaders took a strong stance advocating that their new subway line should run through their existing commercial corridor, rather than along a newly proposed highway. This corridor of transit and commerce formed the basis of the GLUP, and thus the foundation of their smart growth implementation. The vast majority of Arlington's land is under zoning ordinances dictating very-low-density development. This maintains the historic residential neighborhoods, and prevents unplanned development of new commercial centers. When a new development project is proposed, a thorough site plan must be submitted. If the plan is in accordance with the GLUP, including increased public benefits like small parks, then the specified land can be rezoned to allow much taller and dense development.

In addition to small parks, the GLUP calls for safe pedestrian access, public art, wide sidewalks with restaurant seating, street trees, bike lanes, first floor retail, and traffic calming. Although such a plan could be criticized for limiting development, Arlington has enjoyed broad public support through citizen involvement in 40 County Commissions and 60 neighborhood civic associations. As a result of this planning and engagement, Arlington has succeeded in maintaining car ownership rates below the national average and significantly below the rates in neighboring counties. The rates of Metro ridership and walking to work are similarly impressive, particularly when compared to nearby communities such as Fairfax, Virginia. With a very high quality of life and some of the lowest taxes in the area, people have been eager to move there. Maintaining an adequate base of affordable housing has been one of the biggest challenges, but they have taken measures to alleviate the shortage, such as allowing qualifying housing developments a density bonus of 25 percent.

Criticisms of Smart Growth

Critics of smart growth philosophies are also numerous, and their leaders range from libertarians to staunch environmental protection advocates. The most basic argument, often voiced by Wendell Cox and Joshua Utt, is that smart growth tactics simply do not do what they report to do: that denser development does not reduce city expenditures per capita. Cox and Utt's statistical studies of urban growth over the last 50 years are used to support this concept. Automobile interests often support this line of argument, since reducing demand for their product is a core tenant of smart growth.

A second set of arguments centers around quality of life, and justice or equity. Some opponents contend that smart growth advocates are trying to force community members to live a certain lifestyle, one that is very different from the American dream of a single-family home with a white picket fence. This is an image that many citizens hold very dear, so government policies that encourage abandoning that concept as the ideal can be met with firm opposition. This idealized concept is also an impediment to the implementation of smart growth policies in some communities. Building mixed-income housing in a dense neighborhood will not achieve the goals of smart growth if anyone with the means to do so is only buying single-family, detached homes. Conversely, if the planners are able to revitalize an existing neighborhood and create a high demand for local housing, the existing community members can be forced out of the area due to rising housing costs and taxes. This potential for gentrification creates local opposition at the beginning of many proposed smart growth processes.

The final major criticism of smart growth comes from those who believe that no growth is "smart," that the land has already been overdeveloped. Advocates from this line of thought propose a "no growth" strategy and argue that a steady state economy is the only way to avoid massive environmental and human health catastrophes.

See Also: Environmental Impact Statement; Steady State Economy; Sustainability; Sustainable Development.

Further Readings

Brosnan, Robert. "30 Years of Smart Growth." http://www.arlingtonva.us/departments/CPHD/planning/powerpoint/rbpresentation/rbpresentation_060107.pdf (Accessed September 2009).

Duany, Andres, Jeff Speck and Mike Lydon. *The Smart Growth Manual*. New York: McGraw-Hill, 2009.

Dunham-Jones, Ellen and June Williamson. *Retrofitting Suburbia: Urban Design Solutions for Redesigning Suburbs*. Hoboken, NJ: John Wiley & Sons, 2008.

National Audubon Society. "Smart Growth: History, Tools, and Challenges." http://www.audubon.org/campaign/er/library/smart-growth.html#1.1 (Accessed September 2009).

O'Toole, Randy. "The Folly of Smart Growth." http://www.cato.org/pubs/regulation/regv24n3/otoole.pdf (Accessed September 2009).

Smart Growth Network. "About Smart Growth." http://www.smartgrowth.org/about/default.asp?res=1280 (Accessed September 2009).

U.S. Environmental Protection Agency. "About Smart Growth." http://www.epa.gov/dced/about_sg.htm (Accessed September 2009).

U.S. Environmental Protection Agency. "Arlington County, Virginia." http://www.epa.gov/dced/arlington.htm (Accessed September 2009).

Claire Roby
Independent Scholar

SOCIAL ENTREPRENEURSHIP

Social entrepreneurship refers to the use of entrepreneurial methods, such as risk taking, innovation, and team building, to bring about positive social change, typically with extremely scarce resources. The concept of social entrepreneurship has emerged in the private, public, and nonprofit sectors over the last decade, and interest in social entrepreneurship continues to grow. The nonprofit sector—the conventional locus for social change—faces diminishing funding from traditional sources and increased competition for these scarce resources. At the same time, concerns about the concentration of wealth in the private sector have spurred calls for corporate social responsibility and more proactive responses to complex social problems. Meanwhile governments at all levels grapple with multiple demands on public funds.

In light of this, social entrepreneurship has emerged as an innovative approach for dealing with complex social needs. With its emphasis on problem solving and social innovation, social entrepreneurial activities blur the traditional boundaries between the public, private, and nonprofit sectors and emphasize hybrid models of for-profit and nonprofit activities. Within this context, there has been a marked increase in collaboration across sectors. Out of this, a new generation of "social entrepreneurs" has rapidly emerged, driven by innovators increasingly committed to using market-based approaches to solve social problems.

Defining Social Entrepreneurship

Social entrepreneurship is an inherently complex concept about which consensus has been difficult to achieve. While the ideas fueling social entrepreneurship are not new—Victorian private hospitals and the hospice movement are both cited as historic examples of social entrepreneurship—the term as it is used in the academic and popular literature currently encompasses a rather broad range of activities and initiatives.

In spite of the varying definitions of social entrepreneurship, two commonalities emerge in almost every description. The "problem-solving nature" of social entrepreneurship is prominent, as is the corresponding emphasis on initiatives that produce measurable results in the form of changed social outcomes and/or impacts. The main definitional debates are over the locus of social entrepreneurship. Many define it as bringing business expertise and market-based skills to the nonprofit sector in order to help this sector become more efficient providers and deliverers of these services. This category includes social enterprise activities, in which nonprofit organizations develop small, for-profit businesses and channel earnings back into their organizational budgets. It can also include the adoption of private sector management techniques within nonprofit organizations to maximize resource productivity. Some distinguish between for-profit activities that attempt to offset an organization's costs and "social purpose ventures" in which the primary purpose is to make a profit, which can then be used to fund nonprofit ventures.

Others define social entrepreneurship more broadly and argue that social entrepreneurship can occur within the public, private, or nonprofit sectors and is, in essence, a hybrid model involving both for-profit and nonprofit activities as well as cross-sector collaboration. These definitions tend to put more emphasis on the "entrepreneurial" nature of these activities and the creativity and innovation that entrepreneurs bring to solving social problems in unique ways rather than focusing on the social benefits such services can provide. This conceptualization suggests social entrepreneurship can take a variety of forms, including innovative not-for-profit ventures and social purpose business ventures (e.g., for-profit community development banks, and hybrid organizations mixing for-profit and not-for-profit activities). Just as not every new business venture qualifies as "entrepreneurial," not every social venture qualifies as "socially entrepreneurial." This does not depreciate the impact of nonentrepreneurial social initiatives in any way, but merely makes the distinction clear, given the broad use of terminology around social entrepreneurship.

Models and Examples of Social Entrepreneurship

Models of social entrepreneurship in the nonprofit sector abound. As Paul Hawken notes in his book *Blessed Unrest*, it is virtually impossible to count the numbers of social and environmental nonprofit organizations in the world, and equally difficult to keep up with the numbers of new organizations springing up every year. The act of establishing most nonprofit ventures is inherently entrepreneurial, requiring risk taking, innovation, and team building with scarce resources. Among nonprofits, certain organizations such as the Skoll Foundation and its online community, Social Edge, function as accelerators of social entrepreneurship. Created by eBay pioneer Jeff Skoll, the foundation supports social entrepreneurs with awards, grants, recognition, networking, and media exposure.

The microlending industry provides an excellent example of boundary-blurring hybrid social ventures. Microlending was made famous by Nobel laureate Muhammad Yunus and his Grameen Bank in Bangladesh. Specializing in low-interest, uncollateralized loans to the very poor, Grameen Bank reports a high rate of repayment and a high rate of success in helping people raise themselves from abject poverty through entrepreneurship. Innovating and adapting the Grameen model, Matt Flannery, cofounder of Kiva, created a model of microlending that replaces the bank with an online platform connecting individual lenders with individual borrowers. Personal loans to poor entrepreneurs in impoverished nations start for as little as $25 and are used to purchase capital, such as sewing machines or breeding livestock.

From a more purely corporate standpoint, social entrepreneurship has many faces. A common model is for companies to encourage social entrepreneurship among their employees with benefits like release time from work, matching funds for donations and special recognition. For example, in the United Kingdom, Sainsbury's Local Heroes program identifies extraordinary social achievements by employees, honors them at award banquets, and rewards them with educational trips, such as a South Africa trip to experience Fair Trade winemaking.

Another corporate model is to connect customers with causes through guided corporate giving. NAU, a Portland, Oregon–based clothing company, donates 2 percent of the revenue from every sale to one of five Partners for Change, and asks the customer to choose which one to support with each purchase. Partners for Change currently include Mercy Corps, Kiva, Ashoka, Ecotrust, and Breakthrough Institute, most of which also exemplify social entrepreneurship.

Characteristics of Social Entrepreneurs

Interestingly, while many definitions of social entrepreneurship emphasize the "social" rather than the entrepreneurial nature of the activity (e.g., by focusing on nonprofit organizations and their activities), much of the literature on social entrepreneurs emphasizes the "entrepreneurial" characteristics of such individuals. They are often compared to business entrepreneurs with a social mission. Much like their economic counterparts, social entrepreneurs typically do not allow the lack of initial resources to limit their options. In addition, many social entrepreneurs share a strong desire to be in control of their environment with their economic counterparts, the urge to experiment, and a higher-than-average tolerance for uncertainty. Social and economic entrepreneurs also share the same focus on vision and opportunity and the same ability to convince and empower others to help them turn these visions into reality. For social entrepreneurs, however, these characteristics are coupled with a strong desire for social justice.

Somewhat unlike their economic counterparts, social entrepreneurs emerge not only as highly entrepreneurial individuals, but also highly collaborative ones. The ability to develop a network of relationships and contacts is a hallmark of visionary social entrepreneurs, as is the ability to communicate an inspiring vision in order to recruit and motivate staff, partners, and volunteers.

On a final note, it is worth making the distinction between socially entrepreneurial leaders and managers—the former critical as the catalysts for entrepreneurial projects, the latter critical as the individuals who see these initiatives through to fruition. In some cases, one individual will fulfill both roles; in other cases, different individuals will be required for each role.

Issues in Social Entrepreneurship

The majority of social entrepreneurship activities are premised on one form or another of cross-sector collaboration. While there is a lot of support in principle for this, the implementation of collaborative partnerships is much more difficult to achieve. Cross-sector collaborations require not only agreement about basic procedures, but also harmonization of evaluation measures for organizational "success." Finding appropriate performance evaluation measures is critical. For many organizations undertaking socially entrepreneurial activities, there are really two bottom lines to meet—one for economic profit and one for the social mission—and the difficulty of meeting these simultaneously should not be underestimated.

These differences in organizational cultures, while significant in themselves, reflect a much deeper problem for cross-sector collaborations—the lack of a common discursive framework among the public, private, and nonprofit sectors. As such, the challenges of collaboration among these sectors should also not be underestimated, and those advocating cross-sector collaboration will be most likely to succeed if these challenges are understood and accounted for early in the planning processes. However, the term *social entrepreneurship* may also have the advantage of newness—blurring traditional boundaries between sectors in ways that allow new ideas and practices to take root.

See Also: Corporate Social Responsibility; Environmental Marketing; Leadership in Green Business; Sainsbury's; Seventh Generation; Social Marketing; Social Return on Investment.

Further Readings

Bornstein, David. *How to Change the World: Social Entrepreneurs and the Power of New Ideas*, 2nd Ed. New York: Oxford University Press, 2007.

Dees, J. Gregory. "The Meaning of 'Social Entrepreneurship.'" Comments and Suggestions Contributed From the Social Entrepreneurship Working Group, 1998. http://www .caseatduke.org/documents/dees_sedef.pdf (Accessed January 2010).

Hawken, Paul. *Blessed Unrest: How the Largest Social Movement in History Is Restoring Grace, Justice, and Beauty to the World*. New York: Penguin, 2008.

Mair, Johanna and Ignasi Martí. "Social Entrepreneurship Research: A Source of Explanation, Prediction and Delight." *Journal of World Business*, 41/1:36–44 (2005).

Prabhu, Ganesh N. "Social Entrepreneurial Leadership." *Career Development International*, 4/3:140–45 (1999).

Reis, Tom. *Unleashing the New Resources and Entrepreneurship for the Common Good: A Scan, Synthesis and Scenario for Action*. Battle Creek, MI: W. K. Kellogg Foundation, 1999.

Thompson, John, Geoff Alvy and Ann Lees. "Social Entrepreneurship: A New Look at the People and the Potential." *Management Decision*, 38/5:328–38 (2000).

Sherrill Johnson
Independent Scholar

Patricia Ballamingie
Carleton University

Socially Responsible Investing

Socially responsible investing (SRI) is an investment strategy that seeks to optimize environmental and/or social benefits as well as financial returns. In addition to financial indicators such as risk, returns, and asset allocation, socially responsible investors look at nonfinancial factors—social impacts, environmental impacts, and issues of corporate governance—of investments that are often an extension of their own personal life views and values.

Corporations operate in a complex, market-driven environment, and SRI uses market tools to reward and encourage ethical and responsible behaviors. The beginnings of SRI trace back to faith-based groups such as 18th-century Quakers, who avoided any investments associated with war and military weapons. Currently, according to a recent report on U.S. trends by the Social Investment Forum, SRI is growing faster than non-SRI asset rates, with nearly one out of every nine investment dollars now involved in SRI. Between 2005 and 2007, SRI grew by more than 18 percent while the broader investment arena increased less than 3 percent.

Activist-investors are fueling the growth. Investor concerns may be local and/or global. They revolve around issues such as sustainable development, global climate change, environmental degradation, water usage, renewable energy, human rights, labor practices, diversity, addictive substances, and a host of others.

Socially responsible investors include a wide variety of individuals, investment professionals, and activist groups. SRI financial services include mutual funds, banks and credit unions, financial advisers and planners, venture capitalists, mortgage companies, mutual funds, and many others. Nongovernmental organizations (NGOs), such as Social Investment Forum, support SRI with an arsenal of tools and resources including research, conferences, service directories, trend reports, fact sheets, and mutual fund performance charts, which display financial data as well as each fund's company-screening criteria and policies regarding proxy voting.

Screening

SRI relies on both positive and negative screening, that is, examining a company's policies and activities for social impacts that are either particularly positive or particularly negative. Screening begins with the identification of an industry's highest standards—enforced regulations or laws and voluntary industry standards—for social responsibility. Portfolio 21, for example, actively seeks companies making positive impacts in the areas of climate and cleantech; pollution, toxins, and other environmental issues; community development; and board-level governance issues. The fund does negative screening, that is, it restricts investment in companies with poor performance in the areas of diversity, human rights and labor relations; executive pay; and product issues including animal testing, weaponry, gambling, and tobacco.

Many investors and advisers rely on a ratings database of desired measures compiled by KLD, a Boston-based SRI research firm that created one of the first socially responsible investment benchmarks. The database focuses on the return, risk, and financial characteristics of SRI portfolios within unique criteria. Environmental factors include alternative energy, climate change, hazardous waste liabilities, management systems, and regulatory problems. Social considerations include community relations, workforce diversity, employee relations, human rights, product quality, and innovation. Finally, governance factors include accounting, executive compensation, political accountability, transparency, and ownership. Such positive and negative screenings illuminate corporate practices for investors, media, and public audiences and underscore informed choice for investors.

Shareholder Advocacy

As one strategy of SRI, shareholder advocacy works individually as well as collectively. In some instances, investors use political strategies for dialogue—such as massive letter-writing campaigns or meetings with corporate executives—to further their goals in assuring that their values reflect their investment. Other efforts involve filing shareholder resolutions and/ or voting via proxy or resolutions for companies to become more environmentally or social responsible.

One of today's most visible SRI movements is divestment, a combination of advocacy and negative screening. Advocates have urged divestment from companies that do business in oppressive regimes.

Community Investing

Community investing is the fastest-growing SRI choice and brings safe financing to lower-income borrowers and economically distressed communities. It is also one of the most recognized investment movements, with a prominent campaign challenge that echoes the tithing of earlier Christian efforts: investors designate at least 1 percent of their capital for community investment, with a goal of helping U.S. community investment soar to $30 billion by 2010.

Community investing is active globally. For instance, SRI is one of the fastest-growing mechanisms for revitalizing war-torn areas such as Afghanistan, and weather-ravaged areas like Sri Lanka. Within the United States, target urban areas—historically underserved by traditional financial institutions with a bottom line to feed—are revitalized and experience economic growth through small business creation, community-level services, and neighborhood resources in depressed areas.

Community banks—locally owned and operated, and specializing in loans to individuals, small businesses, and nonprofits—are a major force in community investing. In 2009, when large banks were hoarding government bailout money, community banks such as Cross River Bank of Teaneck, New Jersey, were not only making loans, but they were also taking market share from banking giants. Rather than adhering to strict and impersonal formulas for lending, community banks have the ability to meet potential borrowers face to face and use discretion with respect to the many factors that make a community member a good credit risk. Relatedly, microfinance, as developed by Nobel laureate Muhammad Yunus, is another powerful vehicle for community development. Small personal or microenterprise loans, often brokered between private investors and borrowers, offer credit to low-income persons who lack access to traditional banks. Microfinance loans appear to be no more risky than other kinds of loans, and they are widely credited with reducing poverty.

SRI Future

Among the three approaches of SRI—screening, shareholder advocacy, and community investing—the first two concentrate on corporate responsibility while the third produces direct, visible results in areas such as child and healthcare, community facilities, and affordable housing.

One of the challenges facing SRI is the very name itself; after all, the term *responsible* is hard to define. Some refer to the process as "Sustainable and Responsible Investing," or ethical investing, extra-financial factors, and sustainable investing. Just as skeptics of "corporate social responsibility" see such altruism as simply more public relations, some SRI critics question whether the practice can draw the numbers needed for a large global "voice."

So, what makes one investor more attracted to SRI than traditional practices? Competitive corporations want to know what factors might entice new funds; however, empirical evidence proves that there is no guarantee that passionate beliefs translate into action.

Whether driven by environmental or social values, SRI investors ultimately place more importance on financial performance criteria than SRI investment criteria. Plus, many more shareholders have expectations about the larger arena of corporate social responsibility than engage in SRI. For those that do choose SRI, there is a question of how much trust to put in the indexes; that is, could SRI turn into another public relations strategy for corporations hoping to put a shine on their actions?

SRI research is expected to adhere to rigorous research methods and materials. The knowledge base includes direct communication with corporate officers, review of media and public documents, as well as information from nongovernmental and government sources, as well as global SRI research. After finalizing the screening and rating process, SRI research organizations create "best of class" indices that are a focus of companies. Such transparency and credibility advances SRI as an alternative to conventional investment options.

See Also: Cause-Related Marketing; Corporate Social Responsibility; Dow Jones Sustainability Index; Fair Trade; Social Entrepreneurship; Transparency.

Further Readings

Collins, Chuck, Pam Rogers and Joan P. Garner. *Robin Hood Was Right: A Guide to Giving Your Money for Social Change.* New York: W. W. Norton, 2000.

Frynas, Jedrzej George. "The False Development Promise of Corporate Social Responsibility: Evidence From Multinational Oil Companies." *International Affairs* (2005).

Glac, Katherina. "Understanding Socially Responsible Investing: The Effect of Decision Frames and Trade-Off Options." *Journal of Business Ethics*, 87/1 (2009).

Hockerts, Kai and Lance Moir. "Communicating Corporate Responsibility to Investors: The Changing Role of the Investor Relations Function." *Journal of Business Ethics*, 52/1 (June 2004).

Kinder, Peter. "'Socially Responsible Investing': An Evolving Concept in a Changing World." KLD Research & Analytics, Inc. 2005. http://www.kld.com/resources/papers/ SRIEvolving070109.pdf (Accessed April 2009).

Markowitz, Linda. "Can Strategic Investing Transform the Corporation?" *Critical Sociology*, 34/5:681–707 (2008). http://crs.sagepub.com/cgi/content/abstract/34/5/681 (Accessed April 2009).

Social Investment Forum. "2007 Report on Socially Responsible Investing Trends in the United States." http://www.socialinvest.org/pdf/SRI_Trends_ExecSummary_2007.pdf (Accessed April 2009).

Soederberg, Susanne. "The Marketization of Social Justice: The Case of the Sudan Divestment Campaign." Paper presented at International Studies Association, New York City, February 15, 2009. http://www.allacademic.com/meta/p313598_index.html (Accessed April 2009).

Sparkes, Russell and Christopher J. Cowton. "The Maturing of Socially Responsible Investment: A Review of the Developing Link With Corporate Social Responsibility." *Journal of Business Ethics*, 52/1:45–57 (June 2004).

Vyvyan, Victoria, Chew Neg and Mark Brimble. "Socially Responsible Investing: The Green
    Attitudes and Grey Choices of Australian Investors." *Corporate Governance: An
    International Review*, 15/2:370–81 (March 2007).
Waring, Peter and Tony Edwards. "Socially Responsible Investment: Explaining Its Uneven
    Development and Human Resource Management Consequences." *Corporate Governance:
    An International Review*, 16/3:135–45 (May 2008). http://www3.interscience.wiley.com/
    cgi-bin/fulltext/121381605/PDFSTART (Accessed April 2009).

Nancy Van Leuven
Bridgewater State College

SOCIAL MARKETING

Social marketing refers to the use of marketing techniques to motivate behavioral change for the well-being of individuals, communities, or society. Social marketing may be undertaken by governments, nongovernmental organizations (NGOs), or, less commonly, by businesses. Well-known targets of social marketing campaigns include behaviors related to public health and safety, environmental health, and community involvement. Examples include anti-smoking, anti-obesity, and responsible drinking campaigns. Social marketing occasionally addresses topics that are culturally sensitive, such as AIDS prevention, organ donation, and family planning and contraception.

Social marketing is one essential element of sustainable marketing. In their book *Sustainable Marketing*, Diane Martin and John Schouten describe sustainable marketing as both marketing sustainably (i.e., conducting marketing activities according to principles of environmental and social sustainability) and marketing sustainability (e.g., using marketing to influence social behaviors in order to help bring about a culture of sustainability). Social marketing is an indispensable tool for achieving the latter objective.

The academic history of social marketing goes back to the mid-20th century. In 1952, G. D. Wiebe argued that marketing or advertising techniques should be applied to the promotion of good citizenship. In 1971, Philip Kotler and Gerald Zaltman legitimized social marketing as a study and a practice in the marketing discipline. Since 1999, the Social Marketing Institute has been working as a bridge between research and practice. Currently, journals such as *Social Marketing Quarterly* and the *Social Marketing Journal* are devoted to the topic, and more mainstream journals welcome new advances in understanding social marketing.

In practice, social marketing for environmentally related purposes has an even longer history. For example, during World War II social marketing techniques helped promote strategically important civic involvement in practices such as resource conservation and home food production. One poster created to encourage Americans to recycle materials for the war effort read, "Salvage Scrap to Blast the Jap." Another, which encouraged fuel conservation through car sharing, read, "When you ride ALONE, you ride with Hitler! Join a car-sharing club TODAY!" The U.S. government encouraged wartime home food production with a widespread "Victory Garden" campaign, encouraging Americans with posters bearing taglines such as "War Gardens for Victory: Grow Vitamins at Your Kitchen Door," "Sow the Seeds of Victory," and many others.

Beginning roughly in the 1960s, there has been strong societal interest in preserving the natural environment. Early environmentally focused social marketing in the United States included the "Don't Be a Litterbug," "Give a Hoot—Don't Pollute," and "Only You Can Prevent Forest Fires" campaigns.

At the onset of the 21st century, many regarded the task of changing the global consumer culture into a culture of sustainability as one of the most critical challenges facing humanity. Social marketing falls at the center of this task. The underpinnings of culture are held in place by myths and stories that communicate the values by which society lives. Social marketing, done well, has the power to help rewrite the stories that guide global consumer culture. Government, to some extent, and especially nongovernmental organizations (NGOs), continue to market sustainability. However, the real and emerging power of communicating with consumers is in the hands of businesses.

Businesses are beginning to use social marketing (albeit subtle) to link sustainable practices and lifestyles with such deeply held values as health, family, financial security, and, increasingly, fun and fashion. Retailers, such as Sainsbury's, IKEA, and Wal-Mart, use the power of in-store communication and merchandising to persuade shoppers to choose more sustainable products. For example, both Wal-Mart and IKEA made great strides in shifting consumer demand from incandescent lightbulbs to compact fluorescent lightbulbs (CFLs) through sales promotions, favorable shelf space, and in-store customer education. The underlying story: CFLs save you money and make the world better.

Through in-store marketing with "Tip Cards," the UK's Sainsbury's provides consumer education regarding nutritional labeling and food preparation, all aimed at influencing customer choices and lifestyles. The underlying story: local and sustainable foods make your family healthier. These stores use their buying power to influence suppliers to provide more sustainable products and packaging, and to deliver them at prices that are favorable to mass markets. For example, Sainsbury's has developed entire food product lines that increase local content, reduce packaging and additives, and sell for prices that are competitive with less sustainable options. The underlying story: sustainable foods are convenient and affordable. Wal-Mart has developed an entire line of jewelry, called "Love, Earth," that guarantees the chain of custody and relative sustainability of its gold and silver. The underlying story: trace your love to its responsible source. Social marketing reaches beyond marketing communication and product assortments. It also has infrastructural elements that facilitate changed consumer behavior. For example, product take-back programs and recycling collection centers provide the means for consumers to change their behaviors.

See Also: Cause-Related Marketing; Corporate Social Responsibility; Environmental Marketing; Green Retailing; IKEA; Sainsbury's; Sustainable Marketing.

Further Readings

Kotler, Philip and Gerald Zaltmann. "Social Marketing: An Approach to Planned Social Change." *Journal of Marketing*, 35/3:3–12 (1971).

Martin, Diane M. and John W. Schouten. *Sustainable Marketing*. Upper Saddle River, NJ: Pearson Prentice Hall (Forthcoming, 2011).

Sachs, Jonah and Susan Finkelpearl. *From Selling Soap to Selling Sustainability: Social Marketing: State of the World, 2010: Transforming Cultures From Consumerism to Sustainability*. Washington, D.C.: WorldWatch Institute, 2010.

Wiebe, G. D. "Merchandising Commodities and Citizenship on Television." *Public Opinion Quarterly*, 15:679–91 (1952).

John W. Schouten
Diane M. Martin
University of Portland

Social Return on Investment

Creating a socially and environmentally responsible (or sustainable) marketplace is in part hindered by the fact that social or environmental values do not have a price. Nonprofit or nongovernmental organizations (NGOs) in particular find it difficult to express, and receive recognition for, the value of their contributions, since measuring the true value of living in a more just society, or enjoying an art performance, or planting trees is virtually impossible. Increasingly, businesses and corporations that attempt to provide useful social or environmental services are also faced with the challenge of "valuation."

Social return on investment (SROI) is a methodology used by organizations to assign a financial value to the social impact of a project, organization, or policy. By assigning a financial value (in the form of a ratio) to a social value that ordinarily does not have a price or market value, organizers are able to demonstrate that the project will generate a positive impact to individuals or to society as a whole. It is typically used by nonprofit or NGOs, but businesses have been known to use it as well. SROI is a useful tool to identify key value drivers, align stakeholder expectations, and demonstrate social impact, which makes it easier to attract funding. The limits of this method are that some benefits cannot be monetized in a consistent manner, that increased monetization can make the organization lose its social focus, and that intensive efforts are required to prepare a first SROI analysis, as well as lack of recognition by most investors.

Primary Principle

SROI is used by organizations to illustrate the value created by a project or to help decide how to allocated internal resources. This value assessment is based on the perception and experience of participating stakeholders, and facilitates decision makers' understanding of the social, environmental, and economic value they are creating. The objective is to translate into monetary terms actions and initiatives that might not have monetary outcomes. For example, a project that aims to help build self-esteem might not have an economic result, but the benefit for society is indisputable. Therefore, organizations raising funds for these types of projects can use an SROI ratio to showcase the positive economic outcome of their projects, making it easier to convince investment or subsidies.

SROI Value

An SROI analysis is usually illustrated in the form of a ratio, which contrasts the value that has been created with the investment required to attain it. For example, a project presenting an SROI of 4:1 is creating four units of social value for each unit invested. For this reason, SROI can be used by nonprofits to illustrate the positive value created on society by nonmonetized elements (such as a demonstrable cost savings or a positive intervention). However, the application of SROI is highly dependent on stakeholder participation and agreement to appropriately estimate the positive impact created by the project; this means that an SROI value is not fixed, but rather a perception of participating stakeholders; it is possible for two groups evaluating the same activity to arrive at different SROI values.

Doing an SROI Implementation

When using SROI analysis for the first time, an organization will do a mapping process, where it will design an impact map showing the relationship between allocated resources and estimated impact. This map is created through the participation of stakeholders. They are asked to settle on a common value for each of the organization's services. This stakeholder involvement is a key to success, since they are the most likely to know the "value" of a process. Also, different types of stakeholders can supply very different points of views. While stakeholders closer to the organization might provide an opinion based on the efforts to deploy and implement a process, outside stakeholders will be able to supply peripheral opinions since they do not suffer from the tunnel vision of being too close to the organization. For example, they will be best placed to discuss the overall environmental impact of said services. Stakeholders can be implicated either on a one-to-one basis (such as an interview, form letter, or web survey) or in a group setting (such as a general assembly or focus group).

The organization is ultimately responsible for conciliating and reconciling different viewpoints, and must be able to bring the opinions back into a single unified ratio. Following this analysis, the organization will be able to identify which of its processes provides the greatest benefits, and which are the most likely to generate the most positive income. By the same token, when the organization is fundraising, it will be able to showcase processes or projects that generate the most positive ratios.

Potential Benefits

SROI is a useful tool for identifying key value drivers, aligning stakeholder expectations, and demonstrating social impact, which in turn makes it easier to attract funding. By evaluating key drivers, SROI allows participants to make effective organizational decisions. Since indicators and processes are monetized and brought back to a common ratio, it is possible to track and monitor them through time. It also allows managers to focus efforts on important drivers since they have been effectively monetized, and allows them to do a scenario analysis in case there are better means of using internal resources.

By developing a common evaluation framework, SROI is useful for aligning stakeholders' expectations. Hence, the common framework allows stakeholders from different backgrounds or environments to communicate effectively, even if they have different viewpoints (especially if they agreed during the SROI implementation). Also, understanding the overall impact of each process increases the quality of governance, since the organization will be able to respond to stakeholder requirements.

Finally, SROI enables companies to raise funds effectively, since it showcases the positive impact created by the project. It can also illustrate the expected returns of a project, and can help the organization understand grants from the point of view of an investment, not just a loan, shifting to a mentality of opportunity cost.

Potential Limitations of SROI

The potential limits of SROI are that (1) some benefits cannot be monetized in a consistent manner, (2) increased monetization can make the organization lose its social focus, (3) intensive efforts are required to prepare a first SROI, and (4) there may still be lack of recognition by most investors.

One of the key limits of SROI is that some benefits are hard to monetize. Sometimes, stakeholders will be unable to agree on the monetary value of a potential outcome, hence limiting the objectivity of the analysis. For example, monetizing increased self-esteem is a challenging endeavor. Also, there is a social debate whereas monetizing a process might lead the community to lose focus on nonmonetary benefits. Hence, some organizations might lose focus of the values and social impact that they are able to create and solely concentrate on the SROI ratio that they calculated. This in turn leads to short-term thinking, rather than concentrating on the long-term benefits.

Additionally, companies preparing an SROI for the first time might find that it is an intensive venture. With many stakeholders trying to monetize nonmonetary tasks, a large amount of negotiation will be necessary to get all of them to agree. This brings us to the subjectivity of the SROI. If an organization does not use a careful mix of stakeholders, some of its results might lead to ratios that are unacceptable to outsiders. Finally, since nobody officially accredits SROI ratios, and its implementation in not uniformly applied, SROI is viewed as a very imprecise tool. Investing organizations might not recognize the SROI or might not agree to some of the assumptions when it was prepared by the organization.

Practical Applications of SROI

All too often, conventional wisdom has it that economic and environmental goals are in conflict with each other: cleaner production, more efficient use of resources, and less waste, it is said, undermine the bottom line. SROI is in part based on the assumption that social investments can help, rather than hurt, economic investment. As companies like Newman's Own Organics, Stonyfield Farm, and Seventh Generation indicate, making substantial investments in the protection of the environment and the well-being of the larger community can, in fact, also be profitable. To the extent that consumers will continue to demand a higher degree of corporate social responsibility in general, and sustainable environmental stewardship in particular, businesses will find ever more opportunities to receive a return on their social investments.

See Also: Externalities; Newman's Own Organics; Seventh Generation; Social Entrepreneurship; Socially Responsible Investing; Stonyfield Farm; Sustainability; Voluntary Standards.

Further Readings

Brest, Paul and Hal Harvey. *Money Well Spent: A Strategic Plan for Smart Philanthropy*. New York: Bloomberg Press, 2008.

Lawlor, Eilis, Eva Neitzert and Jeremy Nicholls. *Measuring Value: A Guide to Social Return on Investment (SROI)*, 2nd Ed. London: New Economics Foundation, 2008.

The New Economics Foundation. "Measuring Real Value: A DIY Guide to Social Return on Investment." http://www.scribd.com/doc/19678244/NEF-Measuring-Real-Value-Social -Return-on-Investment (Accessed January 2010).

Olsen, Sara. "Social Return on Investment: Standard Guidelines." Working Paper Series, Center for Responsible Business. Berkeley: University of California, Berkeley, 2003.

Scholten, Peter, Jeremy Nicholls, Sara Olsen and Brett Galimidi. *Social Return on Investment: A Guide to SROI Analysis*. Amsterdam, the Netherlands: Lenthe Publishers, 2006.

Jean-François Denault
Université de Montréal

STEADY STATE ECONOMY

Steady state economy (SSE) refers to a theoretical economic system in which the world's population and natural resources remain practically constant through time. A steady state economy would be sustainable in that humans would not extract or deplete natural capital faster than the Earth's systems could replenish it.

Ecological economists, led by Herman E. Daly, posit that classical economic theories, which advocate infinite growth, are inconsistent with scientific reality. Businesses following classical economic theory essentially treat the Earth as an unlimited well of resources and sink for waste. SSE theory holds that since the Earth has limited natural resources, a continuous, unlimited growth of the gross domestic product would lead to the ultimate destruction of the natural resources indispensable for human life. Daly's theory of SSE acknowledges that the economy is a subset of the ecosphere, which is a finite, nongrowing, and closed system with respect to matter. Kenneth E. Boulding's concept of "Spaceship Earth" is an apt metaphor; it maintains that humans must regard the Earth as a spaceship containing limited resources and a limited capacity to process human waste. Also relevant is the concept of carrying capacity, understood as the natural ability of an ecosystem to steadily produce the necessary resources for the living species that live in it.

SSE theory acknowledges that negative environmental externalities, as described by economists Cecil Pigou and Ezra Mishan, are not sustainable. When businesses only take into consideration costs that are directly connected with their own activities, costs of a social nature, such as pollution or the depletion of collective assets (e.g., water, soil, and fisheries), accumulate. Arguing that the laws of economics must conform to the laws of science, Daly maintains that it is urgent to progressively diminish the pressure of human activity until the achievement of a steady state economy, intended as an economy subject to development but not to growth.

A steady state economy is characterized by an equitable distribution of the wealth produced by a careful use of natural resources and by a "constancy" of population and capital stocks. The theory does not deny the necessity for economic development, but it radically changes its meaning: the objectives of the steady state economy are not an accelerated growth and mass production, but rather the production of sufficient wealth for the entire world population, equitably distributed and efficiently placed.

Policy implications suggested by SSE theory commonly include population controls, reduction of per capita consumption, maximum limits on individual income and wealth, and heavy taxes on all resource depletion. Critics of SSE theory tend to focus on the political impossibilities of such proffered solutions.

Sustainable or green business offers a more politically palatable pathway toward a steady state economy. SSE theory has greatly influenced thinking around sustainable business practices.

Lester Brown's *Eco-Economy* begins with and builds on the underlying premises of SSE, and continues with suggestions for a hydrogen-based economy powered by renewable energy sources. The concepts of biomimicry and radical resource utilization discussed in *Natural Capitalism* by Hawken, Lovins, and Lovins also reflect a foundation in SSE theory. The fundamental scientific principles at the heart of SSE theory are the same principles that anchor the strategic framework for sustainability developed by Karl-Henrik Robért and The Natural Step. The work of McDonough and Braungart in cradle-to-cradle design and manufacturing facilitates a move to steady state by keeping resources, such as metals and synthetic substances, in closed technical loops, reducing waste, pollution, and the need for continued resource depletion.

See Also: Biomimicry; Ecological Economics; Environmental Economics; Externalities; Sustainable Development.

Further Readings

Booth, Douglas E. *The Environmental Consequences of Growth: Steady-State Economics as an Alternative to Ecological Decline*. London: Routledge, 1998.

Brown, Lester R. *Eco-Economy: Building a New Economy for the Environmental Age*. New York: W. W. Norton, 2001.

Halevi, Joseph, David Laibman and Edward J. Nell. *Beyond the Steady State: A Revival of Growth Theory*. New York: St. Martin's Press, 1992.

Hawken, Paul, Amory Lovins and L. Hunter Lovins. *Natural Capitalism: Creating the Next Industrial Revolution*. Boston, MA: Little, Brown, 1999.

McDonough, William and Michael Braungart. *Cradle to Cradle: Remaking the Way We Make Things*. New York: North Point Press, 2002.

Mishan, Ezra. "Economic Criteria for Intergenerational Comparisons." *Journal of Economics*, 37/3–4 (September 2007).

Nattrass, Brian. *The Natural Step for Business: Wealth, Ecology and the Evolutionary Corporation*. Gabriola Island, British Columbia, Canada: New Society, 1998.

Pigou, Arthur Cecil. *Essays in Applied Economics*, Reprint Ed. London: Routledge, 1965.

Federico Paolini
University of Siena

STEWARDSHIP

Ecological stewardship refers to the care, protection, and restoration of the Earth and its ecological systems—land, water, and air—that provide life-sustaining ecosystem services. Biologist Gretchen Daily identifies the following ecosystem services:

- Purification of air and water
- Mitigation of droughts and floods
- Generation and preservation of soils and renewal of their fertility
- Detoxification and decomposition of wastes

- Pollination of crops and natural vegetation
- Dispersal of seeds
- Cycling and movement of nutrients
- Control of the vast majority of potential agricultural pests
- Maintenance of biodiversity
- Protection of coastal shores from erosion by waves
- Protection from the sun's harmful ultraviolet rays
- Stabilization of the climate
- Moderation of weather extremes and their impacts
- Provision of aesthetic beauty and intellectual stimulation that lifts the human spirit

Over time, however, human technology and activities have changed the balance of virtually every ecosystem system. Now, with many ecosystems severely stressed, even nearing collapse, the Earth needs the stewardship of those who have benefited most from its ecological services: businesses.

Technological innovations have improved the lives of millions of people. From the industrial revolution to current times, new technologies have provided opportunities for advancements in food production, medicine, transportation, communication, and leisure. The prevailing capitalist economic system has resulted in great wealth for the business owners who provide the goods and services so many have come to depend on. But these lifestyle and economic gains have come with the unaccounted costs of declining resources and increasing environmental degradation.

A History of Environmental Stewardship

Stewardship of resources through conservation has, since the industrial revolution, fallen largely to governments and nongovernmental organizations (NGOs), often with business interests in vocal and well-funded opposition. The earliest environmental regulations in the United States were enacted to protect the unique beauty of several places in the western United States as national parks. By the late 1960s and early 1970s, the U.S. government began to regulate manufacturing waste in an effort to control air and water pollution. The establishment of the U.S. Environmental Protection Agency (EPA) in 1970 and passage of the Clean Water Act in 1972 and the Air Quality Act in 1967 were intended primarily to regulate industrial pollution. Business groups fought these regulations, citing concerns that compliance would raise costs and limit their profits. Federal and state regulation of particular industries followed. For instance, the federal 1992 Energy Policy Act limited the amount of water that showerheads could deliver to 2.5 gallons per minute. Many states have passed automobile fuel efficiency regulations. More recently, efforts to limit emissions of greenhouse gases remain an active issue in the U.S. Congress.

Increasingly, business leaders understand the degree to which their success depends on the health of ongoing ecosystem services. These leaders look beyond the minimum effort and expense needed to remain in compliance with governmental regulations. They see stewardship of the Earth as their responsibility and manage their businesses accordingly, often with the help of NGOs and self-governed industry groups. Not surprisingly, many businesses also find that environmental stewardship pays off in such areas as stakeholder relations, product and process innovations, cost controls, and employee satisfaction and retention.

Much of what is currently going on in the way of stewardship is a function of proactive voluntary measures, and many of those voluntary measures revolve around standards, certifications, and external audits. For example, Worldwide Responsible Accredited Production (WRAP) is an independent organization that certifies and audits factories, which may manufacture many different brands, in the apparel and footwear industries. WRAP conducts audits of working conditions, including environmental factors, and certifies factories for a period of six months at a time. The general effectiveness of voluntary stewardship programs is unclear. A 2008 study from George Mason University found that businesses in voluntary environmental programs actually performed more poorly than nonparticipants within the same industries. This was especially pronounced where companies were left to audit themselves rather than being audited by independent external reviewers.

Land Stewardship

For their survival, many businesses rely directly on ecosystem services provided by the land. Ranchers, foresters, and farmers realize that their business success is intimately connected to the carrying capacity of the land. For example, the National Cattlemen's Association developed an Environmental Stewardship Program to help members learn about conservation and land stewardship practices. Interested members work on mitigating stream bank erosion and enhancing wildlife habitat through innovative irrigation systems, improved seeding operations, and implementing overall grazing plans.

Although timber is a renewable resource, interest in forest stewardship has increased since the early 1990s. Foresters realize the value of maintaining healthy forest ecosystems to ensure reliable yields and long-term health of their industry. One company that understands the relationship between biodiversity and naturally healthy forests is the Collins Company. The company maintains 295,000 acres of biodiverse, multilayered, private forests. The company began operations in the mid-1880s, and over time has developed an organization-wide effort of stewardship and restoration they call their Journey to Sustainability (JTS). These efforts are among some of the voluntary stewardship measures that keep the company in compliance with Forest Stewardship Council (FSC) principles and criteria. The Collins Company enjoys an advantage over less sustainable producers and has the right to market its products as FSC certified.

Water Stewardship

Two major challenges face the stewards of the world's water. First, shortages of clean, fresh water are rapidly becoming a global problem and for years have been serious problems for people in many developing nations. Second, the Earth's oceans and their ecosystems are being systematically decimated. Rising populations exacerbate both of these crises. In 2009, it is estimated that nearly 2 billion people lack access to clean water for drinking or sanitation.

One barrier to government stewardship of water is the often-contested issue of ownership. Many of the world's fisheries exist in international waters making them extremely difficult to regulate or police. Even streams, wetlands, and watersheds that lie wholly inside national boundaries typically face complex issues of ownership and water rights. Here again, it falls increasingly to businesses and consumers to take on the responsibilities of stewardship.

One important movement in ocean stewardship is the development of seafood certification programs, such as the Marine Stewardship Council (MSC), a global organization that administers a seafood certification and labeling program that recognizes and rewards sustainable fishing. The MSC works with fishing industry members throughout the supply chain, including individual fishing boats, seafood companies, and retailers as well as with other stakeholders such as scientists, conservation groups and the public as they promote sustainable seafood production and consumption. The organization helps develop industry-wide standards for sustainable fishing and uses chain-of-custody audits to monitor and enforce them.

According to the U.S. Geological Survey, agriculture is the industry with the greatest impact on fresh water sources, using around 60 percent of all the world's fresh water for irrigation alone. Water stewardship programs in agriculture range from local groups of farmers, such as the California Agricultural Water Stewardship Initiative, to global alliances such as the Alliance for Water Stewardship (AWS), formed at the 2009 World Water Forum in Istanbul. The AWS's charge is to coordinate development of the first worldwide standards for water management around the world.

Water stewardship is a matter of social and environmental responsibility for many firms, including for businesses that don't draw significantly from local freshwater sources. For example, Toyota Logistics Services, which moves thousands of new cars a week through the port of Portland, Oregon, has redesigned it facilities along the Willamette River to reinvest in the local riparian ecosystem. Among many improvements to enhance sustainability, the company added bioswales between its acres of parking lots and the river. The bioswales cool and clean runoff before it flows to the river and provide natural riverbank habitat for wildlife.

Air Stewardship

Much of the progress in the area of air quality stewardship has come from legislation and government enforcement intended to force compliance. For example, in a 2007 Clean Air Act settlement, American Electric Power (AEP) agreed to spend $4 billion to reduce sulfur- and nitrogen-based emissions from their coal-burning power plants. Some industries are becoming more proactive in their treatment of the local atmosphere. For example, associations of hog farmers and dairy farmers, both responsible for enormous amounts of atmospheric methane, have taken on air stewardship through programs like sharing best practices, creating education and certification programs, funding research, and offering awards for innovation or performance.

Product Stewardship

An emerging area in many industries is product stewardship, which is concerned with managing a product's environmental, health, and safety effects throughout its life cycle, from extraction and production to final postconsumer disposition. Product stewardship encompasses every aspect of environmental stewardship, including air, land, and water, and when done properly it can lead to reductions in overall life cycle costs. In one example of product stewardship, Kodak began collecting its one-time-use cameras at the point of film development, and remanufacturing or recycling the plastic camera bodies. Product stewardship goes hand-in-hand with other sustainability-related practices, such as design for disassembly (DfD) and cradle-to-cradle design and manufacturing. One leader in these practices is office furnishings manufacturer Herman Miller. In the automotive sector, BMW has been a leader in product stewardship.

See Also: Clean Fuels; Corporate Social Responsibility; Cradle-to-Cradle; Ecoefficiency; Ecosystem Services; Extended Product Responsibility; Herman Miller; Voluntary Standards.

Further Readings

Chapin, F. Stuart, III, Gary P. Kofinas, Carl Folke and M. C. Chapin. *Principles of Ecosystem Stewardship: Resilience-Based Natural Resource Management in a Changing World*. New York: Springer, 2009.

Daily, Gretchen C. *Nature's Services: Societal Dependence on Natural Ecosystems*. Washington, D.C.: Island Press, 1997.

Darnell, Nicole and Stephen Sides. "Assessing the Performance of Voluntary Environmental Programs: Does Certification Matter?" *Policy Studies Journal*, 36/1 (2008).

Hart, Stuart. L. "A Natural-Resource-Based View of the Firm." *The Academy of Management Review*, 20/4:986–1015 (1995).

Diane M. Martin
John W. Schouten
University of Portland

STONYFIELD FARM

Stonyfield Farm, the world's largest organic yogurt producer, is widely recognized for being a sustainable, mission-driven business model. The company has received national awards for energy efficiency, waste minimization, recycling, corporate environmental leadership, and efforts to address climate change. Stonyfield Farm aims to be ecologically conscious and profitable. The president and "CE-Yo" of Stonyfield Farm, Gary Hirshberg, has described himself as an "ecoentrepreneur" who wanted to change the American diet and save the Earth at the same time. Hirshberg advocates corporate environmental responsibility, maintaining that Earth-friendly practices are also economically sound.

Hirshberg and his partner, Samuel Kaymen, embarked on their venture at a minuscule New Hampshire dairy farm affiliated with a rural homesteading school. Using Kaymen's recipe, they began making quarts of plain organic yogurt from seven cows in 1983. Failed experiments and mounting debt marked the cottage industry's early years. Hirshberg recalled that, at the time, many people dismissed organic food as a "fringy fad." The key challenge to winning acceptance for organic products was taste, and Hirshberg knew that their yogurt would have to be delicious to gain customers. As the business expanded, it began purchasing Jersey milk from local farmers. New filling and capping machines allowed Stonyfield to start offering individual-serving containers in strawberry, cappuccino, peach, and various other flavors.

Stonyfield Farm eschewed producers who used antibiotics, artificial growth hormones, and toxic pesticides. Hirshberg chose to use organic sugar instead of high-fructose corn syrup, agar instead of modified food starch, and organic milk solids instead of gelatin. Stonyfield's slogan became "You just can't fake this stuff." Despite paying for expensive organic ingredients and absorbing higher production costs, the business earned better net

Stonyfield Farm, in Londonderry, New Hampshire, makes organic dairy products such as yogurt and milk.

Source: Mark Warner/Wikipedia

profit margins than its competitors. Hirshberg found that investing in pure, top-quality, organic ingredients was worthwhile because his loyal customers gladly shelled out more for superior yogurt. At the same time, Hirshberg knew that organic food could not be priced at elitist levels if he wanted to "save the world, one yogurt at a time." Stonyfield first sold only to local health food stores. While committed to both the food and politics of the organic movement, Hirshberg also believed that organics would have to accommodate supermarkets and pursue commercial opportunities in order to gain footing. After he commenced handing out free samples at the local Stop & Shop, the chain agreed to carry Stonyfield yogurt at its several hundred supermarkets across New England. In his eagerness to reach mainstream consumers, Hirshberg was integral in turning the organic movement into a potent industry.

For Stonyfield's first eight years, Hirshberg raised money from individual investors and never broke even, despite growing demand for the yogurt. In 1988, the business built a large, modern plant in Londonderry, New Hampshire. Revenues reached $10.2 million in 1992, which was the first profitable year. Stonyfield Farm eventually flourished as the third-largest yogurt operation in the United States, with $330 million in annual sales. By 2003, Hirshberg had sold 80 percent of Stonyfield Farm to the French food-products conglomerate Groupe Danone. Stonyfield remained independent, and Hirshberg retained voting control of the board of directors. Despite criticism that he had "sold out," he asserted that the deal was a "win-win" situation for organic producers and consumers. Buyouts and acquisitions of small, pioneering companies by giant corporations have occurred regularly in the organic industry. While some high-profile undertakings that capitalized on organic food's boom have been the target of harsh critiques, Stonyfield Farm has worked at convincing the public that "Big Organics" can still be sustainable and socially responsible.

Stonyfield Farm tends to assure consumers that they are buying excellent yogurt and are part of a "planet-saving mission." The dairy proclaims that it has been "making healthy, delicious organic yogurts and caring for the Earth" for over 25 years. Yogurt cartons and lids are used for advocating relevant social and ecological issues. Messages include information about Stonyfield's support for organic family farming and its pledge to give 10 percent of annual profits to organizations and projects that "work to protect and restore the Earth." In 2007, the business donated nearly $2 million to nonprofit and educational groups involved with environmental causes.

Stonyfield Farm has been innovative in ecological initiatives. It became the first carbon-neutral business in the United States when it began offsetting all of the carbon dioxide emissions from its facility. It installed New Hampshire's largest solar electric array on the roof of its "Yogurt Works" building in Londonderry in 2003. It utilizes highly efficient lighting systems, motors, compressors, and wastewater treatment. The company believes that carbon-cutting practices, waste reduction, and gains in energy efficiency at its plant have saved over $1.7 million and 46 million kilowatt-hours. Stonyfield Farm created an interest-free loan fund to assist New Hampshire dairy farmers converting to organic production and helped fund the establishment of an organic dairy research farm at the University of New Hampshire. It recycles some used yogurt containers into durable "YoPlanters" pots, which are hand painted by inner-city artists and sold in a partnership with TerraCycle; and recycles others into handles for toothbrushes and razors, in a partnership with Recycline Inc. Through its Planet Protectors program, Stonyfield Farm publicly recognizes people who have contributed significantly to environmental protection. It sponsored Live Earth, a global climate awareness event, in 2007. Hirshberg started a nonprofit called Climate Counts to assess the extent to which other businesses are helping to reduce the effects of global warming.

Stonyfield Farm's line of products includes organic yogurt, smoothies, milk, ice cream, frozen yogurt, YoBaby drinkable yogurts, YoKids squeezable yogurt tubes, O'Soy soy yogurt, and Oikos Greek yogurt. Natural food stores, conventional supermarkets, and mass merchandisers all carry these products. Stonyfield Farm has expanded its customer base by forging emotional connections with consumers, making its label an invitation to an appealing lifestyle. Many of the company's national advertisements focus on the yogurt's healthy qualities, building on its popularity among nutrition-conscious buyers. Connections between human wellness and a healthy planet are highlighted.

A typical advertisement for yogurt caters to ecofriendly inclinations by promising: "You're supporting farmers committed to protecting the environment." Stonyfield professes that its contract with the customer is summed up by the words "We want you to feel good inside." Another key Stonyfield strategy is to cultivate rustic, pastoral images, publicizing its own tale of humble beginnings from "Two Families, Seven Jersey Cows, One Great Yogurt Recipe." The yogurt ingredients are listed as "Our Family Recipe." These promotional efforts at crafting trust have frequently been successful. Loyal organic purchasers tend to deeply embrace the brands they feel connected to. According to a 2007 survey by the Natural Marketing Institute, consumers perceive organic-oriented brands, such as Cascadian Farm, Horizon Organic, and Stonyfield Farm, as small, authentic, and artisanal, despite the fact that larger corporations own these labels.

Stonyfield launched a plan to replace "junk food" vending machines in schools with those featuring nutritious, organic snacks, such as Stonyfield Farm light smoothies, Clif Bars, Stacy's pita chips, Newman's Own Organics pretzels, Earthbound Farm fruit, and Silk soy milk. Hirshberg is also the cocreator of O'Naturals, a fast-food restaurant chain that offers convenient, healthy meals. O'Naturals began with outlets in New England in 2001 and is beginning to expand franchises across the nation. These restaurants proclaim that they do not serve ingredients with "additives, preservatives, artificial flavors, colors, sweeteners and hydrogenated oils." Organic mixed greens, organic beef, organic tofu, and organic ranch dressing are incorporated into tossed salads, flatbread sandwiches, soups, pizza, and noodle dishes. Kids can have organic macaroni and cheese, or hormone-free mini turkey sandwiches; and, of course, Stonyfield Farm organic smoothies are always available.

See Also: Lifestyles of Health and Sustainability (LOHAS); Newman's Own Organics; Organic.

Further Readings

Chomka, Stefan. "Meet America's Green Man." *Grocer*, 230/7806 (May 26, 2007).

Hirshberg, Gary. *Stirring It Up: How to Make Money and Save the World.* New York: Hyperion, 2008.

Hirshberg, Meg Cadoux. "Hitched to Someone Else's Dream." *Inc.*, 30/9 (September 2008).

Lifestyles of Health and Sustainability (LOHAS). "Consumers Prefer Artisanal Foods." http://www.lohas.com/articles/100992.html (Accessed January 2008).

Phillips, David. "Bringing the Cultural Revolution." *Dairy Foods*, 104/12 (December 2003).

Stonyfield Farm. http://www.stonyfield.com (Accessed January 2009).

Robin O'Sullivan
University of Texas at Austin

SunEdison

SunEdison is a leading global solar energy service provider to commercial, government, and utility customers. Since its founding in 2003, SunEdison has engaged in photovoltaic solar projects for states, businesses, and private customers, exceeding business volume of competitors. In April 2009, it marked a milestone in solar energy history by delivering more than 100 gigawatt-hours of electricity from photovoltaics, which essentially can offer electricity to 13,700 residents on an annual basis. SunEdison offers customers the opportunity to purchase solar energy produced by SunEdison's solar equipment to complement the array of energy sources a customer may already enjoy. There are no capital outlays; SunEdison's pricing is competitive. Customers pay only for the energy produced by SunEdison's power plants. If SunEdison does not deliver, customers do not pay. SunEdison's services include renewable power, monitoring, marketing, renewable portfolio standards, and solar tariff services. In its most recent expansion, SunEdison entered into a partnership with Developers Diversified to install solar energy systems at more than 130 shopping centers throughout the United States.

SunEdison is a limited liability corporation that owns and operates 221 solar power plants in North America and Europe. These 221 installations have decreased the carbon footprint by 124 million pounds of carbon dioxide. The company's main offices are in Beltsville, Maryland, with satellite offices in San Clemente, Sacramento, and Ontario, California. Its European offices are located in Spain, Italy, France, and Germany. On its website, SunEdison makes its corporate mission clear: to make solar services available, and to make them available at prices that turn them into rewarding alternatives to traditional energy portfolios. Based on SunEdison's business success in promoting renewable energy, *Fast Company*, a popular trade publication for the business sector, named SunEdison among its annual "Fast 50" listing in 2007.

As the public's demand for sustainable energy sources is growing, and solar energy is one of the few readily available renewable energy technologies, SunEdison works with its customers in assessing the viability of solar power installation. Its most popular customers to date are Whole Foods, Staples office supply store, Sacramento (California) Municipal Utility District, and Kohl's department store. According to the claims of the corporation, SunEdison leads the solar energy industry because of its record of fiscal stability, learned staff, integrity as a corporation, extensive esteemed reputation among a network of suppliers, and a commitment to quality.

Solar energy is marketed to customers generally in one of three ways: integrators, installers, and solar energy providers. A company that engages in solar integration sells solar equipment and necessary products to governments, businesses, and consumers. The system is installed at a given site, but not necessarily by the provider. It may be contracted with another supplier. Assistance with ongoing maintenance is the responsibility of the purchaser who may choose to use the original provider's subcontractors. In this first type of business arrangement, the provider generally steps aside once the sale is made and the customer assumes responsibility for the solar power system. A second type of solar energy commerce focuses mainly on the installation of the system itself. Installers are subcontracted by manufacturers of integrators to install systems. They may have training in installing solar power systems, but generally their expertise is broad, and such installations are one among several other skills they might have.

The last category of solar power commerce is the one that applies to SunEdison. In this type of business arrangement a corporation functions as a holistic provider of solar energy services to its customers, which includes finance, construction, operation, and maintenance. SunEdison business model is partly based on the commitment to provide not only cleaner, but also cheaper energy. Customers of SunEdison's solar energy installations are required to pay only for the amount of energy used. In addition, the company does not charge any up-front costs, rental leases, or equipment purchases. In entering into a business arrangement, SunEdison provides a breakdown of how much money a company can save with an active-install certified solar system. Part of the maintenance plan is SunEdison's commitment to monitor the installation regularly, and to manage the equipment and repairs without interference to the business of the customer. The company also offers its customers a monthly billing that includes a cost savings and a carbon footprint analysis.

SunEdison's Solar Protection Assurance is provided to customers with installed solar power systems. Through this regular upkeep service, SunEdison commits itself to assisting customers through regular cleaning, yearly tune-ups, responding to outages, monitoring, and contractual agreements. Customers have the opportunity to monitor online their solar power usage and receive a user profile through SunEdison's Client Connect. Client Connect is a cyber platform, offering customers critical information, updated every 15 minutes via SunEdison's Energy and Environmental Data System (SEEDS). It details the amount of solar power generated as well as capacities and energy demand offsets. This program details emissions avoided and the health of the environment. It also offers regular displays of messages and alerts.

Though still in its infancy, the solar energy market is steadily growing in the United States. Though solar energy still makes up only a little over 1 percent of total energy production, it is by now a roughly $530 million industry, and the only readily available renewable energy source with virtually limitless growth potential. Manufacturing costs of photovoltaic cells has dropped by 3–5 percent per year since 2000, while at the same time government subsidies have increased. If this trend continues, and especially if fossil fuel prices were to continue to increase, solar energy could very soon reach a stage where it becomes competitively affordable. While China, Europe, and Japan are still far ahead of the United States in total solar cell production and installation, companies like SunEdison have successfully positioned the American market to potentially become very competitive.

Particularly if federal or state government subsidies were to be implemented in the United States, already existing in many other countries, solar energy could quickly become a viable option for businesses that are concerned about their ecological footprint. SunEdison is one of the few solar companies in the United States that seems well positioned to fill a growing demand for clean solar energy.

See Also: Carbon Footprint; Clean Technology; Energy Performance Contracting; Environmental Impact Statement; Pollution Offsets; Sustainability.

Further Readings

Bradford, Travis. *Solar Revolution: The Economic Transformation of the Global Energy Industry*. Cambridge, MA: MIT Press, 2008.
Fast Company. "Fast 50." *Fast Company*, 113 (2007).
Kozlowski, David. "New Thinking Powers Solar." *Building Operation Management*, 55/5 (May 2008).
National Renewable Energy Laboratory. "DOE Solar Energy Technology Program: Overview and Highlights." May 2006. http://www.nrel.gov/docs/fy06osti/39081.pdf (Accessed April 2009).
Pernick, Ron and Clint Wilder. *The Clean Tech Revolution: Discover the Top Trends, Technologies, and Companies to Watch*. New York: Harper, 2008.
Wilson, Marianne. "Staples Aims for a Greener Planet." *Chain Store Age*, 82/13 (December 2006).

Patrick Flanagan
St. John's University

SUPERFUND

Superfund is the popular name for the Comprehensive Environmental Response, Compensation, and Liability Act (CERCLA). CERCLA, passed by Congress in 1980, was aimed at cleaning up the nation's hazardous waste sites. The timing of the legislation was largely in response to media coverage about the troubles of citizens of New York dealing with hazardous waste cleanup at a site known as the Love Canal. As part of CERCLA, Congress created a special fund to cover the cost of hazardous waste cleanup known as the Superfund.

In 1978, residents of the United States became aware of the struggle of residents in the area of Niagara, New York, living close to and on top of a hazardous waste dumpsite. Residents maintained that they were being exposed to dangerous chemicals that caused cancer, miscarriages, and birth defects. The New York Health Commissioner ordered a cleanup of the area. The commissioner did not want to declare the area as uninhabitable, which angered citizens, as they demanded to be relocated. Part of the dispute had to do with who should pay to relocate families and to clean up the area. Although the city owned much of the land, it had previously been owned by a private chemical company. Moreover, city, state, and local governments could not agree on who should be responsible for the cleanup

This Quanta Resources superfund site is located in Edgewater, New Jersey.

Source: U.S. Environmental Protection Agency

costs. The *New York Times* eventually picked up the story, which gave it national attention. Eventually, all of the residents were relocated, and the state government borrowed the money from the federal government for relocation and cleanup.

The Love Canal case brought the issue of toxic waste sites to the foreground of the U.S. policy-making agenda. The site in New York was only one of thousands of potential hazardous waste sites across the nation. Congress quickly realized that, due to the high cost of cleanup, many citizens could be exposed to hazardous wastes. Congress passed CERCLA to address the cost of cleanup. Superfund was financed by a tax charged to chemical and petroleum companies. CERCLA also increased federal government authority to respond directly to the need for cleanup of hazardous waste sites that endangered public health or the environment. Over the first five years, $1.6 billion was collected and put into the Superfund for cleaning up abandoned hazardous waste sites.

The money allocated to the Superfund was only to be drawn on in cases in which the polluter could not be directly indentified. A provision of CERCLA encompassed the "polluters pay" principle, which required companies responsible for hazardous waste pollution to pay for its cleanup. One of the most controversial aspects of the CERCLA was that polluters were responsible to pay for cleanup even if they no longer owned the site. CERCLA also created rules and requirements for closed or abandoned hazardous waste sites.

In order for a site to become eligible for Superfund money, it was required to go through a verification process called the Hazardous Ranking System (HRS). The HRS is a method to score individual sites based on their danger to the surrounding community. It uses a method of assigning numerical values to various factors that are related to specific conditions at a given site. The factors are grouped into categories including the likelihood that a site had released hazardous waste or toxins into the environment, the specific characteristics of the waste, and the proximity of people or easily upset ecosystems. There are four different paths for this waste to be scored in relation to how the toxins may be spread and these include groundwater, surface water, and air migrations, as well as exposure to soil.

Once a site has been scored and assigned a numerical value, it is posted for public comment. The U.S. Environmental Protection Agency (EPA) places an announcement in the Federal Register and delineates the time period during which the agency will accept comments from the public. Once public comments have been collected and incorporated, the site can be placed on the National Priorities List (NPL). The function of the NPL is to guide the EPA in which sites need the most urgent attention. The NPL ascertains which hazardous waste sites warrant further investigation, distinguish which Superfund corrective actions should be undertaken, notify the public of sites that should be further investigated, and contact entities that may be partly responsible for fiscal penalties.

Although communities benefit from the financial aid that may come with designation on the NPL, many communities resist being listed because of the negative association of being categorized as a hazardous waste site. Once designated, CERLA can fund or authorize two different types of actions. The first are labeled removal actions, in which waste needs to be removed. These removal actions can be further designated as emergency, time-critical, or non-time-critical. The second type of action is labeled as remedial. These actions tend to include measures for preventing the migration of pollutants.

There have been over 1,000 sites designated as Superfund sites across the nation. Individuals interested in knowing about sites in their region of the country can look them up on the EPA website, which includes both current sites and delisted sites.

Superfund has been criticized for being an inefficient program because of the time it takes for a site to become listed, and the time it takes some communities to go from designation to cleanup. Some sites have waited decades with no action because of their NPL ranking. Furthermore, remedial and removal actions can be time intensive and extremely costly. Ironically, Superfund actually has inadequate funds to clean up many of the sites on the NPL.

There are some significant challenges that have arisen in the implementation of CERCLA. To start with, local governments are the entities experiencing the problem, but they have to work to get on a national list. Naturally, each community tends to see its own crisis as the most urgent. Second, because EPA actions are subject to lawsuits, it has made the power to enforce regulations, especially the polluters pay principle, very difficult. The process of public comment is important to ensure that the regulations of Superfund sites are not taken over by industry, but can be very time intensive. Finally, the amount of funding provided for by the legislature is not sufficient to fix all of the sites, which leads to public displeasure.

The CERCLA legislation expired in 2004 and the U.S. Congress has not reauthorized Superfund since that time.

See Also: Brownfield Redevelopment; Corporate Social Responsibility; National Priorities List; Persistent Pollutants.

Further Readings

Ferrey, S. *Environmental Law: Examples and Explanations*. New York: Wolters Kluwer: 2007.

Fogelman, V. M. *Hazardous Waste Cleanup, Liability and Litigation: A Comprehensive Guide to Superfund Law*. Westport, CT: Quorum Books, 1992.

Goldsteen, J. B. *The ABCs of Environmental Regulation*. Lanham, MD: Government Institutes, 2005.

U.S. Environmental Protection Agency (EPA). "Superfund." http://www.epa.gov/superfund/ (Accessed January 2010).

Jo A. Arney
University of Wisconsin–La Crosse

Supply Chain Management

Supply chain management (SCM) is a term applied to a management philosophy and to the practices involved in implementing a SCM philosophy, although it can also be used to describe management processes that are not necessarily rooted in a guiding philosophy. SCM, which gained prominence as outsourcing become prevalent in the 1990s, brings together and extends the concerns of total quality management (TQM) and just-in-time (J-I-T) management techniques that preceded it.

Volvo provides customers with environmental information on products and is working with the Swedish and European Union governments to develop "green corridors." The aim is to have these designated areas increase efficiency and safety through specially adapted transport stretches.

Source: Volvo Group

As a management philosophy, SCM is a systems approach that involves assembling a team of upstream (suppliers) and downstream (distributors and end users) partners with a common goal of satisfying a particular customer value or set of values. This outlook recognizes the interdependency of team members and attempts to foster a cooperative effort that will ultimately give the team a competitive advantage in the market place. SCM differs from earlier management philosophies in that it necessarily involves establishing sufficiently trustful relationships with the members of the chain to warrant involving them in strategic decisions. Companies may be involved in multiple supply chains and function in different roles (e.g., partner, supplier, or customer) at any given time. As a guiding management philosophy and implementation technique, SCM is particularly attractive to socially and environmentally responsible businesses, as it gives assurance that standards set to support social welfare and environmental sustainability are being met throughout the life cycle of any given product.

A supply chain, considered as a system, can also be described as a distributed network or channel through which information, materials, labor, equipment, and currency flow. Though supply chains themselves are not new, four major developments forced thinking about them differently than in the past. The first motivating force came from the communications technology advances that forever changed the span of the marketplace for both the producer and the consumer. The second was the trend toward outsourcing in general, and global outsourcing in particular, which the information age enabled. The third was the increasing importance of timeliness in the marketplace. The fourth was the increasing emphasis placed on quality in all aspects of production and delivery.

The Impact of Technology

The development of computer technology and the Internet enabled a near-instantaneous flow of communication and exerted a dual impact on management philosophy and technique, as it addressed deficiencies in both forerunners of SCM. TQM pioneered the practice of viewing each step in a production process as the "customer" of the preceding step, with negotiated product specifications and intermediate quality inspections. TQM was first applied internally, but was later logically extended to suppliers, particularly after the adoption of ISO 9000 standards. J-I-T, which developed in roughly the same timeframe, was concerned with logistics—coordinating production and delivery along the supply chain to effectively manage inventory costs. Both TQM and J-I-T were hampered by communications limitations. Without a method in place for firms to efficiently link separate information systems, production and inventory information were transmitted by telex or fax machines. Computers and the Internet facilitated the flow of information necessary for SCM.

Simultaneously, communications advances lessened the importance of physical location in relationships up and down the supply chain. Information regarding sourcing options was now much more readily available, which facilitated a move away from vertical integration to outsourcing, because suppliers worldwide were now competing against each other. As connectivity to the web became more common, it opened up avenues for business-to-business exchange and electronic commerce. It also facilitated consumer awareness of and information exchange about the standards and practices of businesses, their subsidiaries, and production partners around the globe.

For producers and retailers, the newfound wealth of information allowed them to seek out suppliers from a greatly expanded population. Outsourcing, whether domestic or international, increases the complexity of management and carries with it two sets of costs. The first set is considered "hard costs," because they are quantifiable and generally known. These include acquisition and transportation costs, and any applicable taxes, tariffs, and other import-related costs. The second set comprises "soft costs," so-called because they are harder to quantify and may not be immediately apparent. These include costs related to the following:

- Time and effort involved in identifying and vetting potential supply chain participants
- Communication (e.g., translation costs, and costs arising from miscommunication)
- Differences in cultural norms and/or expectations
- Reconciling differences in legal system requirements
- Currency exchange and fluctuations
- Addressing security concerns (e.g., proprietary information or transmission)

Understanding and containing these soft costs is critical to companies engaged in international outsourcing, and SCM has proven superior to other methods to address these challenges, particularly when consumer demand for timeliness and emphasis on quality are also considered. In this environment, each decision involving procurement becomes a strategic one that involves multiple departments and has the potential to impact the firm's competitive position.

Under the philosophy of SCM, the chain extends to the consumer as the firm's competitive position rests on understanding what its customers value. Two consumer demands alluded to earlier have been instrumental in the development of SCM—timeliness in the market, as it relates to both responsiveness to tastes and availability; and quality, as it relates to both the produced good and the delivered condition of the good. Beyond these aspects, firms must also determine which consumer values they will cater to. Some may choose, as Wal-Mart has, to emphasize price. Amazon, for example, opts to stress a combination of selection and service. Other companies, like Mercedes-Benz, may differentiate themselves from competitors based on quality.

Benefits of Supply Chain Management

The general benefits attributed to SCM include the ability to manage costs by managing logistics (the legacy of J-I-T), reducing risks by fostering stabile relationships, managing labor costs, ensuring sustainability by developing multiple networks, and insuring against market volatility by engaging in several markets. How each of these issues is addressed will depend on the design of a specific supply chain, vis á vis the central value attached to it—minimum cost, minimum social impact, minimum environmental impacts, or maximum resilience.

The list of benefits grows for firms that adopt SCM and implement life cycle cost analysis as part of their vetting process and on-going operations to ensure that social and environmental responsibility extends throughout their supply chain. The added benefits include avoided waste management costs by ensuring universal waste reduction efforts and proper disposal of all wastes; cost savings realized from the use of system-wide water and power conservation measures; and reduced costs and legal liabilities related to the workplace, the workforce, the community-at-large and consumers. Companies can also benefit from positive public reactions to tangible evidence of responsible corporate behavior.

High profile companies that have adopted SCM include Wal-Mart, L'Oreal, and Nike. Nike has outlined a multi-step process for vetting new sources that begins with assembling a detailed dossier on the business. A series of on-site inspections follows, as does an audit by an independent third party. The last steps include an internal review to determine whether the addition to the supply chain is justified before final approval.

Wal-Mart gained notoriety for some of its SCM practices in the early 2000s. Determined to dominate the field of discount general merchandise retailers, the corporation adopted SCM as a cost containment strategy and aggressively sought out suppliers and pitted them against one another, demanding that they meet Wal-Mart's offered price or lose their position in the chain. Wal-Mart was not alone in using this approach, but it was arguably the most successful. The problem with this approach was that it fostered tentative relationships rather than trustful ones. In order to meet the annual demands for cost reductions, suppliers were motivated to substitute cheaper labor or inexpensive materials that failed to meet agreed upon specifications. In several instances, substandard products were delivered that posed consumer safety risks and necessitated recalls.

Greater Responsibility

For companies that adopt SCM, it is a double-edged sword. Although it offers the many advantages listed above, it also carries with it the responsibility for the actions of all supply chain members. The information age that facilitated the development of SCM also empowered the consumer to monitor and comment on product quality and corporate practices. Companies that engage in SCM cannot credibly claim innocence when their suppliers fail to meet standards that their customers demand, but can expect consumer grievances to be aired in a variety of forums. For this reason, those who adopt SCM as a philosophy grounded in corporate social responsibility rather than merely as a practice have a clear advantage.

See Also: Corporate Social Responsibility; Life Cycle Analysis; Responsible Sourcing; Systems Thinking.

Further Readings

Esty, Daniel C. and Andrew S. Winston. *Green to Gold: How Smart Companies Use Environmental Strategies to Innovate, Create Value, and Build Competitive Advantage.* New Haven, CT: Yale University Press, 2006.
Hopkins, M. "Your Next Supply Chain." *MIT Sloan Management Review,* 51/2 (2010).
Mentzer, J., et al. "Defining Supply Chain Management." *Journal of Business Logistics,* 22/2 (2001).
Mentzer, J., ed. *Supply Chain Management.* Thousand Oaks, CA: Sage, 2001.
Myers, M. and M. Cheung. "Sharing Global Supply Chain Knowledge." *MIT Sloan Management Review,* 49/4 (2008).
Sarkis, Joseph, ed. *Greening the Supply Chain.* New York: Springer, 2006.

Susan H. Weaver
Independent Scholar

SUSTAINABILITY

Every economy, on a most basic level, depends on the ecosystem. Without basic resources such as air, water, and soil, and without the equally essential regulating and supporting services of ecosystems such as water and air purification, carbon sequestration, or pollination, there would be no human life, and hence no economy. While the relationship between the economy and the ecosystem has infinitely complex layers of feedbacks and multipliers, several things are largely beyond dispute: (1) through extraction, depletion, permanent alteration, and pollution, current economies are doing grave harm to the natural environment; (2) many, if not most, of the costs of this harm are not borne by those who cause it—rather, they are borne by outsiders and by future generations; and (3) without a well-functioning ecosystem, the very existence of human beings is endangered.

In the last 40 years, sustainability has developed into a widely used and important concept. At its core lies a simple idea: future generations should (or have a right to) find the same opportunities and resources as were provided to the current generation. Critics have dismissed such notion as noble but impossibly vague, and students of the Earth's ecosystem have struggled to generate clearer, more specific, and ultimately operational guidelines for sustainability.

Though uses of the term go back further, sustainability as a central concept of environmental debate gained international traction when the United Nations (UN) commissioned a study in 1983. It was published in 1987 as the Brundtland Commission's report with the title *Our Common Future* (named after former Norwegian Prime Minister Gro Harlem Brundtland, who was Chair of the UN World Commission on Environment and Development). In essence, the report addresses the question of what sustainable development might look like, and details the economic and political changes necessary to make it happen. Its oft-cited definition of sustainable development is "development that meets the needs of the present without compromising the ability of future generations to meet their own needs." The report, based on the work of hundreds of scientists and scholars, summarizes in its "Call for Action" the basic challenge peoples around the world face: "The relationship between the human world and the planet that sustains it has undergone a profound change . . . not only do vastly increased human numbers and their activities have the power [to radically alter planetary systems], but major, unintended changes are occurring in the atmosphere, in soils, in waters, among plants and animals, and in the relationships among all of these. The rate of change is outstripping the ability of scientific disciplines to assess, and . . . frustrating the attempts of political and economic institutions to adapt."

Despite its members from many different nations and cultures, the committee made a number of urgent and sweeping recommendations. Its concluding plea for action underlined their "unanimous conviction that the security, well-being, and very survival of the planet depend on such changes, now."

As sustainability is an inherently transnational challenge, several international organizations and institutions picked up on the call to action in "Our Common Future" and the work of the World Commission on Environment and Development. Within a few years the 1992 Earth Summit emerged, the adoption of the UN sustainability development program Agenda 21, the approval of the Rio Declaration, which spelled out 27 principles geared toward sustainable development, and the establishment of the UN Commission on Sustainable Development. According to critics, however, none of these (or later) efforts fulfilled even the minimal requirement of successfully curbing the exponentially growing ecological footprint of the world's population (and particularly the footprint of the world's worst depleters and polluters—the United States, China, Brazil, and the European Union). As the slow and embattled implementation of the 1997 Kyoto Protocol on global warming, and the 2009 failure in Copenhagen to reach international agreement on basic goals and standards of implementation indicate, broader discussions on sustainability continue to be long on good intentions and short on practical execution. But international politics is not the only arena in which concerns over sustainability play out.

Debates

The idea of sustainability raises fundamental questions about virtually all key components of modern societies. For starters, given that most of humanity's actions alter the natural environment, which of these actions are, and which are not, sustainable? For that matter, what is the maximum population the Earth can sustain? The roughly 6 billion today will grow to 9 billion by 2050, and the number of people going hungry in 2010 is already projected to exceed 1 billion—a world record. Given our fossil-dependent economy that is largely based on nonrenewable energy (and thus, by definition, not sustainable), furthermore, what can and will replace traditional energy resources? And recognizing that, as of today, literally every major economy in the world is based on a system that requires growth (as measured by gross domestic product [GDP]) in order to function, is there such a thing as "sustainable growth"? Or will societies the world over have to construct economic systems that no longer depend on an ever-increasing output (and subsequent waste) of goods? And, as Nobel prize–winning economist Robert Solow has asked, isn't it hypocritical toward both future generations and today's poor if we now suddenly began to limit growth, in effect saying, we had it good, but now there is no more?

Given the profound significance and virtually endless range of questions involved in how to sustain life on Earth, it is not surprising to discover that sustainability is studied and managed across a wide variety of contexts and scales. From the broadest questions about the carrying capacity of Earth to national and international studies of industries, from analyses of cultural traditions to investigations of communal and individual patterns of lifestyles and behaviors, every part of human conduct directly affects, and is in turn affected by, the fundamental question of the underlying ecosystem's sustainability. Yet the people who generally recognize sustainability as a serious challenge are commonly biologists, chemists, and other natural scientists, alongside environmentalists and practitioners such as urban planners, architects, and green business advocates. Traditionally, economists, corporate leaders, or leading politicians have either dismissed the problem or expressed little more than broad and nonbinding intentions. Part of the reason may be that sustainability inherently puts into question basic beliefs and assumptions underlying modern economies from North America to China.

Most traditional economists today, for instance, still don't believe that the world faces any kind of sustainability problem. While they generally agree that an economy cannot survive if it depletes its assets, most neoliberal economists believe that manufactured capital/ assets can readily replace natural capital/assets. And since they also assume that the value and scarcity of a good can be measured by prices, they refer to studies that do indeed suggest that prices for natural commodities have remained quite stable over the last century— ergo, there is no problem of scarcity or sustainability. And of course such economists can point to many examples in which natural resources were in fact replaced by manufactured (and in some cases superior) alternatives, such as when copper wires were replaced by far better and cheaper optical fibers.

But as environmentalists and ecological economists have been pointing out for decades now, there are several obvious flaws in that logic. One is that prices are not a good indication: the market is structurally impaired, for it does not include goods and services that are not priced (nature, fresh air and water, healthy communities, equality, to name a few). Second, markets do not account for costs borne by others, or what is termed *externalities* (making prices a foolishly unreliable indicator). Third, natural capital is not separate from, but rather complements, manufactured capital—the latter cannot exist without, nor replace entirely, the former. No amount of added new logging equipment can replace the forest; no amount of sophisticated pharmaceuticals can make up for polluted air. Indeed, unlike the loss of manufactured capital, the loss of natural capital is permanent, and potentially catastrophic. As environmentalists have repeatedly shown, it is one thing to discuss the relative significance of losing thousands of rare species each year due to pollution and environmental degradation, or to evaluate the ecological consequences of sprawling human habitations. The loss of something like bees, or rainforests, or sufficient fresh water supplies, or the ozone layer, on the other hand, is no more a question of utility maximization or finding technological alternatives as it is a conservationist's romantic hobby. Rather, it is a problem that cold, mature realism requires us to address, for it quickly throws into question the very existence of human beings on earth. It is, to say the least, a basic question of sustainability.

The largest remaining obstacle to translating environmental sustainability standards into meaningful and transparent public policy continues to be the assumption, still largely unquestioned among economists and policy makers alike, that economic well-being depends on traditional economic growth, and that economic growth would be hampered by the costs of serious ecological stewardship. As the aforementioned economist Solow argued, people want consumption, not investment in the future. What has appeared over the last three decades is a deepening chasm between the findings of scholars and scientists on the one hand, and the policy proposals and applications supported by leading economists and politicians on the other. First brought to light by marginal voices such as former World Bank economist Herman Daly, and articulated by largely ignored groups of international scientists like the Club of Rome, what has emerged since the 1970s is an increasingly broad consensus among scholars that (1) the environment is in exponentially mounting distress; (2) that the way modern economies measure growth—as an indiscriminate total of goods and services produced without any qualification as to what is produced—makes no sense and progressively leads us down a very dangerous path; (3) that growth of the economy and growth of well-being have long ceased to move in the same direction; and therefore (4) that leading performance indicators—both economic and environmental—need to be fundamentally changed in order for national and international communities to be able to address radically changing natural and social realities. While there are many continuing debates and disagreements as to how best to move forward, in short, the emerging conclusion is clear: the current path of modern economies is unsustainable.

Sustainability: A Difficult Concept

Even people who generally agree with the proposition that there are limits to the exploitability of the ecosystem nevertheless differ greatly in their focus and emphasis. Two schools of thought in economics, for instance, are represented by "weak" and "strong" sustainability advocates. Both clearly recognize "market failures" and the inherent value of natural ecosystems. Both camps have proposed similar strategies for sustainability enforcements, such as carbon taxes, waste fines, and tradable extraction or pollution permits. Such caps are generally set in two separate categories critical to sustainability: (1) "assimilative" caps (the level of pollution and destruction the ecosystem can presumably assimilate); and (2) "regenerative" caps (the amount of extraction of renewable resources such as timber or fish the ecosystem can presumably regenerate). What proponents of weak and strong sustainability consider tolerable limits, however, varies greatly. The reason is simple: the prime objective of the former continues to be maximization of the value of consumption, taking into account environmental damage. The goal, in short, remains growth (what Daniel Bell called today's secular religion). The prime objective of the latter is maximization of human welfare. Less concerned with narrowly defined maximization, it instead insists on consumption and pollution caps that fully protect the environment and human health (even if it means giving up on growth).

This raises many thorny and complicated questions, chief among them: (1) what are the maximum limits of extraction and pollution that are ecologically sustainable (i.e., roughly sustain current opportunities of life on Earth for future generations); and (2) is economic growth (at least as currently defined) possible within such limits? While research is still a long way from providing conclusive answers to these questions, several theoretical and empirical observations can be made. Logically, production and consumption do not have to be unsustainable (solar energy or education or improved healthcare all have very small ecological footprints). At least since the industrial revolution, however, modern economies have become increasingly resource-intensive and wasteful, rapidly depleting resources, putting billions of pounds of toxic materials into the ground, air, and water, and generally causing irreversible harm to the planet. While modern economies did not set out to wreck the planet (the goal, rather, was to produce and consume the greatest volume of goods as efficiently as possible), the result was rampant exploitation and waste of natural resources (over 90 percent of materials extracted to produce commodities in the United States today become irretrievable waste within the first month).

In short, there are no experiences with sustainable modern economics. What has been demonstrated, however, is that equally high levels of consumption can be achieved with significantly different ecological footprints. Denmark, for instance, has a comparable standard of living to the United States, but its per capita water use and carbon emissions are just a little over half that of its American counterpart, while its per capita amount of waste is less than half, and its amount of recycled and reused materials is more than double. It is important to note that neither model gets close to sustainability. According to ecological footprint estimates, if every earthly inhabitant consumed at the level of the average American, we would need about 3.5 planets; to sustain the consumption patterns of the average Dane, we would need 2 planets.

Scientists who informed international participants at the 2009 UN climate summit in Copenhagen concluded that the world needs to reduce its greenhouse gas emissions by at least 90 percent in order to avoid catastrophe. No enforceable targets were set, but the nonbinding goal articulated was reductions of carbon emissions to avoid more than a 2-degree Celsius increase in temperatures. According to official statements, China, the United States, India, Brazil, and South Africa combined emit more than 80 percent of global emissions. In short, the crisis is recognized, it harms everyone though it is mainly caused by a few, future generations will bear the brunt of the burden, yet so far there has been complete failure to take meaningful action.

Looking at sustainability from the perspective of thermodynamics, physics, and biology, some logical conclusions follow. In a nutshell, the Earth is a finite and closed system. That is, with the critical exception of incoming solar energy, which, in evolutionary terms, constitutes the essential source of the development of ever more complex and structured systems on earth. Taking away or fundamentally altering that closed system leads to a loss of biodiversity and a decline in the Earth's ability to process solar energy. Fossil fuels, for instance, are essentially a highly potent, carbon intensive energy storage system of millions of years of solar-powered decomposition. The reason that oil, gas, and coal are called nonrenewable energy resources is that once we burn them, they are irretrievably gone. They also cause, at current levels of consumption, some estimated 22 billion tons of carbon dioxide emissions each year. While debate goes on about what exactly this means for the future carrying ability of Earth, there is broad consensus that a fossil-based economy is fundamentally unsustainable: if only the average Chinese resident (out of the total 1.5 billion) were to achieve the same level of consumption as the average American (as they are on track to do by about 2030), each day they alone would use 20 million barrels of oil more than the entire world consumes today. As virtually all students of this crisis have concluded, "that kind of demand would stress the earth past its breaking point in an almost endless number of ways." The U.S. National Aeronautics and Space Administration's (NASA) chief climatologist James Hansen put it this way in 2006: if the world does not start using less carbon dioxide within 10 years, we will live on "a different planet." More and more scientists around the world express ever-more dire warnings, not just about sustainability, but about an impending catastrophe.

As such, sustainability would not appear to be an issue that falls along the traditional Right/Left divide, even though it is still often couched in terms of "tree huggers" versus hardworking realists. Conservation of the environment, as the name already suggests, is an inherently conservative issue, as is caring for one's community and for one's children (and their children). Forcing the markets to reflect actual costs of production and consumption more realistically and more responsibly, it would seem, should find support among liberals, conservatives, and progressives alike. The same is true for a multitude of other sustainability strategies: energy independence, and with it less entanglements with foreign dictatorships or the Organization of the Petroleum Exporting Countries (OPEC); support of community-based economics and individual entrepreneurial spirit; corporate social responsibility, which takes into account the health of the community and the environment as much as the proverbial bottom line; the end to a range of government subsidies to large corporations and environmentally unhealthy practices; or the prospect of a much simplified tax system that essentially tracks harm caused to the environment, but also rewards good stewardship. With clearly defined and transparent goals, all such initiatives are likely to find support across the ideological spectrum.

As an issue that concerns social scientists as much as natural scientists, politicians as much as citizens, and as an issue that has a multitude of political, social, and environmental ramifications, the basic question remains: "How can human welfare and economic well-being be achieved without irreversible harm to the planet?"

Sustainability as a Practical Concept

Aside from the theoretical debates within science, politics, and economics, sustainability has a multitude of practical applications. Indeed, to the extent that environmental crises, resource shortages, increasing energy costs, and organized political efforts converge, businesses and consumers alike seek new, more sustainable solutions. According to recent surveys, the percentage of American consumers who are willing to give up time, money, and effort for "green products" has climbed to over 60 percent. The market for organic food alone doubled between 2001 and 2006 and is now a business valued at over $5 billion. Consumers who seek green, sustainable products have considerable power. They can, and increasingly do, demand things such as the following:

- "Extended producer responsibility" (producers taking back their products, reusing as much as possible, and recycling the rest responsibly)
- Radical reductions of resource use, such as water, paper, electricity, or packaging materials (in Germany, for instance, consumers can now deposit all packaging material with the retailer)
- Use of renewable energy resources by producers and retailers
- Organically produced food, cosmetics, apparel
- More community-based production to radically reduce carbon footprint
- Buildings and communities that have environmental ratings
- Green investment portfolios

And businesses throughout the industrialized world respond. According to the UN Centre for Trade and Development (UNCTAD), all of the 25 most globalized transnational corporations (TNCs), with combined revenues exceeding the GDP of all poor and middle-income nations on Earth, now feel obliged to make available sustainability or corporate social responsibility reports. With notable exceptions, however, most of these reports still do not go beyond vague attempts to address an evaluation of what the World Business Council for Sustainable Development has coined *ethical behavior of a company toward society*. Only a handful of TNCs, such as General Electric or the German energy company RWE, follow what analysts call "comprehensive" sustainability strategies. But as some medium-sized enterprises such as Patagonia, Sainsbury's, or Stonyfield Farm have demonstrated, there are business models that successfully pursue both: profitability and environmental stewardship that approximates sustainability.

As the demand for green products grows, however, so does the market for public relations gimmicks and plain manipulation, peddling a wide variety of products under the rubric of "organic," "natural," and "environmentally friendly," even though they fail basic tests of sustainability. As skeptics of corporate sustainability and social responsibility reports have pointed out, furthermore, greenwashing has become a widespread practice to divert attention from the fact that many businesses are essentially incompatible with sustainable practices. Interested primarily in the economic bottom line, these critics argue, most businesses have no long-term commitment to ecosystem preservation, community or place, equity or social justice. As one observer noted, they are more interested in "excessities" than necessities.

If the production, distribution, and consumption of goods were to move in a more sustainable direction, in short, consumers require transparency and readily available detailed information. As many observers have pointed out, however, in order for sustainability to become a widely meaningful reality, there is, above all, a dire need to finally translate the burgeoning findings of scientists into a broad political reality. Without clear goals, much improved performance indicators, and enforceable mandates, sustainability will remain a widely supported but elusive goal.

See Also: Carbon Footprint; Ecological Economics; Ecological Footprint; Externalities; Genuine Progress Indicator; Greenwashing; National Income and Product Accounts; Sainsbury's; Stewardship; Stonyfield Farm.

Further Readings

Edwards, Andres R. *The Sustainability Revolution: Portrait of a Paradigm Shift.* Minneapolis, MN: New Society Publishers, 2005.

Farzin, Y. Hossein. "Is an Exhaustible Resource Economy Sustainable?" *Review of Development Economics,* 8/1:33–46 (2004).

Hicks, John R. *Value and Capital: An Inquiry Into Some Fundamental Principles of Economic Theory.* Oxford, UK: Clarendon Press, 1939.

Marshall, Julian D. and Michael W. Toffel. "Framing the Elusive Concept of Sustainability: A Sustainability Hierarchy." *Environmental Science and Technology,* 39/3: 673–82 (2005).

McDonough, William and Michael Braungart. *Cradle to Cradle: Remaking the Way We Make Things.* New York: North Point Press, 2002.

McKibben, Bill. *Deep Economy: The Wealth of Communities and the Durable Future.* New York: Times Books, 2007.

Solow, Robert M. "Sustainability: An Economist's Perspective." Eighteenth J. Seward Johnson Lecture in Marine Policy. Wood's Hole, MA: Wood's Hole Oceanographic Institution, 1991.

Wackernagel, Matthias and W. Rees. *Ecological Footprint: Reducing Human Impact on the Earth.* Gabriola Island, British Columbia, Canada: New Society Publishers, 1996.

World Commission on Environment and Development. "Our Common Future." New York: Oxford University Press, 1987. http://www.ourcommonfuture.org (Accessed January 2010).

Dirk Peter Philipsen
Virginia State University

SUSTAINABLE DESIGN

Design processes create material culture; that is, the tangible goods, systems, and symbols by which we live communicate and define our places in the world. Sustainable design is the philosophy and activity of designing physical objects that move society toward greater social, economic, and environmental sustainability. Socially, sustainable design has a net positive impact on human capital, that is, on the health and well-being of people and communities. Environmentally sustainable design has a net positive impact on natural capital, meaning that it helps maintain or restore ecosystems.

The Green Building at Macintosh Village, in Manchester, England, includes many features of ecological home design, including a roof-mounted wind turbine.

Source: Terry Whalebone/Wikipedia

Efforts to develop sustainable designs grew out of concerns for limited natural resources, increased nonrecyclable materials in landfills, and increased chemical toxicity in the environment. The earliest efforts include "ecodesign" or "green design," which focused on energy and materials. Sustainable design goes further to include concerns for social and economic sustainability.

Principles of Sustainable Design

One of the best-known approaches to sustainable design was developed by architect Bill McDonough and chemist Michael Braungart. Together they created what is called the cradle-to-cradle design methodology for evaluating chemicals and materials. They modeled their work on the system of natural decomposition, nature's capability to create food or nutrients from waste. Cradle-to-cradle design includes the cyclical use of nonbiodegradable materials; energy from renewable sources such as solar and wind; and safe, nontoxic materials. McDonough and Braungart argue that products can be redesigned and transformed to generate ecological, social and economic value. Sustainable designs are being used by a variety of design professionals including architects, product engineers, and packaging specialists. Design for disassembly (DfD) aids the cradle-to-cradle process by facilitating the return of product components to closed technical cycles of reuse, remanufacturing, and recycling.

Applications of Sustainable Design

Architecture

Sustainably designed buildings save organizations money, have less negative environmental impacts, and make better, healthier workplaces. Sustainable building design grows out of an integrated design process driven by whole-system thinking. It starts from the very beginning of the process with cooperation and decisions from the design team, architects, engineers, and client. Site selection, scheme formation, material selection, material procurement and, finally, project implementation require agreement among all stakeholders. The U.S. Green Building Council's Leadership in Energy and Environmental Design (LEED) building program provides guidelines for sustainability throughout the design and building process. Builders can earn certification based on three levels of sustainability: Silver, Gold, and Platinum. Buildings are graded for energy and water conservation, the reduction of building waste, and the use of renewable and nontoxic materials.

Architecture design firms specializing in LEED-certified buildings enjoy prominence in a growing market. For example, SERA architects in Portland, Oregon, a leader in LEED-certified urban revitalization and sustainable design projects, works to balance urban and natural resource flows. In a recent issue of the *Harvard Business Review*, Charles Lockwood noted that from 2000–07, there was a 232 percent increase in white-collar LEED buildings and 178 percent in blue-collar facilities. He also noted that the Genzyme Center, a 12-story LEED Platinum building in Cambridge, Massachusetts, uses 42 percent less energy and 34 percent less water than traditionally constructed buildings of comparable size.

The future of sustainable building design reaches well beyond the standards for LEED Platinum status to the even more ambitious categories of "living buildings" and "restorative buildings." The Living Building Challenge, issued and administered by the Cascadia Region Green Building Council, is a rigorous set of standards intended to help designers create buildings that generate all their own energy with renewable resources; capture and treat all their wastewater on site; enhance occupants' indoor experiences; and use resources efficiently and for maximum beauty. Restorative buildings would not only be environmentally benign; they would contribute to improving and restoring ecosystems.

Product Design

Sustainable product design offers numerous opportunities for innovation and creativity. Designer who have moved beyond the "form follows function" of modernity and the "form follows fun" of postmodernism find challenges in creating products that perform as well functionally and aesthetically as traditionally made products and are made according to sustainable design principles.

When Nike teamed up with environmentalist and NBA All-Star guard Steve Nash to create Nike Trash Talk, the emphasis was on creating a sustainable, high-performance basketball shoe. Nike used manufacturing waste including leather and synthetic leather waste from the factory floor for the upper. Rather than using toxic adhesives, the upper was sewn together with zigzag stitching. Scrap-ground foam from factory production made up the mid-sole. Nike mixed Nike Grind—recycled waste from their own outsole manufacturing—with a low-toxin, environmentally preferred rubber to make an outsole containing 96 percent fewer toxins. Even the shoelaces and sock liners were made from environmentally preferred materials. Finally, the shoes were packaged in a fully recycled cardboard shoebox. The company continues to push the designers to create more sustainable products. New design specifications are given sustainability scores based on Nike's Considered Index.

Herman Miller, maker of high-end office chairs, has been a champion of the cradle-to-cradle design methods for some time. For example, their Mirra chair is 96 percent recyclable. The chair was designed for disassembly at the end of life. Each connection in the chair takes 30 seconds or less for one person to disassemble, making postconsumer recycling easier and more profitable. When Herman Miller designers start on a new chair, they also consider how materials work together for disassembly; for example, molded steel and plastic won't be integrated into one piece that cannot be dissembled. If Herman Miller designers specify plastics with recycling codes on them, they get a higher score on their internal sustainability rating card than if they choose plastics that may be difficult to identify. Any specified chair component that has to go to the landfill gets a zero on the company's internal score card. The higher the score, the better the chair.

Packaging Design

Packaging provides protection for the product, brand information for the consumer and convenience for everyone in the supply chain. Packaging is also the first thing to be discarded in the consumer use phase of the value chain. The U.S. Environmental Protection Agency (EPA) reports that 90 percent of all fossil fuel–based plastic packaging ends up in a landfill. This translated to over a quarter trillion tons of trash in 2007.

Manufactures are learning to reduce their packaging in an effort to limit the amount of consumer waste that ends up in landfills. The nonprofit Sustainable Packaging Coalition (SPC) advocates for environmentally friendly packaging and innovation in the supply chain. By their definition, sustainable packaging is

- beneficial, safe, and healthy for individuals and communities throughout its life cycle;
- meets market criteria for performance and cost;
- sourced, manufactured, transported, and recycled using renewable energy;
- maximizes the use of renewable or recycled source materials;
- manufactured using clean production technologies and best practices;
- made from materials healthy in all probable end-of-life scenarios;
- physically designed to optimize materials and energy; and
- effectively recovered and utilized in biological and/or industrial cradle-to-cradle cycles.

When Starbucks needed to redesign packaging for a line of chocolates, it was for aesthetic and merchandising reasons. However, using principles of sustainable package design, they were able to create a package that, compared to its unrecyclable predecessor, required 50 to 60 percent less material, weighed less, required fewer folds in the manufacturing process, needed less glue, and eliminated polylaminates, making it entirely recyclable.

Sustainable Design and the Future

In her *Designer's Atlas of Sustainability*, Ann Thorpe places design deep in the heart of ecology, economy, and culture. In discussing the future and frontiers of sustainable design, she points out that design is a potentially powerful change agent. But to effect change, design must be approached and understood in a whole-system context of technology, policy, and behavior. Policy provides the incentive for technology development, and designers create products that change behavior. For example, tax laws may favor the development of renewable energy over fossil fuels, encouraging the development of increasingly effective and versatile photovoltaic technologies. Designers, in turn, incorporate solar electric technology into products—such as clothing, automobiles, homes, and electronic devices—that people want, and that change their behaviors. Design creates the material culture through which a society defines and understands itself. A sustainable society requires increasingly sustainable design.

Further Readings

McDonough, William and Michael Braungart. *Cradle to Cradle: Remaking the Way We Make Things*. New York: North Point Press, 2002.
Sterling, Steve. *Field Guide to Sustainable Packaging*. Chicago, IL: Summit Publishing, 2007.
Sustainable Packaging Coalition. http://www.sustainablepackaging.org/ (Accessed January 2010).
Thorpe, Ann. *The Designer's Atlas of Sustainability: Charting the Conceptual Landscape Through Economy, Ecology, and Culture*. Washington, D.C.: Island Press, 2007.
Williams, Daniel E., David W. Orr and Donald Watson. *Sustainable Design: Ecology, Architecture, and Planning*. Hoboken, NJ: John Wiley & Sons, 2007.

Diane M. Martin
John W. Schouten
University of Portland

SUSTAINABLE DEVELOPMENT

The term *sustainable development* came into use after the publication of the Brundtland Commission's report "Our Common Future" in 1987. Officially named the World Commission on Environment and Development, this panel of experts from different countries was chaired by G. H. Brundtland, and had been given the task in 1983 by the United Nations (UN) General Assembly of defining a new type of global development that reconciled environment and development in both the north and the south. Its famous report paved the way for the major UN Conference on Environment and Development (UNCED, in Rio de Janeiro, Brazil, in 1992), also referred to as the Earth Summit. The Rio Conference was convened 20 years after the UN Conference on the Human Environment (UNCHE, in Stockholm, Sweden, in 1972) and 10 years before the World Summit on Sustainable Development (WSSD, in Johannesburg, South Africa, in 2002). These milestones in the UN sphere have strongly influenced the way that environmental concepts and policies have gradually been integrated into development issues, thereby drawing the contours of a sustainable development perspective.

For years, the overwhelming majority of countries have had sustainable development strategies, developed on a national, state, or local level, or by various public and private institutions, including corporations. At the start of the 21st century, while the idea of sustainable development appears to have broad support, there is much less clarity about either its precise meaning or which steps to take toward its implementation.

More than 30 years after its publication, the Brundtland Report's definition of sustainable development—"Sustainable development is development that meets the needs of the present without compromising the ability of future generations to meet their own needs"—remains by far the most cited. Two key concepts in sustainable development, according to the reports, are

- needs, and in particular the essential needs of the world's poor, to which overriding priority should be given; and
- limitations, and particularly those imposed by the state of technology and social organization on the environment's ability to meet present and future needs.

Broadly defined, an environmental or ecological approach to sustainability tends to stress ecological constraints, or the carrying capacity of a territory, prior to allowing the expansion of development. In the Strategy for Sustainable Living, endorsed in 1991 by the International Union for Conservation of Nature (IUCN), the UN Environment Programme (UNEP), and the World Wildlife Fund (WWF), sustainable development is basically defined as "improving the quality of human life while living within the carrying capacity of supporting ecosystems." The approach taken by most economists, in contrast, focuses on what is necessary to assure further development. A typical expression of this view comes from the economist D. W. Pearce, when he wrote that "sustainable development is readily interpretable as nondeclining human welfare over time—that is, a development path that makes people better off today but makes people tomorrow have a lower 'standard of living' is not 'sustainable.'"

A definition for sustainable consumption used by the UNEP is "the production and use of goods and services that respond to basic needs and bring a better quality of life, while minimizing the use of natural resources, toxic materials and emissions of waste and pollutants over the life cycle, so as not to jeopardize the needs of future generations." The interest in sustainable consumption and production patterns was already visible in the official agreements at the Rio Conference in parallel with business and industry's involvement in sustainability issues. From the start, the formula of sustainable development has set the stage for a possible reconciliation between environmentalism and business, as when the Brundtland Report titled one of its chapters "Producing More with Less." In the same vein, the Business Council for Sustainable Development (BCSD) was founded before the Rio Conference, with the encouragement of its secretary, in order to participate actively in discussions about sustainability.

The Three Dimensions of Sustainable Development

One increasingly predominant notion of sustainable development originated in the mid-1990s from the world of business, and is part of a larger trend that developed in response to increased public and political pressure to respond to the environmental crisis. A crucial component of this development is to see sustainability as resting on three essential pillars, namely the environment, society, and economics (or what literature commonly refers to as the triple bottom line). This type of analysis can be found in corporate reports, variously termed *social responsibility*, *corporate responsibility*, or *sustainable development* reports. Frequently, such reports apply the Global Reporting Initiative (GRI) guidelines—principles and indicators to measure and report economic, environmental, and social performance. Economic indicators include things like financial performance and contributions to sustainability efforts, infrastructure development, or portfolio development. Environmental indicators include carbon emissions development, recycling efforts, energy production and consumption, effects on biodiversity, water use, emission, and waste. Social indicators include amount and type of employment, turnover rate, collective bargaining rights, rates of injury, levels of education, healthcare, and corruption.

Although these reports provide the reader with a wide variety of information, they often lack data and detail, don't explain interaction within and between different categories, do not provide scales of importance, nor make transparent how and why particular criteria developed as they did.

Another version of the "triple bottom line" approach to sustainable development is to design "win-win-win" initiatives leading to combined benefits for the three pillars. A classic example would be for a business to develop successful "green" products that would increase its economic profit and create jobs. Limited "win-win" (two pillar) initiatives are also encouraged in this perspective. A major policy advocated by European Union and Organisation for Economic Co-operation and Development (OECD) countries consists of tax-shifting programs to raise the share of environmental taxes while diminishing taxes on labor. This policy is also known as the "double dividend" (reducing both the amount of pollution and labor costs).

In the context of sustainable development, much has been achieved, specifically regarding interaction between the economic and environmental pillars. The general principle for a synergistic relationship will internalize the costs of environmental externalities, which essentially means that the polluters carry the full cost, in market prices, of the pollution or the degradation of natural resources. The "polluter pays principle," which the OECD advocated in the late 1970s (and was subsequently included in the Rio Declaration) continues to represent a cornerstone of sustainable development policy principles.

Implementing this principle, on the other hand, has turned out to be exceedingly difficult. While there is little difficulty demonstrating the direct benefits to the larger community—both national and international—as well as to future generations, finding parties willing to bear the higher costs has proven almost insurmountable. How to price externalities, and how to discount future costs (e.g., the effects of global warming or the depletion of oil on future generations) is difficult even on a purely conceptual level. Sir Nicolas Stern, former chief economist at the World Bank and lead author of a major report on this problem, has described climate change as "the biggest market failure the world has ever seen."

One can distinguish several approaches toward sustainable development. The first strives for greater equity between populations around the globe (including the Brundtland

Report citing the needs of the world's poor in their definition). This goal is ambitious, and the last decades suggest that achieving it would take structural changes, of which ecological policies would be just a part. Another understanding of the social dimension mixes environmental topics with selected issues pertaining to the social arena. This was the case with the European Union, for instance, when parts of health, social integration, and aid policies were included in the EU Strategy for Sustainable Development (2006). A third understanding of the social aspect of sustainable development requires that attention be paid to the social impacts of environmental policies, such as the impact of a carbon tax on the distribution of revenues within a country.

Future Generations and Global Sustainability

What sets sustainable development criteria apart from other development strategies is that it is primarily concerned with intergenerational equity. As such, it is concerned "future generations," which include, but are not restricted to, the next generation. The idea, not dissimilar from the concept of the "seventh generation," is to evaluate today's actions with an eye toward its consequences on the distant future, and ensure that future generations will live as well as the current generation. The assets and capacities to be sustained in order to meet the needs of the future are not, in the seminal Brundtland conception, limited to the environment, but rather include economic, institutional, and cultural capacities. Nevertheless, as the Brundtland report made very explicit, assuring that the essential vital functions of our ecosystem are sustained is not only significant, it is a prerequisite for all other considerations, be they social, cultural, or economic.

Scientific reports can help assess the range of challenges that the principles of sustainable development have to confront regarding environmental protection. Progress has been made concerning some local issues, especially in developed countries. Cases in point include a reduction of some of the environmental impacts of industrial productions, such as the steel or automobile industry, which have seen significant improvements in water use, toxic runoffs, and soil contamination since the end of the 20th century. Globally, however, the continued exponential increase in consumption of primary resources (energy, land, and water) has not been matched by progress in ecological management.

The "Millennium Ecosystem Assessment," a comprehensive report on the short- and long-term benefits that ecosystems provide for humanity (ecosystem services), concluded in 2005 that "the challenge of reversing the degradation of ecosystems while meeting increasing demands for their services can be partially met under some scenarios . . . but these involve significant changes in policies, institutions and practices that are not currently under way." In an example of bureaucratic understatement, the report conclusion was based on a wealth of data that showed steady or accelerating deterioration of the world's ecosystems due to a continuous depletion of resources, habitat destruction, and an increase in waste production and toxic emissions. The most pressing problem reported was the continued growth of greenhouse gas emissions, which, if not curbed within the very near future, will almost certainly lead to a climate catastrophe.

Despite these dire predictions, or perhaps because of their threat, the adjective *sustainable* has lately been coupled with a variety of concepts, such as growth, cities, agriculture, production, and consumption. Finding more sustainable consumption and production patterns has become increasingly important for the advancement of sustainable development. Transnational corporations, in particular, now consider sustainability important enough to have formed, and support, what has since morphed into the World Business

Council for Sustainable Development, a coalition of major companies exchanging information and developing a common discourse on the topic of sustainability.

This business friendly overture contrasts with the views on economic growth that preceded the debates and conferences on sustainable development. One of the first major international statements on the state of the environment, for instance, the 1972 Club of Rome report "Limits to Growth," very clearly considered economic growth anathema to lasting global development. Many scholarly studies still question the compatibility of growth and sustainable development. Some ecological economists, such as Hermann Daly, consider sustainable development to be an oxymoron, for growth through development that consumes natural resources in a finite world cannot, by definition, be sustainable.

The majority opinion is to make growth "greener" (and for environmentalists is a minimum condition), which in technical terms entails decoupling growth from the use of nonrenewable resources and energy, an objective also highly advocated by OECD countries. The greatest means to achieve this is expected to come from green or ecoefficient technologies. As such technologies are being developed in many contexts and environments, but still comprise only a negligible percentage of the overall economy (only about 7 percent of all energy production in the United States as of 2009, for instance, was from renewable sources), radical shifts in funding priorities and technological developments would have to take place. Expecting to meet sustainability objectives through increasing technological responses, soft institutional changes, and the convergence of demand from various economic players (investors, consumers, and industries) would constitute a type of sustainable development that has been called "ecological modernization."

At the beginning of the 21st century, sustainable development is a widespread area of concern while being a path of hope for the future. It seeks convergence, reconciliation, and equity among multiple strands of development, as well as between current and future generations. Both severe poverty and rampant material excess fundamentally undermine the basic pillars of sustainable development. Scientific forecasts voice concern regarding ecological challenges, suggesting that technological and policy responses have thus far not been able to rise to the challenge of sustainable development—not environmentally, socially, or economically. At the beginning of the 21st century, in short, sustainable development as a concept appears to enjoy almost universal support. Verifiable actions and results on the path toward sustainable development on the part of industries and national economies, however, are increasingly falling behind.

See Also: Clean Technology; Externalities; Factor Four and Factor Ten; Green Design; Green Technology; National Priorities List; Sustainability; Sustainable Design; Triple Bottom Line.

Further Readings

Adams, B. *Green Development*. London: Routledge, 2008.
Daly, Herman E. and Joshua Farley. *Ecological Economics: Principles and Applications*. Washington, D.C.: Island Press, 2003.
Hansen, J., et al. "Target Atmospheric CO_2: Where Should Humanity Aim?" *Open Atmospheric Science Journal*, 2:217–31 (2008).

Hawken, P., A. Lovins and L. Lovins. *Natural Capitalism: Creating the Next Industrial Revolution*. Boston, MA: Little, Brown, 1999.

HM Treasury. "The Stern Review on the Economics of Climate Change." (2006). http://www.hm-treasury.gov.uk/sternreview_index.htm (Accessed May 2009).

Hopwood, B., et al. "Sustainable Development: Mapping Different Approaches." *Sustainable Development*, 13:38–52 (2005).

International Institute for Sustainable Development (IISD). "Sustainable Development Principles." http://www.iisd.org/sd/principle.asp (Accessed May 2009).

International Union for Conservation of Nature (IUCN). *Caring for the Earth: A Strategy for Sustainable Living*. Gland, Switzerland: IUCN, 1991.

Millennium Ecosystem Assessment. "The Millennium Ecosystem Assessment." (2005). http://www.millenniumassessment.org (Accessed May 2009).

Pearce, D., ed. *Blueprint 2: Greening the World Economy*. London: Earthscan, 1991.

Von Weizsäcker, E., A. Lovins and L. Lovins. *Factor Four: Doubling Wealth, Halving Resource Use*. London: Earthscan, 1997.

World Commission on Environment and Development. *Our Common Future*. Oxford, UK: Oxford University Press, 1987.

Zaccai, E., ed. *Sustainable Consumption, Ecology and Fair Trade*. London: Routledge, 2008.

Edwin Zaccai
Université Libre de Bruxelles

SYSTEMS THINKING

The essence of systems thinking is to view a problem as a whole, comprised of parts that interact with and affect one another. A change to one system component has widespread impact, affecting every other part. Thus, solutions to problems are found by considering these impacts. This is in contract to reductionism, which seeks for answers to a problem by considering each individual piece, regardless of its interaction with other pieces of the whole. Systems thinking is a learning process that also enables an individual to simplify a situation perceived as complex. Systems thinking has had a significant influence on the environmental movement through writers ranging from John Muir and Aldo Leopold, to Donella Meadows and Bill McKibben. Application of systems thinking by green businesses requires managers to recognize the complexity, interconnectedness, and sustainability impacts of all their decisions. The complexity in green business arises because multiple competing economic, environmental, technological, and social issues must be considered. The effects of organizations' actions can be unpredictable and have widespread impacts.

Systems

Systems are defined as a set of components comprising ideas, objects, and activities that are interconnected for a purpose. The purpose of the system emerges because systems thinkers and practitioners have an interest in it and want to improve it. With the help of systems diagrams, a situation or system of interest can be explored. Systems of interest are represented by system maps. In contrast to the unbounded complex reality of systems, system maps are bounded. They have an explicit boundary that distinguishes the system of interest from those parts of the situation that are perceived to be of secondary importance. A system of interest

- is an organized assembly of interconnected components;
- consists of a number of subsystems, the behavior of which is affected by being in the system, and which change the behavior of the system by leaving it;
- does something;
- is identified by someone because he is interested in it.

Further, systems are characterized by several concepts. The system environment, comprising the wider context, external influences, and subsystems of secondary importance is distinguished from the system itself by a boundary. The boundary is the result of judgment and simplification processes undertaken by an analyst who is interested in the system. A system transforms inputs into outputs. Subsystems interact within a hierarchy and are interdependent. Communication, feedback, and control take place between subsystems and the system as a whole. Systems are adaptive and dynamic.

Systems Thinking and Systems Practice

No simple or single definition captures all aspects of systems thinking. It involves seeing a situation as a complex system, and holds the central idea that the nature of the system is in the eye of the beholder. Since people have different worldviews comprised of different values and traditions of thinking, there are many ways to see the world through a systems lens. Nevertheless, authors agree upon common features of systems thinking:

- A complex situation produces no solution, but results in multiple comprehensions, all with the aim of acting purposefully within the situation.
- People, and their viewpoints and emotions, belong to the situation.
- Complexity is not a feature of a situation, but arises from the interplay of the observer with the situation.
- The connectedness between events, things, and ideas is as important as the events, things, and ideas themselves.
- Systems thinkers apply techniques that encourage a wide variety of thought in considering a situation, thus not overlooking significant connectedness and important features of the whole situation. One technique is to use multiple partial views by considering the perspectives or viewpoints of various stakeholders in a situation. The purchase of printing paper, for example, can be seen in terms of managing supplier–buyer relationships, financing, or environmentally friendly business behavior. In tackling a situation, the first step is to take a "helicopter view" to see the whole picture and the context of the situation. By doing so, systems thinkers are borrowing ideas from holistic thinking.
- In systems thinking, diagrams play a central role, facilitating learning by representing complex situations in each step of the iterative learning cycle. The learning cycle begins with immersion in the complexity of the situation and ends with an exploration of activities that can improve it. The "rich picture," a situation summary diagram that enables experiencing of complexity, represents as much as possible about a situation including feelings, motivations, pictorial symbols, keywords, and cartoons.

An effective systems practitioner can be described as a reflective thinker who considers four elements continuously: his or her being, his or her engagement with a real-world situation, his or her contextualization of approaches to the real-world situation of interest, and his or her management of his or her involvement in the situation. To consider his or her own being, the systems practitioner must be aware of his or her worldview, beliefs, and traditions of thinking. The practitioner reflects upon whether or not he or she engages a situation as complex. Several systems approaches need to be contextualized within and adapted to a specific situation of interest. Hard Systems Method, Soft Systems Methodology, and Critical Systems Thinking are examples of systems approaches. In a multimethodological mode, various methodologies, methods, and concepts are synthesized and applied in a situation. In managing a situation, the practitioner must decide whether he or she is in or out of the situation.

Implications for Green Business

Systems thinking is at the heart of green business. Organizations that are effective, in terms of sustainable business, collaborate with and learn from a variety of stakeholders. Therefore, effective collaborations and interorganizational relationships are found to be sustainable in systems thinking terms: they are self-organizing. Systems thinking should be used in all aspects of business, from operations management to human resources, and business schools increasingly offer classes in sustainable business and systems thinking applications. Decisions made in each aspect of businesses have far-reaching implications and wide-ranging effects on the global system as a whole, and on its various organizational, economic, social, and ecological subsystems. Varied and sometimes conflicting stakeholder perceptions can result in lack of consensus as to what constitutes a successful green business.

Traditional business asks consumers to think only of the immediate gratification of their individual desires, while environmentalists urge consideration of multiple upstream and downstream impacts to ecological systems. To be fully sustainable, businesses must not only consider the environmental and social impacts of internal decisions, but must help potential customers to be systems thinkers as well. Rather than encouraging customers to consider immediate impacts, such as price and gratification of perceived needs—as their traditional competitors do—green business will educate them as to the multiple indirect impacts of their purchasing decisions. To the extent that consumers become systems thinkers, green businesses will likely flourish in comparison to traditional businesses.

See Also: Balanced Scorecard; Corporate Social Responsibility; Environmental Assessment; Smart Growth; Sustainability.

Further Readings

Checkland, P. B. *Systems Thinking, Systems Practice.* Hoboken, NJ: John Wiley & Sons, [1981] 1998.

Checkland, P. B. and J. Poulter. *Learning for Action: A Short Definitive Account of Soft Systems Methodology and Its Use for Practitioners, Teachers and Students.* Chichester, UK: Wiley, 2006.

McKibben, Bill. *Deep Economy: The Wealth of Communities and the Durable Future.* New York: Holt Paperbacks, 2008.

Meadows, Donella. *Global Citizen*. Washington, D.C.: Island Press, 1991.

Meine, Curt D. and Richard L. Knight. *The Essential Aldo Leopold: Quotations and Commentaries*. Madison: University of Wisconsin Press, 2006.

Muir, John. *John Muir: The Eight Wilderness Discovery Books*. Seattle, WA: The Mountaineers Books, 1992.

Franziska Hasselmann
Swiss Federal Institute of Forest, Snow,
and Landscape Research

Take Back

Consumer societies around the world are producing an ever-growing amount of waste. Resources are being depleted, and communities are beginning to choke from overflowing waste disposal sites. According to studies, well over 95 percent of all materials that go into the production of consumer goods end up as waste within a month. Most materials wasted, furthermore, not only come from nonrenewable resources, but could have been reused or recycled. Ordinarily, producers have little or no responsibility for what happens to the materials that go into producing the goods they sell, or the materials used for packaging. Take-back campaigns and legislations are spreading globally, and all have a simple goal: to reduce the amount of waste. Through both voluntary requests and mandates, producers are asked to take back their products at the end of their life cycles in order to reuse and/or recycle them. Ideally, take-back campaigns and legislations are geared toward a substantial reduction of materials used in the first place.

Take-back schemes have largely come about due to increased pressures on municipal waste site capacity and as a means of reducing the amount of toxic waste sent to landfill or incineration. There is also a growing recognition of the scarcity of natural resources. The electronics waste stream, for instance, has grown rapidly. This has been further exacerbated by a decline in product lifespan (companies increase profits through planned obsolescence) that has brought about an increased frequency of production and consumption. According to estimates, for example, a computer purchased in 1992 would have lasted 4.5 years, whereas the average life span of a computer bought in 2005 was little more than 2 years.

There is also growing pressure to undertake take-back under the premise of Extended Producer/Product Responsibility (EPR). In some countries EPR is mandatory through legislative requirements. According to the latest count, some 30 countries worldwide now have laws that mandate companies to be actively involved in the "end-of-life cycle" of their products. The European Union, for instance, has directives on the books such as the Waste Electronic and Electrical Equipment (WEEE) Directive or the End-of-Life Vehicles Directive (ELV). Other recorded examples of international take-back laws include the following:

- Norway, which requires that up to 90 percent of all electrical and electronic products must be recovered by their producers for reuse or recycling

- The European Union, which has laws that requires automobile manufacturers to take back their old vehicles free of charge, and to reuse or recycle at least 85 percent of their content
- Brazil, which has a National Solid Waste Policy that requires corporations to obtain an "Environmental Operating License" that includes take-back requirements

The fundamental rationale behind take-back laws is not only to enforce corporate responsibility, but rather to create an incentive for producers to pay more attention to the design stage of their product or packaging life cycle. Studies show that most energy and resources are used at the production stage: the more incentive to cut down on materials used at the outset, the fewer problems at the end-of-life cycle of the product.

The United States currently has the distinction of being the only highly developed nation that has no federal take-back laws on the books. Several states, however, have recently passed EPR, or take-back laws, or have EPR legislation pending. Most of those laws, however, are minimal in scope, often restricted to electronics, and frequently demand municipalities to pay for recycling programs. Several transnational companies, some of which are American, on the other hand, have been in the forefront of voluntary take-back campaigns. Indeed, almost all electronics giants, from Motorola to Samsung to Apple, currently have campaigns under way to encourage costumers to return their used cell phones or computers or other electronic devices for resale, reuse, or recycling purposes.

Take-back programs encourage the concept of cradle-to-cradle thinking, as it gives producers the opportunity to accept their original goods back and consequently reuse useful components. Ultimately, successful take-back schemes could, therefore, feed into reuse, remanufacturing, or recycling processes. However, this depends on the capacity of process and product design to enable access to, dismantling of, and reincorporation of used components. Such actions will depend on the quality and status of the product and the value of its components as well as the policies of the Original Equipment Manufacturer (OEM) and could take years to implement since they may also have ramifications further up or down supply chains.

Furthermore, in countries where collective producer responsibility is largely the norm, there is little incentive for producers to go against the grain and invest in mechanisms to take back or recover their individual products. The United Kingdom, for example, offers a collective responsibility system whereby producers pay a proportion of money to a scheme that covers the cost of recycling their goods. Typically, these end-of-life products go to a general collection and recycling plant and require sorting before the materials are crudely separated (usually by machine). To achieve individual take-back in a scenario such as this would require some type of identification system, perhaps radio frequency identification (RFID), or that the products be manually sorted—both of which would be a costly up-front investment. The advantage of a collective system, however, is that it helps overcome issues surrounding "orphan products" that come from producers that no longer exist. Sometimes, retailers may offer take-back programs on consumer goods before sending them either back to OEMs or to an appropriate treatment plant.

There are alternative models for achieving take-back. Instead of changing the product design or collection method, for instance, companies could adopt a holistic business model, making the producer responsible for the entire life cycle of the product. Instead of selling it, the producer retains ownership of the product, as well as increases the incentive to repair or remanufacture the products. This model has been successfully adopted in a business-to-business environment by Xerox, although it may not be applicable across all product types.

With consumer goods, it is essential that consumers are aware of take-back schemes, since without their cooperation and knowledge of such schemes, these goods are unlikely to be returned to OEMs. However, consumers often store unused or old products themselves, perhaps unaware of what can be done with them. Also, some consumers feel that there must be some inherent value to a computer bought years back for double or triple the price that they would pay for a new computer today, and they can be reluctant to get rid of them—in particular, when it might even cost the consumer to have it recycled. Therefore, it is imperative that joint efforts between business, government, and consumer groups seek to educate consumers as to the appropriate options.

Although take-back can help meet the goal of reducing the quantity of hazardous materials going to landfill, it will not in itself succeed in creating more circular materials flows. Until the cost of recycling or the desire to incorporate ecodesign concepts forces a more substantial change in the business culture, it is likely that individual producer responsibility and product take-back will continue to be another retrospective waste management system.

See Also: Closed-Loop Supply Chain; Compliance; Cradle-to-Cradle; E-Waste Management; Extended Producer Responsibility; Extended Product Responsibility; Green Design; Remanufacturing; Resource Management; Reverse Logistics; Stewardship.

Further Readings

Electronics Take Back Coalition. "Producer Responsibility for Electronic Waste—You Make It, You Take It Back." http://www.computertakeback.com/corporate/corporate_main.htm (Accessed September 2009).

Huisman, Jaco and Ab Stevels. "Eco-Efficiency of Take-Back and Recycling: A Comprehensive Approach." In *Proceedings of the 2003 Institute of Electrical and Electronics Engineers (IEEE) International Symposium on Electronics and the Environment.* New York: IEEE, 2003.

Huisman, Jaco, Ab Stevels, Thomas Marinelli and Federico Magalini. "Where Did WEEE Go Wrong in Europe? Practical and Academic Lessons for the U.S." In *Proceedings of the 2003 Institute of Electrical and Electronics Engineers (IEEE) International Symposium on Electronics and the Environment.* New York: IEEE, 2006.

Klausner, Markus and Chris Hendrickson. "Product Takeback Systems Design." http://gdi.ce.cmu.edu/gd/Research/takeback.pdf (Accessed September 2009).

Nakajima, Nina and Willem H. Vanderburg. "A Failing Grade for WEEE Take-Back Programs for Information Technology Equipment." *Bulletin of Science, Technology and Society,* 25/6 (2005).

Pagell, Mark, Zhaohui Wu and Nagesh N. Murthy. "The Supply Chain Implications of Recycling." *Business Horizons,* 50/2 (2007).

Solving the E-Waste Problem (StEP). "E-Waste Take-Back System Design and Policy Approaches." http://www.step-initiative.org/pdf/white-papers/StEP_TF1_WPTakeBack Systems.pdf (Accessed September 2009).

Toffel, Michael W. "Closing the Loop: Product Take-Back Regulations and Their Strategic Implications." *Corporate Environmental Strategy,* 10/9 (2003).

Weee Forum. "About Collective Take-Back Systems." http://www.weee-forum.org/index.php ?section=collective&page=collective_about (Accessed September 2009).

Widmer, Rolf, Heidi Oswald-Krapf, Deepali Sinha-Khetriwal, Max Schnellmann and Heinz Böni. "Global Perspectives on E-Waste." *Environmental Impact Assessment Review,* 25 (2005).

Cerys Anne Ponting
Cardiff University

TESCO

Tesco is a significant international retailer. As a global business, the company perceives itself as having an important role to play in helping to minimize its environmental impacts.

Although based in the United Kingdom, Tesco PLC has developed into an international grocery and general merchandising chain, employing 440,000 staff worldwide and with sales exceeding £59 billion (in the year ending February 2009). Operating income exceeded £3 billion. These figures make Tesco the largest British retailer in terms of both total sales and domestic market share, and the third largest in the world after Wal-Mart (United States) and Carrefour (France). It is claimed that £1 in every £7 of UK retail sales goes the way of Tesco.

Like many companies, the United Kingdom–based Tesco grocery chain now tries to minimize its environmental impact. Here is a Tesco Superstore, in Clapham in London, England.

Source: tescoplc.com

The company was founded in 1919 by Jack (later Sir Jack) Cohen, selling groceries from a stall in London's East End. "Tesco" as a brand first appeared five years later. Growth since the early days has occurred both organically and by acquisition, and today the total number of stores exceeds 3,700.

Today, although Tesco remains focused on grocery items, following the Continental hypermarket model, it has also diversified into consumer electronic goods, clothing, DVDs and CDs, furniture, telecoms, and even insurance and other financial services.

Over the years, Tesco has become increasingly concerned with corporate social responsibility and via its corporate website declares the following overall philosophy:

- Respond to consumer trends such as healthy eating, organics, and Fair Trade
- Treat suppliers fairly so they can "deliver" for customers
- Manage the company's environmental impacts to help reduce costs and inconvenience

- Treat staff as they deserve to be treated so they perform well for customers
- Help staff and customers to support the local organizations and causes they care about

Tesco has identified a number of areas for specific attention in its stated objective of leading its sector toward sustainable consumption by reducing direct environmental impacts and encouraging both suppliers and customers to do the same.

Climate Change

The intention is to lead the way by Tesco's dramatically reducing its own carbon footprint, meanwhile making low-carbon products available to customers at affordable prices. This requires establishing relevant statistics to identify the largest impacts and provide transparency; working with others to achieve change (£25 million was provided to the University of Manchester to finance work in reducing carbon emissions in the supply chain); and empowering customers via product labeling to assist their purchasing decisions. Tesco publicizes its efforts on its corporate website and through news releases; for instance, they report having invested £60 million in low carbon and energy saving technology in 2008 and to have reduced their energy use in the United Kingdom (in 2009) to 50 percent of what it was in 2000. Their efforts focused at enabling consumers to make more environmentally friend choices, to include developing carbon footprint labeling for 100 of their own-branded products, promoting the use of energy-saving lightbulbs by halving their cost, and introducing the Greener Living line of products.

Waste, Recycling, and Packaging

Adoption of the corporate slogan "reduce, reuse, and recycle" has implications in terms of minimizing the amount of waste sent to landfill by reducing the amount of packaging and providing fewer plastic carrier bags. Various initiatives and trials are targeted at helping customers to recycle more and to reduce food waste on the part both of the company and of its customers.

Sustainable Sourcing

Policies are in place to ensure that key resources are sourced responsibly. This has involved work with suppliers and other groups to develop sound policies (regarding such things as using completely legal timber sources, maintaining viable wild fish and shellfish populations, the provision of biofuels, and the appropriate sourcing of palm oil) that are then rigorously implemented.

Responsible Buying and Selling

Tesco is provisioned by more than 5,000 companies around the world, although its stated intention is to place special emphasis on supporting smaller, local businesses, thereby contributing better to their respective communities. Ethical trading, animal welfare, and responsible selling all figure prominently in these aspects of the Tesco philosophy.

Tesco's overarching slogan is, "Every little helps." Its green initiatives, on many fronts, appear to be achieving more than just a small amount of progress.

See Also: Corporate Social Responsibility; IKEA; Responsible Sourcing; Sainsbury's; Waitrose.

Further Readings

Humby, Clive, Terry Hunt and Tim Phillips. *Scoring Points: How Tesco Continues to Win Customer Loyalty*. London: Kogan Page, 2008.

Okoebor, Alfred. *The Big Three: Tesco, Asda and Sainsbury: The Argument Reconsidered*. Saarbrücken, Germany: VDM Verlag, 2009.

Simms, Andrew. *Tescopoly: How One Shop Came Out on Top and Why It Matters*. London: Constable, 2007.

Tesco PLC. "More Than the Weekly Shop: Annual Review and Summary Financial Statement 2008." London: Tesco PLC, 2009. http://www.investis.com/plc/storage/2008_TESCO_REVIEW.pdf (Accessed August 2009).

David G. Woodward
University of Southampton School of Management

Toxics Release Inventory

The Toxics Release Inventory (TRI) is a publicly available database maintained by the U.S. Environmental Protection Agency (EPA) that contains information about toxic chemical releases and waste management activities by federal facilities and certain industries. The TRI program began in 1987 following enactment of the Emergency Planning and Community Right to Know Act (EPCRA) in 1986. That legislation was motivated in part by public outcry following the deaths of thousands in Bhopal, India, following release of methyl isocynate into the air by a Union Carbide plant located there, an incident followed in 1985 by a serious chemical release at another Union Carbide plant located in Institute, West Virginia, in the United States.

The EPA compiles TRI data annually and makes it available through several databases, including Explorer (www.epa.gov/triexplorer) and Envirofacts (www.epa.gov/enviro). Other organizations have also developed interfaces to TRI data, including the Right-to-Know Network and Green Media Toolshed. Material may be accessed in several ways including by geographic area, company, type of pollutant, type of industry, and year.

Currently about 650 chemicals are included in TRI, about twice as many as in 1987. The database has also expanded in terms of the number of industries that are required to submit information, and the thresholds for some chemicals have been lowered. A primary purpose of TRI is to provide information to communities about toxic chemicals being released into their environment, but the reporting process also operates as a type of informational regulation: because companies know they must report releases of toxic chemicals and that the information is accessible by researchers and the general public, they have an incentive to reduce the amount released.

As an example, we can conduct a search by ZIP code using the TRI Explorer interface. Entering in ZIP code 29801 identifies five facilities and their toxic chemical data, including on-site and off-site disposals and release data per facility, as identified by name and an assigned TRI Facility Identification Number (TRIF ID). A uniquely assigned number for the purposes of TRI reporting, facilities usually only obtain one ID number, even if reporting for several buildings within that facility. There are exceptions, however, where one facility will obtain several ID numbers for multiple units belonging to it. A closer review of the data submitted by AGY Aiken, LLC reports 243,026 pounds of on-site releases. Further inspection of the data reveals the origins of these releases—"certain glycol ethers, hydrogen fluoride, methanol."

Using the following formula we can capture the total amount released:

$$\text{land, air, water emissions} + \text{chemicals consumed} + \text{waste stream.}$$

In this case:

$$30{,}782 \text{ (lbs.)} + 197{,}133 \text{ (lbs.)} + 15{,}111 \text{ (lbs.)} = 243{,}026 \text{ (lbs.)}$$

Further emission details become available by clicking on each of the chemical constituents. Clicking on "glycol ethers," for example, leads to the official form capturing the reported data which sheds light on the company's current use of the chemical (glycol ethers), its anticipated future uses, and its source reduction strategies. AGY Aiken, LLC claims to use certain glycol ethers as a manufacturing aid. Air emissions—both point and nonpoint— dominate their emissions reporting, with smaller wastewater releases and disposal quantities comprising the remaining delta.

The TRI is a tool to collect and make available data required to meet the intention of EPCRA, a law that establishes protocols for emergency situations while empowering the public through education.

Only firms in specified sectors (listed below) that have 10 or more full-time workers using a listed toxic chemical above threshold quantities are required to report their emissions to EPA:

- Manufacturing (Standard Industrial Classification (SIC) codes 20 through 39)
- Metal mining (SIC code 10, except for SIC codes 1011, 1081, 1094)
- Coal mining (SIC code 12, except for 1241 and extraction activities)
- Electrical utilities that combust coal and/or oil (SIC codes 4911, 4931, and 4939)
- Resource Conservation and Recovery Act (RCRA) Subtitle C hazardous waste treatment and disposal facilities (SIC code 4953)
- Chemicals and allied products wholesale distributors (SIC code 5169)
- Petroleum bulk plants and terminals (SIC code 5171)
- Solvent recovery services (SIC code 7389)

Federal facilities have also been subject to reporting requirements since 1993 under Executive Order 12856.

Federal facilities have used TRI reporting as a tool to reduce pollution: currently they are required to achieve 50 percent source reduction goals for their reported chemical releases using their first reporting year as the baseline (but no later than 1994). Community and environmental groups use TRI data to track the amounts of toxic chemicals released in their region and to bring pressure on corporations to reduce that amount. TRI data also play a role in disaster response; for instance, they were used after Hurricanes Katrina and Rita to identify factories and industrial sites which might have released toxic chemicals and in 2007 to identify potential contamination risks after the bursting of a holding pond in East Tennessee which collected wastes from a coal-fired electric plant. Private companies have also taken measures to reduce TRI numbers, for instance, by reusing or recycling chemicals that would previously have counted as waste. DuPont reduced iron chloride emissions by developing a product that used the substance in road base (the layer below pavement in a road).

The TRI offers a wealth of information that can help motivate companies to reduce their release of toxic chemicals and provide information to communities and researchers about toxic chemicals in their midst. Overall, the amount of reported chemicals released has been declining: the EPA reported an overall 6 percent reduction in toxic chemicals (by pound) released into the environment in 2008, as compared to 2007, although the amounts of some chemicals increased. The TRI presents an incomplete and potentially misleading picture of the situation because not all companies are required to report to the inventory and not all chemicals are included. Moreover, because release data are reported in pounds, and not toxicity level or human exposure level, the TRI data do not indicate the health or environmental risks from the emissions of a particular business. Industries have successfully lobbied for exemptions to certain kinds of chemical waste, and the database contains no information about the risk a particular chemical poses to humans or the environment. This creates the possibility for deliberately misleading reports: a corporation could publicize that they have substantially reduced their emissions of a chemical that poses low threat to people and the environment while increasing emissions of another chemical that may be highly toxic, even in small amounts.

See Also: Environmental Accounting; Persistent Pollutants; Right to Know; Transparency.

Further Readings

Green Media Toolshed. http://www.scorecard.org (Accessed October 2009).

Hamilton, James T. *Regulation Through Revelation: The Origin, Politics, and Impacts of the Toxics Release Inventory Program.* Cambridge, UK: Cambridge University Press, 2005.

Right-to-Know Network. http://www.rtknet.org/db/tri (Accessed October 2009).

U.S. Environmental Protection Agency. "Toxics Release Inventory (TRI) Program." http://www.epa.gov/tri (Accessed October 2009).

Union Carbide Corporation. http://www.ucarbide.com (Accessed October 2009).

Anju Fritz
University of Maryland

TOYOTA

The Japanese multinational company Toyota is one of the largest automobile manufacturers in the world and produces one of the most popular brands of imported cars in the United States. Toyota introduced the first mass-produced gas/electric hybrid car (the Prius) in 2000

A bus shelter in Chicago with solar panels designed by Toyota. The panels help run fans and circulate air, similar to what Toyota offers in some Prius models, which have fans to draw in outside air to cool the car when parked in the sun.

Source: Toyota

and continues to be an innovation leader in developing energy efficient and alternate fuel vehicles.

Toyota was established in Japan in 1937 as the Toyota Motor Co., Ltd. and began doing business in the United States in 1957 as Toyota Motor Sales, U.S.A., Inc. By 1967 Toyota was third in U.S. sales among imported brands of automobiles and in 1976 became the most popular import brand in the United States, aided in part by the 1973 gas crisis that encouraged many Americans to purchase smaller, more fuel-efficient vehicles. Toyota began manufacturing operations in the United States in 1972 and in 1986 began U.S. vehicle production.

Hybrid, Alternative-Fuel, and Fuel-Efficient Cars

Toyota introduced the Prius in 2000, making it the first company to mass produce and sell hybrid gas/electric cars, or cars that combine an internal combustion engine with an electric battery which can propel the car. The Prius emitted 90 percent less smog-forming gases than conventional cars and had a fuel economy rating of 45 (city) and 51 (highway) mpg (miles per gallon), according to the U.S. Environmental Protection Agency (EPA). In 2010, the Toyota Prius was ranked the most fuel-efficient vehicle overall by the EPA, with 51 mpg for city driving and 48 mpg for highway driving.

The success of the Prius (the most popular hybrid car in the U.S.) encouraged Toyota to expand hybrid technology to its other lines of cars. In 2005, Toyota introduced the Toyota Highlander (a sport utility vehicle [SUV]) and the Lexus RX 400h, the world's first hybrid luxury car. In 2006, the Camry (a sedan) became available as a hybrid, and in 2008 Toyota introduced the Lexus LS 600h, the world's first V8 hybrid. As of 2010, Toyota offers seven hybrid vehicles and has sold over 1 million hybrids in the United States (10 percent of all U.S. Toyota sales are hybrids) and over 2 million worldwide. The company has announced that in the future, it intends to offer hybrid version of every Toyota passenger model.

Toyota conventional fuel cars are also more fuel efficient and produce fewer emissions than the average for U.S. cars. In 2010, the EPA rated the Toyota Yaris, a conventional fuel car, the most efficient subcompact, with 29 mpg city and 36 mpg highway, and a total of seven Toyota cars were rated at 30 mpg or higher for highway driving. Toyota Corporate Average Fuel Economy (CAFE) for cars in 2009 was over 36 mpg (versus the industry average of between 31–32 mpg), Toyota averaged just under 26 mpg for CAFÉ and for trucks versus the industry average of just over 23 mpg. Carbon dioxide (CO_2) emissions for new Toyota cars and trucks were both below the industry average in 2009.

Toyota is currently developing and testing fuel cell and plug-in electric vehicles. In December 2009, it began a global demonstration plan for the Prius Plug-In Hybrid, which is modeled on the third-generation Prius but has a battery that can be recharged by plugging into a 110 or 220 volt outlet. Toyota intends to introduce the FCHV-adv, a fuel cell hybrid vehicle based on the Toyota Highland SUV, in 2015. It uses four compressed hydrogen fuel tanks, an electric motor, a nickel-metal hydride battery, and a power control unit: electricity generated by mixing oxygen and hydrogen gas is used to charge the battery and power the vehicle. The FT-EV II is a battery electric vehicle which is an update of the FT-EV concept introduced in 2009: it uses a lithium-ion battery, can be charged from a 110 or 220 volt outlet, and has a range of about 55 miles on a fully charged battery. Toyota has announced that it will introduce a production electric vehicle in 2012.

Sustainable Operations

Toyota has made numerous reforms to reduce waste and environmental impact, and has an ongoing program to increase environmentally friendly practices. In the operations sphere, Toyota was able to reduce energy consumption at North American nonmanufacturing facilities by 26 percent between 2001–08. Several Toyota facilities have received Leadership in Energy and Environmental Design (LEED) certification (meaning they meet the standards of the U.S. Green Building Council), including a technical training center in Miramar, Florida; a vehicle distribution center in Portland, Oregon, that is powered by 100 percent wind energy and diverts 99 percent of its construction waste from landfill); and several automotive dealerships. In 2008, 93 percent of waste from Toyota's Canadian sales headquarters was recycled, as was 71 percent of waste from the U.S. sales headquarters in FY 2009.

Toyota has been an Energy Star Partner since 2003, and Toyota plants have received 14 Energy Star awards since 2006 (meaning their energy performance is in the top 25 percent of the industry). Water usage per vehicle has been reduced by 20 percent since 2003, and the use of returnable shipping containers within the Toyota parts distribution network has saved over 30 million pounds of wood and 10 million pounds of cardboard since 2003. Toyota has increased the use of materials made from renewable resources in their vehicles; for instance, a soy oil-based foam is used in the Corolla and Lexus RX for passenger seats, and 20 percent of the Lexus HS 250h interior and baggage area is covered with plastics made from renewable resources.

Criticisms

Hybrid vehicles can reduce tailpipe emissions and reduce fuel consumption, but cannot achieve zero emissions because they must burn some fuel to power the internal combustion engine. Hybrid vehicles tend to be more expensive than comparable traditional-fuel cars (and fuel cell cars are predicted to be more expensive), which presents a barrier to purchasing them for less-affluent consumers. Electric vehicles do not burn gasoline, but use electricity that must be generated, and thus indirectly contribute to pollution. Concerns have also been raised about the environmental cost of replacing and disposing of batteries in hybrid and electric vehicles, although Toyota says their hybrid batteries are meant to last for the life of the car and are recyclable.

See Also: Clean Fuels; Green Design; Industrial Ecology; Sustainability; Waste Reduction.

Further Readings

Hasegawa, Yozo. *Clean Car Wars: How Honda and Toyota Are Winning the Battle of the Eco-Friendly Autos.* Tony Kimm, tr. Singapore: John Wiley & Sons (Asia), 2008.

Morgan, James M. and Jeffrey K. Liker. *The Toyota Product Development System: Integrating People, Process and Technology.* New York: Productivity Press, 2006.

Osono, Emi, Nirihiko Shimizu, Hirotaka Takeuchi and John Kyle Dorton. *Extreme Toyota: Radical Contradictions That Drive Success at the World's Best Manufacturer.* Hoboken, NJ: John Wiley & Sons, 2008.

Sperling, Daniel. *Two Billion Cars: Driving Toward Sustainability.* New York: Oxford University Press, 2009.

Union of Concerned Scientists. "Hybridcenter.org: A Project of the Union of Concerned Scientists." http://www.hybridcenter.org (Accessed February 2010).

Sarah Boslaugh
Washington University in St. Louis

TRANSPARENCY

The concern over transparency is not new. Rather, it is an interest that has been modified according to stakeholder interests as well as an evolving economy. The growing acceptance of the association between human activities and events as contributing causes to the adverse impacts on the environment is a recent influence on the growing demand for transparency.

Business transparency generally consists of three parts: the transparency of organizations between each other and related parties; accounting transparency; and transparency to consumers. In each context, the balance between privacy and transparency is manifested in many ways. For example, reasonable disclosure of information to stakeholders is necessary for free market efficiency. The common law prohibiting inflicting harm on others implies recognition of the causes of adverse impacts on people and the environment. Damages caused by activities may be mitigated by disclosure and anticipated by warning notices. In some situations, the onus for disclosure or warning necessarily rests with the source of the emission. For example, some hazardous air pollutants, such as carbon monoxide (CO), that are lethal to human beings at sufficient concentrations are colorless and odorless, which renders them undetectable. Originating sources of such hazardous substances then bear the burden of preventing and mitigating damages.

Regulatory regimes can make it imperative to bring such events to the attention of the public as a safety concern. The same regulations may balance disclosure with limits in recognition that complete transparency is not always appropriate. The exposure of processes to the light of scrutiny could be necessary for systems to operate as designed, while confidentiality can permit discussions that help processes and improve decision making. Meaningful or reasonable transparency in the context of business involves balancing appropriate visibility and opacity to optimize operations and decision making by all stakeholders. Stakeholders in a globalized economy include diverse environmental and social responsibility interests, as well as the usual financial interests of stakeholders in businesses. The resulting regulatory regimes are increasingly complex, and various jurisdictions rely on innovative approaches to transparency to effectively improve access to stakeholders.

Increased transparency has emerged as a means of obtaining improved performance of disclosing entities. Transparency relies on decision makers to obtain the desired improvements and to do so, the approach draws people on both sides of the disclosure boundary into the decision making process. Where transparency is implemented, relevant compliance and enforcement operations may be valuably directed to the availability, accessibility, and adequacy of information with regulation of the use of the information addressed under different regimes. For instance, in response to reports that mistakes in medical procedures resulted in significant fatalities in the United States, mandatory public disclosure of such events was implemented. Similarly, a disclosure of the risk analysis in sport utility vehicle–related injuries was required by regulation rather than regulating the design of the vehicles themselves. Transparency can also influence governments to implement and enforce regulations as seen in the consequences from citizen complaints under the North America Free Trade Agreement (NAFTA) regime.

Business Organizations

In regulated economies, privately owned businesses and closely held corporations are subject to transparency requirements, imposed by the owners or shareholders between themselves as well as by the government. Partners in a business would typically require full transparency between themselves, even if decision making is unequally allocated. The government, employees, and the general public would not be entitled to the same information as partners. Transparency, then, is not uniform, but contextual. Some transparency may be mandated by regulations, or may be voluntarily provided by agreement, or even unilaterally, such as when it is used in business marketing information or in applications for financing.

Publicly traded corporations, those that have securities traded on public stock exchanges, would include additional transparency provisions resulting from securities legislation. In the European Union, January 2008 transparency requirements, as with the Swedish rules of 2007, applied to companies listed on the Luxembourg Stock Exchange. These require periodic publication of information, disclosure of large acquisitions, or dispositions of shares at specified thresholds, and continuing information sufficient to ensure equal treatment of shareholders. In addition to shareholders, other stakeholders in such businesses could include governments, lenders, management, employees, suppliers, and customers. International trade can sometimes involve voluntary disclosure consequences, such as when businesses seek certification to private standards like the ISO 14000 series. Efforts to increase the disclosure of risks have come from each of these groups, as well as from some industries and citizens.

A growing public interest in environmental impacts of business operations brings further pressure for additional transparency. The need for this transparency may make businesses greener, as they may find themselves having to improve and update their products, policies, services, emissions, or waste processes to respond to the demand for corporate accountability. Consumers now have access to information on the hidden ecological costs of doing business with various organizations by using website databases like the GoodGuide.com, which rank the environmental, health, and social impacts of consumer products.

Disclosure regimes can vary according to the entities involved. Securities disclosure information for publicly traded companies can be seen in the Electronic Data Gathering, Analysis and Retrieval system (EDGAR) database of the U.S. Securities and Exchange Commission (SEC), and the equivalent System for Electronic Document Analysis and Retrieval (SEDAR) in Canada. Transparency in some government operations can also be seen in the Environmental Bill of Rights Registry in Ontario, Canada. Information required may be both qualitative and quantitative, and associated with organizational objectives as well as with impacts beyond the organizational boundaries.

Pollutant release and transfer registers also operate in several countries. The mere existence of such registries can only result in beneficial impacts if decision makers act on the information available. Indications of information use become transparent when future impacts of present operations are addressed in management disclosure and analysis requirements, as well as periodic investor guidance in securities filings. Additional meaningful disclosure addressing specific risks, including climate change processes, are not often addressed. For example, few if any businesses disclose the effects climate change will have on their operations.

Demand for more explicit information on identifiable risks associated with climate change impacts and business is increasing. Where it is possible to compute the impact of a business, such as in a carbon footprint or environmental burden calculation, and decision makers are motivated, the transparency of this information could be used to improve efficiencies and reduce costs to the business, as well as to society. Where sustainability is seriously valued, transparency over time, including intergenerational transparency assumes similar significance.

Some existing transparency may be limited by proximity, such as when information is available but is not readily accessible to a wider audience, including the general public. Employees, for instance, may be protected by health and safety protocols, but may in fact be in contact with hazardous substances. However, they may not have ready access to the material information regarding health risks, and may not appreciate the consequences of exposure. In some circumstances, employees that are most vulnerable may not have a commercially reasonable alternative to hazardous substance exposure, in which event, the availability of information and transparency may not be of immediate personal practical relevance. Whether the information is personally significant or not, the information may remain pertinent to decision makers concerned about environmental health.

Beyond the boundaries of businesses, materials handling requires transparency. For instance, regulations applicable to the transportation of dangerous goods and toxic substances may require disclosure of product characteristics in materials-handling data sheets. This disclosure along the supply chain can be negotiated between suppliers, buyers, and contracted service providers, who in turn, as with transportation enterprises, may be subject to similar regulatory disclosure of substances they are transporting. A similar analysis would apply to the sourcing and supply of energy to a business as well as energy sinks associated with the same business. Transparency here may be limited to parties to agreements, or may apply more broadly outside the organization in marketing materials and internally with employees to add value to the workplace.

In a societal context, businesses are typically subject to local zoning and other planning regulations, as well as licensing and permit provisions. Such provisions operate to provide some initial transparency as well as when material changes are made to the business. Communities in the vicinity of facilities may be informed of events either under other bylaws and local regulations or by environmental authorities.

Transparency in operations, as well as in financial reporting, has also become an issue to stakeholders concerned about adverse environmental impacts of human actions and events. Emission trading systems may involve disclosure of environmental impacts and include financial information. Investors continue to call for greater transparency pertinent to their interest in the level of risk associated with environmental impacts of business activities.

Accounting

Transparency, in recent times, has become an issue, as a result of events such as the Enron accounting scandal—coinciding with the progressive adoption of the International Accounting Standards Board (IASB) issued International Financial Reporting Standards (IFRS) worldwide in replacement of the Generally Accepted Accounting Principles (GAAP), the implementation of the Sarbanes-Oxley (SOX) reporting requirements for corporate governance in the United States, and the credit crisis precipitated when transactions appear to have involved misunderstood, misused, and misrepresented financial products.

The IASB is a group of 14 members from nine different countries that works to develop global accounting standards. The goal of this committee is to create global standards that are transparent, enforceable, understandable, and of high quality. The IFRS is a result of this effort, and is endorsed by some countries with notable exceptions. The exceptions include the U.S. Financial Accounting Standards Board, and the Accounting Standards Board (AcSB) of Canada, who do not adhere directly to IFRS but are still involved in discussions. The ongoing task for the IASB is to include these countries in subscribing to coordinated standards.

Consistent global standards could facilitate the presentation of comparative and commonly understood information. This capability is particularly relevant where businesses operate and have stakeholders in multiple jurisdictions. Convergence of various accounting standards, as well as comparable transparency, could assist in achieving this goal.

The credit crisis revealed information about financial transactions between financial institutions and large companies that had either not been visible or not understood by some of the parties to the transactions and their agents. Subsequent inquiries indicate that misrepresentation of finances and accounting may have resulted from reporting practices that had not been designed to disclose pertinent information. The effect of the crisis triggered a call for improved transparency in financial reporting.

Businesses may voluntarily subscribe to industry disclosure provisions, or may be obligated to do so as a condition of membership in an industry association. For instance, transparency is an integral part of the Responsible Care regime, a scheme that chemical producers developed as an industry standard.

Consumer Protection

Customers may also be influenced by the environmental conditions in which a business operates. A food-related business, such as a restaurant that does not have a transparently healthy and sanitary operation, may not remain in business for long for two reasons. For one, it may lose its permit to operate, and for another, it could lose its customers. For consumer and durable goods producers, product life cycle analysis and end user or consumer alternatives need to provide informational transparency for success in product and service delivery.

See Also: Corporate Social Responsibility; Emissions Trading; Informational Regulation; Triple Bottom Line; Voluntary Standards.

Further Readings

Goldschmidt, Mark R. "The Role of Transparency and Public Participation in International Environmental Agreements: The North American Agreement on Environmental Cooperation." *Boston College Environmental Affairs Law Review*, 29/2:343 (Winter 2002).

McFarland, Jeffrey M. "Warming Up to Climate Change Risk Disclosure." *Fordham Journal of Corporate & Financial Law*, 14/2:281–323 (2009).
Weil, David, Archon Fung, Mary Graham and Elena Fagotto. "The Effectiveness of Regulatory Disclosure Policies." *Journal of Policy Analysis & Management*, 25/1:155–81 (Winter 2006).

Lester de Souza
Independent Scholar

TRIPLE BOTTOM LINE

The phrase *triple bottom line* (often abbreviated as TBL or 3BL) refers to an expansion of the factors to be considered in evaluating the success of a corporation or other organization. Besides the traditional bottom line of profitability, the 3BL requires accounting for the effects of the corporation's operations on the physical and social environments (sometimes characterized as natural capital and human capital in analogy with the economic capital that is the focus of the traditional bottom line). The notion of the 3BL was developed in the 1990s as an effort to popularize the notion of corporate responsibility in the ecological and social spheres by broadening the accountability of a corporation beyond the economic interests of the shareholders to include societal stakeholders, who might be affected by the corporation's activities (for instance, a community might be affected by the pollution emitted by a plant and employees might be affected by unsafe working conditions within a plant). These concerns are sometimes expressed with the phrase "people, planet, profit" or "P+," popularized by Royal Dutch/Shell.

Although many would endorse the notion of increased corporate responsibility toward the physical and social environment, there is no obvious way to measure effects in these spheres analogous to the generally accepted standards of financial accounting. For this reason, some critics feel that the very notion of 3BL is misleading because it implies a precision that is not possible to achieve.

In 2003, Wayne Norman and Chris MacDonald charged that 3BL is "inherently misleading" in their seminal paper, "Getting to the Bottom of the 'Triple Bottom Line.'" In their view, linking 3BL to traditional financial accounting is inappropriate as a comprehensive methodology for the evaluation of a company's social or environmental responsibility, and that the term could provide a smoke screen allowing firms to avoid truly effective social, environmental, and economic reporting and performance. Notably, they did not say any supporters of 3BL actually claim to aggregate the data in this way, but only that they would have to do so if their analogy with financial accounting were to have any credibility.

Norman and MacDonald's paper started a debate which continues in academic circles some six years later, but has had less practical effect: the phrase remains in common use in the business world, and consultants are still selling services to help corporations produce nonfinancial reports with a 3BL theme. In many cases, these reports seek to convince stakeholders that they should take seriously the social and environmental "bottom lines" to the same extent that they accept the financial bottom line in the financial statement. Sustainability reports use a wide variety of interpretive tools, including dashboards and scorecards, to augment narrative reporting and graphs, but none of these provide the aggregate measure promised by 3BL. Aside from the problem of there being no universally accepted metrics for environmental and social accounting, the reports are not subject to regulation in the way that financial annual reports are. Few include verification of their claims by outside agencies, making them an excellent opportunity for greenwashing (attempts by a corporation to seem more environmentally responsible than they really are), and given increasing consumer interest in purchasing products which they feel are ecologically and socially responsible, 3BL can easily become just another marketing tool.

Although there is no universally accepted standard for 3BL, a number of approaches have been developed that attempt to account for and report a corporation's environmental and social effects. For instance, the Global Reporting Initiative provides guidelines for reporting on sustainability including 79 indicators, although they do not provide a method to aggregate these into a single score. However, such a method can be added after the fact; for instance, using the scoring method of the Baldrige Performance Excellence Program in the United States, one could assign a numeric score to each indicator by asking four questions of each result:

1. How is the result important to the organization?

2. How has the organization actually planned to achieve that result at the beginning of the year?

3. How does the organization track and trend the result over time?

4. How does the organization benchmark that result against other organizations?

Each lagging indicator is scored, and the scores can be aggregated to a single results score. Of course, each question requires expert judgment, and different individuals might disagree on the answers, and thus, although a numeric score is produced, the meaning of that number is highly open to question (in distinction to the obvious meaning of the accounting bottom line expressed in dollars or other currency).

Another possibility is offered by the ADRI (i.e., Approach, Deployment, Results, and Improvement) method of the Australian Business Excellence Program, which could be adapted to environmental and social accounting. This approach requires that an organization would prepare a written approach and detailed action plan for the deployment of the approach for a given fiscal year, including specifying the metrics to be used to determine the results, indicating how effectively the approach and deployment were working. Each of the three components are scored based on how the documented outcomes compare to the scoped activity. The scores can then be added to obtain a single measure of sustainability for the organization. Of course, this method only measures the organization's success in fulfilling their stated goals, not the total effect on the physical and social environment that is the goal of 3BL, and requires many expert judgments, which could be arguable.

Another approach is to place dollar values on social and environmental results, a practice which is already common in the corporate world for operational, regulatory, and reputational risks. The soon-to-be-released ISO 31000 standard for risk management provides guidelines for risk management that could be helpful in this process. Of course, assignment of monetary values to inherently nonmonetary entities (what is the value of one person's life, or of clean air?) and the corporation is free to choose which effects to include and which to ignore in this process.

The concept of the triple bottom line is very attractive to both corporations and consumers because it seems to promise an accounting system for environmental and social matters that is as transparent and reliable as the system used to report on a corporation's financial results. Although there is nothing even close to this ideal currently in use, there are many alternatives of varying usefulness. The move to develop standardized methods of environmental and social accounting will continue, both because of consumer demand, and because studies have shown that many successful companies have successfully integrated concerns for natural and human capital into their business practices.

See Also: Environmental Accounting; Integrated Bottom Line; Socially Responsible Investing; Transparency.

Further Readings

Australian Business Excellence Program. http://www.saiglobal.com/business-improvement/process/framework/awards.htm (Accessed September 2009).

Baldrige Performance Excellence Program. http://www.quality.nist.gov (Accessed September 2009).

G3 Reporting Guidelines. http://www.globalreporting.org/ReportingFramework/G3Guidelines (Accessed September 2009).

Norman, W. and C. MacDonald. "Getting to the Bottom of the 'Triple Bottom Line.'" *Business Ethics Quarterly*, 14/2:243–62 (2003).

Robert B. Pojasek
Harvard University

U

UPCYCLE

In modern industrial economies, the materials that go into the production, transportation, and consumption of goods are commonly either put into landfills, burned, or recycled. In each case, the material—often based on nonrenewable resources—is either wasted, or its use is downgraded (as in document paper recycled as toilet paper or tires used as playground swings, also referred to as *downcycling*). Upcycling denotes the concept of converting used materials at the end of their original life cycle and turning them into new materials or products of either equal or higher quality (such as creating automobile dashboards from recycled plastic bottles, furniture from discarded wood, or wallets from old tires).

This GE Jenbacher landfill gas engine runs on the offgas produced by landfills, such as methane, carbon dioxide, and nitrogen. Upcycling takes something disposable (material from a landfill) and transforms it into something of greater value (energy).

Source: GE

The term *upcycling* was popularized in the United States by William McDonough and Michael Braungart in their book *Cradle to Cradle: Remaking the Way We Make Things.* In addition to making better use of resources and materials, the book shows that upcycling could also significantly reduce the consumption of nonrenewable raw materials, and, in the production process, result in a reduction of energy use, pollution, and greenhouse gas emissions. As such, upcycling is generally a reinvestment in the environment. "Upcycling is the practice of taking something that is disposable or downcycled and transforming it into something of greater use and value." By contrast, downcycling is the recycling of a material into a material of lesser quality or reuse of a product with crippled functionality for alternative purposes.

Using historical sketches on the roots of the industrial revolution and commentary on science, nature, and society, McDonough and Braungart assert that an industrial system can become a creator of goods and services that generate ecological, social, and economic value. Their vision is based on a system of "life cycle development" initiated by Braungart. The idea is that there is nothing inherently wrong with being "wasteful," as long as the products are completely returned to nature, or are completely reborn as new products. For example, each year a cherry tree dumps a great pile of fruit and leaves on the ground to rot. All of this waste goes back into nature to be reborn as new trees, bacteria, birds, and other parts of the natural ecosystem. According to the authors, humans should try to emulate this natural system instead of trying to do more with less.

The concepts proposed by Braungart and McDonough led to the cradle-to-cradle (C2C) model, where all materials used in industrial or commercial processes—such as metals, fibers, dyes—are seen to fall into one of two categories: "technical" or "biological" nutrients. Technical nutrients are strictly limited to nontoxic, nonharmful synthetic materials that have no negative effects on the natural environment; they can be used in continuous cycles as the same product without losing their integrity or quality. In this manner these materials can be used over and over again instead of being "downcycled" into lesser products, ultimately becoming waste. Biological nutrients are organic materials that, once used, can be disposed of in any natural environment and decompose into the soil, providing food for small life forms without affecting the natural environment.

C2C also seeks to remove dangerous technical nutrients (synthetic materials such as mutagenic materials, heavy metals, and other dangerous chemicals) from current life cycles. If the materials we come into contact with and are exposed to on a daily basis are not toxic and do not have long-term health effects, then the health of the overall system can be better maintained. For example, a fabric factory can eliminate all harmful technical nutrients by carefully reconsidering what chemicals they use in their dyes to achieve the colors they need and attempt to do so with fewer base chemicals.

Upcycling is a concept that can be realized in a multitude of ways, ranging from simple household uses to sophisticated technological and business applications. Simple examples, not unfamiliar to older generations, include turning bacon fat, combined with food crumbs and string, into bird cake (rather than pouring it down the drain and further contaminating waste water); using food trimmings as soup stock, and composting the remainder; and using discarded lumber for furniture or art projects.

But there are also an increasing variety of ever-more creative and sophisticated ways of upcycling, some of which have turned into viable commercial enterprises. Artists have been routine upcyclers for generations, turning everything from broken glass and old fabrics to discarded steel and lumber into projects that can be found around the world (a good example are Derick Melander's massive secondhand clothing sculptures). Architects have begun to develop building projects that fundamentally depend on upcycled materials. Using a combination of traditional stick frame construction and seven cargo containers, for instance, Peter DeMaria built a multifunction church in Los Angeles. The 2010 Vancouver Winter Olympics presented beautiful upcycled medals to its winners. The awards were cast out of materials salvaged from old circuit boards. The all-pervasive billboard in America, aside from its debatable aesthetic, is made out of vinyl, which, according to estimates, leads to roughly 10,000 tons of nonbiodegradable waste ending up in landfills. Two companies, TerraCycle and Yakpak, are set to begin production based on turning discarded vinyl into durable backpacks and messenger bags. Another small but useful upcycling trick was recently discovered by a group of guitar players, who realized that old credit cards make excellent guitar picks. A British company by the name of Elvis & Kresse Organisation (E&KO) uses nothing but industrial waste to make new luxury products, such as turning fire hoses into bags, belts, wallets, and cufflinks.

Even though there are a virtually endless variety of possible green business applications for upcycling, the industry is still in its infancy. While new companies like TerraCycle have demonstrated that upcycling can be both sustainable and profitable, viable business models on a larger scale will still have to be developed. Nevertheless, both as a concept and as a business reality, upcycling squarely focuses public attention on the environmental need for a better, less wasteful use of nonrenewable resources.

See Also: Biomimicry; Corporate Social Responsibility; Cradle-to-Cradle; Ecoeffectiveness; Green Design; Recycling, Business of; Sustainability; Sustainable Design; Sustainable Development.

Further Readings

Benyus, Janine M. *Biomimicry: Innovation Inspired by Nature*. New York: Harper Perennial, 2002.

Brownell, Blaine. *Transmaterial: A Catalog of Materials That Redefine Our Physical Environment*. Princeton, NJ: Princeton Architectural Press, 2005.

Geiser, Kenneth and Barry Commoner. *Materials Matter: Toward a Sustainable Materials Policy (Urban and Industrial Environments)*. Cambridge, MA: MIT Press, 2001.

McDonough, W. and M. Braungart. *Cradle to Cradle: Remaking the Way We Make Things*. New York: North Point Press, 2002.

Rodney Andrew Carveth
Fitchburg State College

Value Chain

Every product moves through a "chain" of activities from resource extraction to production and distribution, while in the process gaining value at each step. Value and cost, in this concept, are not the same: the cost of a step in the chain may be relatively low (as in pumping crude oil out of the ground), but it adds greatly to the value of the good. Conversely, a cost may be relatively high (as in transporting gasoline to the various gas stations), but the value added is very low.

The initial concept of the "value chain" was introduced by Harvard business professor Michael Porter in his 1985 book *Competitive Advantage: Creating and Sustaining Superior Performance*, and further developed in subsequent work. Competitive advantage represents a company's ability to sustain its position in the industry, maintaining levels of profit while outperforming competitors. A perennial topic of discussion in business has been how to develop sustainable competitive advantage—how to remain in that position, without one's competitors duplicating one's value-creating processes.

In the 1990s, a simplified version of the value chain, the value stream, became a popular approach to business process management. The value stream is the point-of-origin-to-point-of-consumption process that brings a product or service to market, composed of process steps that may involve the use or production of intermediate goods or services. Analyzing a business's processes in terms of the value stream can lead to greater efficiency by bringing to light unnecessary or redundant processes, or demonstrating which processes contribute the least value. Such an analysis produces a value stream map, which originated at Toyota under the name "material and information flow map." Though intended for economic efficiency, value stream maps are useful in achieving greater environmental and resource consumption efficiency, and the analysis process has been streamlined through the use of both stand-alone software and software add-ons for packages like Microsoft Visio. In the Toyota method, the model is constructed both horizontally and vertically, with the single horizontal telling the narrative of the product being created and brought to market, and the many verticals telling the story of each process along the way.

Both the value stream and the value chain are compatible with, and in the case of the former explicitly conceived of as part of, "lean production," usually abbreviated (and capitalized for clarity) as simply Lean. Lean is a 1990s generalization of the efficiency-focused production techniques implemented at Toyota, and is a descendant of such management schools of thought as Fordism, scientific management, and the time-motion study of the 1920s (which sought to increase productivity by streamlining the physical motions workers used to do their work, a remarkable goal of efficiency to strive for in a decade still becoming accustomed to the gains of the assembly line).

The core premise of Lean is simple: more value for less work. Lean focuses on "the seven wastes," seven types of *Muda*, a Japanese term for unproductive activity: overproduction, unnecessary transportation, excessive inventory, unnecessary motion (as that targeted by time-motion study), defective product, overprocessing (when more work is done on a product than a customer requires or more is provided than is asked for, or the kind of waste, for instance, that led some airlines to discontinue complimentary snacks), and waiting. Lean differs from the management system used by Toyota in that it is generalized enough to be applied to accounting, to business management, and to services, and not just to manufacturing. Lean also emphasizes profit maximizing less than Toyota does, and the principles of Lean production can in fact be used to optimize a company's business process with ecoefficiency in mind, rather than simply making as much profit as possible from the least amount of effort. The focus on eliminating waste applies not only to wasteful activity, but also to overconsumption of resources.

The value chain defines its constituent processes more specifically than does the value stream. A business adds value by producing products and services, by transforming raw materials, or by creating a brand identity that appeals to customers. When the acclaimed economist Michael Porter published his "Value Chain Analysis," he condensed the various activities a product goes through into five major and four support activities.

Primary Activities

- Inbound logistics: materials handling, warehousing, inventory control, transportation
- Operations: machine operating, assembly, packaging, testing, and maintenance
- Outbound logistics: order processing, warehousing, transportation, and distribution
- Marketing and sales: advertising, promotion, selling, pricing, channel management
- Service: installation, servicing, spare part management

Support Activities

- Firm infrastructure: general management, planning, finance, legal, investor relations
- Human resource management: recruitment, education, promotion, reward systems
- Technology development: research and development, IT, product and process development
- Procurement: purchasing raw materials, lease properties, supplier contract negotiations

This generic value chain provides a model into which to insert processes specific to a given business, and the subsequent mapping of process flows that can be used to identify and isolate value-creating activities. One result of looking at business processes this way has been an increased reliance on outsourcing processes that do not contribute value or that can become more valuable if handled by an outside party. Customer and technical support is a common example, though many companies also outsource their information technology (IT) and at least some portion of their logistics (such as by renting warehouse space) and marketing (through outside marketing firms and advertising agencies).

The competitive advantage gained by value-chain analysis is typically either differentiation (a focus on core competencies, those aspects of the business that make it distinct from and nonimitatable by competitors) or cost advantage (streamlining costs in order to bring goods to market at a lower price than competitors). Either lends itself to running a more sustainable business.

Even among today's business leaders, it is widely acknowledged that making significant progress on reducing the impact of climate change depends to a significant extent on limiting the ecological footprint of supply chains. From design to resource to development, assembly, and distribution, companies worldwide have launched initiatives to study the full life cycle of their products and services—including each activity of the value chain—to identify the opportunities for the greatest reduction of resource use and carbon emissions.

For example, Wal-Mart, the world's largest store chain and the world's largest publicly traded corporation, has set out to establish a sustainability index that will provide detailed information about a product's greenhouse gas emission, solid waste generation, and even workers' wages and rights. As a corporate announcement stated, the "value chain concept gauges how each step in a product's life adds to its worth. But value can be seen from another angle, as embodied in the index: all the environmental, health, and social impacts of a product throughout its life cycle. By creating a single standard for evaluation, Wal-Mart opens a window on products that reveals any negatives—what might be called the 'devalue chain'—and puts them into competitive play." If fully implemented, such an index would provide large-scale transparency, and consequently a wealth of incentives for suppliers and producers to cut environmental and social negatives from value-chain activities within their operations.

Another example is the German-based logistics company DHL, which is aiming for a 30 percent reduction in carbon emissions by 2020. The company plans to accomplish this through optimizing air and ground fleets, improving energy efficiency, and using new, innovative technologies. Again, this is an example of a business going through each step of its value-chain activities in order to identify, and then reduce, waste and pollution.

A third example is the apparel/footwear giant Nike, which recently estimated that the embedded carbon in its products is somewhere around 2.5 times the carbon emitted by the factories making the products. This led the company to invest in efforts to find alternative materials and design processes, and has already resulted in a significant reduction of Nike's carbon footprint.

Value chains, in short, provide a valuable model for examining a business's activities from a sustainability viewpoint, isolating not only inefficient processes or activities that add too little value, but processes with disproportionate environmental impact, materials waste, and energy inefficiencies.

See Also: Best Management Practices; Core Competencies; Economic Value Added; Environmental Assessment; Environmental Impact Statement; Transparency.

Further Readings

Choucri, Nazli, et al., eds. *Mapping Sustainability: Knowledge e-Networking and the Value Chain.* New York: Springer, 2007.

Harvard Business School Press. *Harvard Business Review on Managing the Value Chain.* Cambridge, MA: Harvard Business Press, 2000.

Krajewski, Lee J., Larry P. Ritzman and Manoj K. Malhotra. *Operations Management: Processes and Value Chains*, 8th Ed. New York: Prentice Hall, 2006.

Porter, Michael E. *On Competition*. Cambridge, MA: Harvard Business School Press, 2008.

Bill Kte'pi
Independent Scholar

VOLUNTARY STANDARDS

A voluntary standard is a standard that companies may choose to adopt or not, as opposed to a mandatory standard, which carries the force of law and may be imposed by a governmental or other regulatory agency. Voluntary standards may address products, technical processes, or social processes, and in the past two decades have gained high importance in political, corporate, and technical discourses about how societies should deal with global environmental crises. Although some believe that laws and enforced standards are the best way to deal with environmental problems related to manufacturing and trade, many governments, industry, and business associations are now calling for deregulated, responsibly coordinated action by corporations themselves, arguing that efficient solutions to environmental problems can only be brought about if business actors are included at the table. Along this line, private-sector actors in cooperation with other nonstate-actor-crafted voluntary standards are addressing diverse aspects of greening business, including environmental management systems, products (ecolabeling), reporting, and auditing.

Historical Development of Voluntary Standards

Corresponding to the need for cooperation, humans have standardized artifacts and communication for thousands of years. The industrial revolution brought about mechanized production that created de facto standardized products and processes needing standardized means of production For instance, development of a network of railroads in Great Britain make clear the need for a standard gauge (width between the rails) so that the same rolling stock could be used on different lines. In the late 19th century, standardization became a social activity clearly recognizable as such while since the early 20th century, national standardization organizations emerged in the interaction of industrial associations, engineering bodies, and governmental authorities. Western nation-states supported increasing standardization to improve conditions for trade: common standards generally facilitate international trade while the lack of such standards can be a barrier to it.

Voluntary standards in industry gained significant importance during World War II because of the need to rapidly manufacture large quantities of weapons and munitions: standardization increased efficiency by facilitating mass production of parts, weapons and munitions. National standardization organizations cooperated with nation-states for military goals, leading to the birth of the International Organization for Standardization (ISO) in 1947. ISO is not an acronym, but refers to the Greek concept *isos*, meaning "equal." The aim of ISO is to coordinate the standards developed in the nation-states and to unify standards internationally, with a primary focus on technical standards. Originally, ISO focused on crafting international standards that designed for reissuance at the national level, ideally without further change, but soon moved beyond this. A second major shift took place in the 1980s to 1990s, when ISO standards were expanded to include aspects of corporate culture such as quality assurance. This was the move that finally provided the path to standards for environmental management—a significant terrain of voluntary standards relevant for greening businesses.

However, ISO and the national standardization organizations are not the sole organizations setting standards. Producers often convene to set their own standards, both for single corporations as well as for consortiums. While such standards may be voluntary, they are not necessarily open, meaning available for public scrutiny. However, in the field of voluntary standards regarding the environment, standards are normally open because they were created in reaction to the public's interest in greener products and greener manufacturing and business practices. For this reason, environmental standards related to products are often publicized, and products meeting those standards are identified by means such as labels or certificates that allow consumers to choose products meeting the standards.

Environmental standards may also stem from organizations within movements, such as the organic movement. In such cases, movements may aim to change the structures of a market by creating standards that define a particular type of product. For instance, organic produce is defined by the means by which it is farmed, and is sold with a label differentiating it from nonorganic produce. Standardization efforts growing out of social movements may end up being co-opted by bigger standardization organizations or by a stabilized institutionalization and continued independent existence.

General Characteristics of Voluntary Standards

In contrast to standards enforced by the government, voluntary standards are commonly perceived and constructed as favorable options that organizations can choose from. The general idea runs along the line of stick to the standard, get a certificate, and demonstrate to consumers that this product, or your entire business, adheres to beneficial standards or practices. By voluntarily aligning itself with a standard higher than what is legally required, a company hopes to gain positive public recognition and increased business. Other benefits of adhering to voluntary standards may include incentives from other companies (for instance, lower insurance rates if the voluntary standard reduces company risk), or exemption from otherwise mandatory monitoring by governmental authorities.

At some point, the distinction between mandatory and voluntary standards can become blurry, however, and many see voluntary standards as a form of "soft regulation," which is not, in fact, entirely voluntary, as governments may "suggest" that voluntary standards be met and refusal may not be a practical option. Sometimes self-regulation through voluntary standards may prove insufficient, in which case the standards may be adopted as legally binding.

Voluntary standards are usually set by organizations with expert bodies to discuss and, thereafter, formulate scientifically sound, technically optimal, and practical norms. The acceptance of such standards rests in large part on the assumption that they were formulated by objective experts relying on the best information available. Credibility is also aided by the government and private sector recognition of the organization creating the standard.

Recently, standardization organizations have also sought the participation of social organizations, which are invited to add their moral authority to the development process.

Environmental Standards: How Should They Work?

Western environmental policy depends, to a large degree, on self-regulation of businesses regarding their environmental effects. The term *ecological modernization* refers to a philosophy that for-profit organizations should be allowed to develop environmental standards for themselves. Under this system, corporations advocate for a form of governance, emphasizing self-regulation, and the least-possible enforced regulation by the nation-state. The expected outcome is that corporations will compete to improve their environmental standards, as increasing numbers of consumers seek out green products and services. The outcome of this competition should be high environmental standards that in turn will determine the environmental quality of products and processes.

Examples of environmental standards include ISO 14000 standards. These standards address a variety of issues, including environmental management systems, labeling, life cycle analysis, and auditing. Of course, in some areas several standards may coexist or compete; for instance, in environmental management, the standard "Eco-Management and Audit Scheme" (EMAS) issued by the European Union exists as well as the ISO standard.

Another example comes from the area of ecolabeling. The Forest Stewardship Council (FSC) issues a standard for sustainable forestry, referring to social, economical, ecological, cultural, and spiritual dimensions. While this standard deals very specifically with one class of products, forest products are also eligible for the ecolabel of the European Union. which is designed generally for any products and services.

Voluntary Standards: The Case of Environmental Management Systems

A significant example of environmental voluntary standards is the environmental management system (EMS), a standardized environmental self-governance system that takes responsibility for planning, implementing, and maintaining the environmental plan for an organization. As a management system, the EMS standard focuses on system processes rather than the details at the substantive level. The standard stipulates that the organization meet the standards rather than prescribing exactly how that is to be accomplished. Social scientists and management researchers have found that however rigorous the certification process, standards must still be interpreted by organizations and actors within them, but published standards are not expected to document the process at this level of detail.

Standards are often given credit for bringing about change. This may be contested, as different parties may disagree as to whether substantive environmental change is actually occurring in a particular instance. For example, there may be observable improvements at the technical and administrative level that have little actual effect on the environment. Often employees do not know what an EMS standard stipulates. The certification process requires only that an organization is able to show conformity with the standard during audits, which may be realized through a structure of documentation parallel to the practical workflows or in practices and processes that actually conform to the standard.

Key Issues in Discussions of Voluntary Environmental Standards

Environmental standards can claim a certain amount of respect, based both on the assumed ability of the natural sciences to represent the true state of the environment, and the commonly accepted belief that sustainable development is both practical and morally correct. However, environmental standards are also open to questioning in several fields, including the competence and reliability of the organization that sets the standards, and the usefulness of voluntary standards as an instrument of social change. On the first point, the supposedly exact sciences are enacted within the social framework and are shaped by social interests. Thus, they cannot claim to be entirely neutral or objective, which poses a problem for any claims made for these standards. Taking the discourse of sustainable development in its weak version as a point of reference, standardization organizations often craft standards proving the framework for diminishing the natural stock in favor of human-made capital.

The ability of voluntary standards to bring about change can be very limited. For example, an EMS can only pay attention to what it can understand, that is, processes rather than concrete practices. The standard brings with it a managerial-technocratic language that is able to provide answers limited to the logic of the standard, and which cannot address many concrete issues.

Voluntary standards can also limit reform by implying that if a business meets the accepted standard, it should be immune from further criticism about its environmental practices. The public may not agree, but from the point of the view of the green business, it already managed to accomplish compliance with the standard, and should not be expected to do any more. So standards may have a detrimental effect on the well-being of communities if the standard is insufficient or does not address matters of concern to them.

A tendency emerging in the late 20th century was the increasing inclusiveness of standardization organizations. In the spirit of new governance models, nongovernmental organizations and presumed representatives of civil society became members of committees of standardization organizations to shape standards. While such inclusion may be admirable, in terms of increasing the points of view incorporated in the standards, it also weakens the ability of the included parties to criticize the standard publicly. At the same time, the composition of these committees can become highly political, and other factors, such as the exclusion of certain interests because they can't afford to send their representatives to meetings, may also become relevant. In practice, standardization organizations often depend to a high degree on funding bodies like the nation-state and corporations, which can at least exert indirect influence. To include suitable experts in the committees, members can often make suggestions as to who should be invited, which may limit the diversity of opinion within the organization. These factors may result in the underrepresentation or exclusion of the general public, even though they are stakeholders who will be affected by the standards.

Finally, enforcement of standards depends highly on the auditors who exercise local control. If, as is often the case, organization hires and pays the auditors, then there is an obvious conflict of interest. Another conflict arises as to the role of auditors, who are not elected or otherwise directly accountable to the general public, yet function in a quasi-governmental role.

Overall, then, voluntary environmental standards can be understood as a form of soft regulation of corporate conduct. Standards shape the language in which environmental issues are framed, and are usually stated as manageable within the structures of today's society. Developed in seemingly neutral committees, who in fact may represent a variety of interests, they engender the politics of ecological modernization and decrease the degree to which sustainable development will be applied uniformly on products, process, and services worldwide.

See Also: Certification; Compliance; Environmental Audit; Environmental Management System.

Further Readings

Boiral, Olivier. "Corporate Greening Through ISO 14001: A Rational Myth?" *Organization Science*, 18/1 (2007).

Fiorino, Daniel J. *The New Environmental Regulation*. Cambridge, MA: MIT Press, 2006.

Goetsch, David L. and Stanley Davis. *ISO 14000: Environmental Management*. Upper Saddle River, NJ: Prentice Hall, 2000.

Higgins, Winton and Kristina Tamm Hallström. "Standardization, Globalization and Rationalities of Government." *Organization*, 14/5 (2007).

Prakash, Aseem and Matthew Potoski. *The Voluntary Environmentalists: Green Clubs, ISO 14001, and Voluntary Environmental Regulations*. Cambridge, UK: Cambridge University Press, 2006.

Ingmar Lippert
University of Augsburg

Waitrose

Waitrose supermarket chain is part of the John Lewis Partnership (JLP), principally targeting the UK middle-class market. JLP is held in trust for its approximately 69,000 employees, all of whom are partners. Thus, the 40,000 or so Waitrose employees have a voice in the business. Waitrose operates over 190 supermarkets, four distribution centers, and manages its own farm—the 4,000-acre Leckford Estate in Hampshire, England. It commands approximately 4 percent of the UK market.

Like all major UK supermarket chains, Waitrose is anxious to exhibit its green credentials. Its innovative initiatives range from experiments with "eco-bikes" as an alternative to delivery trucks, to replacing cardboard shipping boxes with reusable plastic. They include locating stores in town centers, using empty delivery vehicles to collect stock from suppliers (backhauling) and encouraging suppliers to deliver Waitrose goods to a store on its return trip from a regional distribution center. Waitrose was the first in the United Kingdom to introduce its own "bag for life" reusable shopping bag, which it sells cheaply and replaces free of charge when damaged or worn out. Returned bags are recycled into "plaswood" furniture that is then provided to store branches, schools, and local authorities and charities. Waitrose also introduced a range of reusable jute bags, experimented with bags made from 33 percent recycled material, and piloted schemes for bag-free green checkout counters at some outlets. Prominent communication in every store encourages customers to reuse existing bags.

One of Waitrose's main programs is setting specific targets against which it subsequently monitors progress. Some of the specific targets to reduce the company's carbon footprint are as follows:

- Reduce carbon dioxide (CO_2) emissions per unit (million £) of sales 10 percent by 2010, 20 percent by 2020, and 60 percent by 2050 (all targets relative to 2001 levels)
- Source 100 percent of electricity from green sources and consider all potential sources of renewable energy for both stores and offices
- Improve the energy efficiency of all facilities 20 percent by 2010 (relative to 2003 levels)
- Reduce energy-related CO_2 emissions from stores' deliveries 15 percent by 2013 (compared with 2005 levels)

Although admitting that its absolute carbon emissions had increased 7 percent by 2008 due to extended store hours and increased store size, in a 2008 report, Waitrose announced that 100 percent of its electricity now comes from renewable sources. One Waitrose store receives power from two local tomato suppliers: the farms' combined heat and power units produce heat for their greenhouses and electricity, which is then purchased by Waitrose. As part of Waitrose's aim to reduce the carbon emissions of its vehicles—one of its biggest concerns—it is investigating many alternative, lower-carbon fuels. In a recent trial, five of its trucks were fueled by rapeseed oil, which has both a cleaner production process and a 20 percent lower carbon footprint than biodiesel. Waitrose is a signatory to the Courtauld Commitment, a UK-based voluntary agreement between grocery retailers, suppliers, manufacturers, and brand owners that encourages less packaging and food waste, and has achieved the first Courtauld target of zero packaging growth, despite increased sales.

According to its website, Waitrose's longtime, consistent goal has been responsible sourcing and strict control of its supply chain. The origin of meat sold in Waitrose supermarkets is carefully tracked and monitored for assurance of quality, sustainability, and humane treatment of animals, some of it coming from the farm at Leckford Estate. Aiming to remove all pesticides from its supply chain, it regularly screens products for pesticide trace residue. Waitrose partners with banana growers on the Windward Islands of the Eastern Caribbean, thus promoting quality and fair labor practices. In 2009, Waitrose signed the Sustainable Palm Oil Pledge, promising that its palm oil will come from suppliers certified by the Roundtable on Sustainable Palm Oil. In the meantime, it is buying Green Palm Certificates equal to its annual palm oil usage. Waitrose is the first supermarket in the United Kingdom to use anaerobic digestion, a form of composting producing renewable energy. It also partners with GROWS (Green Recycling of Waste from Supermarkets) to collect waste and return it to the grower.

Waitrose states that it absolutely complies with legislation where it exists. If there is no legislation, it seeks to develop and implement appropriate standards of its own.

See Also: Clean Fuels; Corporate Social Responsibility; Green Retailing; Responsible Sourcing; Sainsbury's; Tesco; Voluntary Standards.

Further Readings

Hickman, Martin. "Waitrose Makes Sustainable Palm Oil Pledge." *The Independent* (December 3, 2009). http://www.independent.co.uk/environment/green-living/waitrose -makes-sustainable-palm-oil-pledge-1833136.html (Accessed January 2010).

Waitrose. "How We Stack Up: Corporate Social Responsibility Report 2008." Digital Edition. http://publishing.yudu.com (Accessed January 2010).

Waitrose. "Waitrose Corporate and Social Responsibility." http://www.waitrose.presscentre .com/Corporate-Social-Responsibility/Waitrose-Corporate-and-Social-Responsibility-81 .aspx (Accessed January 2010).

David G. Woodward
University of Southampton School of Management

WASTE REDUCTION

The term *waste reduction* includes all efforts to reduce solid waste generation. This includes reusing, recycling, source reduction, designing for disassembly, and dematerialization. Reducing waste is a key imperative in business, not only because it pollutes the ecosystem, but also because it reduces profits. Waste costs money to produce, and it costs money to dispose of it. Waste reduction is the principal focus of *kaizen*, the Japanese principle of improvement made famous in the Toyota Production System.

Early efforts to deal with solid waste focused on reusing products and durable packaging in their original form. As early as the 1900s, Chicago and Cleveland residents were encouraged to find "Waste as Wealth" by sorting and reselling their metal cans for reuse. During the World Wars, citizens were encouraged to reuse rags, wastepaper, foil, books, and newspapers for the war effort. Recycling became easy and popular with the introduction of the all-aluminum can in 1964. Legislation followed one year later when congress passed the Solid Waste Disposal Act. In 1972, the U.S. state of Oregon was the first to encourage more widespread participation in beverage container recycling, and tied refundable cash deposits to returned bottles and cans. Plastic bottle deposits and refunds were added later. The Federal Resource Conservation and Recovery Act of 1976 enacted solid waste legislation and mandated close monitoring of landfills. The 1987 media images of the *Mobro*, a barge carrying garbage from New York, traveling up and down the East Coast for six months looking for a place to dump its load, brought new interest to recycling as an alternative to adding to solid waste.

In an effort to focus on the source of the problem, U.S. Congress passed the Pollution Prevention Act in 1990. In this act, "source reduction" was defined as follows: "any practice which reduced the amount of any hazardous substance, pollutant, or contaminant entering any waste stream or otherwise released into the environment (including fugitive emissions) prior to recycling, treatment or disposal; and reduced the hazards to public health and the environment associated with the release of such substances, pollutants, or contaminants."

Source reduction was made possible through the use of equipment or technology modifications, process or procedure modification, reformulation or redesign of products, substitution or raw materials, and improvements in housekeeping, maintenance, training, or inventory control.

Similar to the term *ecoefficiency*, waste reduction is characterized by the increased efficiency in the use of raw materials, energy, water, or other resources, and by the protection of natural resources by conservation. Despite the use of the word *waste* in the term, it is clear that waste has resource use as its root cause.

There are a myriad of other terms that are derived from ecoefficiency, including waste minimization, waste prevention, pollution prevention, toxic use reduction, and cleaner production. Waste reduction has been used for municipal solid wastes—"reducing the amount of materials entering the municipal waste stream from a specific source by redesigning products or patterns of production or consumption (e.g., using returnable beverage containers).

Source reduction begins with product and packaging design. Designing for disassembly involves designing a product to be disassembled for easier recovery and reuse of components. For example, Nike's "Considered" line of casual footwear is made according to these principles, and contains shoes that have reduced material waste (61 percent), reduced energy use (35 percent), and reduced use of solvents and adhesives (89 percent).

By 2011, Nike plans to have all footwear meeting the Considered line minimum baseline for sustainability.

People are sometimes confused as to whether recycling and reuse are included in the definition of source reduction. A waste management hierarchy was formalized in the Pollution Prevention Act of 1990 as follows:

- Pollution should be prevented or reduced at the source wherever feasible.
- Pollution that cannot be prevented should be reused or recycled in an environmentally safe manner whenever feasible.
- Pollution that cannot be prevented, reused, or recycled should be treated in an environmentally safe manner wherever feasible.
- Disposal or other release into the environment should be employed only as a last resort and should be conducted in an environmentally safe manner.

The difference between reuse and recycling is that recycling requires some conversion of the resource before it can be reused. Reuse is not recycling, because reuse does not alter the physical form of a material. Reuse is preferred to recycling because reuse consumes less energy and less resources than recycling. Think of recycling as a "supporting process" that itself uses and loses resources. It may be that the waste from the recycling also needs some level of waste reduction. This is similar to the phrase, "There is no such thing as a free lunch." When one must recycle, it must be done with the realization that it would be far better if the waste material could be prevented instead, as noted in the waste management hierarchy.

To find the "source," it is important to understand the process responsible for the use and loss of the resources. This is best achieved using a hierarchical process map. The core processes responsible for the product or service are mapped using this method. Then the processes that support the core process are mapped (e.g., steam generation, compressed air, process heating, and so on). Finally, all nonproduct processes (e.g., business processes, infrastructure processes) are mapped. The resources used and lost with each activity in each process are noted and measured.

By using an environmental management system like ISO 14001, it is possible to create an environmental footprint for each of the uses and losses of resources. These uses and losses by activity become the environmental aspects. They can be prioritized for significance by using a risk management system framework. Significant opportunities for waste reduction can now be identified and assigned to employee teams as part of the environmental management program.

Some companies have set zero waste targets and used methods like the systems approach to see how close they can get to zero. Other companies have used other process improvement methods, like the "lean to green" approach, or a Six Sigma approach. Employee involvement is also encouraged with these process improvement methods.

The real key to successful waste reduction is finding the cause of the issue at the source, and then implementing a process improvement program that can help eliminate the cause of the waste reduction using any of the methods presented above. Waste reduction is a fundamental, and is found in every sustainability or corporate responsibility program. A successful approach to waste reduction provides the impetus to address broader issues involving the environmental, social, and economic responsibilities of sustability.

See Also: Environmental Management System; Pollution Prevention; Six Sigma; Sustainability.

Further Readings

Pollution Prevention Act of 1990. http://epw.senate.gov/PPA90.pdf (Accessed August 2009).

Pojasek, R. B. "Framing Your Lean-to-Green Effort." *Environmental Quality Management*, 18/1:85–93 (2008).

Pojasek, R. B. "Risk Management 101: A Primer for EHS Managers." *Environmental Quality Management*, 17/3:95–101 (2008).

Pojasek, R. B. "Selecting Your Own Approach to Pollution Prevention." *Environmental Quality Management*, 12/4:85–94 (2003).

Pojasek, R. B. "Understanding Processes with Hierarchical Process Mapping." *Environmental Quality Management*, 15/2:79–86 (2005).

U.S. Environmental Protection Agency. "An Organizational Guide to Pollution Prevention." EPA/625/C-01/003 (August 2001). http://www.ecy.wa.gov/programs/hwtr/P2/printguid.pdf (Accessed December 2009).

Robert B. Pojasek
Harvard University

Green Business Glossary

A

Acid Mine Drainage: Drainage of water from areas that have been mined for coal or other mineral ores. The water has a low pH because of its contact with sulfur-bearing material and is harmful to aquatic organisms.

Air Pollution Control Device: Mechanism or equipment that cleans emissions generated by a source (e.g., an incinerator, industrial smokestack, or automobile exhaust system) by removing pollutants that would otherwise be released to the atmosphere.

Alternative Compliance: A policy that allows facilities to choose among methods for achieving emission reduction or risk reduction instead of command and control regulations that specify standards and how to meet them. Use of a theoretical emissions bubble over a facility to cap the amount of pollution emitted while allowing the company to choose where and how (within the facility) it complies.

Alternative Fuels: Substitutes for traditional liquid, oil-derived motor vehicle fuels like gasoline and diesel. Includes mixtures of alcohol-based fuels with gasoline, methanol, ethanol, compressed natural gas, and others.

Antarctic "Ozone Hole": Refers to the seasonal depletion of ozone in the upper atmosphere above a large area of Antarctica.

Anti-Degradation Clause: Part of federal air quality and water quality requirements prohibiting deterioration where pollution levels are above the legal limit.

Asbestos: A mineral fiber that can pollute air or water and cause cancer or asbestosis when inhaled. The U.S. Environmental Protection Agency has banned or severely restricted its use in manufacturing and construction.

B

Basalt: Consistent year-round energy use of a facility; also refers to the minimum amount of electricity supplied continually to a facility.

BEN: The U.S. Environmental Protection Agency's computer model for analyzing a violator's economic gain from not complying with the law.

Beryllium: A metal hazardous to human health when inhaled as an airborne pollutant. It is discharged by machine shops, ceramic and propellant plants, and foundries.

Biostabilizer: A machine that converts solid waste into compost by grinding and aeration.

Building-Related Illness: Diagnosable illness whose cause and symptoms can be directly attributed to a specific pollutant source within a building.

By-Product: Material, other than the principal product, generated as a consequence of an industrial process or as a breakdown product in a living system.

C

Carbon Tetrachloride (CC14): Compound consisting of one carbon atom and four chlorine atoms, once widely used as a industrial raw material, as a solvent, and in the production of chlorofluorocarbons. Use as a solvent ended when it was discovered to be carcinogenic.

Categorical Pretreatment Standard: A technology-based effluent limitation for an industrial facility discharging into a municipal sewer system. Analogous in stringency to best availability technology for direct dischargers.

Chlorofluorocarbons (CFCs): A family of inert, nontoxic, and easily liquefied chemicals used in refrigeration, air conditioning, packaging, and insulation, or as solvents and aerosol propellants. Because CFCs are not destroyed in the lower atmosphere, they drift into the upper atmosphere, where their chlorine components destroy ozone.

Class 1 Substance: One of several groups of chemicals with an ozone depletion potential of 0.2 or higher, including CFCs, halons, carbon tetrachloride, and methyl chloroform (listed in the Clean Air Act), as well as hydrobromofluorocarbons and ethyl bromide.

Coke Oven: An industrial process that converts coal into coke, one of the basic materials used in blast furnaces for the conversion of iron ore into iron.

Commercial Waste: All solid waste emanating from business establishments such as stores, markets, office buildings, restaurants, shopping centers, and theaters.

Compliance Schedule: A negotiated agreement between a pollution source and a government agency that specifies dates and procedures by which a source will reduce emissions and, thereby, comply with a regulation.

Cooling Tower: A structure that helps remove heat from water used as a coolant; for example, in electric power–generating plants.

D

Diazinon: An insecticide. In 1986, the U.S. Environmental Protection Agency banned its use on open areas such as sod farms and golf courses because it posed a danger to migratory birds. The ban did not apply to agricultural, home lawn, or commercial establishment uses.

Dioxin: Any of a family of compounds known chemically as dibenzo-p-dioxins. Concern about them arises from their potential toxicity as contaminants in commercial products. Tests on laboratory animals indicate that they are one of the more toxic anthropogenic (man-made) compounds.

Direct Discharger: A municipal or industrial facility that introduces pollution through a defined conveyance or system such as outlet pipes; a point source.

Downstream Processors: Industries dependent on crop production (e.g., canneries and food processors).

E

Emission: Pollution discharged into the atmosphere from smokestacks, other vents, and surface areas of commercial or industrial facilities; from residential chimneys; and from motor vehicle, locomotive, or aircraft exhausts.

Environmental Audit: An independent assessment of the current status of a party's compliance with applicable environmental requirements or of a party's environmental compliance policies, practices, and controls.

Ethylene Dibromide (EDB): A chemical used as an agricultural fumigant and in certain industrial processes. Extremely toxic and found to be a carcinogen in laboratory animals, EDB has been banned for most agricultural uses in the United States.

F

Fluorocarbons (FCs): Any of a number of organic compounds analogous to hydrocarbons in which one or more hydrogen atoms are replaced by fluorine. Once used in the United States as a propellant for domestic aerosols, FCs are now found mainly in coolants and some industrial processes. FCs containing chlorine are called chlorofluorocarbons (CFCs). They are believed to be modifying the ozone layer in the stratosphere, thereby allowing more harmful solar radiation to reach the Earth's surface.

H

Hammer Mill: A high-speed machine that uses hammers and cutters to crush, grind, chip, or shred solid waste.

Hazard Communication Standard: An Occupational Safety and Health Administration regulation that requires chemical manufacturers, suppliers, and importers to assess the hazards of the chemicals that they make, supply, or import, and to inform employers, customers, and workers of these hazards through material safety data sheet information.

I

Indirect Discharge: Introduction of pollutants from a nondomestic source into a publicly owned waste-treatment system. Indirect dischargers can be commercial or industrial facilities whose wastes enter local sewers.

Industrial Pollution Prevention: Combination of industrial source reduction and toxic chemical use substitution.

Industrial Waste: Unwanted materials from an industrial operation; may be liquid, sludge, solid, or hazardous waste.

L

List: Shorthand term for U.S. Environmental Protection Agency list of violating facilities or firms debarred from obtaining government contracts because they violated certain sections of the Clean Air or Clean Water Acts. The list is maintained by the Office of Enforcement and Compliance Monitoring.

M

Maximum Available Control Technology (MACT): The emission standard for sources of air pollution requiring the maximum reduction of hazardous emissions, taking cost and feasibility into account. Under the Clean Air Act Amendments of 1990, the MACT must not be less than the average emission level achieved by controls on the best-performing 12 percent of existing sources, by category of industrial and utility sources.

N

Netting: A concept in which all emissions sources in the same area that are owned or controlled by a single company are treated as one large source, thereby allowing flexibility in controlling individual sources to meet a single emissions standard.

Nonaqueous Phase Liquid (NAPL): Contaminants that remain undiluted as the original bulk liquid in the subsurface; for example, spilled oil.

Nuclear Reactors and Support Facilities: Uranium mills, commercial power reactors, fuel reprocessing plants, and uranium enrichment facilities.

O

Oil Spill: An accidental or intentional discharge of oil that reaches bodies of water. Oil spills can be controlled by chemical dispersion, combustion, mechanical containment, and/or adsorption. Spills from tanks and pipelines can also occur away from water bodies, contaminating the soil, getting into sewer systems, and threatening underground water sources.

P

Performance Standards: 1. Regulatory requirements limiting the concentrations of designated organic compounds, particulate matter, and hydrogen chloride in emissions from incinerators. 2. Operating standards established by the U.S. Environmental Protection Agency for various permitted pollution control systems, asbestos inspections, and various program operations and maintenance requirements.

Plutonium: A radioactive metallic element chemically similar to uranium.

Point Source: A stationary location or fixed facility from which pollutants are discharged; any single identifiable source of pollution (e.g., pipe, ditch, ship, ore pit, factory smokestack).

Polychlorinated Biphenyls: A group of toxic, persistent chemicals used in electrical transformers and capacitors for insulating purposes and in gas pipeline systems as lubricant. The sale and new use of these chemicals, also known as PCBs, were banned by law in 1979.

Q

Quality Assurance/Quality Control: A system of procedures, checks, audits, and corrective actions to ensure that all U.S. Environmental Protection Agency research design and performance, environmental monitoring and sampling, and other technical and reporting activities are of the highest achievable quality.

R

Registration: Formal listing with U.S. Environmental Protection Agency of a new pesticide before it can be sold or distributed. Under the Federal Insecticide, Fungicide, and Rodenticide Act, the U.S. Environmental Protection Agency is responsible for registration (premarket licensing) of pesticides on the basis of data demonstrating no unreasonable adverse effects on human health or the environment when applied according to approved label directions.

Rubbish: Solid waste, excluding food waste and ashes, from homes, institutions, and workplaces.

S

Scrap: Materials discarded from manufacturing operations that may be suitable for reprocessing.

Secondary Materials: Materials that have been manufactured and used at least once and are to be used again.

Sewage: The waste and wastewater produced by residential and commercial sources and discharged into sewers.

Sewer: A channel or conduit that carries wastewater and stormwater runoff from the source to a treatment plant or receiving stream. "Sanitary" sewers carry household, industrial, and commercial waste. "Storm" sewers carry runoff from rain or snow. "Combined" sewers handle both.

Sick Building Syndrome: Building whose occupants experience acute health and/or comfort effects that appear to be linked to time spent therein, but in which no specific illness or cause can be identified. Complaints may be localized in a particular room or zone, or may spread throughout the building.

Strip Mining: A process that uses machines to scrape soil or rock away from mineral deposits just under the Earth's surface.

Suspension: Suspending the use of a pesticide when the U.S. Environmental Protection Agency deems it necessary to prevent an imminent hazard resulting from its continued use. An emergency suspension takes effect immediately; under an ordinary suspension a registrant can request a hearing before the suspension goes into effect. Such a hearing process might take six months.

T

Technology-Based Standards: Industry-specific effluent limitations applicable to direct and indirect sources that are developed on a category-by-category basis using statutory factors, not including water-quality effects.

V

Variance: Government permission for a delay or exception in the application of a given law, ordinance, or regulation.

W

Waste: 1. Unwanted materials left over from a manufacturing process. 2. Refuse from places of human or animal habitation.

Waste Minimization: Measures or techniques that reduce the amount of wastes generated during industrial production processes; term is also applied to recycling and other efforts to reduce the amount of waste going into the waste stream.

Source: U.S. Environmental Protection Agency (http://www.epa.gov/OCEPAterms)

Green Business Resource Guide

Books

Bayon, R., et al. *Voluntary Carbon Markets: An International Business Guide to What They Are and How They Work*. London: Earthscan, 2007.

Beder, Sharon. *Environmental Principles and Policies*. Sydney: UNSW Press.

Beeton, Sue. *Ecotourism: A Practical Guide for Rural Communities*. Collingwood: Landlinks, 1998.

Bellamy Foster, J. *The Vulnerable Planet: A Short History of the Environment*. New York: Monthly Review, 1999.

Bregman, Jacob I. and Kenneth M. Mackenthun. *Environmental Impact Statements*. Boca Raton, FL: Lewis, 1992.

Brohé, Arnaud, et al. *Carbon Markets: An International Business Guide*. London: Earthscan, 2009.

Bullock, Gary. *A Guide to Energy Service Companies*. Upper Saddle River, NJ: Prentice Hall, 2001.

Butterfield, J., et al. *Holistic Management Handbook: Healthy Land, Healthy Profits*. Washington, DC: Island, 2006.

Cairncross, F. *Costing the Earth: The Challenge for Governments, the Opportunities for Business*. Boston, MA: Harvard Business School, 1992.

Chappell, Tom. *The Soul of a Business: Managing for Profit and the Common Good*. New York: Bantam Books, 1996.

Coddington, W. *Environmental Marketing: Positive Strategies for Reaching the Green Consumer*. New York: McGraw-Hill, 1993.

Crane, A. *Marketing, Morality and the Natural Environment*. London: Routledge, 2000.

Feenberg, Andrew. *Questioning Technology*. London: Routledge, 1999.

Fuller, D. A. *Sustainable Marketing: Managerial-Ecological Issues*. Thousand Oaks, CA: Sage, 1999.

Gilpin, Alan. *Environmental Impact Assessment: Cutting Edge for the Twenty-First Century*. Cambridge: Cambridge University Press, 1995.

Goleman, Daniel. *Ecological Intelligence*. New York: Broadway Books, 2009.

Goodman, David and Michael J. Watts, eds. *Globalising Food: Agrarian Questions and Global Restructuring*. London: Routledge, 1997.

Goodstein, E. *The Trade-Off Myth: Fact and Fiction About Jobs and the Environment.* Washington, DC: Island, 1999.

Grant, John. *The Green Marketing Manifesto.* New York: Wiley, 2008.

Harkins, Paul, et al. *Natural Capitalism: Creating the Next Industrial Revolution.* Boston, MA: Little, Brown and Company, 1999.

Heidegger, Martin. *The Question Concerning Technology and Other Essays.* New York: Harper & Row, 1977.

Helm, D. *Economic Policy Towards the Environment.* Oxford: Blackwell, 1991.

Higham, James E. S. *Critical Issues in Ecotourism: Understanding a Complex Tourist Phenomenon.* Woburn, MA: Butterworth-Heinemann, 2007.

Humphrey, Neil and Mark Hadley. *Environmental Auditing.* Bembridge: Palladian Law, 2000.

Kaplan, Robert and Greg Norman. *The Balanced Scorecard: Translating Strategy Into Action.* Cambridge, MA: Harvard Business School, 1996.

Kibert, Charles. *Sustainable Construction: Green Building Design and Delivery.* Hoboken, NJ: John Wiley & Sons, 2008.

Kreske, Diori L. *Environmental Impact Statements: A Practical Guide for Agencies, Citizens, and Consultants.* New York: John Wiley and Sons, 1996.

Lamb, Robert. *The Greening of IT: How Companies Can Make a Difference for the Environment.* New York: IBM, 2009.

Lawrence, David P. *Environmental Impact Assessment: Practical Solutions to Recurrent Problems.* Hoboken, NJ: John Wiley & Sons, 2003.

Makower, J. *The E-Factor: The Bottom-Line Approach to Environmentally Responsible Business.* New York: Plume, 1994.

Marriott, Betty. *Environmental Impact Assessment: A Practical Guide.* New York: McGraw-Hill, 1997.

McDonough, W. and M. Braungart. *Cradle to Cradle: Remaking the Way We Make Things.* New York: North Point, 2002.

McGaw, David. *Environmental Auditing and Compliance Manual.* New York: Van Nostrand Reinhold, 1993.

McKercher, Bob. *Cultural Tourism: The Partnership Between Tourism and Cultural Heritage Management.* New York: Routledge, 2002.

Millennium Ecosystem Assessment. *Ecosystems and Human Well-Being: Synthesis.* Washington, DC: Island, 2005.

Mondt, R. *Cleaner Cars: The History and Technology of Emission Control Since the 1960s.* Warrendale, PA: Society of Automotive Engineers, 2000.

Moore, Emmett B. *The Environmental Impact Statement Process and Environmental Law.* Columbus, OH: Battelle, 2000.

Newman, Nell. *The Newman's Own Organics Guide to a Good Life: Simple Measures That Benefit You and the Place You Live.* New York: Villard, 2003.

Ottman, J. A. *Green Marketing.* Lincolnwood, IL: NTC Business Books, 1993.

Patterson, Walt. *Keeping the Lights On: Towards Sustainable Electricity.* London: Earthscan, 2007.

Patterson, Walt. *Transforming Electricity: The Coming Generation of Change.* London: Earthscan, 1999.

Raffensperger, Carolyn and Joel Tickner. *Protecting Public Health and the Environment.* Washington, DC: Island, 1999.

Raven, P. H., et al. *Environment*. Hoboken, NJ: John Wiley & Sons, 2008.

Raynolds, Laura T., et al. *Fair Trade: The Challenges of Transforming Globalization*. London: Routledge, 2007.

Reinhardt, Forest. *Down to Earth*. Cambridge, MA: Harvard Business School, 2000.

Roberts, P. *The End of Oil: The Decline of the Petroleum Economy and the Rise of a New Energy Order*. London: Bloomsbury, 2004.

Rockwood, L., et al., eds. *Foundations of Environmental Sustainability: The Coevolution of Science and Policy*. New York: Oxford University Press, 2008.

Romm, J. *The Hype About Hydrogen: Fact and Fiction in the Race to Save the Climate*. Washington, DC: Island, 2004.

Samli, A. C. *Social Responsibility in Marketing: A Proactive and Profitable Marketing Management Strategy*. Westport, CT: Quorum Books, 1992.

Savitz, Andrew W. and Karl Weber. *The Triple Bottom Line: How Today's Best-Run Companies Are Achieving Economic, Social, and Environmental Success—and How You Can Too*. Hoboken, NJ: Jossey-Bass, 2006.

Schumacher, E. F. *Small is Beautiful: Economics as if People Mattered*. New York: Harper Perennial, 1973.

Smajgl, A. and S. Larson. *Sustainable Resource Use: Institutional Dynamics and Economics*. London: Earthscan, 2007.

Sorrell, S. and J. Skea, eds. *Pollution for Sale: Emissions Trading and Joint Implementation*. Cheltenham: Edward Elgar, 1999.

Walker, Stuart. *Sustainable by Design: Explorations in Theory and Practice*. London: Earthscan, 2006.

Weaver, David B. *Sustainable Tourism*. Woburn, MA: Butterworth-Heinemann, 2005.

Willard, Bob. *The Sustainability Advantage: Seven Business Case Benefits of a Triple Bottom Line (Conscientious Commerce)*. Gabriola Island, British Columbia, Canada: New Society, 2002.

Wood, Janet. *Local Energy: Distributed Generation of Heat and Power*. London: Institution of Engineering and Technology, 2008.

Yudelson, Jerry. *The Green Building Revolution*. Washington, DC: Island, 2008.

Journals

Agriculture and Human Values (Springer)

Business and Society Review (John Wiley & Sons)
Business Horizons (Elsevier)
Business Strategy and the Environment (John Wiley & Sons)
Business Strategy Review (John Wiley & Sons)

Ecological Economics (Elsevier)
Environmental Science & Technology (American Chemical Society)

Greener Management International (Greenleaf)

Journal of Business Ethics (Springer)
Journal of Cleaner Production (Elsevier)
Journal of Economic Perspectives (American Economic Association)
Journal of Environmental Economics and Management (Elsevier)

Journal of Environmental Planning and Management (Taylor & Francis)
Journal of Industrial Ecology (John Wiley & Sons)
Journal of International Marketing (American Marketing Association)
Journal of Retailing (Elsevier)
Journal of Risk Research (Taylor & Francis)
Journal of Sustainable Development (Canadian Centre of Science and Education)

Management Decision (Emerald)

Nature (Nature Publishing Group)

Organization & Environment (SAGE Publications)

Psychology and Marketing (John Wiley & Sons)

Strategic Change (John Wiley & Sons)
Sustainable Development (John Wiley & Sons)

World Development (Elsevier)

Websites

Business for Social Responsibility Education Fund
    www.bsr.org

Business Social Compliance Initiative
    www.bsci-eu.org

Carbon Monitoring for Action
    carma.org

The Electronic Industry Code of Conduct
    www.eicc.info

Environmental Protection Agency
    www.epa.gov

Ethical Trading Initiative
    www.ethicaltrade.org

Fair Labor Association
    www.fairlabor.org

Greener Design
    www.greenerdesign.com

Green Life
    www.thegreenlife.org

Intergovernmental Panel on Climate Change
    www.ipcc.ch

Light Up the World Foundation
    lutw.org

Sustainability
    www.sustainability.com

United Nations Framework Convention on Climate Change
unfccc.int

U.S. Green Building Council
www.usgbc.org

Village Earth
www.villageearth.org

World Alliance for Decentralized Energy
www.localpower.org

Green Business Appendix

The Association for Sustainable & Responsible Investment in Asia

http://www.asria.org

This is the home page for The Association for Sustainable & Responsible Investment in Asia (ASrIA), a not-for-profit organization dedicated to promoting corporate responsibility and sustainable investment practices in the Asia Pacific region. The ASrIA website, available in English, Japanese, Chinese, Hindi, Indonesian, Korean, and Thai, also acts as a clearing house for information from other sources about sustainable investment in Asia. This website includes information about ASrIA activities (including conferences, seminars, forums, and training), news about sustainable investment in Asia, press releases, links to ASrIA publications (many available for free download), an electronic news bulletin to which anyone can subscribe, a list of available jobs related to sustainable investment in Asia, a posting service for resumes of individuals interested in such jobs, and links to information about ASrIA members. The website also includes portals which provide overviews of general topics such as climate change, toxic chemicals, airlines emissions, and philanthropic investment.

BPI World

http://www.bpiworld.org

BPI World is the home page for the Biodegradable Products Institute (BPI), a professional organization located in New York City which promotes the use and recycling of biodegradable polymeric materials (plastics) via composting. Members include individuals and groups from government, industry, and academia located in the United States, Canada, Europe, and Japan. One aspect of BPI's work is to scientifically determine which products are compostable (meaning that they will biodegrade in a typical municipal or commercial waste facility at a rate similar to yard trimming, food scraps, and similar materials, and must disintegrate leaving no large plastic fragments) and to indicate those which are by the BPI Compostable Logo. The website also includes general information about topics such as composting and the science of biodegradation, a searchable directory of BPI members, and certified compostable products (including bags, foodservice items, resins, and packing materials), and BPI-certified testing labs.

The Center for Responsible Travel

http://www.responsibletravel.org/home/index.html

The Center for Responsible Travel (CREST) is a nonprofit research institution focused on designing, monitoring, evaluating, and improving ecotourism and sustainable tourism.

It promotes societally and environmentally responsible tourism and does research into the potential for ecotourism and sustainable tourism to alleviate poverty and conserve biodiversity. The website includes information about CREST's current research and projects, companies which partner with them to promote eco-friendly travel in Africa and Latin America, their Traveler's Philanthropy project which helps travelers and travel companies learn about ways to improve the lives of local and indigenous communities in the countries they visit, a calendar of events, news about ecotourism and responsible travel, links to related organizations, and CREST research reports and newsletters (downloadable) as well as information about CREST books (available for purchase).

The Consortium on Green Design and Manufacturing

http://cgdm.berkeley.edu

This website, sponsored by the University of California, Berkeley (USA), represents the efforts of the Consortium on Green Design and Manufacturing (CGDM, founded in 1993 and including faculty and students from the University of California, Berkeley, the Lawrence Berkeley National Laboratory, and the Network for Energy, Environment, Efficiency and the Information Economy) to encourage research and education on environmental management, environmental design, and pollution prevention in civil infrastructure, the electronics industry, and service industries. Several software tools are available from the website, including PaLATE (a tool for economic and environmental life-cycle assessment of pavement and roads) and E-COMMUTair (a tool for analyzing the economic and environmental impact of telecommuting). The CGDM website also includes press releases, a bibliography of articles, papers and dissertations by CGDM affiliates, descriptions of ongoing and past research, profiles of people working now or formerly with the CGDM, and links to related sites.

The Global and Forest Trade Network

http://gftn.panda.org

The Global and Forest Trade Network (GFTN) is an initiative within the World Wildlife Fund whose purpose is to create a market for environmentally responsible forest products. Founded in 1991, the GFTN links communities, entrepreneurs, nongovernmental organizations, and over 360 companies in more than 30 countries in the world in order to coordinate national and international efforts to promote responsible forest management and increase the availability of products from sustainably managed forests while ending trade in forest products from illegal sources. The website explains the certification process for forestry products, explains the basics about illegal logging and responsible forest management, and includes information of specific interest to forest managers, buyers of timber products, and financial institutions. It includes several search tools to help locate certified companies and products and links to statistics and reports about the global timber market. The website also includes links to selected news about responsible forestry, arranged by geographic region, press releases, newsletters, and links to factsheets and reports about sustainable forestry.

Green Business Guide

http://www.business.gov/expand/green-business

This website, sponsored by the U.S. government, is devoted to helping small business owners found green businesses or adopt eco-friendly practices for existing businesses. For the former case it includes a checklist of 10 steps for founding ecologically friendly new businesses and a number of hints about starting a green business including links to information about industry partnerships and product stewardship programs with the U.S. Environmental Protection Agency. For existing businesses wishing to "go green," it includes information about energy efficiency, environmental management, available grants, loans and incentives for energy efficient upgrades and green technology development, the process for becoming a government contractor of green products and services, green commuting, green marketing, and green product development. The website also includes case studies of how several companies, including Ben & Jerry's, Whole Foods Market, Patagonia, and McDonald's adopted business practices which lessened their environmental impact and furthered their brand. The site also includes information about Earth Day and green cities in the United States, a podcast series, community forums, and a series of links to U.S. government agencies involved with green business issues.

Sustainable Business

http://www.sustainablebusiness.com

Sustainable Business is the official website of an organization of the same name located in Huntington, New York, whose purpose is to encourage businesses that contribute to an equitable and ecologically sustainable economy. The website offers a wealth of information for people interested in founding or operating a green business as well as people interested in working for or investing in green businesses. The website includes a statement of the organization's beliefs; news relevant to sustainable businesses; financial information and news of interest to investors in sustainable businesses; an employment service to connect job seekers with open positions in environmentally conscious businesses, nonprofits, and government agencies; a networking service to connect green businesses with potential investors and partners; an events calendar of workshops and conferences; and a resource directory of websites, databases, and other resources relating to sustainable businesses.

Sarah Boslaugh
Washington University in St. Louis

Index

Article titles and their page numbers are in **bold.**